FUNDAMENTALS OF NATURAL COMPUTING

Basic Concepts, Algorithms, and Applications

CHAPMAN & HALL/CRC
COMPUTER and INFORMATION SCIENCE SERIES

Series Editor: Sartaj Sahni

PUBLISHED TITLES

HANDBOOK OF SCHEDULING: ALGORITHMS, MODELS, AND PERFORMANCE ANALYSIS
Joseph Y.-T. Leung

THE PRACTICAL HANDBOOK OF INTERNET COMPUTING
Munindar P. Singh

HANDBOOK OF DATA STRUCTURES AND APPLICATIONS
Dinesh P. Mehta and Sartaj Sahni

DISTRIBUTED SENSOR NETWORKS
S. Sitharama Iyengar and Richard R. Brooks

SPECULATIVE EXECUTION IN HIGH PERFORMANCE COMPUTER ARCHITECTURES
David Kaeli and Pen-Chung Yew

SCALABLE AND SECURE INTERNET SERVICES AND ARCHITECTURE
Cheng-Zhong Xu

HANDBOOK OF BIOINSPIRED ALGORITHMS AND APPLICATIONS
Stephan Olariu and Albert Y. Zomaya

HANDBOOK OF ALGORITHMS FOR WIRELESS NETWORKING AND MOBILE COMPUTING
Azzedine Boukerche

HANDBOOK OF COMPUTATIONAL MOLECULAR BIOLOGY
Srinivas Aluru

FUNDEMENTALS OF NATURAL COMPUTING: BASIC CONCEPTS, ALGORITHMS, AND APPLICATIONS
Leandro Nunes de Castro

FUNDAMENTALS OF NATURAL COMPUTING
Basic Concepts, Algorithms, and Applications

Leandro Nunes de Castro
Catholic University of Santos (UniSantos)
Brazil

 Chapman & Hall/CRC
Taylor & Francis Group
Boca Raton   London   New York

Chapman & Hall/CRC is an imprint of the
Taylor & Francis Group, an informa business

Published in 2006 by
Chapman & Hall/CRC
Taylor & Francis Group
6000 Broken Sound Parkway NW, Suite 300
Boca Raton, FL 33487-2742

International Standard Book Number-10: 1-58488-643-9 (Hardcover)
International Standard Book Number-13: 978-1-58488-643-3 (Hardcover)
Library of Congress Card Number 2006006350

Library of Congress Cataloging-in-Publication Data

de Castro, Leandro N., 1974-
    Fundamentals of natural computing : basic concepts, algorithms, and applications / author, Leandro Nunes de Castro.
        p. cm. -- (Chapman & Hall/CRC Computer and information science series)
    Includes bibliographical references and index.
    ISBN 1-58488-643-9 (9781584886433 : alk. paper)
    1. Evolutionary programming (Computer science) 2. Neural networks (Computer science) 3. Quantum computers. 4. Molecular computers. I. Title. II. Series.

QA76.618.D42 2006
005.1--dc22   2006006350

Taylor & Francis Group
is the Academic Division of Informa plc.

Visit the Taylor & Francis Web site at
http://www.taylorandfrancis.com

and the CRC Press Web site at
http://www.crcpress.com

To Nature,

Lásara, José and Elizabete

This new science started from the assumption that in the end it is possible to bring together completely different fields of investigation.

FOREWORD

The idea of writing this book was based on two main motivations. First, to bring together several topics relating computing and nature in a single volume. Some of these topics have been dealt with by specific texts for some time (e.g., neural networks and evolutionary algorithms), but some are quite young (e.g., swarm intelligence, artificial immune systems, DNA and quantum computing). A second motivation was because all natural computing topics are, by themselves, sources of endless fascination and usefulness, and I would like to bring such excitement to a broad audience and to potentially new researchers in the field.

Although there are some popular science books that discourse about all or most of these subjects in an informal and illustrative way (see Appendix C), I wanted to do it more formally, providing the reader with the conceptual and algorithmic fundaments of natural computing, together with some design principles, exercises, illustrations and lists of potential applications.

This book has started from my own research background over ten years ago, but mostly from lectures given at the University of Kent at Canterbury (UK) on a final year undergraduate course called, at that time, Novel Computing Paradigms. This same course has been adapted and is currently called Natural Computing, and has been further extended to become a graduate course in Computer Science and Engineering at the Universities of Campinas (Unicamp) and of Santos (UniSantos) in Brazil. But certainly these are not the only courses related to the subject matters presented in this book. Nowadays, there are several courses throughout the world on bio-inspired computing, bioinformatics, artificial life, fractal geometry, molecular computing, etc., that could benefit from this text, mainly graduate and final year undergraduate courses.

The writing of this text started on June 2002. Any lengthy document that involves so many concepts and ideas, and that takes so long to be written contains errors and omissions, and this book is surely no exception to this rule. If you find errors or have other comments to improve this volume, please e-mail them to me at lnunes@unisantos.br. The errata will be added to the book's website: http://lsin.unisantos.br/lnunes/books.

Leandro Nunes de Castro

PREFACE

Natural computing is a new and broad field of investigation that brings together not only words but also subjects that were once believed to be mutually exclusive: nature and computing. Some problems strike researchers for the difficulty in providing (feasible) solutions. They cannot be resolved using the standard computing and problem solving techniques. They stimulate hard thinking and the search for solutions in the best problem-solver ever known – *Nature*.

The hallmark of any new discipline or field of investigation is that it must be able to provide successful solutions and models to old, yet unresolved, problems and to new ones as well. But this is only one of the benefits of natural computing. It also gives us a new form of understanding and interacting with nature; new methods of simulating, emulating and even realizing natural phenomena have been proposed. Natural computing simulations of nature are now being used as realistic models of plant growth, development, landscapes, complex behaviors, etc.

If we acknowledge the fact that computing may need a change in paradigm soon, when current computer technology reaches its limit in processing power and information storage, then natural computing has closed its loop and completed all its three main branches:

a) The use of nature to inspire the development of novel computing techniques for solving complex problems.

b) The use of computers to recreate natural phenomena and organisms.

c) The use of natural materials (e.g., biomolecules) to compute.

What follows is a tour of fields never gathered between the same book jacket covers. We will look at the fundaments of the three main branches of natural computing: computing inspired by nature, the simulation and emulation of nature in computers, and computing with natural materials. What unites these different systems, processes and theories is a recurring inspiration, use and interaction with nature. At each scale you will be able to see the imprint of nature and how it can be used for computational purposes.

Over the last few decades, there has been a considerable expansion of interest in the exchange of ideas between the natural sciences and computing. Numerous conferences have been organized, books edited and written, and journals started (cf. Appendix C). In summary, this book is an attempt to consolidate and integrate the basic concepts, algorithms, and applications across a wide range of interrelated fields of research and disciplines into a single coherent text. At one level the result can be viewed as just that - an integration and consolidation of knowledge. Nevertheless, I have felt that the process of putting all these ideas together into a single volume has led to the emergent phenomenon in which the whole is more than the sum of its parts.

Main Features

The contents of this book are very attractive for undergraduate students, lecturers, academic researchers (mainly newcomers) and industry parties in Biology, Computing, Engineering, (Applied) Mathematics and Physics, interested in developing and applying alternative problem solving techniques inspired by nature, knowing the most recent computing paradigms, and discovering how computers can aid the understanding and simulation of nature. The book is written in a textbook style with illustrative and worked examples, and provides different types of exercises to assess the comprehension of the contents presented and stimulate further investigation into each of the subjects. The many questions, computational exercises, thought exercises and projects that appear at the end of each chapter are aimed at complementing, solidifying, practicing and developing the readers' ability to explore the many concepts and tools introduced. With few exceptions, such as the thought questions, projects and challenges, these should be solved with a fairly small amount of work, what may depend on the readers' background.

Most chapters contain a brief description of the main ideas and algorithms about selected applications from the literature. In most cases either real-world or benchmark problems were chosen for presentation. These examples serve the purpose of illustrating some types of problems that can be solved using the methods introduced, not to provide an in depth description of how the problems were effectively solved using the techniques. They are illustrations of potential applications instead of case studies, what allows the presentation of a larger number of applications in detriment of a detailed, lengthier description.

The innovative concepts and tools described in this volume have already become very good candidates to be applied as solution tools to real-world problems, and have also been incorporated as part of academic degrees in disciplines related to computing, biological modeling, biomedical engineering, and bioinformatics. Although there are books in the literature dealing with each of these parts separately, there is no single volume dealing with these subjects altogether. Similarly to the standard well-known Artificial Intelligence books, this volume was primarily written to support first courses on the many subjects covered.

Structure of the Book

The book is composed of eleven chapters divided into three main parts, plus three appendices. Each of its three main parts is composed of a set of topics which, by themselves, would constitute a discipline on their own. It would be impracticable to cover all subjects in detail, include the most recent advances in each field, and provide the necessary biological and mathematical background in the main text. The approach taken was, thus, to present the fundaments of each technique; that is, the simplest and pioneer proposals in each field. Recent developments can be found in the cited specialized literature, and are being proposed each day.

The eleven chapters composing the main body of the text are:

The motivation from nature is stressed in all chapters independently and an appendix (Appendix A) was created to support the search and quick access for the meaning of the biological, chemical and physical terminology. Another appendix (Appendix B) was especially created to provide the reader with some minimal background knowledge and reference to the mathematics and basic theoretical foundations necessary to understand the contents of the book. The reader is pointed to the appendices when necessary. One last appendix (Appendix C) was written to guide the reader, mainly newcomers, to the literature. Appendix C contains my personal comments on selected bibliography (reason why the list is incomplete and others' comments may differ from mine), and a list of journals and conferences in which to publish and find information on the subjects.

A quick glance through the book is enough to note that emphasis was given to Part I of the text. The reasons are twofold. First, some fields of investigation composing this part are older than those composing parts II and III or are used as a basis for Part II and, thus, have a wider bibliography and set of well-established techniques. Second, the first part could serve as a general reference for the many existing courses on bio-inspired computing, but for which there is no single volume available to date. For Part II there are several books on Fractals and on Artificial Life that could be used as standard references. In Part III there are not only specific books on DNA and Quantum computing, but also books involving both paradigms. See Appendix C and the respective chapters for a list of these references.

How to Use this Book

Natural computing is a very broad area and, therefore, this is a big book. The text has been written to be used by newcomers to the field, lecturers (as a textbook) and independent readers. Each one of these may approach the text differently. This section provides a brief guidance on how to use this book.

For independent students and newcomers

This book is supposed to be a self-contained and accessible introduction to natural computing, even to the independent reader. The readers are assumed to have some background knowledge on linear algebra, discrete mathematics, notions from computing and basic programming skills. If this is not the case, they are strongly encouraged to go through Appendix B thoroughly before start reading the main text. Even if the reader is familiar with all these concepts, he/she is encouraged to have a quick look at Appendix B and certify the familiarity with all the concepts presented, because these were selected due to their importance to a basic understanding of natural computing. As a guide, an independent reader may want to go through the suggested programs for a one-term or a full-year course, as described below.

The appendices and exercises were designed so as to help the reader get to know and understand each field independently and as a single discipline. The Table of Contents and introductions to each chapter provide a brief description of their contents that may be useful while choosing what topics to study. The Natural Computing Virtual Laboratory (LVCoN http://lsin.unisantos.br/lvcon) can also be used as a platform to play with many of the techniques discussed in this text. The reader is also invited to download the book slides from the website: (http://lsin.unisantos.br/lnunes/books).

For instructors

As this book covers a diverse range of topics, it may be used in various different courses and in distinct ways. An ideal course would take two semesters, but an introductory (general overview) natural computing course could be taught in a single semester.

In a one-semester general introductory course about the whole book, the following chapters/sections are suggested: Chapters 1 and 2 (all sections); Chapter 3 (Sections 3.2, 3.4 and 3.5); Chapter 4 (Sections 4.2, 4.3, 4.4.2, 4.4.3, 4.4.5); Chapter 5 (Sections 5.2, and 5.4); Chapter 6 (Sections 6.2, 6.3, 6.5, 6.6, 6.7.1); Chapter 7 (Sections 7.2, 7.3, 7.7); Chapter 8 (Sections 8.2, 8.3.1, 8.3.2, 8.3.3, 8.3.9, 8.3.10, 8.3.11); Chapter 9 (Sections 9.2 and 9.3); Chapter 10 (Sections 10.2, 10.3, and 10.4). In addition to these, the introductory sections of each chapter are mandatory, and the sections that make a parallel between the natural and the computational systems are highly recommended so that the student can understand the loop 'nature ↔ computing'.

In a one-year course, the suggestion is to divide the subjects into two terms as follows: Chapters 1, 2 and Part I to be lectured in the first term, and Parts II and III to be lectured in the second term together with a brief review of Chapter 2. This way, the first term would be a course on bio-inspired computing or nature-inspired computing. The second term would cover the remaining. The following material may require additional references for a one-year introductory course or for a specific, more in-depth, course on a given subject: Chapter 3 (Section 3.6); Chapter 6 (Sections 6.6 and 6.7); Chapter 7 (Sections 7.4, 7.5, 7.6 and 7.8);

Chapter 8 (Sections 8.2 and 8.3); Chapter 9 (Section 9.4); Chapter 10 (Section 10.6).

This volume can also be used as a basic text for an introductory course on any of the subjects covered here. In this case, additional sources of reference should be searched, including conference proceedings. The most recent advances and variations of the algorithms presented could be added so as to complement the contents presented here.

Concerning Appendix B, dedicated to the Theoretical Background for the book, the required background for each chapter is as follows:

Chapter 3: Section B.2.1, B.2.4, B.4.1, and B.4.4.

Chapter 4: Sections B.1, B.4.2, B.4.3, B.4.4 and B.4.5.

Chapter 5: Sections B.4.1, B.4.4 and B.4.5.

Chapter 6: Sections B.4.1, B.4.4 and B.4.5.

Chapter 7: Sections B.1, B.4.3, B.4.5 and B.4.6.

Chapter 8: Part I of the book, with emphasis on Chapters 3 and 4.

Chapter 9: Section B.3.

Chapter 10: Sections B.1, B.2, B.3 and B.4.7.

Also, take the attention of the students for Appendices A and C, written as a general glossary to the terminology and a guide to the literature, respectively.

Notation

While writing such a wide-ranging volume, some major concerns are uniformity and notation. Special care was taken to provide a concise and uniform notation to the text. The readers are invited to go through this Notation section before commencing reading the book. Appendix B complements this notation by providing a brief review of the theoretical background necessary for a thorough understanding of the book using the standard notation style of the whole text.

- o   Scalars are denoted by italic lowercase Roman letters: e.g., a, b, x, y, z.

- o   Vectors are denoted by boldface lowercase Roman letters: e.g., $\mathbf{x}, \mathbf{y}, \mathbf{z}$.

- o   Matrices are denoted by boldface uppercase Roman letters: e.g., $\mathbf{X}, \mathbf{Y}, \mathbf{Z}$.

- o n-dimensional Euclidean space: $\Re^n$.

- o   Matrix with n rows and m columns, and real-valued entries: $\mathbf{X} \in \Re^{n \times m}$.

- o   Vector with n entries (always assumed to be a column vector, unless otherwise specified): $\mathbf{x} \in \Re^n$.

- o   The transpose of a vector or matrix is denoted by an italic uppercase letter 'T': e.g., $\mathbf{x}^T, \mathbf{X}^T$.

- o   Problem of minimizing function $f(\mathbf{x})$ over choices of $\mathbf{x}$ that lie in the set of feasible solutions: $\min f(\mathbf{x})$.

- o   Problem of maximizing function $f(\mathbf{x})$ over choices of $\mathbf{x}$ that lie in the set of feasible solutions: $\max f(\mathbf{x})$.

- Set of all the minimizers of $\min f(\mathbf{x})$; that is, the value(s) of the argument $\mathbf{x}$ that minimize the objective: $\arg\ \min f(\mathbf{x})$.
- Set of all the maximizers of $\max f(\mathbf{x})$; that is, the value(s) of the argument $\mathbf{x}$ that maximize the objective: $\arg\ \max f(\mathbf{x})$.

List of Abbreviations

ACA	Ant-Clustering Algorithm
ACO	Ant-Colony Optimization
AI	Artificial Intelligence
AIS	Artificial Immune System
ALife	Artificial Life
ANN	Artificial Neural Networks
APC	Antigen-Presenting Cell
API	Application Programmers Interface
BZ	Beloussov-Zhabotinsky Reaction
BIC	Biologically Inspired Computing
CA	Cellular Automata
CAM	Content-Addressable Memory
CAS	Complex Adaptive System
CI	Computational Intelligence
CQED	Cavity Quantum Electrodynamics
DIFS	Deterministic Iterated Fuction System
DLA	Diffusion limited aggregation
DNA	Deoxyribonucleic Acid
DOF	Degrees of Freedom
DR	Detection Rate
EA	Evolutionary Algorithm
ES	Evolution Strategies
FAR	False Alarm Rate
FN	False Negative
FP	False Positive
FSM	Finite-State Machine
H	Hydrogen
HPP	Hamiltonian Path Problem
ICARIS	International Conference on Artificial Immune Systems
IFS	Iterated Function System
IR	Infrared

JSS	Job Shop Scheduling
LAN	Local Area Network
LMS	Least Mean Squares
LTD	Long-Term Depression
LTP	Long-Term Potentiation
MSE	Mean-Squared Error
MHC	Major Histocompatibility Complex
MLP	Multi-Layer Perceptron
MRCM	Multiple Reduction Copy Machine
MST	Minimal Spanning Tree
NMR	Nuclear Magnetic Resonance
ODE	Ordinary Differential Equation
OH	Hydroxyl Group
PAM	Parallel Associative Memory
PCR	Polymerase Chain Reaction
PDS	Parallel-Distributed System
PS	Particle Swarm
QAP	Quadratic Assignment Problem
QED	Quantum Electrodynamic
QFT	Quantum Fourier Transform
RDM	Reaction-Diffusion Model
RIFS	Random Iterated Function System
RL	Reinforcement Learning
RNA	Ribonucleic Acid
SAT	Satisfiability Problem
SC	Soft Computing
SOM	Self-Organizing Map
SOS	Self-Organizing System
SPIR	Two-Spirals Problem
SSE	Sum-Squared Error
TM	Turing Machine
TN	True Negative
TP	True Positive
TSP	Traveling Salesman Problem
VRP	Vehicle Routing Problem
VSM	Virtual State Machine

Pseudocodes

Instead of providing codes in a specific programming language, this book provides a set of pseudocodes, which I hope are self-explanatory, as one convenient way of expressing algorithms. These codes are included to clarify the techniques and thus are neither optimized for speed nor complete. Some routines are omitted and others are left as exercises, in which cases the readers are encouraged to try and implement them using any computer language they are familiar with.

The standard pseudocode used here follows the one illustrated in Algorithm 0.1, which corresponds to the training algorithm of a simple perceptron neural network (Algorithm 4.2). In this procedure, line 1 contains the output of the algorithm within brackets, which is a matrix $\mathbf{W}$, the function's name – perceptron – and the input parameters within parentheses $(\text{max\_it}, \alpha, \mathbf{X}, \mathbf{D})$. The repetition structures, such as **for** and **while** are standard and terminated with an **end**. Comments are indicated by double slashes '//'. All codes contain internal functions (e.g., sum in line 12 that sums all elements e_{ji}) that are, in most cases, common in computing. Other functions are used, but when they are not internal, their role is explained in the main text (e.g., initialize).

It is important to realize that the whole book uses matrix notation. Therefore, when a notation like $\mathbf{Wx}_i$ is found (line 8), it means that matrix $\mathbf{W}$ multiplies vector $\mathbf{x}_i$, and this is only possible if the number of columns of $\mathbf{W}$ is the same as the number of rows of $\mathbf{x}_i$. The notation used in line 10, $\mathbf{e}_i\mathbf{x}_i^T$, represents the outer product between the vector of errors $\mathbf{e}_i$ and the vector of inputs $\mathbf{x}_i$ (see Appendix B for a complement on notation).

```
1.   procedure [W] = perceptron(max_it,α,X,D)
2.      initialize W    //set it to zero or small random values
3.      initialize b    //set it to zero or small random values
4.      t ← 1; E ← 1
5.      while t < max_it & E > 0 do,
6.          E ← 0
7.          for i from 1 to N do,      //for each training pattern
8.              yᵢ ← f(Wxᵢ + b)          //network outputs for xᵢ
9.              eᵢ ← dᵢ - yᵢ             //determine the error for xᵢ
10.             W ← W + α eᵢ xᵢᵀ          //update the weight matrix
11.             b ← b + α eᵢ             //update the bias vector
12.             E ← E + sum( e²ⱼᵢ )       //j = 1,...,o
13.             veᵢ ← eᵢ
14.         end for
15.         t ← t + 1
16.     end while
17. end procedure
```

Algorithm 0.1: Learning algorithm for the perceptron with multiple outputs.

Acknowledgments

While writing this book I have benefited from the support of many colleagues, students, researchers and institutions. This support has ranged from the permission to use copyrighted material to technical discussions and the review of chapters and sections. I have made a great effort to find the owners of copyrighted material, but in a few cases this has not been possible. I would like to take this opportunity to apologize to any copyright holder whose rights were unwittingly infringed.

I am particularly indebted to the following people for their varied contributions to this volume: André Homeyer, André L. Vizine, André P. L. F. de Carvalho, Ângelo C. Loula, Craig Reynolds, Danilo M. Bonfim, Demetri Terzopoulos, Eduardo R. Hruschka, Eduardo J. Spinosa, Elizabete L. da Costa, Eric Klopfer, Fabrício O. de França, Fernando J. Von Zuben, George B. Bezerra, Hazinah K. Mammi, Helder Knidel, Jon Timmis, Jungwon Kim, Karl Sims, Katrin Hille, Ken Musgrave, Lalinka C. T. Gomes, Laurent Keller, Leonid Perlovsky, Maciej Komosinski, Norhaida M. Suaib, Osvaldo Pessoa, Peter J. Bentley, Rodrigo Pasti, Ronald Kube, Sandra Cohen, Sartaj Sahni, Tiago Barra, Vitorino Ramos, Witold Pedrycz, and Yupanqui J. Muñoz. The following institutions kindly provided me permission to reproduce part of their copyrighted materials: Elsevier, Gameware Development, The MIT Artificial Intelligence Laboratory, Springer-Verlag, The IEEE Press, The MIT Press, and W. H. Freeman and Company. I also would like to thank all CRC Press staff for their great work in publishing this book; particular thanks go to Robert Stern, Jay Margolis and Marsha Pronin. A special thanks also goes to my brother, Sandro de Castro, who managed to find time to prepare the cover of the book. Last, but not least, all the institutional and financial support I received from UniSantos, Unicamp, CNPq and FAPESP were crucial for the development of this project.

Leandro Nunes de Castro

CONTENTS

PART I – COMPUTING INSPIRED BY NATURE

PART II – THE SIMULATION AND EMULATION OF NATURAL PHENOMENA IN COMPUTERS

7. FRACTAL GEOMETRY OF NATURE

PART III – COMPUTING WITH NEW NATURAL MATERIALS

9. DNA COMPUTING

CHAPTER 1

FROM NATURE TO NATURAL COMPUTING

"... science has often made progress by studying simple abstractions when more realistic models are too complicated and confusing."
(I. Stewart, Does God Play Dice, Penguin Books, 1997, p. 65)

"Often the most profound insights in science come when we develop a method for probing a new regime of Nature."
(M. Nielsen and I. Chuang, Quantum Computation and Quantum Information, Cambridge University Press, 2000, p. 3)

1.1 INTRODUCTION

During the early days of humanity natural resources were used to provide shelter and food. We soon learned to modify and manage nature so as to breed crops and animals, build artifacts, control fire, etc. We then started to observe and study biological, chemical, and physical phenomena and patterns in order to better understand and explain how nature works. As examples, by learning about the physical laws of motion and gravity it became possible to design aircrafts; and by understanding some basic principles of life it is now possible to manage nature in various levels, from the creation of transgenic food to the control of diseases.

With the advent of computers, the way human beings interact with nature changed drastically. Nature is now being used as a source of inspiration or metaphor for the development of new techniques for solving complex problems in various domains, from engineering to biology; computers can simulate and emulate biological life and processes; and new material and means with which to compute are currently being investigated. *Natural computing* is the terminology introduced to encompass these three types of approaches, named, respectively: 1) *computing inspired by nature*; 2) *the simulation and emulation of natural phenomena in computers*; and 3) *computing with natural materials*. This book provides an introduction to the broad field of natural computing. It constitutes a textbook-style treatment of the central ideas of natural computing, integrated with a number of exercises, pseudocode, theoretical and philosophical discussions, and references to the relevant literature in which to gather further information, support, selected websites, and algorithms involving the topics covered here. This introductory chapter provides some motivations to study natural computing, challenges the student with some sample ideas, discusses its philosophy and when natural computing approaches are necessary, provides a taxonomy and makes a brief overview of the three branches of the proposed taxonomy for natural computing.

1.1.1. Motivation

Why should we study natural computing and why should research in this broad area be supported? There are many reasons for doing so; from the engineering of new computational tools for solving complex problems whose solutions are so far unavailable or unsatisfactory; to the design of systems presenting nature-like patterns, behaviors and even the design of new forms of life; and finally to the possibility of developing and using new technologies for computing (new computing paradigms). Although still very young in most of its forms, the many products of natural computing are already available in various forms nowadays, in washing machines, trains, toys, air conditioning devices, motion pictures, inside computers as virtual life, and so forth. Some of these applications will be reviewed throughout this book with varying levels of details.

Natural phenomena (e.g., processes, substances, organisms, etc.) have long inspired and motivated people to mimic, design, and build novel systems and artifacts. For many centuries, the observation of the natural world has allowed people to devise theories about how nature works. For example, physics is abounded with laws describing electromagnetism (Maxwell's equations), thermodynamics (first law: conservation, second law: entropy, and third law: absolute zero), motion (Newton's laws), and so forth. Artifacts, such as sonar echolocation, chemical substances used for pharmaceutical purposes, infrared imaging systems, airplanes, submarines, etc., were all developed by taking inspiration from nature, from animals (bats, birds, etc.) to chemical substances.

Natural computing is the computational version of this process of extracting ideas from nature to develop 'artificial' (computational) systems, or using natural media (e.g., molecules) to perform computation. The word artificial here means only that the systems developed are human-made instead of made by nature. While not the rule, in some cases, the products of natural computing may turn out to be so life-like that it becomes difficult to tell them apart from natural phenomena. Natural computing can be divided into three main branches (Figure 1.1) (de Castro and Von Zuben, 2004; de Castro, 2005):

1) *Computing inspired by nature*: it makes use of nature as inspiration for the development of problem solving techniques. The main idea of this branch is to develop computational tools (algorithms) by taking inspiration from nature for the solution of complex problems.

2) *The simulation and emulation of nature by means of computing*: it is basically a synthetic process aimed at creating patterns, forms, behaviors, and organisms that (do not necessarily) resemble 'life-as-we-know-it'. Its products can be used to mimic various natural phenomena, thus increasing our understanding of nature and insights about computer models.

3) *Computing with natural materials*: it corresponds to the use of natural materials to perform computation, thus constituting a true novel computing paradigm that comes to substitute or supplement the current silicon-based computers.

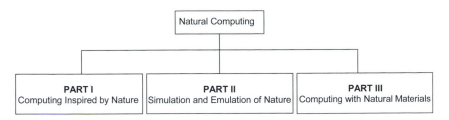

Figure 1.1: The three main branches of natural computing and their order of appearance in the book.

Therefore, natural computing can be defined as the field of research that, based on or inspired by nature, allows the development of new computational tools (in software, hardware or 'wetware') for problem solving, leads to the synthesis of natural patterns, behaviors, and organisms, and may result in the design of novel computing systems that use natural media to compute.

Natural computing is thus a field of research that testimonies against the specialization of disciplines in science. It shows, with its three main areas of investigation - *computing inspired by nature, the simulation and emulation of nature by means of computing*, and *computing with natural materials* - that knowledge from various fields of research are necessary for a better understanding of life, for the study and simulation of natural systems and processes, and for the proposal of novel computing paradigms. Physicists, chemists, engineers, biologists, computer scientists, among others, all have to act together or at least share ideas and knowledge in order to make natural computing feasible.

It is also important to appreciate that the development and advancement of natural computing leads to great benefits to the natural sciences, like biology, as well. Many computational tools developed using ideas from nature and the biological sciences are applied to create models and solve problems within the biosciences. This application domain is becoming even more important over the last few years with the emergence of new fields of investigation like computational biology and bioinformatics (Attwood and Parry-Smith, 1999; Baldi and Brunak, 2001; Waterman, 1995). Natural computing has also proven to be useful for a better understanding of nature and life processes through the development of highly abstract models of nature. Sometimes natural computing techniques can be directly aimed at being theoretical models of nature, providing novel insights into how nature works.

1.2 A SMALL SAMPLE OF IDEAS

The history of science is marked by several periods of almost stagnation, intertwined with times of major breakthroughs. The discoveries of Galileo, Newtonian mechanics, Darwin's theory of evolution, Mendel's genetics, the development of quantum physics, and the design of computers are just a small sample of the scientific revolutions over the past centuries. We are in the midst of another technological revolution - the *natural computing age*; a time when the interaction and the similarity between computing and nature is becoming each day greater. The transformation may be revolutionary for all those involved in the development of natural computing devices, but, if they do their job well, it will not necessarily make much difference for the end users. We may notice our spreadsheets recalculating faster, our grammar checker finally working, several complex problems being solved, robots talking naturally to humans, cars driving themselves, new forms of life, and patterns emerging in a computer screen in front of us, computers based on biomolecules, etc. But we will all be dealing with the end results of natural computing, not with the process itself. However, will ordinary people and end-users get a chance to experiment and play with natural computing? In fact, we can get our hands dirty already. And we will start doing this just by testing our ability to look at nature and computing in different ways.

Below are discussions about some natural phenomena and processes involving natural means: 1) clustering of dead bodies in ant colonies; 2) bird flocking; and 3) manipulating DNA strands. All of them have already served as inspiration or media for the development of natural computing techniques and will be presented here as a first challenge and motivation for the study of natural computing. Read the descriptions provided and try to answer the following questions.

To clean up their nests, some ant species group together corpses of ants or parts of dead bodies, as illustrated in Figure 1.2. The basic mechanism behind this type of clustering or grouping phenomenon is an attraction between dead items mediated by the ants. Small clusters of items grow by attracting more workers to deposit more dead bodies. This grouping phenomenon can be modeled using two simple rules:

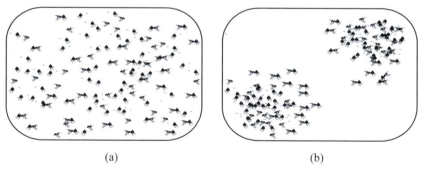

(a)   (b)

Figure 1.2: Clustering of dead bodies in an ant colony. (a) Initial distribution of ants. (b) Clustered bodies.

Pick up rule: if an ant finds a dead body, it picks it up and wanders around the arena until it finds another dead body. The probability or likelihood that an ant picks up a dead body is inversely proportional to the number of items in that portion of the arena; that is, the more dead bodies around, the smaller the probability it is picked up, and vice-versa.

Dropping rule: while wandering around, the loaded ant eventually finds more dead bodies in its way. The more dead bodies are found in a given region of the arena, the higher the probability the ant drops the dead body it is carrying at that location of the arena, and vice-versa.

As a result of these very simple behavioral rules, all dead items will eventually be brought together into a single group, depending on the initial configuration of the arena and how the rules are set up.

Question 1: what kind of problem could be solved inspired by this simple model of a natural phenomenon?

Question 2: how would you use these ideas to develop a computing tool (e.g., an algorithm) for solving the problem you specified above?   ■

Figure 1.3 illustrates a bird flock. When we see birds flocking in the sky, it is most natural to assume that the birds 'follow a leader'; in this picture, the one in front of the flock. However, it is now believed (and there are some good evidences to support it) that the birds in a flock do not follow any leader.

There is no 'global rule' that can be defined so as to simulate a bird flock. It is possible, however, to generate scripts for each bird in a simulated flock so as to create a more realistic group behavior (for example, in a computer simulation). Another approach, one that is currently used in many motion pictures, is based on the derivation of generic behavioral rules for individual birds. The specification of some simple individual rules allows realistic simulation of birds flocking. The resultant flock is a result of many birds following the same simple rules.

Question 1: describe (some of) these behavioral rules that, when applied to each bird in the flock, result in an emergent group behavior that is not specifically defined by the individual rules. It means that such rules, together with the interactions among individual birds, result in a global behavior that cannot often be predicted by simply looking at the rules.

Figure 1.3: Illustration of a flock of birds.

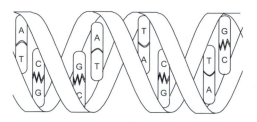

Figure 1.4: Double strand of DNA.

Question 2: can you extend these rules to herds of land animals and schools of fish? That is, is there a significant qualitative difference between these various types of group behavior?   ∎

Figure 1.4 depicts a double strand of DNA. The DNA molecules contain the genetic information of all living beings on earth. It is known that this genetic information, together with the environmental influences, determines the phenotype (expressed physical characteristics) of an individual.

Roughly, DNA molecules are composed of four bases which bind to each other exclusively in a complementary fashion: A binds with T, and C binds with G. Genetic engineering techniques can nowadays be used to artificially manipulate DNA so as to alter the genetic information encoded in these molecules. For instance, DNA molecules can be *denatured* (separated into single strands), *annealed* (single strands can be 'glued' together to form double strands of DNA), *shortened* (reduced in length), *cut* (separated in two), *multiplied* (copied), *modified* (e.g., new sequences inserted), etc.

Question 1: if the information that encodes an organism is contained in DNA molecules and these can be manipulated, then life can be seen as information processing. Based on your knowledge of how standard computers (PCs) work, propose a new model of computer based on DNA strands and suggest a number of DNA manipulation techniques that can be used to compute with molecules.

Question 2: what would be the advantages and disadvantages of your proposed DNA computer over the standard computers?   ∎

If you have not tried to answer these questions yet, please take your time. They may give you some flavor of what researchers on some branches of natural computing do. If you want to check possible answers to these questions, please refer to Chapters 5, 8, and 9 respectively

    . . .

So, how did you get on?

If your answers were too different from the ones presented in this volume, do not worry; they may constitute a potentially new algorithm or computing paradigm!

1.3 THE PHILOSOPHY OF NATURAL COMPUTING

One important question this book tries to answer is how researchers discover the laws and mechanisms that are so effective in uncovering how nature functions and how these can be used within and for computing. A natural result of this line of investigation is the proposal of novel ways of computing, solving real-world problems, and synthesizing nature. Scientific explanations have been dominated by the formulation of principles and rules governing systems' behaviors. Researchers usually assume that natural systems and processes are governed by finite sets of rules. The search for these basic rules or fundamental laws is one of the central issues covered in this book. It is not easy to find such rules or laws, but enormous progress has been made. Some examples were provided in Section 1.2.

Most of the computational approaches natural computing deals with are based on highly simplified versions of the mechanisms and processes present in the corresponding natural phenomena. The reasons for such simplifications and abstractions are manifold. First of all, most simplifications are necessary to make the computation with a large number of entities tractable. Also, it can be advantageous to highlight the minimal features necessary to enable some particular aspects of a system to be reproduced and to observe some emergent properties. A common question that may arise is: "if it is possible to do something using simple techniques, why use more complicated ones?"

This book focuses on the extraction of ideas and design aspects of natural computing, in particular the teaching of modeling, how to make useful abstractions, and how to develop and use computer tools or algorithms based on nature. In contrast to some books on the technological and advanced aspects of specific topics, this text outlines the relations of theoretical concepts or particular technological solutions inspired by nature. It is therefore important to learn how to create and understand abstractions, thus making a suitable simplification of a system without abolishing the important features that are to be reproduced.

Which level is most appropriate for the investigation and abstraction depends on the scientific question asked, what type of problem one wants to solve, or the life phenomenon to be synthesized. As will be further discussed, simple behavioral rules for some insects are sufficient for the development of computational tools for solving combinatorial problems and coordinate collective robotic systems. These are also useful for the development of computer simulations of biological systems in artificial life, and the creation of abstract models of evolution, and the nervous and immune systems, all aimed at solving complex problems in various domains.

Natural computing usually integrates experimental and theoretical biology, physics and chemistry, empirical observations from nature and several other sciences, facts and processes from different levels of investigation into nature so as to design new problem solving techniques, new forms of mimicking natural phenomena, and new ways of computing, as summarized in Figure 1.5.

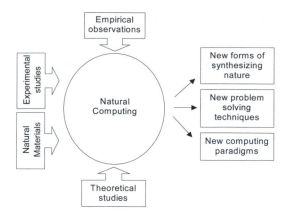

Figure 1.5: Many fields of investigation have to be integrated for the study and development of natural computing. As outcomes, new ways of computing, new problem solving techniques, and possible forms of synthesizing nature result.

1.4 THE THREE BRANCHES: A BRIEF OVERVIEW

This section provides a very brief overview of the three branches of natural computing and their main approaches. The bibliography cited is basically composed of references from pioneer works and books where further and didactic information can be found on all topics discussed. Instead of trying to cover the areas reviewed in detail, general comments about most natural computing tools, algorithms, techniques and their potential application areas are provided, together with a discussion of how the resultant computational tools or systems relate (interact) with nature and the natural sciences. The reader will be pointed to the chapters that deal specifically with each of the approaches discussed.

1.4.1. Computing Inspired by Nature

The main motivation for this part of the book is that nature has greatly enriched computing. More importantly, nature has been very successful in solving highly complex problems. In a very low level, there is an urge for survival in living organisms: they have to search for food, hide from predators and weather conditions, they need to mate, organize their homes, etc. All this requires complex strategies and structures not usually directly modeled or understood. But, for instance, viewing a colony of ants foraging for food as an 'intelligent behavior' is not always very intuitive for us, used to attribute 'intelligent behaviors' to 'intelligent beings'. What if I tell you that the way ants forage for food has inspired algorithms to solve routing problems in communication networks? Can you imagine how this is done?

Among all natural computing approaches, computational algorithms and systems inspired by nature are the oldest and most popular ones. They arose with two main objectives in mind. First, researchers were interested in the modeling of natural phenomena and their simulation in computers. The common goal in this direction is to devise theoretical models, which can be implemented in computers, faithful enough to the natural mechanisms investigated so as to reproduce qualitatively or quantitatively some of their functioning. Theoretical models are supposed to provide a deeper insight and better understanding of the natural phenomena being modeled, to aid in the critical analysis and design of experiments, and to facilitate the recovery of results from laboratory experimentation or empirical observations. There is a vast number of theoretical models available in the literature concerning all natural sciences, including biology, ethology, ecology, pharmacology, nutrition and health care, medicine, geophysics, and so forth.

However, the focus of computing inspired by nature, under the umbrella of natural computing, is most often on problem solving instead of on theoretical modeling, and this leads to the second objective of computing based on nature. The second objective, thus, involves the study of natural phenomena, processes and even theoretical models for the development of computational systems and algorithms capable of solving complex problems. The motivation, in this case, is to provide (alternative) solution techniques to problems that could not be (satisfactorily) resolved by other more traditional techniques, such as linear, non-linear, and dynamic programming. In such cases, the computational techniques developed can also be termed *bio-inspired computing* or *biologically motivated computing* (Mange and Tomassini, 1998; de Castro and Von Zuben, 2004), or *computing with biological metaphors* (Paton, 1994).

As computing inspired by nature is mostly aimed at solving problems, almost all approaches are not concerned with the creation of accurate or theoretical models of the natural phenomena being modeled. In many situations highly abstract models, sometimes called metaphors (Paton, 1992), are proposed mimicking particular features and mechanisms from biology. What usually happens is that a natural phenomenon, or a theoretical model of it, gives rise to one particular computational tool and this is then algorithmically or mathematically improved to the extent that, in the end, it bears a far resemblance with the natural phenomenon that originally motivated the approach. Well-known examples of these can be found in the fields of artificial neural networks and evolutionary algorithms, which will be briefly discussed in the following.

A landmark work in the branch of bio-inspired computing was the paper by McCulloch and Pitts (1943), which introduced the first mathematical model of a neuron. This neuronal model, also known as artificial neuron, gave rise to a field of investigation of its own, the so-called *artificial neural networks* (Fausett, 1994; Bishop, 1996; Haykin, 1999; Kohonen, 2000). Artificial neural networks, to be discussed in Chapter 4, can be defined as information processing systems designed with inspiration taken from the nervous system, in most cases the human brain, and with particular emphasis on problem solving. There are several types of artificial neural networks (ANNs) and learning algorithms used to set up (train) these networks. ANNs are distinct from what is currently known as *computational neuroscience* (O'Reilly and Munakata, 2000; Dayan and Abbot, 2001; Trappenberg, 2002), which is mainly concerned with the development of biologically-based computational models of the nervous system.

Another computing approach motivated by biology arose in the mid 1960's with the works of I. Rechenberg (1973), H. P. Schwefel (1965), L. Fogel, A. Owens and M. Walsh (Fogel et al., 1966), and J. Holland (1975). These works gave rise to the field of *evolutionary computing* (Chapter 3), which uses ideas from evolutionary biology to develop *evolutionary algorithms* for search and optimization (Bäck et al., 2000a,b; Fogel, 1998; Michalewicz, 1996). Most evolutionary algorithms are rooted on the neo-Darwinian theory of evolution, which proposes that a population of individuals capable of reproducing and subject to genetic variation followed by natural selection result in new populations of individuals increasingly more fit to their environment. These simple three processes, when implemented in computers, result in evolutionary algorithms (EAs). The main types of EAs are the *genetic algorithms* (Mitchell, 1998; Goldberg, 1989), *evolution strategies* (Schwefel, 1995; Beyer, 2001), *evolutionary programming* (Fogel, 1999), *genetic programming* (Koza, 1992, 1994; Bahnzaf et al., 1997), and *classifier systems* (Booker et al., 1989; Holmes et al., 2002).

The term *swarm intelligence* (Chapter 5) was coined in the late 1980's to refer to cellular robotic systems in which a collection of simple agents in an environment interact based on local rules (Beni, 1988; Beni and Wang, 1989). Nowadays, the term is being used to describe any attempt to design algorithms or problem-solving devices inspired by the collective behavior of social organisms, from insect colonies to human societies. Swarm intelligence has two main frontlines: algorithms based on the collective behavior of social insects (Bonabeau et al., 1999), and algorithms based on cultures or sociocognition (Reynolds, 1994; Kennedy et al., 2001). In the first case, the collective behavior of ants and other insects has led to the development of algorithms for solving combinatorial optimization, clustering problems, and the design of autonomous robotic systems. Algorithms based on cultures and sociocognition demonstrated effectiveness in performing search and optimization on continuous and discrete spaces.

Artificial immune systems (AIS) or *immunocomputing* (Chapter 6), borrow ideas from the immune system and its corresponding models to design computational systems for solving complex problems (Dasgupta, 1999; de Castro and Timmis, 2002; Timmis et al., 2003). This is also a young field of research that emerged around the mid 1980's. Its application areas range from biology to robotics. Similarly to ANNs, EAs and swarm intelligence, different phenomena, processes, theories and models resulted in different types of immune algorithms, from evolutionary-like algorithms to network-like systems. Several other (emerging) types of algorithms inspired by nature can be found in the literature. For instance, it is possible to list the *simulated annealing* algorithm, the systems based on *growth* and *development*, and the *cells and tissues* models (Kirkpatrick et al., 1983; Aarts and Korst, 1989; Paton et al., 2004; Kumar and Bentley, 2003; Glover and Kochenberger, 2003; de Castro and Von Zuben, 2004).

Figure 1.6 summarizes the main components of computing inspired by nature to be discussed in this book and the respective chapters.

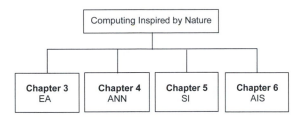

Figure 1.6: The main components of computing inspired by nature to be discussed in this book: ANN: artificial neural networks (neurocomputing); EA: evolutionary algorithms (evolutionary computing); SI: swarm intelligence; AIS: artificial immune systems (immunocomputing).

1.4.2. The Simulation and Emulation of Nature in Computers

While biologically inspired computing is basically aimed at solving complex problems, the second branch of natural computing provides new tools for the synthesis and study of natural phenomena that can be used to test biological theories usually not passive of testing via the traditional experimental and analytic techniques. It is in most cases a synthetic approach aimed at synthesizing natural phenomena or known patterns and behaviors. There is also a complementary relationship between biological theory and the synthetic processes of the simulation and emulation of nature by computers. Theoretical studies suggest how the synthesis can be achieved, while the application of the theory in the synthesis may be a test for the theory. There are basically two main approaches to the simulation and emulation of nature in computers: by using *artificial life* techniques or by using tools for studying the *fractal geometry of nature* (Figure 1.7).

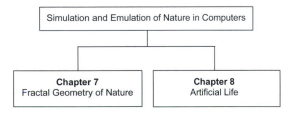

Figure 1.7: The two main approaches for the simulation and emulation of nature in computers and the chapters in which they are going to be presented.

Recent advances in computer graphics have made it possible to visualize mathematical models of natural structures and processes with unprecedented realism. The resulting images, animations, and interactive systems are useful as scientific, research, and educational tools in computer science, engineering, biosciences, and many other domains. One major breakthrough in the modeling and synthesis of natural patterns and structures is the recognition that nature is fractal in the sense that it can be successfully emulated by *fractal geometry* (Mandelbrot, 1983; Peitgen et al., 1992; Flake, 2000; Lesmoir-Gordon et al., 2000). In a simplified form, *fractal geometry* (Chapter 7) is the geometry of nature, with all its irregular, fragmented, and complex structures. In general, fractals are characterized by infinite details, infinite length, self-similarity, *fractal dimensions*, and the absence of smoothness or derivative. Nature provides many examples of fractals, for instance, ferns, coastlines, mountains, cauliflowers and broccoli, and many other plants and trees are fractals. Moreover, organisms are fractals; our lungs, our circulatory system, our brains, our kidneys, and many other body systems and organs are fractal.

There are a number of techniques for modeling fractal patterns and structures, such as *cellular automata* (Ilachinski, 2001; Wolfram, 1994), *L-systems* or *Lindenmayer systems* (Lindenmayer, 1968; Prusinkiewicz and Lindenmayer, 1990), *iterated function systems* (Hutchinson, 1981; Barnsley and Demko, 1985; Barnsley, 1988), *particle systems* (Reeves, 1983), *Brownian motion* (Fournier et al., 1982; Voss, 1985), and others. Their applications include computer-assisted landscape design, the study of developmental and growth processes, and the modeling and synthesis (and corresponding analysis) of an innumerable amount of natural patterns and phenomena. But the scope and importance of fractals and fractal geometry go far beyond these. Forest fires have fractal boundaries; deposits built up in electro-plating processes and the spreading of some liquids in viscous fluids have fractal patterns; complex protein surfaces fold up and wrinkle around toward three-dimensional space in a fractal dimension; antibodies bind to antigens through complementary fractal dimensions of their surfaces; fractals have been used to model the dynamics of the AIDS virus; cancer cells can be identified based on their fractal dimension; and the list goes on.

Artificial life (ALife) will be discussed in Chapter 8. It corresponds to a research field that complements traditional biological sciences concerned with the analysis of living organisms by trying to synthesize life-like behaviors and creatures in computers and other artificial media (Langton, 1988; Adami, 1998; Levy, 1992). Differently from nature-inspired computing, approaches in the ALife field are usually not concerned with solving any particular problem. ALife has, as major goals, to increase the understanding of nature (life-as-it-is), enhance our insight into artificial models and possibly new forms of life (life-as-it-could-be), and to develop new technologies such as software evolution, sophisticated robots, ecological monitoring tools, educational systems, computer graphics, etc.

ALife systems have, thus, been designed to simulate and emulate behaviors or organisms in order to allow the study or simulation of natural phenomena or processes. In most cases it emphasizes the understanding of nature, and applications as a problem solver are left in second plan. For instance, ALife systems have been created to study traffic jams (Resnick, 1994); the behavior of synthetic biological systems (Ray, 1994); the evolution of organisms in virtual environments (Komosinski and Ulatowski, 1999); the simulation of collective behaviors (Reynolds, 1987); the study and characterization of computer viruses (Spafford, 1991); and others. Its major ambition is to build living systems out of non-living parts; that is, to accomplish what is known as 'strong ALife' (Sober, 1996; Rennard, 2004).

1.4.3. Computing with Natural Materials

Computing with natural materials is concerned with new computing methods based on other natural material than silicon. These methods result in a non-standard computation that overcomes some of the limitations of standard, sequential John von Neumann computers. As any mathematical operation can be broken down into bits, and any logical function can be built using an AND and a NOT gate, any computable 'thing' can be worked out by appropriately wired AND and NOT gates. This independence of a specific representation makes it possible to use new concepts for the computational process based on natural materials, such as, chemical reactions, DNA molecules, and quantum mechanical devices.

The history of computer technology has involved a sequence of changes from one type of realization to another; from gears to relays to valves to transistors to integrated circuits. Nowadays, a single silicon chip can contain millions of logic gates. This miniaturization of the most basic information processing elements is inevitably going to reach a state where logic gates will be so small so as to be made of atoms. In 1965 G. Moore (1965) observed that there is an exponential growth in the number of transistors that can be placed in an integrated circuit. According to what is now known as the "Moore's law", there is a doubling of transistors in a chip every couple of years. If this scale remains valid, by the end of this decade, silicon-based computers will have reached their limits in terms of processing power. One question that remains thus, is "What other materials or media can be used to perform computation in place of silicon?". Put in another form, after certain level of miniaturization of the computing devices, the standard physical laws will be no longer applicable, because quantum effects will begin to take place. Under this perspective, the question that arises refers to "How should we compute under quantum effects?".

Computing with natural materials is the approach that promises to bring a major change in the current computing technology in order to answer the questions above. Motivated by the need to identify alternative media for computing, researchers are now trying to design new computers based on molecules, such as membranes, DNA and RNA, or quantum theory. These ideas resulted in what is now known as *molecular computing* (Păun et al., 1998; Gramß et al., 2001; Calude and Păun, 2001; Păun and Cutkosky, 2002; Sienko et al., 2003) and *quantum computing* or *quantum computation* (Hirvensalo, 2000; Nielsen and Chuang, 2000; Pittenger, 2000), respectively. Figure 1.8 summarizes the main components of computing with natural materials.

Molecular computing is based upon the use of biological molecules (biomolecules) to store information together with genetic engineering (biomolecular) techniques to manipulate these molecules so as to perform computation. It constitutes a powerful combination between computer science and molecular biology. The field can be said to have emerged due to the work of L. Adleman who, in 1994, solved an NP-complete problem using DNA molecules and biomolecular techniques for manipulating DNA (Adleman, 1994). Since then, much has happened: several other 'molecular solutions' to complex problems have been proposed and molecular computers have been shown to perform universal computation. The main advantages of molecular computing are its high speed, energy efficiency, and economical information storage. An overall, striking observation about molecular computing is that, at least theoretically, there seem to be many diverse ways of constructing molecular-based universal computers. There are, of course, the possibility of errors and difficulties in implementing real molecular computers. When compared with the currently known silicon-based computers, molecular computers offer some unique features, such as the use of molecules as data structures and the possibility of performing massively parallel computations. In Chapter 9, this book reviews one particular molecular computing approach, namely, *DNA computing*.

When the atomic scale of logic gates is reached, the rules that will prevail are those of *quantum mechanics*, which are quite different from the classical rules that determine the properties of conventional logic gates. Thus, if computers are to become even smaller in the (not so far) future, quantum technology must complement or supplement the current technology. *Quantum computation* and *quantum information*, introduced in Chapter 10, is the study of the information processing tasks that can be accomplished using quantum mechanical systems (Nielsen and Chuang, 2000). In quantum computers information is stored at the microphysical level where quantum mechanisms prevail.

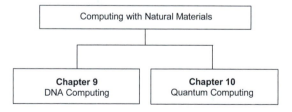

Figure 1.8: The two main branches of computing with natural materials or media.

In such cases, a bit could represent both zero and one simultaneously, and measurements and manipulations of these quantum bits are modeled as matrix operations. The seminal paper by R. Feynman (1982) introduced a computer capable of simulating quantum physics. A little later on, D. Deutsch (1984) published a paper where he demonstrated the universal computing capability of such quantum computers. Another seminal work that served to boost the field was the paper by P. Shor (1994) introducing the first quantum algorithm capable of performing efficient factorization, something that only a quantum computer could do. What is important to remark about quantum computing, though, is that it can provide entirely novel types of computation with qualitatively new algorithms based on quantum mechanics; quantum technology thus offers much more than simply adding the capability of processing more bits using the current silicon-based computers. Therefore, quantum computing aims at nontraditional hardware that would allow quantum effects to take place.

1.5   WHEN TO USE NATURAL COMPUTING APPROACHES

While studying this book, the reader will be faced with a diverse range of problems to be solved, phenomena to be synthesized and questions to be answered. In all cases, one or more of the natural computing approaches briefly reviewed above will be used to solve the problem, synthesize the phenomenon or answer the question. However, it is important to acknowledge that natural computing is not the only field of investigation that provides solutions to these, nor is it always the most suitable and efficient approach. To clarify when natural computing should be used, let us present some examples and arguments in each of the three branches. Let us assume that you have just finished your undergraduation course and now have an engineering or science degree in hand. In your first interview for a job in a major company, you are posed with three problems and given some time to provide solutions to them.

Problem 1: the company is expanding rapidly and now wants to build a new factory in a country so far unattended. The site where the factory is to be built has already been chosen as long as the cities to be attended by this factory. Figure 1.9 depicts the scenario. The problem is: given all the cities in the map find the smallest route from the factory to all cities passing by each city exactly once and returning to the departure city (factory). This problem is well-known from the literature and is termed *traveling salesman problem* (TSP).

Such problems have several practical applications, from fast food delivery to printed circuit board design. Although this problem may be simple to state it is hard to solve, mainly when the number of cities involved is large. For example, if there is one factory plus three other cities, then there are 6 possible routes; if there is one factory plus four other cities, then there are 24 possible routes; if there is one factory plus five other cities, then there are 120 possible routes; and so on. This corresponds to a factorial growth in the number of possible routes in relation to the number of cities to be attended.

Figure 1.9: Map of the new country to be attended by the company. The city where the factory is going to be built is detached together with the 27 cities to be attended.

The most straightforward solution you could provide to this problem is to suggest the testing of all possible routes and the choice of the smallest one; an approach we usually call 'brute force' or 'exhaustive search'. Although simple, this approach is only efficient for a small number of cities. Assuming your computer is capable of analyzing 100 routes per second, it would take much less than a second to solve a three cities instance of this problem and a bit more than a second to solve a five cities instance of the problem. For the problem presented, however, the scenario is much different: there are 27! possible routes, and this corresponds to approximately 1.1×10^{28} possible routes to be tested. In your computer, the exhaustive search approach would take, approximately, 3.0×10^{23} hours, or 1.3×10^{22} days, or 3.5×10^{19} years of processing time to provide a solution.   ■

Problem 2: the expansion plans of the company include the development of motion picture animations. The first animation to be designed involves a herd of zebras running away from a few hungry lions. Your task is to propose a means of realistically and effectively simulating the collective behavior of the zebras.

The first proposal you may come out with is to write a *script* for each zebra; that is, a computer program that fully describes the action of the zebras in the field. This, of course, would seem easier than writing a single script to coordinate the whole herd. Again, though simple, this approach has some drawbacks when the number of zebras involved is large and when more realistic scenarios are to be created. First, as the script is static, in the sense that the behaviors modeled do not change over time, the resultant collective behavior is the same every time the simulation is run. Second, the number of scripts to be written grows with the size of the herd and the complexity of the scripts grows in a much larger scale, because the more zebras, the more difficult it becomes to coordinate the behavior of all animals so as to avoid collisions, etc., and all aspects of the collective behavior have to be accounted for within the script.

In such cases, it is much more efficient and realistic to try and find a set of simple rules that will be responsible for guiding the individual behavior of the zebras across the field, similarly to the discussion presented in Section 1.2 for a bird flock. More details on how to solve this type of problem will be provided in Chapter 8.

To make your task more complete and challenging, the interviewer also asks how you would create the surrounding environment in which the animals will be placed. The simplest approach, in this case, would be to draw the scenarios or maybe photograph a real scenario and use it as a background for the animation, as sometimes done in cartoons. The drawing or pictures have to be passed, frame by frame, and the animals placed on this environment for running the simulation. A problem with this solution is the amount of memory required to store all these images (drawings or photographs). For instance, a digital photo of medium quality requires around 500KB of memory to be stored.

An efficient way of reproducing natural scenarios and phenomena is by using techniques from fractal geometry (Chapter 7). Brownian motion, cellular automata, L-systems, iterated function systems, and particle systems can be used to generate scenarios of natural environments; simulate fire, chemical reactions, etc., all with unprecedented realism and, in most cases, at the expense of a small amount of memory usage. These techniques usually employ a small number of rules or equations that are used to generate patterns or simulate phenomena *on line*, thus reducing the need for huge storage devices. ∎

Problem 3: in the current computer technology, bits constitute the most basic unit of information. They are physically stored in integrated circuits or chips based on silicon technology. The need to increase the memory capacity and processing speed of computers forced the chips to be able to accommodate more and more bits, thus promoting a miniaturization of the electronic devices. To have an idea of the dramatic decrease in size of electronic devices, by the year 2000 approximately 10^9 atoms were necessary to represent a single bit of information. It is estimated that by the year 2020 a single atom will be used to represent one bit of information. The question is: if we will unavoidably reach the miniaturization limit of current computer technology, what are the other types (than silicon) of materials that could be used for computing?

Note that the answer to this question may involve a change of computing *paradigm*. That is, if we are to investigate the possibility of computing with a material that is different from silicon, then a completely new type of storing and manipulating information may be used. An example of a potential novel material for computing was given in Section 1.2 with the use of DNA. In that case, information was stored in DNA molecules and genetic engineering techniques were proposed as means to manipulate DNA for computing (Chapter 9). ∎

To summarize, natural computing approaches almost invariably correspond to alternative solution techniques to all the problems discussed in this volume and many others to which they are applicable. In many situations there are other solution techniques for a given problem and these may even provide superior results. Thus, it is important to carefully investigate the problem to be solved before choosing a specific solution method, be it a natural computing tool or a different one. To give some hints on when natural computing should be used, consider the following list. Natural computing could be used when:

- The problem to be solved is complex, i.e., involves a large number of variables or potential solutions, is highly dynamic, nonlinear, etc.

- It is not possible to guarantee that a potential solution found is optimal (in the sense that there is no better solution), but it is possible to find a quality measure that allows the comparison of solutions among themselves.

- The problem to be solved cannot be (suitably) modeled, such as pattern recognition and classification (e.g., vision) tasks. In some cases, although it is not possible to model the problem, there are examples (samples) available that can be used to 'teach' the system how to solve the problem, and the system is somehow capable of 'learning from examples'.

- A single solution is not good enough; that is, when diversity is important. Most standard problem-solving techniques are able to provide a single solution to a given problem, but are not capable of providing more than one solution. One reason for that is because most standard techniques are deterministic, i.e., always use the same sequence of steps to find the solution, and natural computing is, in its majority, composed of probabilistic methods.

- Biological, physical, and chemical systems and processes have to be simulated or emulated with realism. Euclidean geometry is very good and efficient to create man-made forms, but has difficulty in reproducing natural patterns. This is because nature is fractal, and only fractal geometry provides the appropriate tools with which to model nature.

- Life behaviors and phenomena have to be synthesized in artificial media. No matter the artificial media (e.g., a computer or a robot), the essence of a given natural behavior or pattern is extracted and synthesized in a, usually, much simpler form in artificial life systems.

- The limits of current technology are reached or new computing materials have to be sought. Nature abounds with information storage and processing systems, and the scientific and engineering aspects of how to use these natural materials to compute are the main challenges of the third branch of natural computing: computing with natural materials.

These prerequisites may not sound so clear yet, but you will naturally have a better idea of why, when, and how natural computing should or have been used as you progress in the book.

1.6 SUMMARY

Natural computing is the terminology used to refer to three types of systems: 1) computational algorithms for problem-solving developed by taking inspiration from natural phenomena; 2) computational systems for the simulation and/or emulation of nature; and 3) novel computing devices or paradigms that use media other than silicon to store and process information.

Although all these branches are quite young from a scientific perspective, several of them are already being used in our everyday lives. For instance, we now have 'intelligent' washing machines, games, (virtual) pets, control systems, etc., all based on, or using, computing devices inspired by nature; research on ALife and the fractal geometry of nature has allowed the creation of realistic models of nature and the simulation and emulation of several plants and animal species, and has aided the study of developmental processes and many other natural phenomena; and computing with natural materials has given new insights into how to complement or supplement computer technology as known currently.

Natural computing is highly relevant for today's computer scientists, engineers, biologists, and other professionals because it offers alternative, and sometimes brand new, solutions to problems yet unsolved or poorly resolved. It has also provided new ways of seeing, understanding, and interacting with nature. There is still much to come and much to do in such a broad and young field of investigation as natural computing. However, we now know for sure that this is not only a promising field of research; its many applications, outcomes, and perspectives have been affecting our lives, even if this is not perceived by most people. And this is just the beginning. Welcome to the natural computing age!

1.7 QUESTIONS

1. Find evidences in the literature that support the idea that birds in a flock do not follow a leader.

2. The movie "A Bug's Life" by Disney/Pixar starts with an ant colony harvesting for food. But instead of harvesting food for the colony, the ants were harvesting for the grasshoppers, who obliged them to do so. The long line of ants carrying food is stopped when a leaf falls down on their way, more precisely on their *trail*. The ants in front of the line just interrupted by the leaf start despairing, claiming they do not know what to do; that is, where to go. Another 'more instructed' ant suggests that they might walk around the leaf thus keeping their harvesting. One ant even claims that such catastrophe was not nearly as bad as the "twig of 93".

    Based upon this brief summary of the beginning of the movie "A Bug's Life", what can you infer about the way ants forage and harvest food, more specifically, why did the ants loose their direction when the leaf fell in their trail?

3.   What is the Occam's Razor (also spelled Ockham's Razor)? What is its relationship with the approach usually adopted in natural computing (Section 1.3)?

4.   Name some limitations of the current silicon-based computing paradigm.

5.   Name one natural phenomenon, process, or system that is a potential candidate to become a new natural computing technique. Include the biological and theoretical background, and the utility of the approach proposed.

1.8   REFERENCES

[1]   Aarts, E. and Korst, J. (1989), *Simulated Annealing and Boltzman Machines – A Stochastic Approach to Combinatorial Optimization and Neural Computing*, John Wiley & Sons.

[2]   Adami C. (1998), *An Introduction to Artificial Life*, Springer-Verlag/Telos.

[3]   Adleman, L. M. (1994), "Molecular Computation of Solutions to Combinatorial Problems", *Science*, **226**, November, pp. 1021–1024.

[4]   Attwood, T. K. and Parry-Smith, D. J. (1999), *Introduction to Bioinformatics*, Prentice Hall.

[5]   Bäck, T., Fogel, D. B. and Michalewicz, Z. (2000), *Evolutionary Computation 1 Basic Algorithms and Operators*, Institute of Physics Publishing (IOP), Bristol and Philadelphia.

[6]   Bäck, T., Fogel, D. B. and Michalewicz, Z. (2000), *Evolutionary Computation 2 Advanced Algorithms and Operators*, Institute of Physics Publishing (IOP), Bristol and Philadelphia.

[7]   Bahnzaf, W., Nordin, P., Keller, R. E. and Frankone, F. D. (1997), *Genetic Programming: An Introduction*, Morgan Kaufmann.

[8]   Baldi, P. and Brunak, S. (2001), *Bioinformatics: The Machine Learning Approach*, 2nd ed., Bradford Books.

[9]   Barnsley, M. F. (1988), *Fractals Everywhere*, Academic Press.

[10]   Barnsley, M. F. and Demko, S. (1985), "Iterated Function Systems and the Global Construction of Fractals", *Proc. of the Royal Soc. of London*, **A339**, pp. 243-275.

[11]   Beni, G. (1988), "The Concept of Cellular Robotic Systems", *Proc. of the IEEE Int. Symp. on Intelligent Control*, pp. 57–62.

[12]   Beni, G. and Wang, J. (1989), "Swarm Intelligence", *Proc. of the 7th Annual Meeting of the Robotics Society of Japan*, pp. 425–428.

[13]   Bentley, P. J. (2001), *Digital Biology*, Headline.

[14]   Beyer, H.-G. (2001), *Theory of Evolution Strategies*, Springer-Verlag.

[15]   Bishop, C. M. (1996), *Neural Networks for Pattern Recognition*, Oxford University Press.

[16]   Bonabeau E., Dorigo, M. and Theraulaz, T. (1999), *Swarm Intelligence: From Natural to Artificial Systems*, New York: Oxford University Press.

[17]   Booker, L. B., Goldberg, D. E. and Holland, J. H. (1989), "Classifier Systems and Genetic Algorithms", *Artificial Intelligence*, **40**, pp. 235–282.

[18]   Calude, C. S. and Păun, G. (2001), *Computing with Cells and Atoms: An Introduction to Quantum, DNA, and Membrane Computing*, Taylor & Francis.

[19]   Dasgupta, D. (1999), *Artificial Immune Systems and Their Applications*, Springer-Verlag.

[20] Dayan, P. and Abbot, L. F. (2001), *Theoretical Neuroscience: Computational and Mathematical Modeling of Neural Systems*, The MIT Press.

[21] de Castro, L. N. (2005), "Natural Computing", In M. Khosrow-Pour, *Encyclopedia of Information Science and Technology*, Idea Group Inc.

[22] de Castro, L. N. and Timmis, J. I. (2002), *Artificial Immune Systems: A New Computational Intelligence Approach*, Springer-Verlag.

[23] de Castro, L. N. and Von Zuben, F. J. (2004), *Recent Developments in Biologically Inspired Computing*, Idea Group Publishing.

[24] Deutsch, D. (1985), "Quantum Theory, the Church-Turing Principle and the Universal Quantum Computer", *Proc. of the Royal Soc. of London*, **A 4000**, pp. 97–117.

[25] Fausett, L. (1994), *Fundamentals of Neural Networks: Architectures, Algorithms, and Applications*, Prentice Hall.

[26] Feynman, R. P. (1982), "Simulating Physics with Computers", *Int. Journal of Theor. Physics*, **21**(6/7), pp. 467–488.

[27] Flake, G. W. (2000), *The Computational Beauty of Nature*, MIT Press.

[28] Fogel, D. B. (1998), *Evolutionary Computation: Toward a New Philosophy of Machine Intelligence*, IEEE Press.

[29] Fogel, L. J. (1999), *Intelligence through Simulated Evolution: Forty Years of Evolutionary Programming*, Wiley-Interscience.

[30] Fogel, L. J., Owens, A. J. and Walsh, M. J. (1966), *Artificial Intelligence through Simulated Evolution*, Wiley, New York.

[31] Fournier, A., Fussell, D. and Carpenter, L. (1982), "Computer Rendering of Stochastic Models", *Comm. of the ACM*, **25**, pp. 371–384.

[32] Glover, F. W. and Kochenberger, G. A., (2003), *Handbook of Metaheuristics*, Springer.

[33] Goldberg, D. E. (1989), *Genetic Algorithms in Search, Optimization, and Machine Learning*, Addison-Wesley Pub Co.

[34] Gramß, T., Bornholdt, S., Groß, M., Mitchell, M. and Pellizzari, T. (2001), *Non-Standard Computation*, Wiley-VCH.

[35] Haykin, S. (1999), *Neural Networks: A Comprehensive Foundation*, 2nd ed., Prentice Hall.

[36] Hirvensalo, M. (2000), *Quantum Computing*, Springer-Verlag.

[37] Holland, J. H. (1975), *Adaptation in Natural and Artificial Systems*, MIT Press.

[38] Holmes, J. H., Lanzi, P. L., Stolzmann, W. and Wilson, S. W. (2002). "Learning Classifier Systems: New Models, Successful Applications", *Information Processing Letters*, **82**(1), pp. 23–30.

[39] Hutchinson, J. (1981), "Fractals and Self-Similarity", *Indiana Journal of Mathematics*, **30**, pp. 713–747.

[40] Ilachinski, A. (2001), *Cellular Automata: A Discrete Universe*, World Scientific.

[41] Kennedy, J., Eberhart, R. and Shi. Y. (2001), *Swarm Intelligence*, Morgan Kaufmann Publishers.

[42] Kirkpatrick, S., Gerlatt, C. D. Jr. and Vecchi, M. P. (1983), "Optimization by Simulated Annealing", *Science*, **220**, 671–680.

[43] Kohonen, T. (2000), *Self-Organizing Maps*, Springer-Verlag.

[44] Komosinski, M. and Ulatowski, S. (1999), "Framsticks: Towards a Simulation of a Nature-Like World, Creatures and Evolution", In D. Floreano and F. Mondada, *Lecture Notes in Artificial Intelligence 1674* (Proc. of the 5th European Conf. on Artificial Life), pp. 261–265.

[45]   Koza, J. R. (1992), *Genetic Programming: On the Programming of Computers by Means of Natural Selection*, MIT Press.

[46]   Koza, J. R. (1994), *Genetic Programming II: Automatic Discovery of Reusable Programs*, MIT Press.

[47]   Kumar, S. and Bentley, P. J. (2003), *On Growth, Form and Computers*, Academic Press.

[48]   Langton, C. (1988), "Artificial Life", In C. Langton (ed.), *Artificial Life*, Addison-Wesley, pp. 1-47.

[49]   Lesmoir-Gordon, N., Rood, W. and Edney, R. (2000), *Introducing Fractal Geometry*, ICON Books UK.

[50]   Levy, S. (1992), *Artificial Life*, Vintage Books.

[51]   Lindenmayer, A. (1968), "Mathematical Models for Cellular Interaction in Development, Parts I and II", *Journal of Theoretical Biology*, **18**, pp. 280–315.

[52]   Mandelbrot, B. (1983), *The Fractal Geometry of Nature*, W. H. Freeman and Company.

[53]   Mange, D. and Tomassini, M. (1998), *Bio-Inspired Computing Machines: Towards Novel Computational Architecture*, Presses Polytechniques et Universitaires Romandes.

[54]   McCulloch, W. and Pitts, W. H. (1943), "A Logical Calculus of the Ideas Immanent in Nervous Activity", *Bulletin of Mathematical Biophysics*, **5**, pp. 115–133.

[55]   Michalewicz, Z. (1996), *Genetic Algorithms + Data Structures = Evolution Programs*, Springer-Verlag, 3rd Ed.

[56]   Mitchell, M. (1998), *An Introduction to Genetic Algorithms*, The MIT Press.

[57]   Moore, G. E. (1965), "Cramming More Components into Integrated Circuits", *Electronics*, **38**(8).

[58]   Nielsen, M. A. and Chuang, I. L. (2000), *Quantum Computation and Quantum Information*, Cambridge University Press.

[59]   O'Reilly, R. C. and Munakata, Y. (2000), *Computational Explorations in Cognitive Neuroscience: Understanding the Mind by Simulating the Brain*, The MIT Press.

[60]   Paton, R. (1992), "Towards a Metaphorical Biology", *Biology and Philosophy*, **7**, pp. 279–294.

[61]   Paton, R. (Ed.) (1994), *Computing with Biological Metaphors*, Chapman & Hall.

[62]   Paton, R., Bolouri, H. and Holcombe, M. (2004), *Computing in Cells and Tissues: Perspectives and Tools of Thought*, Springer-Verlag.

[63]   Păun, G. and Cutkosky, S. D. (2002), *Membrane Computing*, Springer-Verlag.

[64]   Păun, G., Rozenberg, G. and Saloma, A. (1998), *DNA Computing*, Springer-Verlag.

[65]   Peitgen, H.-O, Jürgens, H. and Saupe, D. (1992), *Chaos and Fractals: New Frontiers of Science*, Springer-Verlag.

[66]   Pittenger, A. O. (2000), *An Introduction to Quantum Computing Algorithms*, Birkhäuser.

[67]   Prusinkiewicz, P. and Lindenmayer, A. (1990), *The Algorithmic Beauty of Plants*, Springer-Verlag.

[68]   Ray, T. S. (1994), "An Evolutionary Approach to Synthetic Biology", *Artificial Life*, **1**(1/2), pp. 179–209.

[69]   Rechenberg, I. (1973), *Evolutionsstrategie: Optimierung Technischer Systeme Nach Prinzipien der Biologischen Evolution*, Frommann-Holzboog, Stuttgart.

[70]   Reeves, W. T. (1983), "Particle Systems – A Technique for Modeling a Class of Fuzzy Objects", *ACM Transactions on Graphics*, **2**(2), pp. 91–108.

[71]   Rennard, J.-P. (2004), "Perspectives for Strong Artificial Life", In L. N. de Castro and F. J. Von Zuben (Eds.), *Recent Developments in Biologically Inspired Computing, Idea Group Publishing*, Chapter 12, pp. 301–318.

[72]   Resnick, M. (1994), *Turtles, Termites, and Traffic Jams: Explorations in Massively Parallel Microworlds*, MIT Press.

[73]   Reynolds, C. W. (1987), "Flocks, Herds, and Schools: A Distributed Behavioral Model", *Computer Graphics*, **21**(4), pp. 25–34.

[74]   Reynolds, R. G. (1994), "An Introduction to Cultural Algorithms", In A. V. Sebald and L. J. Fogel (Eds.), *Proceedings of the Third Annual Conference on Evolutionary Programming*, World Scientific, River Edge, New Jersey, pp. 131–139.

[75]   Schwefel, H. –P. (1965), *Kybernetische Evolutionals Strategie der Experimentellen Forschung in der Stromungstechnik*, Diploma Thesis, Technical University of Berlin.

[76]   Schwefel, H. –P. (1995), *Evolution and Optimum Seeking*, John Wiley & Sons.

[77]   Shor, P. W. (1994), "Algorithms for Quantum Computation: Discrete Logarithms and Factoring", *Proc. of the 35th Annual Symposium on Foundations of Computer Science*, Santa Fe, NM, IEEE Computer Society Press, pp. 124–134.

[78]   Sienko, T., Adamatzky, A. and Rambidi, N. (2003), *Molecular Computing*, MIT Press.

[79]   Sober, E. (1996), "Learning from Functionalism – Prospects for Strong Artificial Life". In M. A. Boden (Ed.), *The Philosophy of Artificial Life*, Oxford: Oxford University Press, pp. 361–378.

[80]   Spafford, E. H. (1991), "Computer Viruses – A Form of Artificial Life?", in C. G. Langton, C. Taylor, J. D. Farmer (eds.), *Artificial Life II*, Addison-Wesley, pp. 727–746.

[81]   Timmis, J., Bentley, P. J. and Hart, E. (Eds.) (2003), *Artificial Immune Systems*, Proc. of the 2nd International Conference on Artificial Immune Systems (ICARIS 2003), Springer-Verlag.

[82]   Trappenberg, T. (2002), *Fundamentals of Computational Neuroscience*, Oxford University Press.

[83]   Voss, R. F. (1985), "Random Fractals Forgeries", In R. A. Earnshaw (ed.), Fundamental Algorithms for Computer Graphics, Springer-Verlag: Berlin, pp. 805–835.

[84]   Waterman, M. S. (1995), *Introduction to Computational Biology: Maps, Sequences, and Genomes*, Chapman & Hall.

[85]   Wolfram, S. (1994), *Cellular Automata and Complexity*, Perseus Books.

[86]   Zurada, J. M., Robinson, C. J. and Marks, R. J. (1994), *Computational Intelligence: Imitating Life*, IEEE Press.

CHAPTER 2

CONCEPTUALIZATION

"Adaptationism, the paradigm that views organisms as complex adaptive machines whose parts have adaptive functions subsidiary to the fitness-promoting function of the whole, is today about as basic to biology as the atomic theory is to chemistry. And about as controversial."
(D. Dennett, Darwin's Dangerous Idea: Evolution and the Meanings of Life, Penguin Books, 1995, p. 249)

"Nothing is wrong with good metaphors so long as we don't take them as reality."
(C. Emmeche, 1997, Aspects of complexity in life and science, Philosophica, 59(1), pp. 41-68.)

"The complexity of a model is not required to exceed the needs required of it."
(B. S. Silver, The Ascent of Science, 1998, Oxford University Press, p. 79)

2.1   INTRODUCTION

Perhaps the most remarkable characteristic of natural computing is the encompassing of a large variety of disciplines and fields of research. Ideas, principles, concepts, and theoretical models from biology, physics, and chemistry are most often required for a good understanding and development of natural computing. This interdisciplinarity and multidisciplinarity also involves the use of concepts that aid in the description and understanding of the underlying phenomenology. However, most of these concepts are made up of 'slippery' words; words whose meaning can be different under different domains and words whose definitions are not available or agreed upon yet.

Therefore, this chapter presents, in a descriptive manner, what is meant by words such as adaptation, complexity, agents, self-organization, emergence, and fractals. Very few closed definitions will be provided. Instead, descriptive explanations sufficient for the conceptualization and comprehension of the basic ideas incorporated in natural computing will be given. Some popular science books will be cited because they introduce important topics using a terminology with little formalism and mathematics. Some more technical literature may eventually be cited, but most of it can be readily reached with a quick web search.

For those readers anxious to put their hands at work with some computational and algorithmic description of natural computing, this chapter can be temporarily skipped. However, readers are strongly encouraged to go through this chapter thoroughly, because it will give him/her better philosophical basis and insights into concepts that will be widely used throughout the book. This chapter will also provide a better understanding of what is meant by and what is behind of many of the phenomena, processes, and systems studied in natural computing. Finally, this chapter makes use of the many concepts reviewed to start our discussion of nature, which led to the development of natural computing. Several natural systems and processes are used to illustrate the meaning of the concepts investigated here. Most of these examples from nature will be returned to in later chapters.

2.1.1. Natural Phenomena, Models, and Metaphors

All the approaches discussed in this volume are rooted on and sometimes enhanced by their natural plausibility and inspiration. The focus is on how nature has offered inspiration and motivation for their development. However, all these approaches are very appealing to us mainly for computational reasons, and they also may help us study and understand the world we inhabit, and even to create new worlds, new forms of life, and new computing paradigms. They hold out the hope of offering computationally sufficient accurate mechanistic accounts of the natural phenomena they model, mimic, or study, almost always with a view of computing, understanding, and problem solving. They have also radically altered the way we think of and see nature; the computational beauty and usefulness of nature.

Modeling is an integral part of many scientific disciplines and lies behind great human achievements and developments. Most often, the more complex a system, the more simplifications are embodied in its models. The term *model* can be found in many different contexts and disciplines meaning a variety of different things. Trappenberg (2002) has defined "models [as] abstractions of real world systems or implementations of a hypothesis in order to investigate particular questions or to demonstrate particular features of a system or a hypothesis." (Trappenberg, 2002; p. 7) It corresponds to a (schematic) description of a system, theory, or phenomenon, which accounts for its known or inferred properties and that may be used for further study of its characteristics. Put in a different form, models can be used to represent some aspect of the world, some aspect of theories about the world, or both simultaneously. The representative usefulness of a model lies in its ability to teach us something about the phenomenon it represents. They mediate between the real world and the theories and suppositions about that world (Peck, 2004).

The critical steps in constructing a model are the selection of salient features and laws governing the behavior of the phenomena under investigation. These steps are guided by *metaphor* and knowledge transfer (Holland, 1998). However, for purely practical reasons, many details are usually discarded. In the particular case of natural computing, models are most often simple enough to understand, but rich enough to provide (*emergent*) behaviors that are surprising, interesting, useful, and significant. If all goes well, what may commonly be the case, the result may allow for the prediction and even reproduction of behaviors observed in nature, and the achievement of satisfactory performances when a given function is required from the model.

The word *metaphor* comes from the Greek for 'transference'. It corresponds to the use of language that assigns one thing to designate another, in order to characterize the latter in terms of the former. Metaphors have traditionally been viewed as implicit comparisons. According to this view, metaphors of the form *X is a Y* can be understood as *X is like Y*. Although metaphors can suggest a comparison, they are primarily attributive assertions, not merely comparisons (Wilson and Keil, 1997; p. 535-537). For example, to name a computational tool developed with inspiration in the human brain an 'artificial neural network' or a 'neurocomputing device' corresponds to attributing salient properties of the human brain to the artificial neural network or neurocomputing device. The first part of this book, dedicated to computing inspired by nature, is sometimes referred to as *computing with biological metaphors* (Paton, 1994) or *biologically inspired computing* (de Castro and Von Zuben, 2004).

The use of metaphors from nature as a means or inspiration to develop computational tools for problem solving can also be exemplified by the development of 'artificial immune systems' for computer protection against viruses. One might intuitively argue: "if the human immune system is capable of protecting us against viruses and bacteria, why can't I look into its basic functioning and try to extract some of these ideas and mechanisms to engineer a computer immune system?" Actually, this type of metaphor has already been extracted, not only by academic research institutions, but also by leading companies in the computing area, such as IBM (Chapter 6). There is, however, an important difference between a metaphor and a model. While models are more concerned with quantitatively reproducing some phenomena, metaphors are usually high-levels abstractions and inspirations taken from a system or process in order to develop another. Most metaphors are basically concerned with the extraction or reproduction of qualitative features.

A simple formula, a computer simulation, a physical system; all can be models of a given phenomenon or process. What is particularly important, though, is to bear in mind what are the purposes of the model being created. In theoretical biology and experimental studies, models may serve many purposes:

- Through modeling and identification it is possible to provide a deeper and more quantitative description of the system being modeled and its corresponding experimental results.
- Models can aid in the critical analysis of hypotheses and in the understanding of the underlying natural mechanisms.
- Models can assist in the prediction of behaviors and design of experiments.
- Models may be used to simulate and stimulate new and more satisfactory approaches to natural systems, such as the behavior of insect societies and immune systems.
- Models may allow the recovery of information from experimental results.

An *experiment* can be considered as a procedure performed in a controlled environment for the purpose of gathering observations, data, or facts, demonstrating known facts or theories, or testing hypotheses or theories. Most biological experiments are usually made *in vivo*, within a living organism (e.g., rats and mice), or *in vitro*, in an artificial environment outside the living organism (e.g., a test tube).

There is a significant conceptual difference between experiment, simulation, realization, and emulation. In contrast to experiments, *simulations* and *realizations* are different categories of models (Pattee, 1988). Simulations are metaphorical models that 'stand for' something else, and may cover different levels of fidelity or abstraction. They can be performed by physical modeling, by writing a special-purpose computer program, or by using a more general simulation package that is usually still aimed at a particular kind of simulation. They can be used, for instance, to explore theories about how the real-world functions based on a controlled medium (e.g., a computer). As an example, the simulation of a car accident can be performed by specifying the place and conditions in which the car was driven and then using a given medium (e.g., the computer) to run the simulation. Computer simulation is pervasive in natural computing. It has been used to design problem-solving techniques that mimic the behavior of several biological phenomena (Chapter 3 to Chapter 6), it has served to drive synthetic environments and virtual worlds (Chapter 7 and Chapter 8), and it has been used to simulate DNA computers (Chapter 9).

The *realization* of a system or organism corresponds to a literal, material model that implements certain functions of the original; it is a substantive functional device. Roughly speaking, a realization is evaluated primarily by how well it can function as an implementation of a design specification, and not in relation to the goodness of the measurements (mappings) they perform. A system or function is used to realize another when one performs in exactly the same way as another (Pattee, 1988; Mehler, 2003). To *emulate* a system is to imitate or reproduce its functions using another system or medium. The emulating system has to perform the same functions of the emulated system in the way the latter does. A typical example in computer science is the emulation of one computer by (a program running on) another computer. You may emulate a system as a replacement for the system, whereas you may simulate a system if the goal is, for instance, simply to analyze or study it.

Natural computing approaches are aimed at simulating, emulating, and sometimes realizing natural phenomena, organisms, and processes with distinct goals. The metaphorical representation of simulations makes them suitable for designing problem solving techniques and mimics of nature. Realizations of nature, on the contrary, would be the primary target of the so-called strong artificial life (Chapter 8). It is also important to acknowledge that, as most natural computing approaches to be studied here usually have not the same goals as models, they have the advantages of being explicit about the assumptions and relevant processes incorporated, allowing for a closer control of the variables involved, and providing frameworks to explain a wide range of phenomena.

Due mainly to these differences in goals and levels of details incorporated, most of the highly simplified models discussed in this volume are usually treated as metaphors, simulations, or simple abstractions of natural phenomena or processes. In addition, natural computing techniques are usually based upon a different modeling approach. The theoretical models used in biological sciences are based, in most cases, on ordinary differential equations (ODE) or Monte Carlo simulations. For example, when theoretical biologists want to create a model of an army ant, they use some rule of thumb such as "the more phero-mone (a chemical released by ants) an ant detects, the faster it runs". This rule would be translated into an equation of the type $dx/dt = kP$, where dx/dt is the speed (distance change, dx, divided by time change, dt) of the ant, k is a constant of proportionality, and P is the pheromone level. This simple formula captures the essence of ant movement as described.

Despite the differences in approach, level of details, and accuracy, it is unde-niable, and this will become clearer throughout the text, that the inspiration from nature and the relationship with it is the core of natural computing. Metaphors are important approaches not only for the creation of useful and interesting tools, but they may also aid the design of more accurate models and a better un-derstanding of nature. Thus, it is not surprising that many researchers in natural computing call their products models instead of metaphors.

2.1.2. From Nature to Computing and Back Again

In most cases, the first step toward developing a natural computing system is to look at nature or theoretical models of natural phenomena in order to have some insights into how nature is, works, and how it behaves. In other cases, it might happen that you have a given problem at hand, and you know some sort of natu-ral system solves a similar problem. A good example is the immune system metaphor for computer security mentioned above. Another classical example is the neural network metaphor: if there are brains that allow us to reason, think, process visual information, memorize, etc., why can I not look into this system and try to find its basic functioning mechanisms in order to develop an (intelli-gent) 'artificial brain'?

The problem with the extraction of metaphors and inspiration from nature is that it is usually very difficult to understand how nature works. In the particular case of the brain, though some basic signal transmission processes might be al-ready known (and many other facts as well), it is still out of human reach to fully uncover its mysteries, mainly some cognitive abilities such as hate and love. The use of technological means (e.g., computers) to simulate, emulate or reproduce natural phenomena may also not be the most suitable approach. Would computers, such as the ones we have nowadays, be suitable to build an 'artificial brain' or an 'artificial organism'? Can we simulate 'wetware' with the current 'hardware'? Furthermore, even if we do know how some natural proc-esses work, would it still be suitable to simply reproduce them the way they are? For example, we know that most birds are capable of flying by flapping wings, however airplanes fly using propellers or turbines. Why do airplanes not fly by flapping wings?

Last, but not least, sometimes looking at nature or theoretical studies may not be sufficient to give us the necessary insight into what could be done in computing and engineering with these phenomena. We have already seen, in Chapter 1, that the clustering of dead bodies in ants may result in computer algorithms for solving clustering problems, and that simple behavioral rules applied to many virtual birds result in flock-like group behaviors. What if I tell you that the behavior of ants foraging for food resulted in powerful algorithms for solving combinatorial optimization problems? Also, what if I tell you that the behavior of ant prey retrieval has led to approaches for collective robotics? Can you have an idea of how these are accomplished without looking at the answers in Chapter 5?

Due to all these aspects, designing novel natural computing systems may not be a straightforward process. But this book is not about how to design new natural computing devices, though some insights about it will certainly be gained. Instead, it focuses on how the main natural computing systems available nowadays were motivated, emerged, and can be understood and designed. Designing natural computing systems is basically an engineering task; that is, physical, mechanical, structural, and behavioral properties of nature are made useful to us in computational terms. They can become new problem-solving techniques, new forms of (studying) nature, or new forms of computing. Each part of natural computing, and its many branches, is rooted in some specific feature(s):

- Evolutionary algorithms were inspired by evolutionary biology.
- Artificial neural networks were inspired by the functioning of the nervous system.
- Swarm systems are based on social organisms (from insects to humans).
- Artificial immune systems extract ideas from the vertebrate immune system.
- Fractal geometry creates life-like patterns using systems of interactive functions, L-systems, and many other techniques.
- Artificial life is based on the study of life on Earth to simulate life on computers and sometimes develop synthetic forms of life.
- DNA computing is based on the mechanisms used to process DNA strands in order to provide a new computing paradigm.
- Quantum computing is rooted on quantum physics to develop another new computing paradigm.

Although it is generally difficult to provide a single engineering framework to natural computing, some of its many branches allow the specification of major structures and common design procedures that can be used as frameworks to the design of specific natural computing techniques. For instance, evolutionary algorithms can be designed by specifying a representation for candidate solutions to a problem, some general-purpose operators that manipulate the candidate solutions, and an evaluation function that quantifies the goodness or quality of each candidate solution (Chapter 3). In artificial life, however, it is much harder to provide such a framework. It will be seen that most artificial life (ALife) approaches reviewed here are based on the specification of usually simple sets of rules describing the behavior of individual agents. The remaining of the ALife project will involve the modeling of the agents, environment, etc., which are not part of the scope of this book.

2.2 GENERAL CONCEPTS

2.2.1. Individuals, Entities, and Agents

There is a body of literature about *agents* and agent-based systems. One of the main themes of this book is collectivity; populations of individuals, insect societies, flocks of bird, schools of fish, herds of land animals, repertoires of immune cells and molecules, networks of neurons, and DNA strands. What all these systems have in common is the presence of a number of individual *entities* or *components*. When we model or study these systems, the *individuals* may go by the generic name of *agents*. However, the words individuals, entities, components, and agents are sometimes used interchangeably and with no distinction throughout the text.

The term agent is currently used to mean anything between a mere subroutine of a computer program and an intelligent organism, such as a human being. Intuitively, for something to be considered an agent, it must present some degree of autonomy or identity; that is, it must, in some sense, be distinguishable from its environment by some kind of spatial, temporal, or functional boundary. Traditionally, agent-based models are drawn on examples of biological phenomena and processes, such as social insects and immune systems (Rocha, 1999). These systems are formed by distributed collections of interacting elements (agents) that work under no central control. From simple agents, who interact locally following simple rules of behavior and responding to environmental stimuli, it is possible to observe a synergistic behavior that leads to higher-level behaviors that are much more intricate than those of individuals.

Agent-based research has a variety of definitions of what is an agent, each hoping to explain one particular use of the word. These definitions range from the simplest to the lengthiest ones. Here are some examples:

"An agent is anything that can be viewed as perceiving its environment through sensors and acting upon that environment through effectors." (Russell and Norvig, 1995)

"Perhaps the most general way in which the term agent is used is to denote a hardware or (more usually) software-based computer system that enjoys the following properties: autonomy, social ability, reactivity, and proactiveness." (Wooldridge and Jennings, 1995) [Summarized definition, for the full version please consult the cited reference]

"An autonomous agent is a system situated within and part of an environment that senses that environment and acts on it, over time, in pursuit of its own agenda and also so as to effect what it senses in the future." (Franklin and Graesser, 1997)

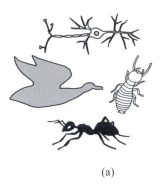
(a)

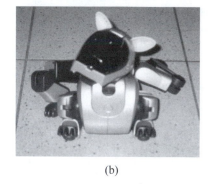
(b)

Figure 2.1: Examples of agents. (a) Pictorial representation of biological agents (bird, neuron, termite, and ant). (b) Physical agent (the AIBO ERS-210 robot by Sony®).

Therefore, an agent can be understood as an entity endowed with a (partial) representation of the environment, capable of acting upon itself and the environment, and also capable of communicating with other agents. Its behavior is a consequence of its observations, knowledge, and its interactions with other agents and the environment. Agents can be of many types, including biological (e.g., ants, termites, neurons, immune cells, birds, etc.), physical (e.g., robots), and virtual agents (e.g., a computer algorithm, Tamagotchi, etc.), as illustrated in Figure 2.1.

2.2.2. Parallelism and Distributivity

There are several well-known examples involving the capability of processing more than one thing at the same time. In the natural world, *parallel processing* is evident in insect societies, brain processing, immune functioning, the evolution of species, and so forth. All these examples will be studied in this book.

In order for evolution to occur, there must be a number of individuals in a population competing for limited resources. These individuals suffer genetic variation and those more fit (adapted) to the environment have higher probabilities of survival and reproduction. All the individuals in the population play important roles in exploring the environment and sometimes exchanging (genetic) information, thus producing progenies more adapted to the life in a spatial location.

In insect societies, in particular in ant colonies, a colony of ants has individuals assigned to various tasks, such as harvesting food, cleaning the nest, and caring for the queen. Termites, bees, and wasps also perform similar tasks in a distributed way; there are individuals allocated for different tasks. All insects in a colony work in parallel in a given task, but they may switch tasks when needed. For instance, some worker ants may be recruited for battle when the nest is being invaded.

In immune systems, a large variety and number of cells are involved in an immune response. When a virus infects a cell, some specialized immune cells, named T-cells, recognize fragments of this virus presented by a molecular complex of another specialized antigen presenting cell. This recognition triggers the action of many other immune cells to the site of infection. In addition, several other cells are performing the same and other processes, all at once in a distributed and parallel form.

In the human nervous system, a huge number of neurons are involved in processing information at each time instant. Talking while driving, watching TV while studying, hearing a name while having a conversation with someone in the middle of a party (a phenomenon called the 'cocktail party effect'), all these are just samples of a kind of parallel processing. Who does not know about the joke of not being able to walk while chewing gum? The vast number of neurons we have endow us with this capability of processing multiple information from multiple sensors at the same time.

What is surprising about each of the individual processes from the examples above is that they are all a product of a large number of elements and processes occurring in parallel. At the lowest level of analysis, evolution requires a large number of individuals to allow for a genetic variety and diversity that ultimately result in a higher adaptability; insect colonies are composed of thousands, sometimes millions, of insects that work in concert to maintain life in the colony; immune systems are composed of approximately 10^{12} lymphocytes (a special type of immune cell); and the human brain contains around 10^{11} nervous cells. Each of these individual agents contributes its little bit to the overall global effect of evolution, maintenance of life in the colony (insects), and the body (immune systems), and thought processes and cognition (nervous system).

From a biological and computational perspective, all the end results discussed are going to be emergent properties of the parallel and distributed operations of individual entities. All these systems can be termed *parallel-distributed systems* (PDS). Rumelhart and collaborators (Rumelhart et al., 1986; McClelland et al., 1986) have coined the term *parallel distributed processing* (PDP) to describe parallel-distributed systems composed of processing elements, in particular neurons. They used this terminology to refer to highly abstract models of neural function, currently known as *artificial neural networks* (ANN). These will be discussed in more detail in Chapter 4 under the heading of Neurocomputing. PDP networks are thus a particular case of parallel-distributed systems.

2.2.3. Interactivity

A remarkable feature of natural systems is that individual agents are capable of interacting with one another or the environment. Individual organisms interact with one another in variety of forms: reproductively, symbiotically, competitively, in a predator-prey situation, parasitically, via channels of communication, and so on. At a macro level, an important outcome of these interactions is a struggle for limited resources and life. Individuals more adapted to the (local) environment tend to survive and mate thus producing more progenies and propagating their genetic material. Genetic variation together with the selection of the fittest individuals leads to the creation of increasingly fitter species. Besides, interactivity allows for the emergence of self-organized patterns.

Interactivity is an important mean nature has to generate and maintain life. Complex systems, organisms, and behaviors emerge from interacting components. For instance, take the case of genes, known to be the basic functional elements of life. Researchers have created genetically modified organisms in which a single gene has been deleted or blocked, a process known as *knockout*. In some situations, these researchers have been surprised to find that some other gene(s) can take over its whole function or at least part of it. Similar cases are constantly being reported in the news where people with damaged brains, from accidents for example, are capable of recovering some of their lost functions after often long periods of treatment and recovery. It is observed, in most of these cases, that other portions of the brain assume the functions previously performed by the damaged areas. Interactions, thus, are not only necessary for the complexity, diversity, and maintenance of life, but it also leads to emergent phenomena and behaviors that cannot be predicted by simply looking at discrete components.

In all the main systems studied in this book, several types of interactions can be observed. For instance, immune cells and molecules communicate with one another and foreign agents through chemical messengers and physical contact; insects may also communicate with one another via chemical cues, dancing (e.g., bees dance to indicate where there is food to the other bees in the nest) or physical contact (e.g., antennation); and neurons are known to be connected with one another via small portions of its axons known as synapses. All these communication and contact means allow for the interaction of individual agents in the many systems. The interactions between individuals can be basically of two types: direct and indirect. One important example of direct interaction, namely *connectivity*, and one important example of indirect interaction, namely *stigmergy*, will be discussed in the next two sections. Other important examples of direct interaction are reproduction and molecular signaling, and these will be specifically discussed in the next few chapters.

Connectivity

Connectionist systems employ a type of representation whereby information is encoded throughout the nodes and connections of a network of basic elements, also called units. Their content and representational function is often revealed only through the analysis of the activity patterns of the internal units of the system. Although the term *connectionism* appeared in the mid 1980s to denote network models of cognition based on the spreading activation of numerous simple units (cf. Rumelhart et al., 1986; McClelland et al., 1986), it can refer to any approach based on interconnected elements. These systems are sometimes referred to as *networks*.

The peculiarity of connectionist systems is due to several factors. The connections establish specific pathways of interaction between units; two units can only interact if they have a connection linking them. The connection is also in most cases an active element of interaction, i.e., it not only specifies who interacts with whom, but it also quantifies the degree of this interaction by weighting the signal being transmitted. The direct interaction via connections also results in a structured pattern for the system that may, for instance, reflect the structural organization of the environment in which the network is embedded. Networks are also very successful examples of parallel-distributed processors, for instance, neural networks and immune networks. These two types of networks will be fully explored in this book in Chapter 4 and Chapter 6, respectively.

Stigmergy

Grassé (1959) introduced the concept of *stigmergy* as a means to refer to how the members of a termite colony of the genus *Macrotermes* coordinate nest building. He realized how individual termites could act independently on a structure without direct communication or interactions. This process was termed *indirect social interactions* to describe the same mechanism of indirect communication among bees in a bee colony (Michener, 1974).

The concept of stigmergy provides a general mechanism that relates individual and colony-level behaviors: individual behaviors modify the environment, which in turn modifies the behavior of other individuals. The environment thus mediates the communication of individuals, i.e., there is an indirect communication, instead of direct, by means such as antennation, trophalaxis (food or liquid exchange), mandibular contact, visual contact, and so on (Bonabeau et al., 1999). Self-organization is thus made possible due to the intensity of the stigmergic interactions among termites that can adopt a continuum of interactions.

Grassé (1959) gave the original example to illustrate stigmergy involving nest building in termite colonies (Figure 2.2). He observed that termite workers are stimulated to act during nest building according to the configuration of the construction and of other workers. Termite workers use soil pellets, which they impregnate with a chemical substance known as *pheromone*, to build pillars. Initially, termites deposit pellets in a random fashion until one of the deposits reaches a critical size. Then, if the group of builders is large enough and the pillars start to emerge, a coordination phase begins. The accumulation of pellets reinforces the attractivity of deposits due to the diffusing pheromone emitted by the pellets. Therefore, the presence of an initial deposit of soil pellets stimulates workers to accumulate more pellets through a *positive feedback* or self-reinforcing mechanism (Dorigo et al., 2000).

It is possible to extend the idea of stigmergy to other domains (Holland and Melhuish, 1999). It can be seen as an even more impressive and general account of how the interaction of simple entities, such as ants or termites, can produce a wide range of highly organized and coordinated behaviors and behavioral outcomes, simply acting and exploiting the influence of the environment. By exploiting the stigmergic approach to coordination, researchers have been able to design a number of successful algorithms and systems that can be applied to several domains, such as discrete optimization, clustering, and robotics.

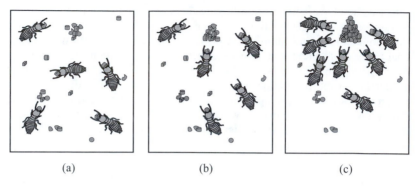

(a)   (b)   (c)

Figure 2.2: Termite mound building. (a) Pellets are initially deposited randomly in space. If the group of builders is large enough and pillars start to emerge (b), then a coordinated building phase starts (c).

Chapter 5 reviews some of these applications focusing on those systems inspired by the behavior of ants. Chapter 8 also provides some examples of stigmergic interactions such as the wasp nest building behavior.

2.2.4. Adaptation

Adaptation can be defined as the ability of a system to adjust its response to stimuli depending upon the environment. Something, such as an organism, a device, or a mechanism, that is changed (or changes) so as to become more suitable to a new or a special application or situation, becomes more adapted to the new application or situation. The use of the word adaptation is, in many cases, related with *evolution* (cf. Wilson and Keil, 1999; p. 3–4). However, many other important concepts in natural computing, such as learning and self-organization, can also be viewed as types of, or resulting from, adaptation mechanisms.

Learning

Learning may be viewed as corresponding to the act, process, or experience of gaining knowledge, comprehension, skill, or mastery, through experience, study, or interactions. Learning systems are those able to change their behavior based on examples in order to solve information-processing demands. An important virtue of adaptation in learning is the possibility of solving information processing tasks and the ability to cope with changing (dynamic) environments.

A consideration of what it takes to learn reveals an important dependence on gradedness (the passing through successive stages of changes) and other aspects of natural mechanisms (O'Reilly and Munakata, 2000). Learning, or more generally adapting, can be viewed as a synonym for *changing* with the end result of knowledge (memory) acquisition. When a system learns it changes its pattern of behavior (or another specific feature), such as the way information is processed.

It is much easier to learn if the system responds to these changes in a graded, proportional manner, instead of radically altering the way it behaves.

These graded changes allow the system to try out a number of different patterns of behavior, and get some kind of graded proportional indication of how these changes are affecting the system's interaction with the environment. By exploring several little changes, the system can evaluate and strengthen those that improve performance, while abandoning or weakening those that do not. There are, however, other types of learning procedures in nature that are more discrete than the graded one just described. For instance, it is believed that there are some specialized areas in the brain particularly good at 'memorizing' discrete facts or events.

In contrast to some beliefs, learning does not depend purely on consciousness and also does not require a brain. Insect societies learn how to forage for food, and our immune systems learn how to fight against disease-causing agents - a principle explored in the vaccination procedures. Even evolution can be viewed as resulting in learning, though evolutionary systems are more appropriately characterized as adaptive systems in the context of natural computing.

Neurocomputing models, and others based on computational neuroscience, provide useful accounts of many forms of learning, such as graded learning and memorization. Chapter 4 reviews some of the standard and most widely spread neurocomputing techniques, and provides a discussion about the main learning paradigms in this field, namely *supervised*, *unsupervised*, and *reinforcement learning*.

Evolution

In its simplest form, the *theory of evolution* is just the idea that life has changed over time, with younger forms descending from older ones. This idea existed well before the time of Charles Darwin, but he and his successors developed it to explain both the diversity of life and the adaptation of living things to their environment (Wilson and Keil, 1999; p. 290–292).

In contrast to learning, evolution requires some specific processes to occur. First, evolution involves an individual or a population of individuals that reproduce and suffer genetic variation followed by natural selection. Without any one of these characteristics, there is no evolution. Therefore, there cannot be evolution if there is a single individual, unless this individual is capable of asexually reproducing. Also, some variation has to occur during reproduction so that the progeny brings some 'novelty' that allows it to become more adapted to the environment. Finally, natural selection is responsible for the maintenance of the genetic material responsible for the fittest individuals to the environment; these will have survival and reproductive advantages over the others, less fit individuals. The outcome of evolution, like the outcome of learning, is a better adaptability to life and the environment.

Both, the evolved genetic configuration of organisms together with their learning capabilities make important contributions to their adaptability to the environment. But perhaps only in the context of learning the genetic encoding can be fully understood, much as the role of DNA itself in shaping the pheno-type must be understood in the context of emergent developmental processes.

2.2.5. Feedback

Essentially, *feedback* occurs when the response to a stimulus has an effect of some kind on the original stimulus. It can be understood as the return of a por-tion of the output of a process or system to the input, especially when used to maintain performance or to control a system or process. The nature of the re-sponse determines how the feedback is labeled: *negative feedback* is when the response diminishes the original stimulus (they go in the opposite direction); and *positive feedback* is when the response enhances the original stimulus (they go in the same direction). An important feature of most natural systems described in this text is that they rely extensively on *feedback*, both for growth and self-regulation.

Take the case of the human brain as an example of extensive feedback loops and their importance. The brain can be viewed as a massive network of neurons interconnected via tiny gaps known as synapses. Any brain activity, such as thinking of a word or recognizing a face, triggers a vast array of neural circuitry. Each new brain activity triggers a new array, and an unimaginably large number of possible neuronal circuits go unrealized during the lifetime of an individual. Beneath all that apparent diversity, certain circuits repeat themselves over and over again. All these feedback and reverberating loops are believed to be neces-sary for learning, and are consequences of the high interconnectivity of the brain.

Positive Feedback

Positive feedback is a sort of self-reinforcing (growth) process in which the more an event occurs, the more it tends to occur. Take the case of the immune system as an example. When a bacterium invades our organism, it starts repro-ducing and causing damage to our cells. One way the immune systems find to cope with these reproducing agents is by reproducing the immune cells capable of recognizing these agents. And the more cells are generated, the more cells can be generated. Furthermore, the immune cells and molecules release chemicals that stimulate other immune cells and molecules to fight against the disease-causing agent. Therefore, the response of some immune cells provides some sort of *positive feedback* to other immune cells reproduce and join the pool of cells involved in this immune response.

The termite mound building behavior discussed previously is another example of a positive feedback mechanism. The more soil pellets are deposited in a given portion of the space, the more pellets tend to be deposited in that portion be-cause there is more pheromone attracting the termites (Figure 2.3). But these self-reinforcing (positive feedback) processes have to be regulated by *negative feedback* processes, otherwise the systems would go unstable or the resources would be depleted.

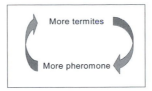

Figure 2.3: Example of positive feedback.

There are several other examples of positive feedback in nature:

- *Human breeding*: the more humans reproduce, the more humans exist to reproduce.

- *Feeding the baby*: a baby begins to suckle her mother's nipple and a few drops of milk are released, stimulating the production and release of more milk.

- *Avalanche*: an avalanche starts at rest and, when disturbed, accelerates quickly towards its end point at the base of a slope.

- *Autocatalysis*: autocatalysis occurs in some digestive enzymes such as pepsin. Pepsin is a protein-digesting enzyme that works in the stomach. However, the stomach does not secrete pepsin; it secretes an inactive form, called pepsinogen. When one pepsinogen molecule becomes activated, it helps to activate other pepsinogens nearby, which in turn can activate others. In this way, the number of active pepsin molecules can increase rapidly by using positive feedback.

- *Giving birth*: while giving birth, the more uterine contractions a mother has, the more it is stimulated to have, until the child is born.

- *Scratching an itch*: scratching an itch makes it more infected and damaged, and thus more itchy.

- *Ripening fruits*: a ripening apple releases the volatile plant hormone ethylene, which accelerates the ripening of unripe fruit in its vicinity; so nearby fruits also ripen, releasing more ethylene. All the fruits become quickly ripe.

Negative Feedback

Negative feedback by contrast, plays the role of regulating positive feedback so as to maintain a(n) (dynamic) equilibrium of the medium. It refers to change in the opposite direction to the original stimulus. The thermostat is one of the most classic examples of negative feedback. It takes the reading of a room's temperature, measures that reading according to a desired setting, and then adjusts its state accordingly. If the room's temperature is too low, more hot air is allowed to flow into the room; else if the temperature is too high, then more cold air flows into the room (Figure 2.4).

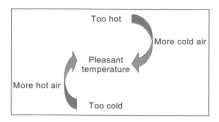

Figure 2.4: Example of negative feedback.

Negative feedback is at the heart of every stable, self-regulating system. If a company raises prices too high, people stop buying, and soon the company cuts the price to increase sales. In the immune system example given above, after the infection is successfully eliminated, specific immune cells are stimulated to release other chemical substances that suppress the replication of immune cells, thus ceasing the immune response. Without this negative feedback mechanism, death by uncontrolled cell reproduction would be inevitable. And without the positive feedback, death from infection would be inevitable.

There are also plenty of examples of negative feedback in nature:

- *Ecosystems*: in an ecosystem composed of, say rabbits and grass, when there is plenty of grass to feed the rabbits, they tend to reproduce with greater rates. But as there are more rabbits in the environment, the more grass will be eaten, and the less grass will be left as food; the amount of grass provides a feedback to the rabbits birth rate.

- *Homeostasis*: blood glucose concentrations rise after eating a meal rich in sugar. The hormone insulin is released and it speeds up the transport of glucose out of the blood and into selected tissues, decreasing blood glucose concentrations.

- *Metabolism*: exercise creates metabolic heat that raises the body temperature. Cooling mechanisms such as vasodilatation (flushed skin) and sweating begin, decreasing the body temperature.

- *Climate theory*: the curvature of the earth helps making it so that continental glaciers expanding equator ward experience strong sunlight and tend to melt. Another example is the tendency for continental glaciers to make cold, high-pressure regions, which do not favor further snowfall.

2.2.6. Self-Organization

An important question in biology, physics, and chemistry is "Where does order come from?" The world abounds with systems, organisms, and phenomena that maintain a high internal energy and organization in seeming defiance of the laws of physics (Decker, 2000). Water particles suspended in air form clouds; a social insect grows from a single celled zygote into a complex multicellular organism and then participates in a structured social organization; birds gather together in a coordinated flock; and so forth. What is so fascinating is that the organization seems to emerge spontaneously from disordered conditions, and it does not appear to be driven solely by known physical laws or global rules. Somehow, the order arises from the multitude of interactions among the simple parts. Self-organization may also go by the name 'emergent structuring', 'self-assembly', 'autocatalysis', and 'autopoiesis', though most of these concepts have some slight differences to the self-organization concept provided here (see Project 3).

Self-organization refers to a broad range of pattern-formation processes in both physical and biological systems, such as sand grains assembling into rippled dunes, chemical reactants forming swirling spirals, cells making up highly structured tissues, and fishes joining together in schools. A basic feature of these diverse systems is the means by which they acquire their order and structure. In self-organizing systems, pattern formation occurs through interactions internal to the system, without intervention by external directing influences. As used here, a *pattern* corresponds to a particular, organized arrangement of objects in space or time. Examples of biological behavioral patterns include a school of fish, a raiding column of army ants, the synchronous flashing of fireflies, and the complex architecture of a termite mound (Camazine et al., 2001). But self-organization does not only affect behavioral patterns, it is also believed to play a role in the definition of patterns, such as shapes, and the coating of several animals (see Figure 2.5). In these cases, it is believed that not only the genetic code of these animals determine their physical expressed characteristics, some self-organized processes may also be involved.

The concept of self-organization can also be conveyed through counterexamples. A system can form a precise pattern receiving instructions from outside, such as a blueprint, recipe, orders, or signals. For instance, the soldiers marching form a neat organized process that is not self-organized. Their sequence of steps, direction of movement, velocity, etc., are all dictated by specific instructions. In such cases, the process is organized but not self-organized. It is less obvious, however, to understand how a definite pattern can be produced in the absence of such instructions.

The self-organized pattern formation in social systems is one of the main themes of this book, and will become clearer and more exemplified along the chapters. It will be seen that many social insects, such as ants, termites, and bees, are capable of building extremely complex nests without following any blueprint, recipe, leader, or template. It is interesting to note that, although all of them have a queen (the queen ant, the queen termite, and the queen bee), queens are basically involved in reproduction, mainly when the colony size is very large.

There is even an interesting historical fact about the queen bee. Until the late 19th century, in a time when men were considered superior to women, queen bees were called kings, because people could not accept that anything so well organized could be run by females. Although females could actually run the colony, it is now known that, in most cases, the main role of the queens is to lay eggs. Even more surprising, such complex patterns of behavior, architecture designs, and foraging and hunting strategies, do not require any global control or rule whatsoever; they are amazing self-organized undertakings.

(a)

(b)

(c)

Figure 2.5: Animal patterns believed to involve self-organized pattern formation. (a) Polygonal shapes on the shell of a turtle. (b) Stripes coating the tiger skin. (c) Stripes in the iguana's tail.

Characteristics of Self-Organization

Self-organization refers to spontaneous ordering tendencies sometimes observed in certain classes of complex systems, both natural and artificial. Most self-organized systems present a number of features:

- *Collectivity and interactivity*: self-organizing systems (SOS) are usually composed of a large number of elements that interact with one another and the environment.

- *Dynamics*: the multiplicity of interactions that characterize self-organizing systems emphasize that they are dynamic and require continual interactions of lower-level components to produce and maintain structure.

- *Emergent patterns*: SOS usually exhibit what appears to be spontaneous order; the overall state of a self-organized system is an emergent property.

- *Nonlinearities*: an underlying concept in self-organization is nonlinearity. The interactions of components result in qualitatively new properties that cannot be understood as the simple addition of the individual contributions.

- *Complexity*: most self-organizing systems are complex. The very concepts of complexity and emergence are embodied in SOS. However, it is more accurate to say that complex systems can be self-organizing systems.

- *Rule-based*: most SOS are rule-based, mainly biological self-organizing systems. Examples of rules governing natural self-organized systems were already reviewed, such as the ones that result in dead body clustering in ant colonies. Further examples will be given in the following chapters.

- *Feedback loops*: positive and negative feedback contribute to the formation of self-organized processes by amplifying and regulating fluctuations in the system.

Alternatives to Self-Organization

Self-organization is not the only means responsible for the many patterns we see in nature. Furthermore, even those patterns that arise through self-organization may involve other mechanisms, such as the genetic encoding and physical constraints (laws). Camazine et al. (2001) provide four alternatives to self-organization:

- *Following a leader*: a well-informed leader can direct the activity of the group, providing each group member with detailed instructions about what to do. For example, a captain in a battle field gives order to each soldier relating to where to attack, etc.

- *Building a blueprint*: a blueprint is a compact representation of the spatial or temporal relationships of the parts of a pattern. For instance, each musician of an orchestra receives a musical score that fully specifies the pattern of notes within the composition and the tonal and temporal relationships among them.

- *Following a recipe*: each member of the group may have a recipe, i.e., a sequential set of instructions that precisely specify the spatial and temporal actions of the individual's contribution to the whole pattern. For example, you tell someone how to get to your place by specifying the precise sequence of streets he/she must follow. A blueprint is different from a recipe because it does not specify how something is done, only what is to be done.

- *Templates*: a template is a full-size guide or mold that specifies the final pattern and strongly steers the pattern formation process. For example, a company that makes car parts uses a template in which raw material is poured in order to make the desired parts; each part has its own template.

2.2.7. Complexity, Emergence, and Reductionism

Complexity and emergence are some of the most difficult terms to conceptualize in this chapter. Viewpoints and definitions of complexity (complex systems) and emergence vary among researchers and disciplines. This section discusses only some of the many perspectives; further and more complete studies can be found, for instance, in (Emmeche, 1997; Baas and Emmeche, 1997), the special issue on Complex Systems of the Science magazine (Science, 1999), on the Santa Fe volumes on Artificial Life and Complex Systems (e.g., Cowan et al., 1994; Morowitz and Singer, 1995), and on the Artificial Life and Complex Systems journals (see Appendix C).

Complexity

To start the discussion, let us present a very simplistic idea that fits into the context of natural computing: a *complex system* is a system featuring a large number of interacting components whose aggregate activity is nonlinear (not derivable by summing the behavior of individual components) and typically exhibit self-organization (Morowitz and Singer, 1995; Holland, 1995; Gallagher and Appenzeller, 1999; Rocha, 1999). Consider the case of an organism; say the human body. Can you fully uncover how it works by looking at its major systems and organs? The answer is no. Take an even more reductionist approach, and try to understand the organism by analyzing all its cells and molecules. Can you understand it now? Not still. Some limitations of this more traditional *reductionist* way of thinking will be discussed later. What is important here is the fact that for complex systems we are unable to understand/explain their behavior by examining its component parts alone.

The studies on *complexity* suggest that not only the internal organization (e.g., the genetic code of a biological organism) of a system is sufficient for its full understanding, but also how the system itself and its component parts interact with one another and the environment. The internal microstructure, self-organizing capabilities, and natural selection are part of the most important aspects necessary for a better understanding of complex systems.

Perhaps the most remarkable contribution of complexity to science was the perception that many natural phenomena and processes can be explained and sometimes reproduced by following some basic and simple rules. For instance, the most striking aspects of physics are the simplicity of its laws. Maxwell's equations, Schrödinger's equations, and Hamiltonian mechanics can each be expressed in a few lines. Many ideas that form the foundations of nature are also very simple indeed: nature is lawful, and some basic laws hold everywhere. Nature can produce complex structures even in simple situations and can obey simple laws even in complex situations (Goldenfeld and Kadanoff, 1999).

The five basic forms of investigating complex systems have already been discussed (Section 2.1.1), namely, experimentation, simulation, theoretical modeling, emulation, and realization. Experiments are necessary for raising a range of information about how the natural organisms and processes behave. Simulations are often used to check the understanding, validate experimental results, or simulate a particular system or process. Theoretical models are useful for the understanding, complementation, prediction, critical analysis, and quantitative and qualitative description of natural phenomena. Finally, realizations and emulations are fundamental for the possibility of creating and studying (new) life-like patterns and forms of life.

In order to explore the complexity inherent in nature, one must focus on the right level of description. In natural computing, higher-levels of description are usually adopted. Most systems developed are highly abstract models or metaphors of their biological counterparts. The inclusion of too many processes, details, and parameters, can obscure the desired qualitative understanding and can also make the creation of computational systems based on nature unfeasible. For instance, the 'artificial' neural networks are based upon very simple mathematical models of neural units structured in a network-like architecture and subjected to an iterative procedure of adaptation (learning). Despite all this simplicity, the approaches to be presented still capture some important features borrowed from nature that allow them to perform tasks and solve problems that, in most cases, could not be solved satisfactorily with the previously existing approaches.

In *Hidden Order*, J. Holland (1995) starts with a discussion of how natural (biological and social) systems are formed and self-sustained. Among the several instances discussed, there is the case of the immune system with its numerous cells, molecules and organs. Other examples range from the New York City to the central nervous system. These systems are termed *complex adaptive systems* (CAS), in which the (complex) behavior of the whole is more than a simple sum of individual behaviors. One of the main questions involved in complex adaptive systems is that of how a decentralized system - with no central planning or control - is self-organized.

Despite the differences among all complex adaptive systems and organizations, in most cases, the persistence of the system relies on some main aspects: 1) *interactions*, 2) *diversity*, and 3) *adaptation*. Adaptability allows a system or organism to become better fit to the environment or to learn to accomplish a given task. Adaptability also has to do with the system's capability of processing information (or computing); another important feature of a complex adaptive system. The major task of surviving involves the gathering of information from the environment, its processing and responding accordingly. It is clear, thus, that computing or processing information does not require a brain; ants, immune systems, evolutionary processes, flocks of birds, schools of fish, and many other complex adaptive systems present the natural capability of processing information.

According to Holland, the choice of the name *complex adaptive systems* is more than a terminology "It signals our intuition that general principles rule CAS behavior, principles that point to ways of solving attendant problems." (Holland, 1995; p. 4). This turns back to the idea that there are general rules or principles governing natural systems. The question, thus, can be summarized as how to extract these general principles. This is also one of the main aims of this book and will be illustrated and more clearly identified in all chapters.

Emergence

Important questions about complex adaptive systems rely upon the understanding of *emergent* properties. At a very low level, how do living systems result from the laws of physics and chemistry? That is, how do the genes specify the unfolding processes of biochemical reactions and interactions that result in the development of an organism? Are the genes the necessary and sufficient ingredients to development? At higher levels, how insect societies are organized? How do brains process information? How does the immune system cope with disease-causing agents? Why does a flock of bird present such a coordinated behavior? None of these questions can be answered without having in mind the concept of *emergence*; that is to say, the properties of the whole are not possessed by, nor are they directly derivable from, any of the parts - a water particle is not a cloud, and a neuron is not conscious.

All the systems and processes discussed above present behaviors by drawing upon populations of relatively 'unintelligent' individual agents, rather than a single, 'intelligent' agent. They are *bottom-up* systems, not *top-down*. They are complex adaptive systems that display emergent behaviors. Emergence is a concept tightly linked with complex systems. In these systems, agents residing on one scale or level start producing behaviors that lie scale(s) above them: social insects create colonies; social animals create flocks, herds, and schools; immune cells and molecules compose the immune system; neurons form brains, and so forth. The movement from low-level rules to higher-level sophistication is what we call emergence (Johnson, 2002).

One important feature most emergent systems and processes discussed in this book share is that they are rule-governed. Remember, nature is lawful! It means that it is possible to describe them in terms of a usually simple set of laws or rules that govern the behavior of individual agents. This book presents various instances of small numbers of rules or laws generating (artificial) systems of surprising complexity and potential for problem solving, computing, and the simulation of life-like patterns and behaviors. It will be interesting to note that even with a fixed set of rules, the overall emergent behaviors are *dynamic* (they change over time) and, in most cases, unpredictably.

Holland (1998) underlines a number of features of emergent (or complex) systems and suggests that emergence is a product of coupled, context-dependent interactions. These interactions, together with the resulting system are nonlinear, where the overall behavior of the system cannot be obtained by summing up the behavior of its constituent parts. Temperature and pressure are easy to grasp examples of emergent properties. Individual molecules in motion or closely placed result in temperature and pressure, but the molecules by themselves do not present a temperature or a pressure.

For short, the complex and flexible behavior of the whole depends on the activity and interactions of a large number of agents described by a relatively small number of rules. There is no easy way to predict the overall behavior of the system by simply looking at the local rules that govern individual behaviors, though we may have some intuition about it. The difficulty increases even further when the individual agents can adapt; that is, when the complex systems become complex adaptive systems. Then, an individual's strategy (or the environment) is not only conditioned by the current state of the system, it can also change over time. Even when adaptation may not affect an individual directly, it may still affect it indirectly. For instance, stigmergic effects result in the adaptation of the environment as a result of the action of individual agents. This, in turn, results in different patterns of behavior for the agents. What is important to realize, thus, is that the increase in complexity results in an increase of the possibilities for emergent phenomena.

To make the conceptual idea of emergence even clearer, consider the following example. Imagine a computer simulation of highway traffic. Simulated 'semi-intelligent' cars are designed to drive on the highway based on specific interactions with other cars following a finite set of rules. Several rules can be defined, such as:

- Cars should drive on the high speed lane if their speed is over 100Km/h, otherwise they should drive on the lower speed lanes.
- Cars should only take over using the higher speed lane.
- Avoid collisions.
- Stop in the red light.
- Vary speed smoothly.

As the cars can vary speed, change lanes, and are subjected to several constraints (speed, traffic lights, road limits, etc.), their programming enables them to have unexpected behaviors. Such a system would define a form of complex behavior, for there are several agents dynamically interacting with one another and the environment in multiple ways, following local rules and unaware of any higher-level instructions. Nevertheless, the behavior of such a system can only be considered emergent if discernible macro-behaviors are observed. For instance, these lower-level rules may result in the appearance of traffic jams and a higher flux of cars on the high-speed lane than on the lower speed ones. All these are emergent phenomena that were not programmed in the local rules applied to the cars. They simply emerged as outcomes of the lower-level interactions, even though some of them could be predicted.

Several features and behaviors of natural systems are not so obviously emergent; only a detailed analysis of complex interactions at the organism level can show that they are genuinely new features that only appear at the higher levels. However, emergent behaviors of natural systems can be observed through the creation of models. Even highly abstract and simplified models, such as the ones that will be reviewed here, allow us to simulate, emulate, observe, and study several emergent behaviors of natural complex adaptive systems. In particular, DNA strands, chromosomes, ant colonies, neurons, immune systems, flocks of birds, and schools of fish, will all be seen to present a large variety of emergent properties.

Reductionism

The classical *vitalist* doctrines of the 18th century are based on the idea that all life phenomena are animated by immaterial life spirits. These life spirits determine the various life phenomena, but are themselves unexplainable and indescribable from a physical perspective. By contrast, the *reductionist* position, also in the 18th century, insisted that a large part, if not all, of the life phenomena can be reduced to physics and chemistry (Emmeche et al., 1997).

For long, scientists have been excited about the belief that natural systems could be understood by reductionism; that is, by seeking out their most fundamental constituents. Physicists search for the basic particles and forces, chemists seek to understand chemical bonds, and biologists scrutinize DNA sequences and molecular structures in an effort to understand organisms and life. These reductionist approaches suggest that questions in physical chemistry can be answered based on atomic physics, questions in cell biology can be answered based on how biomolecules work, and organisms can be understood in terms of how their cellular and molecular systems work (Gallagher and Appenzeller, 1999; Williams, 1997).

However, apart from a few radicals, the reductionists do not claim that the higher psychological functions can be reduced to physics and chemistry. As an outcome of the scientific development in many areas, such as cytology, neuroanatomy, immunology, and neurophisiology, it became very difficult to maintain the more classical positions (Emmeche et al., 1997).

Advances in science and technology have led to transformations in the vitalists' and reductionists' positions as well. After a number of scientific discoveries in the early 19th century, the vitalists gradually limited their viewpoints to a narrower field. They now insisted that only the higher psychological functions were irreducible, but admitted that a large range of biological phenomena could be described scientifically. Reductionists now claimed that every phenomenon in the world, including the highest psychological ones, could be reduced to physics and chemistry (Emmeche et al., 1997).

Although the reductionist approaches work to some extent, scientists are now beginning to realize that reductionism is just one of the many tools needed to uncover the mysteries of nature. There might be additional principles to life and nature embodied in properties found at higher levels of organization, and that cannot be seen in lower levels. These properties are known as emergent properties. For instance, it is still not possible to understand higher psychological phenomena, such as love and hate, by simply looking at how neurons work.

2.2.8. Bottom-up vs. Top-down

Broadly speaking, there are two main approaches to addressing the substantive question of how in fact nature works. One exploits the model of the familiar serial, digital computer, where representations are symbols and computations are formal rules (algorithms) that operate on symbols. For instance, 'if-then' rules are most often used in formal logic and circuit design. The second approach is rooted in many natural sciences, such as biology, neuroscience, and evolutionary theory, drawing on data concerning how the most elementary units work, interact, and process information. Although both approaches ultimately seek to reproduce input-output patterns and behaviors, the first is more *top-down*, relying heavily on computer science principles, whereas the second tends to be more *bottom-up*, aiming to reflect relevant natural constraints.

Bottom-Up

Most complex systems exhibiting complex autonomous patterns and behaviors are parallel and distributed systems of interacting components taking decisions that directly affect their state and the state of some other components (e.g., their neighbors). Each component's decisions are based on information about its own local state. Bottom-up systems and models are those in which the global behavior emerges out of the aggregate behavior of an ensemble of relatively simple and autonomous elements acting solely on the basis of local information.

The reductionist approach is to some extent a bottom-up approach. Reductionism assumes that the study of individual components is sufficient for a full understanding of the organism as a whole. Bottom-up approaches also seek for the study of component parts, but do not necessarily claim that the whole is just the sum of its parts; it allows for emergent phenomena as well. Biology is in most cases a reductionist scientific enterprise (it has been changing over the last few years though). Natural computing, although highly inspired or based on biology, is more rooted on bottom-up approaches than on purely reductionist techniques. Theories about complexity and emergence are pervasive in natural computing, and may be of primary importance for the study and formalization of natural computing techniques.

Another way of viewing bottom-up systems or approaches is related to how they develop or evolve; a design perspective: the most deeply embedded structural unit is created first and then combined with more of its parts to create a larger unit, and so on. For instance, if a system develops or evolves by constantly adding new parts to the system until a given criterion or state of maturity is achieved, this system is said to follow a bottom-up design.

A classical example of bottom-up design in natural computing involves artificial neural networks whose architecture is not defined *a priori*; they are a result of the network interactions with the environment. Assume, for instance, that you are trying to design an artificial neural network capable of recognizing a set of patterns (e.g., distinguish apples from oranges). If the initial network has a single neuron and more neurons are constantly being added until the network is capable of appropriately recognizing the desired patterns, this network can be said to follow a bottom-up architecture design. In nature, a classical example of a bottom-up system is the theory of evolution, according to which life on Earth is a result of a continuous and graded procedure of adaptation to the environment.

Top-Down

In the early days of artificial intelligence, by the time of the Dartmouth summer school in 1956, most researchers were trying to develop computer programs capable of manipulating symbolic expressions. These programs were developed in a *top-down* manner: by looking at how humans solve problems and trying to 'program' these problem-solving procedures into a computer. Top-down approaches assume that it is possible to fully understand a given system or process by looking at its macro-level patterns.

Most artificial intelligence techniques based upon the top-down paradigm are known as knowledge-based or expert systems. They rely upon the existence of a knowledge base or production system containing a great deal of knowledge about a given task provided by an expert. In these systems, an action is taken if a certain specific condition is met. A set of these production systems results in an intelligent system capable of inferring an action based on a set of input conditions. Knowledge-based systems were rooted on a philosophy inspired by cognitive theories of the information processes involved in problem solving.

Top-down systems and approaches can also be studied under the perspective of how the system develops with time - the design perspective: an over-complicated system is initially designed and then parts of it are eliminated so as to result in a satisfactorily more parsimonious structure still capable of performing its task. We have discussed, in the previous section, that it is possible to design an artificial neural network for pattern recognition by simply adding neurons until the desired patterns are recognized. The opposite direction could also be adopted: an over-sized network could be initially designed and neurons would then be pruned until a network of reasonable size remained, still capable of meeting its goal.

To make the distinction even clearer, consider the case of building a sand castle. A bottom-up approach is the one in which you keep pouring sand and molding it, and a top-down approach is the one in which you initially pour a lot of sand, making a big mountain of sand, and then you start molding the castle from the mountain of sand.

2.2.9. Determinism, Chaos, and Fractals

One of the classical positions in the theory of science is that scientific theories are capable of providing deterministic relations between the elements being investigated. A *deterministic system* can have its time evolution predicted precisely; all events are inevitable consequences of antecedent sufficient causes. The main characteristic of this type of deterministic system is *predictability*. When it is possible to predict the development or time evolution of a system from some predefined conditions there is a deterministic relation between the elements that constitute the system. This classical perspective demands the capacity of predicting the time evolution of a system, thus precluding the appearance of new and emergent phenomena.

One of the most interesting and exciting results of recent scientific development, mainly in physics, is the remodeling of the relation between determinism and prediction. It is now evident that there are many systems that can be described adequately as being strictly deterministic but that still remain unpredictable. The impossibility of predicting the properties arising within many systems considered totally deterministic is the consequence of the well-known Poincaré's treatment of the three-body problem and the Hadamard's investigation of the sensitivity to initial states - insights from the latter half of the 19th century that have recently given rise to the *chaos theory*. Several processes in physics and biology are deterministic but unpredictable. Thus, one of the very important theoretical consequences of chaos theory is the divorce between determinism and predictability.

Before chaos theory, scientists (mainly physicists) were suffering from a great ignorance about disorder in the atmosphere, in the turbulent sea, in the fluctuations of wildlife populations, in the oscillations of the heart and the brain. The irregular side of nature, the discontinuous and erratic side, has been a puzzle to science. The insights from chaos theory led directly into the natural world - the shapes of clouds, the paths of lightening, the microscope intertwining of blood vessels, the galactic clustering of stars, the coast lines. Chaos has created special techniques of using computers and special kinds of graphic images, pictures that capture fantastic and delicate structures underlying complexity. The new science has spawned its own language, such as the word *fractals*.

The word fractal comes to stand for a way of describing, calculating, and thinking about shapes that are irregular and fragmented, jagged and broken-up - shapes like the crystalline curves of snowflakes, coast shores, mountains, clouds, and even the discontinuous dusts of galaxies (see Figure 2.6). A fractal curve implies an organizing structure that lies hidden among the hideous complication of such shapes.

(a)

(b)

Figure 2.6: Examples of the fractal geometry of nature. (a) Clouds. (b) Top of mountains in Alaska.

Fractals - the term used to describe the shape of chaos - seem to be everywhere: a rising column of cigarette smoke breaks into swirls; a flag snaps back and forth in the wind; a dripping faucet goes from a steady pattern to a chaotic one; and so forth (Gleick, 1997; Stewart, 1997).

Chaos theory is often cited as an explanation for the difficulty in predicting weather and other complex phenomena. Roughly, it shows that small changes in local conditions can cause major perturbations in global, long-term behavior in a wide range of 'well-behaved' systems, such as the weather. Therefore, chaos embodies three important principles: sensitivity to initial conditions, cause and effect are not proportional, and nonlinearities. This book talks very little about chaos; however Chapter 7 describes the Fractal Geometry of Nature as the main branch of the study of biology by means of computers aimed at creating life-like shapes or geometrical patterns.

2.3 SUMMARY

This chapter started with a discussion of what are models, metaphors, experiments, simulations, emulations, and realizations. These concepts are important for they allow us to distinguish natural computing techniques from theoretical models, computer simulations from realizations, and so on. Some comments about the difficulty in creating a general framework to design natural computing systems were also made. However, it was argued that each approach, with the exception of some topics in the second part of this book, do have specific frameworks for their design. Also, it was emphasized that this is a book about how existing techniques can be understood, reproduced, and applied to particular domains, not a book about how to engineer new techniques. Of course, from reading this text the reader will certainly get the feeling of natural computing and will then find it much easier to go for his/her own personal explorations and design of novel natural computing approaches.

If the contents of this chapter were to be summarized in a single word, this word would be *complexity* or *complex system*. Complexity encompasses almost all the terminology discussed here. It may involve a large number of interacting individuals presenting or resulting in emergent phenomena, self-organizing processes, chaos, positive and negative feedback, adaptability, and parallelism. Although this book is not about the *theory of complex systems*, it provides design techniques, pseudocode, and applications for a number of complex systems related with nature.

2.4 EXERCISES

2.4.1. Questions

1. Provide alternative definitions for all the concepts described in Section 2.2.

2. List ten journals that can be used as sources of information about natural systems and processes that could be useful for natural computing and explain your choice.

3. Name two connectionist systems in nature in addition to the nervous system and the immune network. Explain.

4. Section 2.2.5 presented the concepts of positive and negative feedback. In Section 2.2.3, the presence of an initial deposit of soil pellets was demonstrated to stimulate worker termites to accumulate more pellets through a positive feedback mechanism. It is intuitive to think that if no negative feedback mechanism existed in this process, the process could go uncontrolled. Name three negative feedback mechanisms involved in the termite mound building process.

5. Name a natural system or process that involves both, positive and negative feedback, and describe how these are observed.

6. Exemplify some natural systems (excluding those involving humans) that exhibit alternatives to self-organization. That is, provide an example of a natural system whose parts behave in a follow a 'leader' manner, one that follows a 'blueprint', one that follows a 'recipe', and one that follows a 'template'. The resultant pattern or process may vary from one case to another. Provide a list of references.

7. Discuss the advantages and disadvantages of self-organization over its alternatives.

8. The discussion about self-organization shows it is potentially relevant to fundamental questions about evolution. One issue is whether life itself can come into existence through a self-organizing process. A second issue is the relationship between natural selection and self-organization once life is up and running. Discuss these two possible implications of self-organization to life.

9. Two other important concepts in natural computing are *competition* and *cooperation* among agents. Provide definitions for both concepts under the perspective of agent-based theory.

2.4.2. Thought Exercise

1. An example of vehicular traffic flow was given as an instance of an emergent system (Section 2.2.7). Assume that there is a road with a single lane, a radar trap is installed in this road and a number of cars are allowed to run on this road. Embody the following rules in each car:

   - If there is a car close ahead of you, slow down.

   - If there isn't any car close ahead of you, speed up (unless you are already at maximum speed).

   - If you detect a radar trap, then slow down.

   What types of emergent behavior would you expect from this system? Will there be any traffic jam?

   (After answering the question, see Chapter 8 for some possible implications.)

2.4.3. Projects and Challenges

1. Write a computer program to simulate the traffic flow described in the exercise above. Compare the results obtained with your conclusions and the discussion presented in Chapter 8.

2. It was discussed in Section 2.2.6 that some patterns coating the skin of many animals are resultant from self-organizing processes. Perform a broad search for literature supporting this claim. Name, summarize, and discuss the most relevant works, and give your conclusive viewpoint.

3. Write an essay discussing the similarities and differences between the concepts *autopoiesis*, *autocatalisys*, and *self-organization*. The essay should be no longer than 10 pages. Provide a list of references.

4. Name other alternatives to self-organization and give examples.

5. From our math courses we know that a point is a zero-dimensional geometrical form, a line is a one-dimensional form, a square is a two-dimensional form, and so on. Can you determine the dimension of a snowflake? Justify your answer.

2.5 REFERENCES

[1] Baas, N. A. and Emmeche, C. (1997), "On Emergence and Explanation", *Intellectica*, **1997/2**(25), pp. 67–83.

[2] Bonabeau, E., Dorigo, M. and Théraulaz, G. (1999), *Swarm Intelligence from Natural to Artificial Systems*, Oxford University Press.

[3] Camazine, S., Deneubourg, J. -L., Franks, N. R., Sneyd, J., Theraulaz, G. and Bonabeau, E. (2001), *Self-Organization in Biological Systems*, Princeton University Press.

[4] Cowan, G., Pines, D. and Meltzer, D. (Eds.) (1994), *Complexity: Metaphors, Models, and Reality*, Proc. Volume XIX, Santa Fe Institute, Studies in the Sciences of Complexity. Addison-Wesley Pub. Company.

[5] de Castro, L. N. and Von Zuben, F. J. (2004), *Recent Developments in Biologically Inspired Computing*, Idea Group Inc.

[6] Decker, E. H. (2000), "Self-Organizing Systems: A Tutorial in Complexity", *Vivek: A Quarterly in Artificial Intelligence*, **13**(1), pp. 14–25.

[7] Dorigo, M., Bonabeau, E. and Theraulaz, G. (2000), "Ant Algorithms and Stigmergy", *Future Generation Computer Systems*, **16**, pp. 851–871.

[8] Emmeche, C. (1997), "Aspects of Complexity in Life and Science", *Philosophica*, **59**(1), pp. 41–68.

[9] Emmeche, C., Koppe, S. and Stjernfelt, F. (1997), "Explaining Emergence: Towards an Ontology of Levels", *Journal for General Philosophy of Science*, **28**, pp. 83–119.

[10] Franklin, S. and Graesser, A. (1997), "Is It an Agent, or Just a Program?: A Taxonomy for Autonomous Agents", in J. P. Muller, M. J. Wooldridge, and N. R. Jennings (eds.), *Intelligent Agents III: Agents Theories, Architectures, and Languages*, Springer-Verlag, pp. 21–35.

[11] Gallagher, R. and Appenzeller, T. (1999), "Beyond Reductionism", *Science*, **284**(5411), pp. 79.

[12] Gleick, J. (1997), *Chaos: Making a New Science*, Vintage.

[13] Goldenfeld, N. and Kadanoff, L. P. (1999), "Simple Lessons from Complexity", *Science*, **284**(5411), pp. 87–89.

[14] Grassé, P. -P. (1959), "La Reconstruction du Nid et Les Coordinations Interindividuelles Chez *Bellicositermes Natalensis* et *Cubitermes sp*. La Théorie de la Stigmergie: Essai d'interprétation du Comportement des Termites Constructeurs", *Insect. Soc.*, **6**, pp. 41–80.

[15] Holland, J. H. (1995), Hidden Order: How Adaptation Builds Complexity, Addison-Wesley.

[16] Holland, J. H. (1998), *Emergence: From Chaos to Order*, Oxford University Press.

[17] Holland, O. and Melhuish, C. (1999), "Stigmergy, Self-Organization, and Sorting in Collective Robotics", *Artificial Life*, **5**(2), pp. 173–202.

[18] Johnson, S. (2002), Emergence: The Connected Lives of Ants, Brains, Cities and Software, Penguin Books.

[19] Kennedy, J., Eberhart, R. and Shi. Y. (2001), *Swarm Intelligence*, Morgan Kaufman Publishers.

[20] Lewin, R. (1999), *Complexity: Life at the Edge of Chaos*, 2nd. Ed., Phoenix.

[21] McClelland, J. L., Rumelhart, D. E. and the PDP Research Group (1986), Parallel Distributed Processing: Explorations in the Microstructure of Cognition, Volume 2: Psychological and Biological Models, The MIT Press.

[22] Mehler, A. (2003), "Methodological Aspects of Computational Semiotics", *S.E.E.D. Journal (Semiotics, Evolution, Energy, and Development)*, **3**(3), pp. 71–80.

[23] Michener, C. D. (1974), *The Social Behavior of Bees: A Comparative Study*, Harvard University Press.

[24] Morowitz, H. and Singer, J. L. (Eds.) (1995), *The Mind, The Brain, and Complex Adaptive Systems*, Addison-Wesley Publishing Company.

[25] O'Reilly, R. C. and Munakata, Y. (2000), Computational Explorations in Cognitive Neuroscience: Understanding the Mind by Simulating the Brain, The MIT Press.

[26] Pattee, H. H. (1988), "Simulations, Realizations, and Theories of Life", in C. Langton (ed.), *Artificial Life*, Addison-Wesley, pp. 63–77.

[27] Paton, R. (1994), *Computing with Biological Metaphors*, International Thomson Computer Press.

[28] Peck, S. L. (2004), "Simulation as Experiment: A Philosophical Reassessment for Biological Modeling", *Trends in Ecology and Evolution*, **19**(10), pp. 530–534.

[29] Rocha, L. M. (1999), "Complex Systems Modeling: Using Metaphors from Nature in Simulation and Scientific Models", *BITS: Computer and Communications News*, Computing, Information, and Communications Division, Los Alamos National Laboratory, November.

[30] Rumelhart, D. E., McClelland, J. L. and the PDP Research Group (1986), Parallel Distributed Processing: Explorations in the Microstructure of Cognition, Volume 1: Foundations, The MIT Press.

[31] Russell, S. J. and Norvig, P. (1995), *Artificial Intelligence: A Modern Approach*, Prentice Hall.

[32] Science (1999), Special Issue on Complex Systems, *Science Magazine*, **284**(5411).

[33] Solé, R. and Goodwin, B. (2002), *Signs of Life: How Complexity Pervades Biology*, Basic Books.

[34] Stewart, I. (1995), *Nature's Numbers*, Phoenix.

[35] Stewart, I. (1997), Does God Play Dice?: The New Mathematics of Chaos, Penguin Books.

[36] Trappenberg, T. (2002), *Fundamentals of Computational Neuroscience*, Oxford University Press.

[37] Williams, N. (1997), "Biologists Cut Reductionist Approach Down to Size", *Science*, **277**(5325), pp. 476–477.

[38] Wilson, R. A. and Keil, F. C., (eds.) (1999), *The MIT Encyclopedia of the Cognitive Sciences*, The MIT Press.

[39] Wooldridge, M. J. and Jennings, N. R. (1995), "Intelligent Agents: Theory and Practice", *Knowledge Engineering Reviews*, **10**(2), pp. 115–152.

PART I

COMPUTING INSPIRED BY NATURE

CHAPTER 3

EVOLUTIONARY COMPUTING

"... one general law, leading to the advancement of all organic beings, namely, multiply, vary, let the strongest live and the weakest die."
(C. Darwin, The Origin of Species, 1859; Wordsworth Editions Limited (1998), p. 186).

"The work done by natural selection is R and D, so biology is fundamentally akin to engineering, a conclusion has been deeply resisted out of misplaced fear for what it might imply. In fact, it sheds light on some of our deepest puzzles. Once we adopt the engineering perspective, the central biological concept of function and the central philosophical concept of meaning can be explained and united. Since our own capacity to respond to and create meaning – our intelligence – is grounded in our status as advanced products of Darwinian processes, the distinction between real and artificial intelligence collapses. There are important differences, however, between the products of human engineering and the products of evolution, because of differences in the processes of evolution into focus, by directing products of our own technology, computers, onto the outstanding questions."
(D. Dennett, Darwin's Dangerous Idea: Evolution and the Meanings of Life, Penguin Books, 1995, p. 185–186)

"... nothing in biology makes sense, except in the light of evolution."
(T. Dobzhansky, The American Biology Teacher, 35, 1973, p. 125–129)

3.1 INTRODUCTION

Evolutionary computing, also called *evolutionary computation*, is the field of research that draws ideas from evolutionary biology in order to develop search and optimization techniques for solving complex problems. Most *evolutionary algorithms* are rooted on the Darwinian theory of evolution. Darwin proposed that a population of individuals capable of reproducing and subjected to (genetic) variation followed by selection result in new populations of individuals increasingly more fit to their environment. Darwin's proposal was very radical at the time it was formalized, in the late 1850s, because it suggested that a simple algorithmic process of reproduction plus variation and natural selection was sufficient to produce complex life forms.

This simple theory for the origin and diversity of life resulted in the development of one of the most useful natural computing approaches to date, namely, the evolutionary algorithms (EAs). There are several types of EAs, among which the most classical ones are *genetic algorithms*, *evolution strategies*, *evolutionary programming*, and *genetic programming*. This chapter starts by describing what is problem-solving as a search task, and follows with an introduction to *hill-climbing* and the *simulated annealing* algorithm, some traditional search techniques. The motivation behind the simulated annealing algorithm is provided and it is theoretically compared with hill-climbing and some of its variations. This paves the ground for a better understanding of EAs. The focus of the chapter is on the *standard genetic algorithm*, but an overview of the other main evolutionary algorithms is also provided. To appropriately describe the inspiration behind all evolutionary algorithms, with particular emphasis on the genetic algorithm, some background on evolutionary genetics is provided.

3.2   PROBLEM SOLVING AS A SEARCH TASK

Under the evolutionary perspective to be studied in this chapter, a *problem* may be understood as a collection of information from which something (e.g., knowledge) will be extracted or inferred. For instance, consider the cases of a numeric function to be maximized, and the problem of allocating a number of classes to some set of students, known as a timetabling problem. In the first problem, some knowledge (information) about the function to be optimized is available (e.g., the function itself, $f(x) = x^3 + x + 3$), and the objective is to determine the values of x that maximize this function. In the timetabling problem, a lot of information might be available, such as the number of students, classrooms, teachers, and so forth. The objective may be, for example, to make a timetable for all the classes at a college in a semester, given the information available.

The process of *problem solving* corresponds to taking actions (steps), or sequences of actions (steps), that either lead to a desired performance or improve the relative performance of *individuals*. This process of looking for a desired performance or improved performances is called *search*. A search algorithm will take a problem as input and return a solution to it. In this case, one or more individuals, which could be viewed as *agents*, will be used as *candidate solutions* to the problem. Some knowledge about the desired performance of a given individual is not always available, but it might still be possible to evaluate the relative quality of individuals that are being used as means to solve the problem.

The first step in problem solving is the problem formulation, which will depend on the information available. Three main concepts are then involved in problem solving (Michalewicz and Fogel, 2000):

1) *Choice of a representation*: encoding of alternative candidate solutions (individuals) for manipulation. For each problem, the representation of a candidate solution is of paramount importance, and its corresponding interpretation implies the *search space* and its size. The search (or state) space is defined by the initial *state* (*configuration*) and the set of possible states (configurations) of the problem.

2) *Specification of the objective*: description of the purpose to be fulfilled. This is a mathematical statement of the task to be achieved. It is not a function, but rather an expression. For instance, if the objective is to minimize the function $f(x) = x^3 + x + 3$ given above, then the objective can be stated as follows: min $f(x)$.

3) *Definition of an evaluation function*: a function that returns a specific value indicating the quality of any particular candidate solution (individual), given the representation. In cases when no knowledge about the desired performance of individuals is available, the evaluation function may be used as a means to evaluate the relative quality of individuals, thus allowing for the choice of one or more high quality individuals within a set of candidate solutions.

3.2.1. Defining a Search Problem

Given a search space S, assume the existence of some *constraints* that, if violated, avoid the implementation of a solution. In the search for improved solutions to a problem, we have to be able to move from one solution to the next without violating any of the constraints imposed by the problem formulation, that is, we need operators that generate or select *feasible* solutions. See Appendix B.4.1 for a brief review of optimization problems.

It is now possible to define a search (or optimization) problem (Michalewicz and Fogel, 2000). Given the search space S, together with its feasible part F, $F \subseteq S$, find $x^* \in F$ such that $eval(x^*) \leq eval(x), \forall x \in F$.

In this case, the evaluation function that returns smaller values for x is considered better. Thus, the problem is one of *minimization*. However, we could just as easily use an evaluation function for which larger values are favored, thus turning the search problem into one of *maximization*. To maximize something is equivalent to minimize the negative of the same thing ($\max f(x) = \min -f(x)$). It is important to stress that the search process itself does not know what problem is being solved, as will be further discussed in Section 3.9.1. All it knows is the information provided by the evaluation function, the representation used, and how the possible solutions are sampled. If the evaluation function does not correspond with the objective, we will be searching for the right answer to the wrong problem!

The point x that satisfies the above condition is called a *global solution* or *global optimum*. Finding such a global solution to a problem might be very difficult, but sometimes finding the best solution is easier when it is possible to concentrate on a small portion of the search space. Effective search techniques have to provide a mechanism for balancing two apparently conflicting objectives: *exploiting* the best solutions found so far and at the same time *exploring* the search space. Among the search techniques to be studied, some (e.g., hillclimbing) exploit the best available solution but neglect exploring a large portion of the search space, while others (e.g., simulated annealing and genetic algorithms) combine exploration with exploitation of the search space. It is important to note, however, that it has been proved mathematically that there is no way to choose a single search method that can have an average superior performance in all runs for all problems; a theorem known as 'No Free Lunch' (Wolpert and Macready, 1997).

It is possible, on the other side, to assess and compare the performance of different algorithms in specific problem domains and, therefore, it is possible to look for a technique that provides the best performance, on average, in this domain.

In contrast to the global optimum, a *local optimum* is a potential solution $x \in F$ in respect to a neighborhood N of a point y, if and only if $eval(x) \leq eval(y)$, $\forall\, y \in N(x)$, where $N(x) = \{y \in F : dist(x,y) \leq \varepsilon\}$, *dist* is a function that determines the distance between x and y, and ε is a positive constant. Figure 3.1 illustrates a function with several local optima (minima) solutions, a single global optimum (minimum), and the neighborhood of radius ε around one of the local optima solutions.

The evaluation function defines a response surface, which will later be termed *fitness landscape* (Section 3.4.4), that is much like a topography of hills and valleys. The problem of finding the (best) solution is thus the one of searching for a peak, assuming a maximization problem, in such a fitness landscape. If the goal is that of minimization, then a valley has to be searched for. Sampling new points in this landscape is basically made in the immediate vicinity of current points, thus it is only possible to take local decisions about where to search next in the landscape. If the search is always performed uphill, it might eventually reach a peak, but this might not be the highest peak in the landscape (global optimum). The search might sometimes go downhill in order to find a point that will eventually lead to the global optimum. Therefore, by not knowing the fitness landscape and using only local information, it is not possible to guarantee the achievement of global optima solutions.

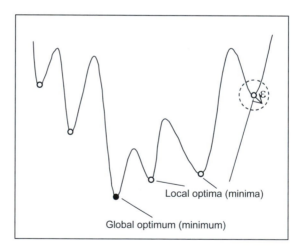

Figure 3.1: Illustration of global and local minima of an arbitrary function. Dark circle: global minimum; White circles: local minima.

3.3   HILL CLIMBING AND SIMULATED ANNEALING

This section introduces two standard search techniques: *hill-climbing* and *simulated annealing*. The simulated annealing algorithm can be seen as a sort of probabilistic hill-climber, and it can also be viewed as a special case of an evolutionary algorithm, which is the main subject of this chapter. Additionally, the study of these two procedures serves as a preparation for the study of evolutionary algorithms, a better understanding of their behavior, and for performance comparisons as well.

3.3.1. Hill Climbing

Hill climbing is a local search method that uses an *iterative improvement* strategy. The strategy is applied to a single point - the current point (or state) - in the search space. At each iteration, a new point x' is selected by performing a small displacement or perturbation in the current point x, i.e., the new point is selected in the neighborhood of the current point: $x' \in N(x)$. Depending on the representation used for x, this can be implemented by simply adding a small random number, Δx, to the current value of x: $x' = x + \Delta x$.

If that new point provides a better value for the evaluation function, then the new point becomes the current point. Else, some other displacement is promoted in the current point (a new neighbor is chosen) and tested against its previous value. Termination occurs when one of the following *stopping criteria* is met:

- No further improvement can be achieved.
- A fixed number of iterations have been performed.
- A goal point is attained.

Let x be the current point, g the goal point (assuming it is known), and max_it a maximum number of iterations allowed. Algorithm 3.1 contains the pseudocode of a standard (simple) hill-climbing algorithm.

```
procedure [x] = hill-climbing(max_it,g)
  initialize x
  eval(x)
  t ← 1
  while t < max_it & x != g & no_improvement do,
      x' ← perturb(x)
      eval(x')
      if eval(x') is better than eval(x),
            then x ← x'
      end if
      t ← t + 1
  end while
end procedure
```

Algorithm 3.1: A standard (simple) hill-climbing procedure.

```
procedure [best] = IHC(n_start,max_it,g)
  initialize best
  t1 ← 1
  while t1 < n_start & best != g do,
    initialize x
    eval(x)
    x ← hill-climbing(max_it,g) //Algorithm 1
    t1 ← t1 + 1
    if x is better than best,
          then best ← x
    end if
  end while
end procedure
```

Algorithm 3.2: An iterated hill-climbing procedure.

Hill climbing algorithms have some weaknesses. First, they usually terminate at local optima solutions. Second, there is no information about the distance from the solution found and the global optimum. Third, the optimum found depends on the initial configuration. Finally, it is generally not possible to provide an upper bound for the computational time of the algorithm.

An important aspect of this algorithm, however, is that it performs a *blind search*, i.e., the current point is randomly perturbed and evaluated to check for an improvement of the objective function (see discussion in Section 3.9.1). Note also that hill-climbing methods can only converge to local optima solutions, and these values are dependent upon the starting point. Furthermore, as the global optimum is usually unknown, there is no general procedure to bound the relative error with respect to it. As the algorithm only provides local optima solutions, it is reasonable to start hill-climbing methods from a large variety of points. The hope is that at least one of these initial locations will have a path leading to the global optimum. The initial points might be chosen randomly, or by using a regular grid or pattern, or even by using other types of information. In the case the standard hill-climbing algorithm allows for multiple initializations and maintains a 'memory' of the best solution found so far, we have the so-called *iterated hill-climbing*, as described in Algorithm 3.2 (The input variable n_start is the number of different starting points to be used).

As an alternative stopping criterion, it is possible to verify if the algorithm reached a local optima solution by comparing the current value of **x** (or *best*) with some of its previous values. If the value of **x** (*best*) does not change significantly after a number of iterations, then the algorithm can be said to have converged to a local optimum and the process can be halted.

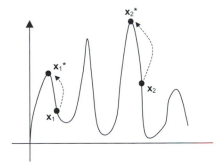

Figure 3.2: An illustrative example of the behavior of the basic hill-climbing algorithm.

There are a number of variations for the hill-climbing algorithm, and the ones discussed here are the simplest cases. Figure 3.2 illustrates the performance of the iterated hill-climbing algorithm assuming only two different initial points, x_1 and x_2. This function is uni-dimensional and contains only four peaks, with the third peak from left to right corresponding to the global optimum (the other peaks are all local optima solutions). If the first start point is at x_1, then the inner loop of the algorithm converges to the local optimum x_1^*. In the second start, point x_2, the global optimum will be determined and *best* = x_2^*.

Note that in practical applications the number of local optima solutions is very high, making such pure a trial and error strategy almost always unfeasible. Nevertheless, the good exploitation capability of hill-climbing has been used in combination with many other techniques capable of performing a better exploration of the search space.

The standard hill-climbing procedure can be modified in order to accommodate a probabilistic selection of the perturbed point x'. The probability that x' is selected depends upon the relative merit of x and x', i.e., the difference between the values returned by the evaluation function for these two points. This modification leads to a new algorithm called *stochastic hill-climbing* procedure, as described in Algorithm 3.3, with the capability of escaping from local optima solutions. The parameter T is a control parameter for the decay of the exponential function, and it remains constant along the iterations.

In Algorithm 3.3, by looking at the probability P of accepting the new point x', $P = 1/(1+\exp[(\text{eval}(x)-\text{eval}(x'))/T])$, it is possible to note that the greater the value of T, the smaller the importance of the relative difference between the evaluation of x and x'. If T is very large, then the search becomes similar to a random search.

```
procedure [x] = stochastic hill-climbing(max_it,g)
  initialize x
  eval(x)
  t ← 1
  while t < max_it & x != g do,
     x' ← perturb(x)
    eval(x')
    if random[0,1) < (1/(1+exp[(eval(x)-eval(x'))/T])),
          then x ← x'
    end if
    t ← t + 1
  end while
end procedure
```

Algorithm 3.3: A stochastic hill-climbing procedure.

3.3.2. Simulated Annealing

The *simulated annealing* (SA) algorithm, proposed by Kirkpatrick et al. (1983), was inspired by the *annealing* process of physical systems. Annealing corresponds to subjecting a material (e.g., glass or metal) to a process of heating and slow cooling in order to toughen and reduce brittleness.

The SA algorithm is based upon that of Metropolis et al. (1953), which was originally proposed as a means of finding the equilibrium configuration of a collection of atoms at a given temperature. The basic ideas of SA were taken from *statistical thermodynamics*, the branch of physics that makes theoretical predictions about the behavior of macroscopic systems (both, liquid and solid) on the basis of laws governing its component atoms.

Although the connection between the Metropolis algorithm and optimization problems had already been noted by Pincus (1970), it was Kirkpatrick et al. (1983) who proposed that it could form the basis of a general-purpose search (optimization) technique to solve combinatorial problems. The authors realized that there is a useful connection between statistical thermodynamics and multi-variate combinatorial optimization. A detailed analogy with annealing in solids could provide a framework to develop an optimization algorithm capable of escaping local optima solutions (configurations).

A fundamental question in statistical thermodynamics was raised concerning what happens to a system in the limit of low temperature. Do the atoms remain fluid or solidify? If they solidify, do they form a crystalline structure or a glass? The SA algorithm was instantiated with a procedure used to take a material to its ground state; that is, to a state of lowest energy in which a crystal, instead of a glass, can be grown from a melt. The material is first heated to a high temperature so that it melts and the atoms can move freely. The temperature of the melt is then slowly lowered so that at each temperature the atoms can move enough to begin adopting a more stable orientation. If the melt is cooled down slowly enough, the atoms are able to rest in the most stable orientation, producing a crystal. This heating followed by a slow cooling process is known as *annealing*.

Kirkpatrick et al. (1983) proposed that when it is possible to determine the energy of a system, the process of finding its low temperature is akin to combinatorial optimization. Nevertheless, the concept of temperature of a physical system had no obvious equivalent in a problem being solved (optimized). By contrast, any cost function could play the role of the energy of the system. It remained thus important to propose an equivalent to the temperature parameter when dealing with optimization problems.

Basic Principles of Statistical Thermodynamics

In *statistical thermodynamics* large systems at a given temperature approach spontaneously the equilibrium state, characterized by a mean value of energy, depending on the temperature. By simulating the transition to the equilibrium and decreasing the temperature, it is possible to find smaller and smaller values of the mean energy of the system (Černý, 1985).

Let $\mathbf{x}$ be the current configuration of the system, $E(\mathbf{x})$ the energy of $\mathbf{x}$, and T the temperature. The equilibrium is by no means a static situation. In equilibrium, the system randomly changes its state from one possible configuration to another in such a way that the probability of finding the system in a particular configuration is given by the Boltzmann-Gibbs distribution:

$$P(\mathbf{x}) = K.\exp(-E(\mathbf{x})/T) \qquad (3.1)$$

where K is a constant. If we assume a system with a discrete number of possible states, then the mean energy E_m of the system in equilibrium is given by

$$E_m = \left[\sum_{conf} E_{conf} \exp(-E_{conf}/T) \middle/ \sum_{conf} \exp(-E_{conf}/T) \right] \qquad (3.2)$$

The numerical calculation of E_m might be quite difficult if the number of configurations is high. However, it is possible to employ a Monte Carlo simulation of the random changes of state from one configuration to another such that, in equilibrium, Equation (3.1) holds. One such algorithm is the Metropolis et al. (1953) procedure.

The Simulated Annealing Algorithm

Assuming the general problem of minimizing a function, the simulated annealing algorithm works as follows. Having a current configuration $\mathbf{x}$, this is given a small random displacement, resulting in $\mathbf{x}'$, where the point $\mathbf{x}'$ is chosen in the neighborhood of $\mathbf{x}$. The energy $E(\mathbf{x}')$ of this new configuration is computed, and one has to decide whether to accept $\mathbf{x}'$ as the new configuration, or to reject it. The resulting change in the energy of the system, $\Delta E = E(\mathbf{x}') - E(\mathbf{x})$, is calculated. If $\Delta E \leq 0$, then the displacement is accepted and the configuration $\mathbf{x}'$ is used as the starting point for the next iteration step. If $\Delta E > 0$, the probability that the configuration $\mathbf{x}'$ is accepted is given by Equation (3.3), which is a particular case of the Boltzmann-Gibbs distribution (Equation (3.1)).

$$P(\Delta E) = \exp(-\Delta E/T) \qquad\qquad (3.3)$$

To implement the random portion of the procedure, one can simply generate a random number $r = \texttt{random}[0,1)$, sampled over the interval $[0,1)$, and compare it with $P(\Delta E)$. If $r < P(\Delta E)$, the new configuration $\mathbf{x}'$ is retained, else it is discarded and $\mathbf{x}$ is maintained in the next iteration of the algorithm.

By repeating the basic steps described in the paragraph above, one is simulating the thermal motion of atoms in thermal contact with a heat bath at temperature T. By using $P(\Delta E)$ as described by Equation (3.3), the system evolves into a Boltzmann distribution.

To apply this algorithm as a general-purpose optimization procedure, the energy of the system is replaced by an evaluation function and the configurations are defined by a set of variables for this function. The temperature is a *control parameter* in the same units as the evaluation function. As annealing corresponds to first heating and then slowly cooling down the system, the temperature T in the simulation usually starts with a high value and is decreased during the iterative procedure. At each temperature, the simulation must proceed long enough for the system to reach a steady state. The sequence of temperatures and the number of rearrangements of the parameters attempted to reach equilibrium at each temperature are termed *annealing schedule* (Kirkpatrick et al., 1983).

Let $\mathbf{x}$ be the current configuration of the system, $\mathbf{x}'$ be the configuration $\mathbf{x}$ after a small random displacement, and T the temperature of the system. The standard simulated annealing algorithm for solving minimization problems is described in Algorithm 3.4. The function $g(T, t)$ is responsible for reducing the value of the temperature. Usually a geometrical decrease is employed, e.g., $T \leftarrow \beta.T$, where $\beta < 1$.

```
procedure [x] = simulated_annealing(g)
   initialize T
   initialize x
   eval(x)
   t ← 1
   while not_stopping_criterion do,
       x' ← perturb(x)
       eval(x')
       if eval(x') is less than eval(x),
            then x ← x'
       else if random[0,1) < exp[(eval(x)-eval(x'))/T],
            then x ← x'
       end if
       T ← g(T,t)
       t ← t + 1
   end while
end procedure
```

Algorithm 3.4: The simulated annealing procedure.

```
Step 1:      initialize T
             initialize x
Step 2:      x' ← perturb(x)
             if eval(x') is less than eval(x),
                   then x ← x'
             else if random[0,1) < exp[(eval(x)-eval(x'))/T]
                   then x ← x'
             end if
             repeat this step k times
Step 3:      T ← β.T
             if T ≥ Tmin
                   then goto Step 2
             else goto Step 1
```

Algorithm 3.5: Typical implementation of the simulated annealing procedure.

The Metropolis algorithm can be viewed as a *generalized iterative improvement* method, in which controlled uphill steps can also be incorporated into the search. Also, this algorithm is very similar to the stochastic hill-climber, with the difference that it allows changes in the parameter T during the run. Most implementations of the simulated annealing algorithm follow the simple sequence of steps (Michalewicz and Fogel, 2000) presented in Algorithm 3.5.

In this typical implementation (Algorithm 3.5) the algorithm is initialized a number of times, whenever the temperature reaches is minimal value (frozen temperature). Also, for each temperature value the algorithm is run a number k of times.

From Statistical Thermodynamics to Computing

In the previous sections the simulated annealing algorithm was described together with its inspiration in physical systems. Table 3.1 summarizes how to interpret the terminology from the physics domain into the one used in the simulated annealing algorithm.

Table 3.1: Interpretation from the physics terminology into the computational domain.

Physics	Simulated Annealing Algorithm
State	(Feasible) solution to the problem, also called point in the search space
Energy	Value returned by the evaluation function
Equilibrium state	Local optimum
Ground state	Global optimum
Temperature	Control parameter
Annealing	Search by reducing T
Boltzmann-Gibbs distribution	Probability of selecting a new point

3.3.3. Example of Application

Consider the uni-dimensional function $g(x)$ presented in Figure 3.3. The variable x is defined over the interval $[0,1]$, $x \in [0,1]$, and assume the global maximum of this function is unknown. It was discussed above that to optimize this function it is necessary to have a well-defined representation, objective function, and evaluation function.

Representation. The most straightforward representation, in this case, is to use the real value of variable x to represent the candidate solutions to the problem. Another representation can be obtained by using a bitstring data structure, as will be described in detail in Section 3.5.2. In order to perturb the current point, i.e., to generate a candidate point in the neighborhood of the current point, a Gaussian random noise of zero mean and small variance, $G(0,\sigma)$, can be added to the current point. If the added noise results in an individual that lies within the domain of x, $x \in [0,1]$, then accept it; else, discard it: $x' = x + G(0,\sigma)$, where σ is a small positive constant. The same could be done with a uniform, instead of Gaussian, distribution.

Objective. The objective for this problem is to find the maximal value of the function $g(x)$; that is, $\max g(x)$.

Evaluation. The evaluation function used to determine the relative quality of candidate solutions is obtained by simply evaluating the function $g(x)$ for the values of x.

By using real values over the interval $[0,1]$ to represent the variable x, all hill-climbing procedures described (simple, iterated, and stochastic) and the simulated annealing algorithm can be implemented and applied to find the global maximum of $g(x)$. For the simple hill-climbing, different initial configurations can be tried as attempts at finding the global optimum.

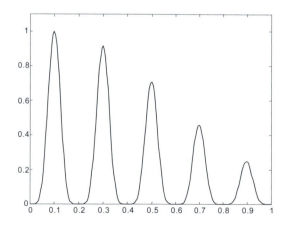

Figure 3.3: Graph of the function $g(x) = 2^{-2((x-0.1)/0.9)^2}\left(\sin(5\pi x)\right)^6$ to be maximized.

3.4   EVOLUTIONARY BIOLOGY

Evolutionary biology is a science concerned, among other things, with the study of the diversity of life, the differences and similarities among organisms, and the adaptive and non-adaptive characteristics of organisms. Its importance are manifold, from the health sciences to the understanding of how the living organisms adapt to the environment they inhabit. For instance, evolutionary biology helps in the understanding of disease epidemics, population dynamics, and the production of improved cultures. Over roughly the last 60 years, computer scientists and engineers realized that evolutionary biology has various interesting ideas for the development of theoretical models of evolution (some of them being rather abstract models) that can be useful to obtain solutions to complex real-world problems.

The word *evolution* is originated from the Latin *evolvere*, which means to unfold or unroll. Broadly speaking, evolution is a synonym for 'change'. But what type of change? We do not usually employ the word evolution to refer to the changes suffered by an individual during its lifetime. Instead, an evolving system corresponds to the one in which there is a descent of entities over time, one generation after the other, and in which characteristics of the entities differ across generations (Futuyma, 1998). Therefore, *evolution* can be broadly defined as *descent with modification* and often *with diversification*. Many systems can be classified as evolutionary: languages, cellular reproduction in immune systems, cuisines, automobiles, and so on.

Any evolutionary system presents a number of features:

- *Population(s)*: in all evolutionary systems there are populations, or groups, of entities, generally termed *individuals*.

- *Reproduction*: in order for evolution to occur, the individuals of the population(s) must reproduce either sexually or asexually.

- *Variation*: there is variation in one or more characteristics of the individuals of the population(s).

- *Hereditary similarity*: parent and offspring individuals present similar characteristics. Over the course of generations, there may be changes in the proportions of individuals with different characteristics within a population; a process called *descent with modification*.

- *Sorting of variations*: among the sorting processes, it can be emphasized *chance* (random variation in the survival or reproduction of different variants), and *natural selection* (consistent, non-random differences among variants in their rates of survival and reproduction).

Adaptation as a result of variation plus natural selection leads to improvement in the function of an organism and its many component parts. "Biological *or* organic evolution *is change in the properties of populations of organisms, or groups of such populations, over the course of generations*." (Futuyma, 1998; p. 4). Note that according to this definition of evolution, individual organisms do not evolve and the changes of a population of individuals that are assumed to be evolutionary are those resultant from inheritance, via the genetic material, from one generation to the other.

The history of evolutionary biology is marked by a number of hypotheses and theories about how life on earth appeared and evolved. The most influential theory to date is the one proposed by Charles Darwin and formalized in his book *On the Origins of Species by Means of Natural Selection, or the Preservation of Favoured Races in the Struggle for Life* (Darwin, 1859). Historically, Alfred Wallace is also one of the proponents of the theory of evolution by means of natural selection, but it was Darwin's book, with its hundreds of instances and arguments supporting natural selection, the landmark for the theory of evolution.

Among the many preDarwinian hypotheses for the origin and development of beings, the one proposed by Jean Baptist Pierre Antoine de Monet, chevalier de Lamarck, was the most influential. According to Lamarck, every species originated individually by spontaneous generation. A 'nervous fluid' acts within each species, causing it to progress up the chain over time, along a single predetermined path that every species is destined to follow. No extinction has occurred: fossil species are still with us, but have been transformed. According to Lamarck, species also adapt to their environments, the more strongly exercised organs attract more of the nervous fluid, thus getting enlarged; conversely, the less used organs become smaller. These alterations, acquired during an individual's lifetime through its activities, are inherited. Like everyone at that time, Lamarck believed in the so-called *inheritance of acquired characteristics*.

The most famous example of Lamarck's theory is the giraffe: according to Lamarck, giraffes need long necks to reach the foliage above them; because they are constantly stretching upward, the necks grow longer; these longer necks are inherited; and over the course of generations the necks of giraffes get longer and longer. Note that the theory of inheritance of acquired characteristics is not Lamarck's original, but an already established supplement to his theory of 'organic progression' in which spontaneous generation and a chain of beings (progression from inanimate to barely animate forms of life, through plants and invertebrates, up to the higher forms) form the basis. Lamarck's theory may also be viewed as a *transformational theory*, in which change is programmed into every member of the species.

3.4.1. On the Theory of Evolution

Darwin's studies of the natural world showed a striking diversity of observations over the animal and vegetal kingdoms. His examples were very wide ranging, from domestic pigeons, dogs, and horses, to some rare plants. His research that resulted in the book *Origin of Species* took literally decades to be concluded and formalized.

In contrast to the Lamarckian theory, Darwin was certain that the direct effects of the conditions of life were unimportant for the variability of species.

"Seedlings from the same fruit, and the young of the same litter, sometimes differ considerably from each other, though both the young and the parents ... have apparently been exposed to exactly the same conditions of life; and this shows how unimportant the direct effects of the conditions of life are in comparison with the laws of reproduction, and of growth, and of inheritance; for had the action of the conditions been direct, if any of the young had varied, all would probably have varied in the same manner." (Darwin, 1859; p. 10)

Darwin starts his thesis of how species are formed free in nature by suggesting that the most abundant species (those that range widely over the world) are the most diffused and which often produce well-marked varieties of individuals over the generations. He describes some basic rules that promote improvements in organisms: reproduce, change and compete for survival.

Natural selection was the term used by Darwin to explain how new characters arising from variations are preserved. He starts thus paving the ground to his theory that slight differences in organisms accumulated over many successive generations might result in the appearance of completely new and more adapted species to their environment. As defended by himself

"... as a general rule, I cannot doubt that the continued selection of slight variations ... will produce races differing from each other ..." (Darwin, 1859; p. 28) and "... I am convinced that the accumulative action of Selection, whether applied methodically and more quickly, or unconsciously and more slowly, but more efficiently, is by far the predominant Power." (Darwin, 1859; p. 35)

In summary, according to Darwin's theory, evolution is a result of a population of individuals that suffer:

- Reproduction with inheritance.
- Variation.
- Natural selection.

These very same processes constitute the core of all evolutionary algorithms. Before going into the details as to how reproduction and variation happen within individuals and species of individuals, some comments about why Darwin's theory was so revolutionary and 'dangerous' at that time (and, to some people, until nowadays) will be made.

3.4.2. Darwin's Dangerous Idea

Darwin's theory of evolution is controversial and has been refuted by many because it presents a sound argument for how a "Nonintelligent Artificer" could produce the wonderful forms and organisms we see in nature. To D. Dennett (1991), *Darwin's dangerous idea* is that evolution, thus life, can be explained as the product of an *algorithmic process*, not of a superior being (God) creating everything that might look wonderful to our eyes. But the reason there is a section on Dennett's book here is not to discuss particular beliefs. Instead, to discourse about some key interpretations of evolution, from a computational perspective, presented by D. Dennett in his book *Darwin's Dangerous Idea: Evo-*

lution and the Meanings of Life. These are not only interesting, but also useful for the understanding of why the theory of evolution is suitable for the comprehension and development of a class of search techniques known as evolutionary algorithms.

Dennett defines an algorithm as a certain sort of formal process that can be counted on (logically) to yield a certain sort of result whenever it is run or instantiated. He emphasizes that *evolution* can be understood and represented in an abstract and common terminology as an *algorithmic process*; it can be lifted out of its home base in biology. Evolutionary algorithms are thus those that embody the major processes involved in the theory of evolution: a population of individuals that *reproduce with inheritance*, and suffer *variation* and *natural selection*.

Dennett also discusses what can be the outcomes of evolution and its probable implications when viewed as an *engineering process*. He stresses the importance of genetic variation and selection, and quotes an interesting passage from M. Eigen (1992).

> "Selection is more like a particularly subtle demon that has operated on the different steps up to life, and operates today at the different levels of life, with a set of highly original tricks. Above all, it is highly active, driven by an internal feedback mechanism that searches in a very discriminating manner for the best route to optimal performance, not because it possesses an inherent drive towards any predestined goal, but simply by virtue of its inherent non-linear mechanism, which gives the appearance of goal-directedness." (Eigen, 1992; quoted by Dennett, 1991, p. 195)

Another important argument is that evolution requires *adaptation* (actually it can also be seen as adaptation plus selection, as discussed in the previous chapter). From an evolutionary perspective, adaptation is the reconstruction or prediction of evolutionary events by assuming that all characters are established by direct natural selection of the most adapted state, i.e. the state that is an 'optimum solution' to a 'problem' posed by the environment. Another definition is that under adaptation, organisms can be viewed as complex adaptive systems whose parts have (adaptive) functions subsidiary to the fitness-promoting function of the whole.

The key issue to be kept in mind here is that evolution can be viewed as an algorithmic process that allows - via reproduction with inheritance, variation and natural selection - the most adapted organisms to survive and be driven to a state of high adaptability (optimality) to their environment. These are the inspiring principles of evolutionary algorithms; the possibility of modeling evolution as a search process capable of producing individuals (candidate solutions to a problem) with increasingly better 'performances' in their environments.

3.4.3. Basic Principles of Genetics

The theory of evolution used in the development of most evolutionary algorithms is based on the three main aspects raised by Darwin as being responsible for the evolution of species: reproduction with inheritance, variation, and selection. However, the origins of heredity along with variations, which were some of the main ingredients for the natural selection theory, were unknown at that time. This section explores the genetic basis of reproduction and variation in order to provide the reader with the necessary biological background to develop and understand evolutionary algorithms, in particular *genetic algorithms*. The union of genetics with some notions of the selection mechanisms, together with Darwin's hypotheses led to what is currently known as neo-Darwinism.

Gregor Mendel's paper establishing the foundations of *genetics* (a missing bit for a broader understanding of the theory of evolution) was published only in 1865 (Mendel, 1865), but it was publicly ignored until about the 1900. He performed a series of careful breeding experiments with garden peas. In summary, Mendel selected strains of peas that differed in particular *traits* (characteristics). As these differences were clearly distinguishable, their *phenotypes* (measurable attributes, or observable physical or biochemical characteristics of an organism) were identified and scored. For instance, the pea seeds were either smooth or wrinkled, the pod shape was either inflated or constricted, and the seed color was either yellow or green. Then, Mendel methodically performed crosses among the many pea plants, counted the progeny, and interpreted the results. From this kind of data, Mendel concluded that phenotypic traits were controlled by *factors*, later called *Mendelian factors*, and now called *genes*. *Genotype* is the term currently used to describe the genetic makeup of a cell or organism, as distinguished from its physical or biochemical characteristics (the phenotype). Figure 3.4 summarizes the first experiment performed by Mendel.

The basic structural element of all organisms is the *cell*. Those organisms whose genetic material is located in the *nucleus* (a discrete structure within the cell that is bounded by a nuclear membrane) of the cells are named *eukaryotes*. *Prokaryotes* are the organisms that do not possess a nuclear membrane surrounding their genetic material. The description presented here focuses on eukaryotic organisms.

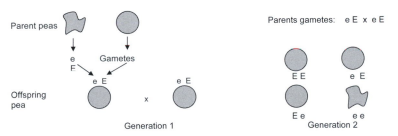

Figure 3.4: First experiment of Mendel. When crossing a normal pea with a wrinkled pea, a normal pea was generated (generation 1). By crossing two daughters from generation 1, three normal peas were generated plus one wrinkled pea. Thus, there is a recessive gene (e) that only manifests itself when there is no dominant gene together. Furthermore, there is a genetic inheritance from parents to offspring; those offspring that carry a factor that expresses a certain characteristic may have offspring with this characteristic.

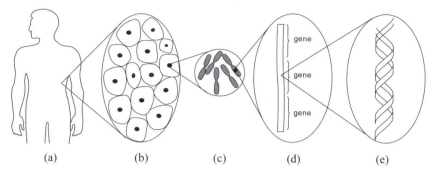

(a)   (b)   (c)   (d)   (e)

Figure 3.5: Enlargement of an organism to focus the genetic material. (a) Human organism. (b) Cells composing the organism. (c) Each cell nucleus contains chromosomes. (d) Each chromosome is composed of a long DNA segment, and the genes are the functional portions of DNA. (e) The double helix of DNA. (Modified with permission from [Griffiths et al., 1996], © W. H. Freeman and Company.)

In the cell nucleus, the genetic material is complexed with protein and is organized into a number of linear structures called *chromosomes*, which means, 'colored body', and is so named because these threadlike structures are visible under the light microscope only after they are stained with dyes. A *gene* is a segment of a helix molecule called *deoxyribonucleic acid*, or *DNA* for short. Each eukaryotic chromosome has a single molecule of DNA going from one end to the other. Each cell nucleus contains one or two sets of the basic DNA complement, called *genome*. The genome itself is made of one or more chromosomes. The *genes* are the functional regions of DNA. Figure 3.5 depicts a series of enlargements of an organism to focus on the genetic material.

It is now known that the DNA is the basis for all processes and structures of life. The DNA molecule has a structure that contributes to the two most fundamental properties of life: reproduction and development. DNA is a double helix structure with the inherent feature of being capable of replicating itself before the cell multiplication, allowing the chromosomes to duplicate into *chromatids*,

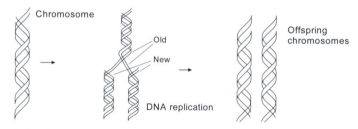

Figure 3.6: When new cells are formed, the DNA replication allows a chromosome to have a pair of offspring chromosomes and be passed onto the offspring cells.

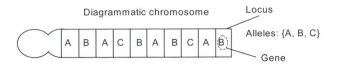

Figure 3.7: Diagrammatic chromosome depicting the locus, genes and three alleles {A, B, C}.

which eventually become *offspring chromosomes* that are transmitted to the *offspring cells*. The DNA replication process is essentially the same for sexual and asexual reproduction and is depicted in Figure 3.6.

It is the DNA replication property that allows the replication of cells and organisms during generations. Therefore, the DNA can be viewed as the 'string' that connects any organism to its descendants (Griffiths et al., 1996).

Genetics is a science named after its main object of study, namely, the genes. Studies in genetics demonstrated that many differences among organisms are results of differences in the genes they carry. Therefore, a gene can also be defined as the genetic factor that controls one trait or one characteristic of the organism. Together with the environmental influences, the genotype determines the phenotype of an organism. The different forms of a gene that determine alternate traits or characteristics are called *alleles*. The specific place on a chromosome where a gene is located is termed *locus*. Figure 3.7 presents a diagrammatic chromosome depicting the locus, genes, and alleles.

The genetic material of eukaryotes is distributed among multiple chromosomes, whose number usually varies according to the characteristics of the species. Many eukaryotes have two copies of each type of chromosome in their nuclei, so their chromosome complement is said to be *diploid*. In diploids, the members of a chromosome pair are called *homologous chromosomes*. Diploid eukaryotes are produced by the fusion of two *gametes* (mature reproductive cell specialized for sexual fusion), one from the female parent and another from the male parent. The fusion produces a diploid cell called *zygote*, which then undergoes embryological development. Each gamete has only one set of chromosomes and is said to be *haploid*.

Eukaryotes can reproduce by asexual or sexual reproduction. In *asexual reproduction*, a new individual develops from either a single cell or from a group of cells in the absence of any sexual process. It is found in multicellular and unicellular organisms. Single-celled eukaryotes grow, double their genetic material, and generate two progeny cells, each of which contains an exact copy (sometimes subjected to a small variation) of the genetic material found in the parent cell. The process of asexual reproduction in haploids is illustrated in Figure 3.8.

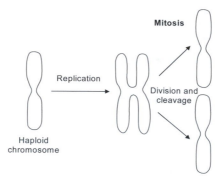

Figure 3.8: Asexual reproduction in haploids. The chromosome replicates itself, the cell nucleus is divided through a process named mitosis, and then the cell is divided into two identical progeny.

Sexual reproduction is the fusion of two *haploid gametes* (sex cells) to produce a single *diploid zygote* cell. An important aspect of sexual reproduction is that it involves *genetic recombination*; that is, it generates gene combinations in the offspring that are distinct from those in the parents. Sexually reproducing organisms have two sorts of cells: *somatic* (body) cells, and *germ* (sex) cells. All somatic cells reproduce by a process called *mitosis* that is a process of nuclear division followed by cell division. Figure 3.9 illustrates the process of sexual reproduction.

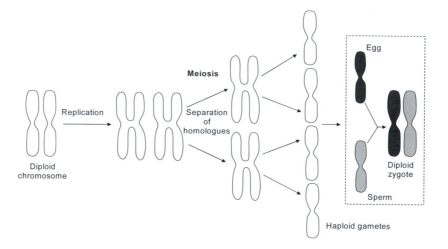

Figure 3.9: Sexual reproduction. A diploid chromosome replicates itself, then the homologues are separated generating haploid gametes. The gametes from each parent are fused to generate a diploid zygote.

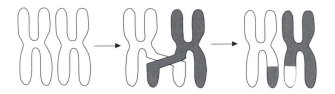

Figure 3.10: Crossing over between two loci in a cell undergoing the first meiotic division. Of the four chromatids, two will have new combinations and two will retain the parental combination of alleles.

In the classical view of the meiosis process in sexual reproduction, homologous chromosomes first undergo the formation of a very tight association of homologues, and then the reciprocal physical exchange of chromosome segments at corresponding positions along pairs of homologous chromosomes, a process termed *crossover* (Russel, 1996). Crossing-over is a mechanism that can give rise to *genetic recombination*, a process by which parents with different genetic characters give birth to progeny so that genes are associated in new combinations. Figure 3.10 depicts the crossing-over process.

The differences among organisms are outcomes of the evolutionary processes of *mutation* (a change or deviation in the genetic material), *recombination* or *crossover* (exchange of genetic material between *chromosomes*; see Figure 3.10), and *selection* (the favoring of particular combinations of genes in a given environment). With the exception of gametes, most cells of the same eukaryotic organism characteristically have the same number of chromosomes. Further, the organization and number of genes on the chromosomes of an organism are the same from cell to cell. These characteristics of chromosome number and gene organization are the same for all members of the same species. *Deviations* are known as *mutations*; these can arise spontaneously or be induced by chemical or radiation mutagens. Several types of mutation exist, for instance point mutation, deletion, translocation, and inversion. Point mutation, deletion and inversion are illustrated in Figure 3.11.

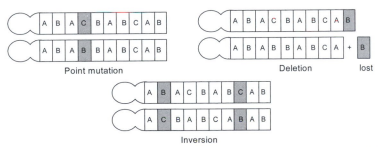

Figure 3.11: Some types of chromosomal mutation, namely, point mutation, deletion, and inversion.

3.4.4. Evolution as an Outcome of Genetic Variation Plus Selection

So far we have seen the two types of reproduction, sexual and asexual, and some of the main mechanisms that alter the genetic makeup of a population of individuals, emphasizing crossover and mutation. It still remains to discuss the process by which these altered individuals survive over the generations.

Populations of individuals change over time. The number of individuals may increase or decrease, depending on food resources, climate, weather, availability of breeding areas, predators, and so forth. At the genetic level, a population may change due to a number of factors, such as mutation and selection. These processes not only alter allele frequencies, but also result in changes in the adaptation and diversity of populations, thus leading to the evolution of a species (Gardner et al., 1991).

The viability and fertility of an individual are associated with *fitness*, a term that is used to describe the overall ability of an organism to survive and reproduce. In many populations, survival and reproductive ability are variable traits. Some individuals die before they have a chance to reproduce, whereas others leave many progeny. In a population of stable size, the average number of offspring produced by an individual is one.

Variation in fitness is partially explained by the underlying genetic differences of individuals. The crossing-over of parental genetic material and mutation can increase or decrease fitness, depending on their effects on the survival and reproductive capabilities of the individuals. Thus, genetic recombination and mutation can create phenotypes with different fitness values. Among these, the most fit will leave the largest number of offspring. This differential contribution of progeny implies that alleles associated with superior fitness will increase in frequency in the population. When this happens, the population is said to be undergoing *selection*.

As Darwin made a series of observations of domestic animals and plants, and also those existing free in nature, he used the term natural selection to describe the latter in contrast to men's selection capabilities of domestic breeds. To our purposes, the more general term *selection* is assumed in all cases, bearing in mind that selection under nature has been originally termed natural selection, and selection made by men has been sometimes termed artificial selection.

Under the evolutionary biology perspective, *adaptation* is the process by which traits evolve making organisms more suited to their immediate environment; these traits increase the organisms' chances of survival and reproduction. Adaptation is thus responsible for the many extraordinary traits seen in nature, such as eyes that allow us to see, and the sonar in bats that allow their guidance through the darkness. Note however, that, more accurately speaking, adaptation is a result of the action of both, variation and selection. Variation by itself does not result in adaptation; there must be a way (i.e., selection) of promoting the maintenance of those advantageous variations.

S. Wright (1968-1978) introduced the concept of *adaptive landscapes* or *fitness landscapes*, largely used in evolutionary biology. In his model, each population of a *species* (reproductively isolated group) is symbolized by a point on a *topographic map*, or *landscape*. The contours of the map represent different levels of adaptation to the environment (fitness). Populations at high levels (peaks) are more adapted to the environment, and populations at low levels (valleys) are less adapted. At any one time, the position of a population will depend on its genetic makeup. Populations with alleles that improve fitness will be at a higher peak than populations without these alleles. Consequently, as the genetic makeup of a population changes, so will its position on the adaptive landscape. Figure 3.12 depicts a landscape representing the different levels of adaptation of the populations in relation to the environment.

The adaptive (fitness) landscape corresponds to the response surface discussed in Section 3.2.1 in the context of problem solving via search in a search space. Note that, under the evolutionary perspective, the search performed is for individuals with increased survival and reproductive capabilities (fitness) in a given environment (fitness landscape).

A *niche* can thus be defined as the region consisting of the set of possible environments in which a species can persist; members of one species occupy the same ecological niche. In natural ecosystems, there are many different ways in which animals may survive (grazing, hunting, on water, etc.), and each survival strategy is called an ecological niche. However, it is generally recognized that the niche of a single species may vary widely over its geographical range. The other fundamental concept of niche was proposed by Elton (1927) "The niche of an animal means its place in the biotic environment, its relations to food and enemies;" where the term *biotic* refers to life, living organisms. Thus, niche in this case is being used to describe the role of an animal in its community (Krebs, 1994).

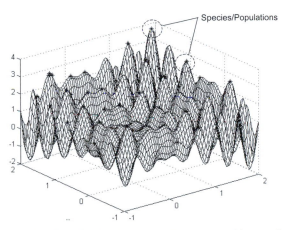

Figure 3.12: An example of an adaptive landscape. The topographic map (landscape or surface) corresponds to the different levels of adaptation of the populations (points in the landscape) to the environment. The populations or individuals at each peak are assumed to be reproductively isolated, i.e., they only breed with individuals in the same peak, thus forming species inhabiting distinct niches.

It has been discussed that evolution is an outcome of genetic variation plus selection. In order for continuing evolution to occur, there must be mechanisms that increase or create genetic variation and mechanisms that decrease it. We have also seen that recombination and mutation cause differences among (variations in) organisms. Other two important mechanisms of evolution are the so-called *genetic drift* (chance fluctuations that result in changes in allele frequencies) and *gene flow* (spread of genes among populations via *migration*). It is known that selection and genetic drift decrease variation, while mutation, recombination, and gene flow increase genetic variation (Colby, 1997).

Natural selection sifts through the genetic variations in the population, preserving the beneficial ones and eliminating the harmful ones. As it does this, selection tends to drive the population uphill in the adaptive or fitness landscape. By contrast, the random genetic drift will move the population in an unpredictable fashion. The effect of all these mechanisms (mutation, recombination, genetic drift, gene flow, plus selection) will bring the population to a state of 'genetic' equilibrium, corresponding to a point near or at a peak on the adaptive landscape. Actually, the population will hover around a peak because of fluctuations caused by genetic drift. Note also, that, under nature, the environment is constantly changing, hence the population is also adapting to the new landscape resultant from the new environment, in a never-ending process of variation and selection.

3.4.5. A Classic Example of Evolution

A classic example of evolution comes from species that live in disturbed habitats. In the particular example of the evolution of melanic (dark) forms of moths, human activity has altered the environment and there has been a corresponding change in the species that inhabit this environment. The peppered moth, *Biston betularia*, is found in wooded areas in Great Britain, where it exists in two color forms, light and dark; light being the typical phenotype of this species. The difference between the two forms is believed to involve a single gene. Since 1850, the frequency of the dark form has increased in certain areas in England, in particular in industrialized parts of the country. Around heavily industrialized cities, such as Manchester and Birmingham, the frequency of the dark form has increased drastically from 1 to 90% in less than 100 years. In other areas of England, where there is little industrial activity, the dark form has remained very rare (Gardner et al., 1991).

The rapid spread of the dark form in industrialized areas has been attributed to natural selection. Both, light and dark forms are active at night. During the day, the moths remain still, resting on tree trunks and other objects in the woodlands. Since birds may find the moths and eat them during their resting period, camouflage is their only defense against predation. On white or gray tree bark, the light moths are protectively colored, especially if the bark is overgrown with lichens. However, in industrialized areas most of the lichens have been killed by pollution and the tree bark is oftenest darkened by soot. Such conditions offer little or no cover for the light moths, but make ideal resting spots for the dark ones. Predatory birds have difficulties in seeing the dark moths on the darkened barks, similarly to what happens with light moths on light barks. The spread of the dark moths in industrialized areas thus appears to be a result of its selective advantage on a sooty background. Because the dark moth survived more in polluted woods, the gene responsible for the dark color increased in frequency over industrialized regions.

3.4.6. A Summary of Evolutionary Biology

The three basic principles of Darwin's theory of evolution, together with the genetics introduced by Mendel and some ideas about the natural selection process, form what is currently known as neo-Darwinism; for Darwin had no knowledge of genetics, a missing bit of his theory.

We have seen that evolution is a result of a reproducing population(s) of individuals, which suffer genetic variation followed by selection. The variation affects the genetic makeup of individuals (genotype), which will present a selective advantage over the others if their phenotypes confer them a better adaptability to the environment they inhabit. This degree of adaptability to the environment (capability of surviving and reproducing) is broadly termed fitness.

The genetic entities that suffer variation are located within the cell nucleus and are named chromosomes, whose basic functional units are the genes. In both types of reproduction, asexual and sexual, there may occur a deviation (mutation) in one or more alleles of a chromosome, allowing the appearance of a new character in the phenotype (physical and chemical characteristics) of the offspring individual. In sexual reproduction, in addition to mutation, there is the recombination (crossing-over) of parental genetic material, resulting in offspring that present features in common with both parents.

The survival and reproductive advantage of an individual organism, its fitness, endows it with a selective advantage over the others, less fit individuals. Those individuals whose genetic makeup result in a phenotype more adapted to the environment have higher probabilities of surviving and propagating their genotypes. It can thus be seen that there is a competition for survival among the many offspring generated. As argued by Darwin

"Every being, which during its natural lifetime produces several eggs or seeds, must suffer destruction during some period of its life, and during some season or occasional year, otherwise, on the principle of geometrical increase, its numbers would quickly become so inordinately great that no country could support the product. Hence, as more individuals are produced than can possibly survive, there must in every case be a struggle for existence, either one individual with another of the same species, or with the individuals of distinct species, or with the physical conditions of life." (Darwin, 1859; p. 50)

An adaptive landscape (or surface) is a topographic map used to represent the degree of adaptation of individuals in a given environment. Individuals that only reproduce with each other are part of the same species, which occupies one or more biological niche. As the fittest individuals of the population have higher chances of surviving and reproducing, the outcome of evolution is a population increasingly more fit to its environment.

Viewed in this manner, evolution is clearly a search and optimization problem solving process (Mayr, 1988). Selection drives phenotypes as close as possible to the optimum of a fitness landscape, given initial conditions and environmental constraints. Evolution thus, allows the discovery of functional solutions to particular problems posed by the environment in which some organisms live (e.g., a more appropriate color for moths living in a sooty area). This is the perspective assumed when using evolutionary biology as a source of inspiration for the development of evolutionary computation. In such a case, an evaluation function will define a *fitness landscape* for the problem, and finding the best solution to the problem (most adapted population or individual in a given environment) will correspond to the search for a peak (assuming a maximization problem) of this landscape by subjecting individuals to genetic variation (e.g., crossover and mutation) operators.

3.5 EVOLUTIONARY COMPUTING

Evolution can be viewed as a search process capable of locating solutions to problems offered by an environment. Therefore, it is quite natural to look for an algorithmic description of evolution that can be used for problem solving. Such an algorithmic view has been discussed even in philosophy (Section 3.4.2). Those iterative (search and optimization) algorithms developed with the inspiration of the biological process of evolution are termed *evolutionary algorithms* (EAs). They are aimed basically at problem solving and can be applied to a wide range of domains, from planning to control. *Evolutionary computation* (EC) is the name used to describe the field of research that embraces all evolutionary algorithms.

The basic idea of the field of evolutionary computation, which came onto the scene about the 1950s to 1960s, has been to make use of the powerful process of natural evolution as a problem-solving paradigm, usually by simulating it on a computer. The original three mainstreams of EC are *genetic algorithms* (GAs), *evolution strategies* (ES), and *evolutionary programming* (EP) (Bäck et al., 2000a,b). Another mainstream of evolutionary computation that has been receiving increasingly more attention is *genetic programming* (Koza, 1992; Koza, 1994a). Despite some differences among these approaches, all of them present the basic features of an evolutionary process as proposed by the Darwinian theory of evolution.

3.5.1. Standard Evolutionary Algorithm

A standard evolutionary algorithm can be proposed as follows:

- *A population of individuals that reproduce with inheritance.* Each individual represents or encodes a point in a search space of potential solutions to a problem. These individuals are allowed to reproduce (sexually or asexually), generating offspring that inherit some features (traits) from their parents. These inherited traits cause the offspring to present resemblance with their progenitors.

- *Genetic variation.* Offspring are prone to genetic variation through mutation, which alters their genetic makeup. Mutation allows the appearance of new traits in the offspring and, thus, the exploration of new regions of the search space.

- *Natural selection.* The evaluation of individuals in their environment results in a measure of adaptability, quality, or fitness value to be assigned to them. A comparison of individual fitnesses will lead to a competition for survival and reproduction in the environment, and there will be a selective advantage for those individuals of higher fitness.

The standard evolutionary algorithm is a generic, iterative, and probabilistic algorithm that maintains a population **P** of N individuals, $\mathbf{P} = \{\mathbf{x}_1, \mathbf{x}_2, \ldots, \mathbf{x}_N\}$, at each iteration t (for simplicity of notation the iteration index t is suppressed). Each individual corresponds to (represents or encodes) a potential solution to a problem that has to be solved. An individual is represented using a *data structure*. The individuals $\mathbf{x}_i$, $i = 1, \ldots, N$, are evaluated to give their measures of adaptability to the environment, or *fitness*. Then, a new population, at iteration $t + 1$, is generated by *selecting* some (usually the most fit) individuals from the current population and *reproducing* them, sexually or asexually. If employing sexual reproduction, a *genetic recombination* (*crossover*) operator may be used. Genetic variations through *mutation* may also affect some individuals of the population, and the process iterates. The completion of all these steps: reproduction, genetic variation, and selection, constitutes what is called a *generation*. An initialization procedure is used to generate the initial population of individuals. Two parameters *pc* and *pm* correspond to the genetic recombination and variation probabilities, as will be further discussed.

Algorithm 3.6 depicts the basic structure of a standard evolutionary algorithm. Most evolutionary algorithms can be implemented using this standard algorithm, with some differences lying on the representation, selection, reproduction, variation operators, and in the order these processes are applied. The stopping criterion is usually a maximum number of generations, or the achievement of a prespecified objective.

Note that all evolutionary algorithms involve the basic concepts common to every algorithmic approach to problem solving discussed in Section 3.1: 1) representation (data structures), 2) definition of an objective, and 3) specification of an evaluation function (fitness function). Although the objective depends on the problem to be solved, the representation and the evaluation function may depend on the designers' experience or expertise.

```
procedure [P] = standard_EA(pc,pm)
   initialize P
   f ← eval(P)
   P ← select(P,f)
   t ← 1
   while not_stopping_criterion do,
        P ← reproduce(P,f,pc)
        P ← variate(P,pm)
        f ← eval(P)
        P ← select(P,f)
        t ← t + 1
   end while
end procedure
```

Algorithm 3.6: A standard evolutionary algorithm.

3.5.2. Genetic Algorithms

All evolutionary algorithms embody the basic processes of the Darwinian theory of evolution, as described in Algorithm 3.6. However, the *genetic algorithms* (GAs) are those that use a vocabulary borrowed from natural genetics. This method has been offered by a number of researchers (e.g., Fraser, 1959; Friedberg, 1958; Anderson, 1953; Bremermann, 1962), but its most popular version was presented by J. Holland (1975). This is the one to be introduced here.

The data structures representing the *individuals* (*genotypes*) of the population are often called *chromosomes*; these are one-chromosome individuals, i.e., *haploid* chromosomes. In standard genetic algorithms the individuals are represented as strings of binary digits {0,1}, or *bitstrings*. In accordance with its biological source of inspiration (i.e. genetics) each unit of a chromosome is a *gene*, located in a certain place in the chromosome called *locus*. The different forms a gene can assume are the *alleles*. Figure 3.13 illustrates a bitstring of length $l = 10$ corresponding to a chromosome in a genetic algorithm. Note the similarity of this representation with the one used for the diagrammatic chromosome of Figure 3.7.

Figure 3.13: Bitstring of length $l = 10$ representing a chromosome in a standard genetic algorithm. Each position (locus) in the chromosome can assume one of the two possible alleles, 0 or 1.

As discussed in Section 3.1, the problem to be solved is defined and captured in an *objective function* that allows to evaluate the *fitness* (procedure `eval` of Algorithm 3.6) of any potential solution. Each genotype, in this case a single chromosome, represents a potential solution to a problem. The meaning of a particular chromosome, its *phenotype*, is defined according to the problem under study. A genetic algorithm run on a population of chromosomes (bitstrings) corresponds to a search through a space of potential solutions.

As each chromosome x_i, i, ... , N, often corresponds to the encoded value of a candidate solution, it often has to be decoded into a form appropriate for evaluation and is then assigned a fitness value according to the objective. Each chromosome is assigned a probability of reproduction, p_i, i, ... , N, so that its likelihood of being selected is proportional to its fitness relative to the other chromosomes in the population; the higher the fitness, the higher the probability of reproduction, and vice-versa. If the fitness of each chromosome is a strictly positive number to be maximized, selection (procedure `select` of Algorithm 3.6) is traditionally performed via an algorithm called *fitness proportional selection* or *Roulette Wheel selection* (Fogel, 2000a).

The assigned probabilities of reproduction result in the generation of a population of chromosomes probabilistically selected from the current population. The selected chromosomes will generate offspring via the use of specific *genetic operators*, such as *crossover*, and *bit mutation* might be used to introduce genetic variation in the individuals. Crossover is the genetic recombination operator embodied in the procedure `reproduce` of Algorithm 3.6. Mutation is the genetic variation mechanism of procedure `variate` of Algorithm 3.6. The iterative procedure stops when a fixed number of iterations (generations) has been performed, or a suitable solution has been found.

Roulette Wheel Selection

In the *fitness proportional* or *Roulette Wheel* (RW) selection method, the probability of selecting a chromosome (individual of the population) is directly proportional to its fitness value. Each chromosome x_i, $i = 1$, ... , N, of the population is assigned to a part of the wheel whose size is proportional to its fitness value. The wheel is then tossed as many times as parents (N) are needed to create the next generation, and each winning individual is selected and copied into the parent population. Note that this method allows an individual to be selected more than once and the 'death' of some individuals as well.

Figure 3.14 depicts the RW applied to a population composed of four individuals. To play the roulette might correspond to obtaining a value from a random number generator with uniform distribution in the interval [0,1] (see Section 3.5.3 for another form of implementing the RW selection). The value obtained is going to define the chromosome to be chosen, as depicted in Figure 3.14. The higher the fitness of an individual, the larger the portion of the roulette to be assigned to this individual, thus the higher its probability of being selected.

Ind.	Chromosome	Fitness	Degrees
1	0001100101010	16	240
2	0101001010101	4	60
3	1011110100101	2	30
4	1010010101001	2	30

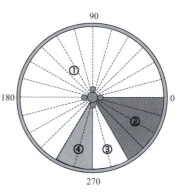

Figure 3.14: Roulette Wheel selection. (reproduced with permission from [de Castro and Timmis, 2002]. © L. N. de Castro and J. Timmis, 2002).

Crossover

In biological systems, crossing-over is a process that yields the recombination of alleles via the exchange of segments between pairs of chromosomes (see Figure 3.10). As discussed in Section 3.4.3, it occurs only in sexually reproducing species. This process can be abstracted as a general operator to the level of the data structures discussed, i.e., bitstrings. Crossing-over proceeds basically in three steps (Holland, 1975):

- Two strings $\mathbf{x} = x_1x_2 \ldots x_l$ and $\mathbf{y} = y_1y_2 \ldots y_l$ are selected from the current population $\mathbf{P}$.

- A number r indicating the crossover point is randomly selected from $\{1, 2, \ldots, l–1\}$.

- Two new strings are formed from $\mathbf{x}$ and $\mathbf{y}$ by exchanging the set of attributes to the right of position r, yielding $\mathbf{x'} = x_1 \ldots x_i y_{i+1} \ldots y_l$ and $\mathbf{y'} = y_1 \ldots y_i x_{i+1} \ldots x_l$.

The two new chromosomes (strings), $\mathbf{x'}$ and $\mathbf{y'}$, are the offspring of $\mathbf{x}$ and $\mathbf{y}$. This single-point crossover is illustrated in Figure 3.15 for two strings of length $l = 8$.

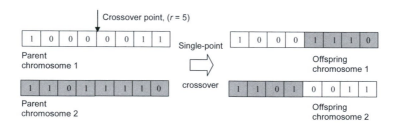

Figure 3.15: Single-point crossover for a pair of chromosomes of length $l = 8$.

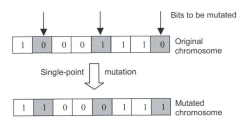

Figure 3.16: Point mutations for a chromosome of length *l* = 8 (three points are selected for being mutated).

Mutation

In genetics (Section 3.4.3), point mutation is a process wherein one allele of a gene is randomly replaced by (or modified to) another to yield a new chromosome (structure). Generally, there is a small probability of mutation at each gene in the structure (in the GA context, a bitstring). It means that, each bitstring in the population *P* is operated upon as follows (Holland, 1975):

- The numbers $r,...,u$ indicating the positions to undergo mutation are determined by a random process where each position has a small probability *pm* of undergoing mutation, independently of the other positions.

- A new string $\mathbf{x'} = x_1...x_r...x_u...x_l$ is generated where $x_r...x_u$ are drawn at random from the set of alleles for each gene. In the case of bitstrings, if a position has an allele '0', then it becomes '1', else if it is originally '1', then it becomes '0'.

The parameter *pm* is the mutation probability at each position. Figure 3.16 illustrates the process of point or bit mutation.

3.5.3. Examples of Application

Algorithm 3.6 shows the standard evolutionary algorithm, which is the same as a standard genetic algorithm, and follows the Darwinian theory of evolution. In the standard GA, selection - `select` - is performed via Roulette Wheel, reproduction - `reproduce` - is performed by crossing-over pairs of the selected individuals, and genetic variation - `variate` - is performed via mutation. Fitness evaluation - `eval` - depends on the problem under study.

To illustrate how to put all these procedures together in practical problems, consider the following examples. The first example - pattern recognition - is aimed at illustrating the basic steps and implementation of a standard genetic algorithm. In this particular case, the GA capability of learning a known pattern will be tested. The second example - numeric function optimization - the GA potential of determining the (unknown) global optimal (maximum) to a numeric function will be assessed. The focus of this latter example is in the encoding (representation) scheme used; the basic selection and genetic operators are the same as those used in the first example.

$$\begin{bmatrix} 0 & 1 & 0 \\ 0 & 1 & 0 \\ 0 & 1 & 0 \\ 0 & 1 & 0 \end{bmatrix}$$

(a)   (b)

Figure 3.17: Character to be recognized using a standard genetic algorithm. (a) Pictorial view. (b) Matrix representation.

A Step by Step Example: Pattern Recognition (Learning)

As a first example of how the standard GA works, consider the problem of evolving a population of individuals toward recognizing a single character; the number '1' depicted in Figure 3.17(a). The resolution of this picture is *rows* = 4 and *columns* = 3. This character is drawn by using two colors, light and dark gray, and can be represented by a matrix of '0s' and '1s', where each '0' corresponds to the dark gray, while each '1' corresponds to the light gray (Figure 3.17(b)).

In the genetic algorithm a small population **P** with only $N = 8$ individual chromosomes will be chosen to be evolved toward recognizing the character '1'. This population **P** will correspond to a matrix with $N = 8$ rows (the individuals) of $l = 12$ columns. To generate a single individual we simply place one row of the character after the other downward. For example, the target individual of Figure 3.17 is represented by the vector $x_1 = [010010010010]$. (Compare Figure 3.17(b) with x_1.)

The standard evolutionary algorithm presented in Algorithm 3.6 also corresponds to the standard genetic algorithm. What differs the standard GA from the other evolutionary algorithms (*evolution strategies* and *evolutionary programming*) is basically the data representation, the operators, the selection method, and the order in which these operators are applied; the basic idea remains. The standard GA has the following main features:

- Binary representation of the data structures.
- Fixed population size.
- Single-point crossover.
- Point mutation.
- Roulette Wheel selection.

Initialization

Figure 3.18 depicts a randomly initialized population **P** to be evolved by Algorithm 3.6. The matrix **P** corresponding to the individuals of Figure 3.18 is:

$$\mathbf{P} = \begin{bmatrix} 0 & 0 & 0 & 0 & 1 & 0 & 1 & 1 & 1 & 1 & 0 & 1 \\ 1 & 0 & 0 & 1 & 1 & 1 & 1 & 1 & 1 & 0 & 1 & 1 \\ 1 & 0 & 0 & 0 & 0 & 1 & 1 & 0 & 0 & 1 & 0 & 1 \\ 1 & 0 & 0 & 1 & 0 & 0 & 0 & 1 & 0 & 0 & 0 & 0 \\ 1 & 1 & 0 & 1 & 0 & 1 & 0 & 0 & 0 & 0 & 1 & 0 \\ 1 & 1 & 0 & 1 & 1 & 0 & 0 & 0 & 1 & 0 & 1 & 0 \\ 1 & 0 & 0 & 1 & 0 & 1 & 0 & 1 & 0 & 1 & 1 & 0 \\ 1 & 1 & 1 & 0 & 1 & 1 & 0 & 0 & 0 & 1 & 0 & 1 \end{bmatrix}$$

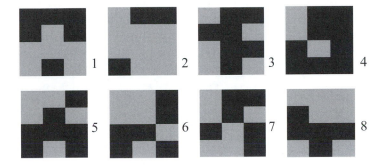

Figure 3.18: Randomly initialized population *P* for the standard GA.

Fitness Evaluation

After initializing the population of candidate solutions, the next step in the algorithm (Algorithm 3.6) is to evaluate the quality of each individual in relation to the problem to be solved, i.e., to calculate the fitness of each individual. As the goal is to generate individuals that are as similar as possible to the target individual (Figure 3.17), a straightforward way of determining fitness is by counting the number of similar bits between each individual of the population and the target individual. The number of different bits between two bitstrings is termed *Hamming distance*. For instance, the vector **h** of Hamming distances between the individuals of **P** and the target individual is **h** = [6,7,9,5,5,4,6,7].

The fitness of the population can be measured by subtracting the length $l = 12$ of each individual by its respective Hamming distance to the target individual. Therefore, the vector of fitness becomes $\mathbf{f} = [f_1, \dots, f_8] = [6,5,3,7,7,8,6,5]$.

The ideal individual is the one whose fitness is $f = 12$. Therefore, the aim of the search to be performed by the GA is to maximize the fitness of each individual, until (at least) one individual of **P** has fitness $f = 12$.

Roulette Wheel Selection

To perform selection via Roulette Wheel, each individual will be assigned a portion of the Roulette proportional to its fitness; the higher the fitness, the higher the slice of the roulette, and vice-versa. Roulette Wheel can be implemented as follows:

- Sum up the fitness of all individuals, $f_T = \Sigma_i f_i$, $i = 1, \dots, N$.

- Multiply the fitness of each individual by 360 and divide it by the total fitness: $f_i' = (360. f_i)/f_T$, $i = 1, \dots, N$, determining the portion of the wheel to be assigned for each individual.

- Sort a random number in the interval $(0,360]$ and compare it with the interval belonging to each individual. Select the individual whose interval contains the sorted number and place it in the new population **P**.

The whole process is illustrated in Figure 3.19, including the size of each portion and its interval (Portion). (Note that the Roulette Wheel can also be implemented within the interval $[0,1]$ instead of $(0,360]$, as discussed previously.)

Individual	Chromosome	Fitness	Degree	Portion of the Roulette
1	000010111101	6	46	(0,46]
2	100111111011	5	38	(46,84]
3	100001100101	3	23	(84,107]
4	100100010000	7	54	(107,161]
5	110101000010	7	54	(161,215]
6	110110001010	8	61	(215,276]
7	100101010110	6	46	(276,322]
8	111011000101	5	38	(322,360]

(a)

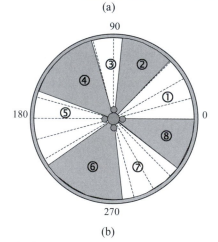

(b)

Figure 3.19: Roulette Wheel selection. (a) Descriptive table. (b) Roulette Wheel. (*Note: the degrees were rounded for didactic purposes.*)

Assuming that **s** = [230,46,175,325,275,300,74,108] were the sorted random numbers, the population **P** will be composed of the individuals **P** = [**x**₆, **x**₁, **x**₅, **x**₈, **x**₆, **x**₇, **x**₂, **x**₄]^T, where ^{T} corresponds to the matrix transposition operator (Appendix B.1.4).

It can be noticed that the less fit individual of the population (**x**₃) went extinct, while the most fit individual (**x**₆) was selected twice. This can be interpreted as individual **x**₆ having had a higher probability of leaving progenies, while **x**₃ having left no progeny at all.

Reproduction

In the standard GA, reproduction is accomplished by selecting pairs of individuals and performing the crossing-over of their genetic material with a crossover probability *pc*. Basically, two approaches can be taken. In the first one, both offspring replace the parents, and in the second one a single offspring replace a single parent and the process is repeated *N* times. Let us assume the first case in which both offspring will replace both parents. Thus, for the matrix **P**, *N*/2 random numbers *r* over the interval [0,1] will be chosen and if $r \leq pc$, then the crossing-over is performed according to the procedure described in Section 0; else, the original parents are repeated in the next generation.

To illustrate this procedure, let **r** = [0.5,0.7,0.3,0.9] be the four values randomly chosen for *r*, and *pc* = 0.6. There are four pairs of parents in the population **P**: **x**₆ and **x**₁; **x**₅ and **x**₈; **x**₆ and **x**₇; **x**₂ and **x**₄. In this case, the first and third pairs were selected for crossing over. Thus, **x**₆ will be crossed over with **x**₁, and **x**₆ will again be crossed over with **x**₇. What remains to be done is to define the crossover point between the strings in both cases. Assume that the randomly chosen crossover points, *cp*, from the interval [1,*l*–1] are *cp* = 5 and *cp* = 9, respectively, for each pair. Figure 3.20 illustrates the crossing-over between **x**₆ and **x**₁, and **x**₆ and **x**₇, for *cp* = 5 and *cp* = 9, respectively.

Figure 3.20: Crossing-over between *x*₆ and *x*₁, and *x*₆ and *x*₇.

The new population **P** is thus:

$$\mathbf{P} = \begin{bmatrix} 1 & 1 & 0 & 1 & 1 & 0 & 1 & 1 & 1 & 1 & 0 & 1 \\ 0 & 0 & 0 & 0 & 1 & 0 & 0 & 0 & 1 & 0 & 1 & 0 \\ 1 & 1 & 0 & 1 & 0 & 1 & 0 & 0 & 0 & 0 & 1 & 0 \\ 1 & 1 & 1 & 0 & 1 & 1 & 0 & 0 & 0 & 1 & 0 & 1 \\ 1 & 1 & 0 & 1 & 1 & 0 & 0 & 0 & 1 & 1 & 1 & 0 \\ 0 & 0 & 0 & 0 & 1 & 1 & 0 & 1 & 0 & 0 & 1 & 0 \\ 1 & 0 & 0 & 1 & 1 & 1 & 1 & 1 & 1 & 0 & 1 & 1 \\ 1 & 0 & 0 & 1 & 0 & 0 & 0 & 1 & 0 & 0 & 0 & 0 \end{bmatrix}.$$

Mutation (Genetic Variation)

Assume a mutation probability $pm = 0.02$. For each position in matrix $\mathbf{P}$, a random number r is generated and compared with pm. If $r \leq pm$, then the corresponding bit is selected for being mutated. For the bits selected, as depicted in $\mathbf{P'}$ below (underlined and boldface numbers), the population $\mathbf{P}$ before and after mutation is:

$$\mathbf{P} = \begin{bmatrix} 1 & 1 & 0 & 1 & 1 & 0 & 1 & 1 & 1 & 1 & 0 & 1 \\ 0 & 0 & 0 & 0 & 1 & 0 & 0 & 0 & 1 & 0 & 1 & 0 \\ 1 & 1 & 0 & 1 & \underline{\mathbf{0}} & 1 & 0 & 0 & 0 & 0 & 1 & 0 \\ 1 & 1 & 1 & 0 & 1 & 1 & 0 & 0 & 0 & 1 & 0 & 1 \\ 1 & 1 & 0 & 1 & 1 & 0 & 0 & 0 & 1 & 1 & 1 & 0 \\ 0 & 0 & 0 & 0 & 1 & 1 & 0 & 1 & 0 & 0 & 1 & 0 \\ 1 & 0 & 0 & 1 & 1 & 1 & \underline{\mathbf{1}} & 1 & 1 & 0 & 1 & 1 \\ 1 & 0 & 0 & 1 & 0 & 0 & 0 & 1 & 0 & 0 & 0 & 0 \end{bmatrix};$$

$$\mathbf{P} = \begin{bmatrix} 1 & 1 & 0 & 1 & 1 & 0 & 1 & 1 & 1 & 1 & 0 & 1 \\ 0 & 0 & 0 & 0 & 1 & 0 & 0 & 0 & 1 & 0 & 1 & 0 \\ 1 & 1 & 0 & 1 & \underline{\mathbf{1}} & 1 & 0 & 0 & 0 & 0 & 1 & 0 \\ 1 & 1 & 1 & 0 & 1 & 1 & 0 & 0 & 0 & 1 & 0 & 1 \\ 1 & 1 & 0 & 1 & 1 & 0 & 0 & 0 & 1 & 1 & 1 & 0 \\ 0 & 0 & 0 & 0 & 1 & 1 & 0 & 1 & 0 & 0 & 1 & 0 \\ 1 & 0 & 0 & 1 & 1 & 1 & \underline{\mathbf{0}} & 1 & 1 & 0 & 1 & 1 \\ 1 & 0 & 0 & 1 & 0 & 0 & 0 & 1 & 0 & 0 & 0 & 0 \end{bmatrix}$$

Evaluation

The last step in Algorithm 3.6 before selection is applied again is to evaluate the fitness of the population at the end of each generation. This will result in the following vector of fitnesses: $\mathbf{f} = [5,9,7,5,7,10,6,6]$, and $f_T = 55$.

If the total fitness is divided by the number of individuals in the population, the average fitness can be obtained $f_{av} = f_T/N$. The average fitness at the previous generation is $f_{av} = 47/8 = 5.875$, while the average fitness at the present generation is $f_{av} = 55/8 = 6.875$. If one looks at the Hamming distance instead of the fitness, it can be noticed that the average Hamming distance is being reduced towards zero. It can thus be noticed that by simply performing selection, crossover, and mutation, the population of individuals becomes increasingly more capable of recognizing the desired pattern. Note that in this particular example, minimizing the **h** is equivalent to maximizing the evaluation function chosen $(f_i = l - h_i)$.

Figure 3.21(a) depicts the final population evolved by the GA after 171 generations. Figure 3.21(b) depicts the average Hamming distance, and the Hamming distance of the best individual of the population (individual x_3) at the end of the evolutionary search process.

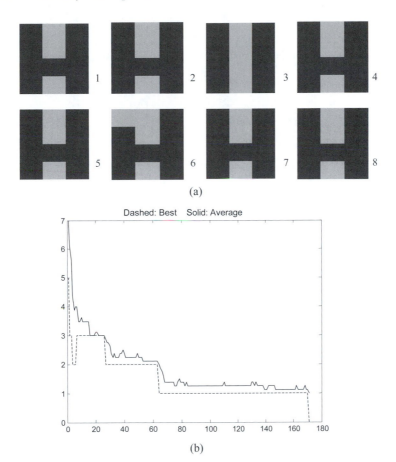

(a)

(b)

Figure 3.21: Final population and evolution of the population over the generations. (a) Final population. (b) Evolution of the population. Dashed line: Hamming distance of the best individual in relation to the target one. Solid line: Average Hamming distance of the whole population.

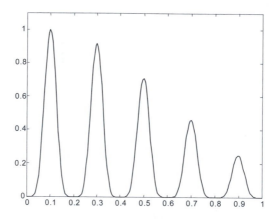

Figure 3.22: Graph of the function $g(x) = 2^{-2((x-0.1)/0.9)^2} \left(\sin(5\pi x)\right)^6$ to be maximized with a standard GA.

Numerical Function Optimization

The second problem to be explored is the one of maximizing a simple uni-dimensional function presented in Section 3.3.3. The function is defined as

$$g(x) = 2^{-2((x-0.1)/0.9)^2} \left(\sin(5\pi x)\right)^6,$$

and is depicted in Figure 3.22 (Goldberg, 1989). The problem is to find **x** from the range [0..1] which maximizes $g(\mathbf{x})$, i.e., to find **x*** such that

$$g(\mathbf{x}^*) \geq g(\mathbf{x}), \ \forall \ \mathbf{x} \in [0..1].$$

The first aspect that has to be accounted for is the representation. How to represent the real value x using a binary string? Although it is not necessary to use a binary representation for x, it is our intent to employ the standard GA and its basic operators, as described in the previous example. Actually, some authors (e.g., Michalewicz, 1996), argue that a float-point representation is usually the most suitable one for numeric function optimization problems. Evolution strategies (Section 3.6.1) have also been very successful in solving numeric function optimization problems.

The length l of the bitstring depends on the numeric precision required. Assuming a bitstring of length l, the mapping from a binary string $\mathbf{x} = \langle x_l, ..., x_2, x_1 \rangle$ into a real number z is completed in two steps (Michalewicz, 1996):

- Convert the binary string $\mathbf{x} = \langle x_l, ..., x_2, x_1 \rangle$ from base 2 to base 10: $\left(\langle x_l, ..., x_2, x_1 \rangle \right)_2 = \left(\sum_{i=0}^{l-1} x_i . 2^i \right)_{10} = z'$; and

- Find the corresponding real value for z: $z = z_{min} + z' \cdot \dfrac{z_{max} - z_{min}}{2^l - 1}$; where

  z_{max} is the upper bound of the interval in which the variable is defined, z_{min} is the lower bound of this interval, and l is the string length.

To evaluate each individual of the population one has to decode the binary chromosomes **x** into their real values and then evaluate the function for these values. The remaining steps of the algorithm (selection, recombination, and mutation) are the same as the ones described for the previous example.

3.5.4. Hill-Climbing, Simulated Annealing, and Genetic Algorithms

All the algorithms discussed so far, hill-climbing (HC), simulated annealing (SAN), and genetic algorithms (GAs), have a number of versions. The focus of this text was on the standard versions of all of them. The major difference among these three approaches is that the evolutionary algorithms are population-based search techniques, while HC and SAN work with a single individual.

In the case of the standard hill-climbing search (Algorithm 3.1), its success or failure in determining the global optimum of a given function (problem) depends on its starting point. It is clear that if the starting point is not on a hill that will lead to the global optimum, the algorithm will never be able to reach this peak in a single run. Additional runs, with different starting conditions, will have to be performed. This is automatically performed in the iterated hill-climbing procedure (Algorithm 3.2). The stochastic hill-climbing algorithm (Algorithm 3.3) incorporates random variation into the search for optimal solutions, just as the simulated annealing (Algorithm 3.4) and genetic algorithm (Algorithm 3.6). In these algorithms, no two trials could be expected to take exactly the same course (sequence of steps), i.e., to do the same search in the search space.

The stochastic hill-climber, simulated annealing, and genetic algorithms are capable of escaping local optima solutions due to their stochastic nature. The simulated annealing algorithm allows the acceptance of a new point, given a Boltzmann probability, which might result in the exploration of a different region of the search space. The broader exploration of the search space in evolutionary algorithms is accomplished by the use of multiple individuals and genetic operators, which allow the creation of individuals with new features and also the sharing of features with their parents.

In the genetic algorithms, which are population-based approaches, there is the concept of competition (via a selective mechanism that privileges better – more fit – individuals) between candidate solutions to a given problem. It is also interesting to note that while hill-climbing and simulated annealing algorithms generate new candidate points in the neighborhood of the current point, evolutionary algorithms allow the examination of points in the neighborhood of two (or even more) candidate solutions via the use of genetic operators such as crossover.

To conclude this section, let us cite a passage quoted by Michalewicz (1996), which serves as a metaphor for comparing hill-climbing, simulated annealing, and genetic algorithms:

"Notice that in all [hill-climbing] methods discussed so far, the kangaroos can hope at best to find the top of a mountain close to where he starts. There's no guarantee that this mountain will be Everest, or even a very high mountain. Various methods are used to try to find the actual global optimum.

In simulated annealing, the kangaroo is drunk and hops around randomly for a long time. However, he gradually sobers up and tends to hop up hill.

In genetic algorithms, there are lots of kangaroos that are parachuted into the Himalayas (if the pilot didn't get lost) at random places. These kangaroos do not know that they are supposed to be looking for the top of Mt. Everest. However, every few years, you shoot the kangaroos at low altitudes and hope the ones that are left will be fruitful and multiply." (Sarle, 1993; quoted by Michalewicz, 1996; p. 30)

3.6   THE OTHER MAIN EVOLUTIONARY ALGORITHMS

This section provides a brief overview of the other three main types of evolutionary algorithms: evolution strategies (ES), evolutionary programming (EP) and genetic programming (GP).

3.6.1. Evolution Strategies

The *evolution strategies* (ES) are a class of evolutionary algorithms, developed by Rechenberg (1965), Schwefel (1965), and Bienert, employed mainly to solve parameter optimization problems (Schwefel, 1995; Beyer, 2001). They were initially used to tackle problems on fluid mechanics and later generalized to solve function optimization problems, focusing the case of real-valued functions. The first ESs operated with a single individual in the population subjected only to mutation.

The data structure of an ES corresponds to real-valued vectors of candidate solutions in the search space. Let us consider the most general case of an ES. An individual $\mathbf{v} = (\mathbf{x}, \sigma, \theta)$ may be composed of the attribute vector $\mathbf{x}$, and the sets of the strategy parameters σ and θ. The adaptation procedure is usually implemented performing first (the recombination and) the mutation - according to some probability density function - of the parameter vectors σ and θ, resulting in σ' and θ', and then using the updated vectors of the strategy parameters to (recombine and) mutate the attribute vector $\mathbf{x}$, resulting in $\mathbf{x}'$. The pseudocode to implement an ES is the same as the one described in Algorithm 3.6, with the difference that the strategy parameters suffer genetic variation before the attribute vector.

Note that this new type of representation incorporates the strategy parameters into the encoding of the individuals of the population. This is a very powerful idea, because it allows the evolutionary search process to adapt the strategy parameters together with the attribute vector. This process is known as *self-adaptation*. Despite self-adaptation, recombination operators are not always used. Recombination was placed within parenthesis in the paragraph above because most ESs only use mutation; no crossover between individuals is performed. Assuming this simpler case where there is no recombination, it remains to the designer only to define how the mutation and selection operators are performed.

Selection

There are basically two types of selection mechanisms in the standard ES: $(\mu + \lambda)$ selection and (μ, λ) selection. In the $(\mu + \lambda)$-ES, μ parents generate λ offspring and the whole population $(\mu + \lambda)$ is then reduced to μ individuals. Thus, selection operates in the set formed by parents and offspring. In this case, the parents survive until the offspring become better adapted to the environment (with higher fitness) than their parents. In the (μ, λ)-ES, μ parents generate λ offspring ($\lambda \geq \mu$) from which μ will be selected. Thus, selection only operates on the offspring set.

Both $(\mu + \lambda)$-ES and (μ, λ)-ES have the same general structure. Consider thus the most general case of a (μ, λ)-ES. An individual $\mathbf{v} = (\mathbf{x}, \sigma, \theta)$ is composed of three elements: $\mathbf{x} \in \mathfrak{R}^l$, $\sigma \in \mathfrak{R}^{l\sigma}$, and $\theta \in (0, 2\pi]^{l\theta}$, where l is the dimension of $\mathbf{x}$, $l\sigma \in \{1, \dots, l\}$, and $l\theta \in \{0, (2l - l\sigma)(l\sigma - 1)/2\}$ (Bäck et al., 2000b). The vector $\mathbf{x}$ is the attribute vector, σ is the vector of standard deviations, and θ is the vector of rotating angles (see further explanation).

Crossover

There are a number of recombination operators that can be used in evolution strategies, either in their usual form, producing one new individual from two randomly selected parents, or in their global form, allowing attributes to be taken for one new individual potentially from all individuals available in the population. Furthermore, recombination is performed on strategy parameters as well as on the attribute vector. Usually, recombination types are different for the attribute vector and the strategy parameters. In the following, assume $\mathbf{b}$ and $\mathbf{b}'$ to stand for the actual parts of $\mathbf{v}$. In a generalized manner, the recombination operators can take the following forms:

$$\mathbf{b}_i' = \begin{cases} \mathbf{b}_{S,i} & (1) \\ \mathbf{b}_{S,i} \text{ or } \mathbf{b}_{T,i} & (2) \\ \mathbf{b}_{S,i} + u(\mathbf{b}_{T,i} - \mathbf{b}_{S,i}) & (3) \\ \mathbf{b}_{Sj,i} \text{ or } \mathbf{b}_{Tj,i} & (4) \\ \mathbf{b}_{Sj,i} + u(\mathbf{b}_{Tj,i} - \mathbf{b}_{Sj,i}) & (5) \end{cases} \qquad (3.4)$$

where S and T denote two arbitrary parents, u is a uniform random variable over [0,1], and i and j index the components of a vector and the vector itself, respectively. In a top-down manner we have (1) no recombination, (2) discrete recombination (or uniform crossover), (3) intermediate recombination, (4) global discrete recombination, and (4) global intermediate recombination.

Mutation

The mutation operator in ESs works by adding a realization of an l-dimensional random variable with uniform distribution $X \sim N(0,C)$, 0 as the expectation vector and co-variance matrix given by

$$C = (c_{ij}) = \begin{cases} \text{cov}(X_i, X_j) & i \neq j \\ \text{var}(X_i) & i = j \end{cases} \tag{3.5}$$

C is symmetric and positive definite, and has the probability density function given by

$$f_X(x_1,...,x_n) = \sqrt{\frac{\det(C)}{(2\pi)^l}} \exp\left(\frac{1}{2} x^T C^{-1} x\right) \tag{3.6}$$

where the matrix C is the co-variance matrix described by the mutated values σ' and θ' of the strategy parameters. Depending on the number of parameters incorporated in the representation of an individual, the following types of self-adaptation procedures are possible.

A Single Standard Deviation for all Attributes of x

$l\sigma = 1, l\theta = 0, X \sim \sigma' N(0,I)$.

In this case, the standard deviation of all attributes of x are identical, and all attributes are mutated as follows:

$$\sigma^{t+1} = \sigma^t \exp(\tau_0.N(0,1))$$
$$x_i^{t+1} = x_i^t + \sigma^{t+1}.N_i(0,1) \tag{3.7}$$

where $\tau_0 \propto l^{-1/2}$ and $N_i(0,1)$ indicates that the random variable is sampled independently for each i. The lines of equal probability density of the normal distribution are hyperspheres in an l-dimensional space.

Individual Standard Deviations for each Attribute of x

$l\sigma = l, l\theta = 0, X \sim N(0, \sigma'.I)$.

Each attribute of x has its own standard deviation σ_i that determines its mutation as follows:

$$\sigma_i^{t+1} = \sigma_i^t \exp(\tau'.N(0,1) + \tau.N_i(0,1))$$
$$x_i^{t+1} = x_i^t + \sigma_i^{t+1}.N_i(0,1) \tag{3.8}$$

where $\tau' \propto (2.l)^{-1/2}$ and $\tau \propto (2.l^{1/2})^{-1/2}$.

The lines of equal probability density of the normal distribution are hyperellipsoids.

Individual Standard Deviations for each Attribute of **x** *and Correlated Mutation*

$l\sigma = l, l\theta = l.(l-1)/2, X \sim N(0, C)$.

The vectors σ and θ represent the complete covariance matrix of an l-dimensional normal distribution, where the co-variances c_{ij} ($i \in \{1, \dots, l-1\}$, and $j \in \{i+1, \dots, l\}$) are represented by a vector of rotation angles θ_k that describe the coordinate rotations necessary to transform an uncorrelated mutation vector into a correlated mutation vector. Thus, the mutation is performed as follows:

$$\sigma_i^{t+1} = \sigma_i^t \exp(\tau'.N(0,1) + \tau.N_i(0,1))$$
$$\theta_i^{t+1} = \theta_i^t + \beta.N_j(0,1) \qquad (3.9)$$
$$x^{t+1} = x^t + N(0, C(\sigma^{t+1}, \theta^{t+1}))$$

where $N(0, C(\sigma^{t+1}, \theta^{t+1}))$ corresponds to the vector of correlated mutations and $\beta = 0.0873$ (5°).

This way, the mutation corresponds to hyper-ellipsoids capable of rotating arbitrarily, and θ_k characterizes the rotation angles in relation to the coordinate axes.

The last step to implement the (μ, λ)-ES or $(\mu + \lambda)$-ES with correlated mutation is to determine $C(\sigma^{t+1}, \theta^{t+1})$. Rudolph (1992) demonstrated that a symmetric matrix is positive definite if, and only if, it can be decomposed as $C = (ST)^T ST$, where S is a diagonal matrix with positive values corresponding to the standard deviations ($s_{ii} = \sigma_i$) and

$$T = \prod_{i=1}^{l-1} \prod_{j=i+1}^{l} R_{ij}(\theta_{ij}) \qquad (3.10)$$

Matrix T is orthogonal and is built by a product of $l(l-1)/2$ rotation matrices R_{ij} with angles $\theta_k \in (0, 2\pi]$. An elementary rotation matrix $R_{ij}(\theta)$ is the identity matrix where four specific entries replaced by $r_{ii} = r_{jj} = \cos\theta$ and $r_{ij} = -r_{ji} = -sen\theta$.

3.6.2. Evolutionary Programming

Fogel and collaborators (Fogel et al., 1966) introduced *evolutionary programming* (EP) as an evolutionary technique to develop an alternative form of *artificial intelligence*. An intelligent behavior was believed to require two capabilities: 1) the prediction of a certain environment, and 2) an appropriate response according to this prediction. The environment was considered as being described by a finite sequence of symbols taken from a finite alphabet. The evolutionary task was then defined as the evolution of an algorithm capable of operating in the sequence of symbols generated by the environment in order to produce an output symbol that maximizes the performance of the algorithm in relation to the next input symbol, given a well-defined cost function. *Finite state machines* (FSM) were considered suitable to perform such task.

Although the first evolutionary programming techniques were proposed to evolve finite state machines, they were later extended by D. Fogel (1992) to operate with real-valued vectors subject to Gaussian mutation, in a form similar to the one used in evolution strategies.

In a later development of evolutionary programming, called meta-evolutionary programming (meta-EP), an individual of the population is represented by $\mathbf{v} = (\mathbf{x}, \mathbf{var})$, where $\mathbf{x} \in \Re^l$ is the attribute vector, $\mathbf{var} \in \Re^{lvar}$ is the variance vector ($var = \sigma^2$) of the mutation variables, l is the dimension of $\mathbf{x}$, and $lvar \in \{1, \dots, l\}$. The variance is also going to suffer mutation in a self-adaptive manner, similarly to the evolution strategies. In another approach, Fogel (1992, pp. 287-289) introduced the *Rmeta-PE*, which incorporates the vector of the correlation coefficients into the set of parameters that will suffer mutation. It was implemented in a form essentially identical to the evolution strategies, but few experiments were performed with this version of EP.

Selection

The selection mechanism used in evolutionary programming is essentially similar to the *tournament selection* sometimes used in genetic algorithms. After generating μ offspring from μ parents and mutating them (see next section), a stochastic tournament will select μ individuals from the set formed by the parents plus offspring. It thus corresponds to a stochastic $(\mu + \lambda)$ selection in evolution strategies.

For each individual $\mathbf{v}_i \in \mathbf{P} \cup \mathbf{P'}$, where $\mathbf{P}$ denotes the whole population and $\mathbf{P'}$ denotes the offspring population, q individuals are randomly selected from the set $\mathbf{P} \cup \mathbf{P'}$ and compared with $\mathbf{v}_i$ in relation to their fitness values. Given $\mathbf{v}_i$, we count how many individuals have a fitness value less than $\mathbf{v}_i$, resulting in a score w_i ($i \in \{1, \dots, 2\mu\}$), and the μ individuals with greater score w_i are selected to form the population for the next generation:

$$w_i = \sum_{j=1}^{q} \begin{cases} 1 & \text{if } fit(\mathbf{v}_i) \le fit(\mathbf{v}_{\chi_j}) \\ 0 & \text{otherwise} \end{cases} \qquad (3.11)$$

where $\chi_j \in \{1, \dots, 2\mu\}$ is an integer and uniform random variable sampled for each comparison.

Mutation

In the standard EP, a Gaussian mutation with independent standard deviation for every element x_i of $\mathbf{x}$ is obtained as the square root of a linear transformation applied to the fitness value of $\mathbf{x}$ (Bäck and Schwefel, 1993):

$$x_i^{t+1} = x_i^t + \sigma_i.N_i(0,1)$$
$$\sigma_i = \sqrt{\beta_i.fit(\mathbf{x}) + \gamma_i} \qquad (3.12)$$

where the parameters β_i and γ_i must be adjusted according to the problem under study.

To solve the parameter adjustment problems of EP, the meta-EP self-adapts l variances by individual, similarly to what is performed with the ES:

$$x_i^{t+1} = x_i^t + \sqrt{var_i} \ .N_i(0,1)$$

$$var_i^{t+1} = var_i^t + \sqrt{\alpha.var_i} \ .N_i(0,1)$$

(3.13)

where α is a parameter that guarantees that var_i is nonnegative, and $N_i(0,1)$ indicates that the random variable is sampled independently for each i. Fogel (1992) proposed the simple rule $\forall var_i \leq 0$, $var_i \leftarrow \xi$, where ξ is a value close to zero, but not zero.

Although this idea is similar to the one used in the ES, the stochastic process behind the meta-EP is fundamentally different from the one behind ES (Bäck, 1994). The procedure to implement EP is the same as the one used to implement ES, but with $\lambda = \mu$, and is just a slight variation of the standard evolutionary algorithm procedure (Algorithm 3.6).

3.6.3. Genetic Programming

Genetic programming (GP), proposed by Cramer (1985) and further developed and formalized by Koza (1992, 1994a), constitutes a type of evolutionary algorithm developed as an extension of genetic algorithms. In GP, the data structures that suffer adaptation are representations of computer programs, and thus the fitness evaluation involves the execution of the programs. Therefore, GP involves a search based upon the evolution of the space of possible computer programs such that, when run, they will produce a suitable output, which is usually related to solving a given problem.

The main question that motivated the proposal of genetic programming was "How can computers learn to solve problems without being explicitly programmed to do so?" In its original form, a computer program is basically the application of a sequence of functions to arguments: *functional paradigm*. Implementing GP is conceptually straightforward when associated with programming languages that allow the manipulation of a computer program in the form of data structures, such as the LISP programming language. However, it is possible to implement GP using virtually any programming language that can manipulate computer programs as data structures and that can then compile, link and execute the new data structures (programs) that arise. As LISP was the original language used to study and apply GP and is easily understandable, it will be briefly reviewed in the following.

In LISP there are only two types of entities: *atoms* (constants and variables) and *lists* (ordered set of items delimited by parentheses). A symbolic expression, or *S*-expression, is defined as an atom or a list, and it constitutes the only syntactic form in LISP; that is, all programs in LISP are *S*-expressions. In order to evaluate a list, its first element is taken as a function to be applied to the other elements on the list. This implies that the other elements from the list must be evaluated before they can be used as arguments for the function, represented by the first element of the list. For example, the *S*-expression $(- 4\ 1)$

requires function '−' to be applied to two arguments represented by the atoms '4' and '1'. The value returned by evaluating this expression is '3'.

A list may also contain other lists as arguments, and these are evaluated in a recursive, depth-first way; for instance, the S-expression (− (× 2 4) 6) returns '2' when evaluated. Therefore, the programs in LISP can be seen as compositions of functions. It is also important to note that these functions must not be restricted to simple arithmetic operators. It is possible to implement all types of constructs that can be used in a computer program (sequence, selection, and repetition). An important feature of LISP is that all LISP programs have a single S-expression that corresponds to the parse tree of the program. Figure 3.23 illustrates the parse tree representations of both S-expressions exemplified above.

As with all other evolutionary algorithms, in GP a problem is defined by its representation and the specification of an objective function that will allow the definition of a fitness function. The representation in GP consists of choosing an appropriate *function set F* and an appropriate *terminal set T*. The functions chosen are those that are *a priori* believed to be useful and sufficient for the problem at hand, and the terminals are usually variables or constants. The computer programs in the specified language are then the individuals of the population that can be represented in a data structure such as a tree. These individuals have to be executed so that the phenotype of the individual is obtained. While defining the representation in GP, two important aspects must be taken into account:

- *Syntactic closure*: it is necessary to examine the values returned by all functions and terminals such that all of them will accept as argument any value and data type that may be returned by a function from *F* or a terminal from *T*.

- *Sufficiency*: defining the search space corresponds to defining *F* and *T*. It is intuitive thus that the search space has to be large enough to contain the desired solution. Besides, the functions and terminals chosen have to be adequate to the problem domain.

Several types of functions and terminals can be used, such as arithmetic functions (e.g., +, ×, −, /), logic functions (e.g., AND, OR, NAND), and standard programming functions. The terminals may be variables, constants, or functions that do not receive arguments.

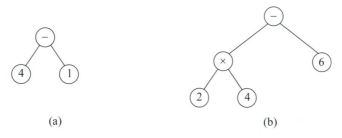

(a)   (b)

Figure 3.23: Parse tree representation of S-expressions in LISP. (a) (− 4 1). (b) (− (× 2 4) 6).

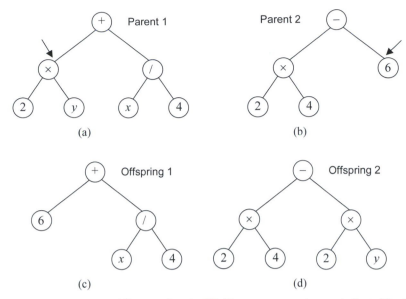

Figure 3.24: Crossover of *S*-expressions in GP. The crossover points are indicated by the arrows. (a) (+ (× 2 *y*) (/ *x* 4)). (b) (− (× 2 4) 6). (c) (+ 6 (/ *x* 4)). (d) (− (× 2 4) (× 2 *y*)).

Crossover

One interesting feature of genetic programming is that the individuals of the population may have different sizes. The recombination operator (crossover) is used to generate individuals (computer programs) for the next generation from parent programs chosen according to their fitness value. The offspring programs are composed of sub-trees from the parent programs, and may have different sizes and shapes from those of their parents. Similarly to the genetic algorithms, the application of crossover in GP requires two *S*-expressions as inputs. However, recombination cannot be random in this case, for it has to respect the syntax of the programs. This means that different crossover points may be chosen for each of the parent *S*-expressions. The crossover point isolates a sub-tree from each parent *S*-expression that will be exchanged to produce the offspring. Crossover in GP is illustrated in Figure 3.24. In this example, $F = \{+, -, \times, /\}$ and $T = \{1,2,3,4,5,6,x,y\}$.

Mutation

Mutation in GP may be useful but is not always used. Mutation may be important for the maintenance and introduction of diversity and also to maintain a dynamic variability of the programs being evolved. Mutation in GP is illustrated in Figure 3.25.

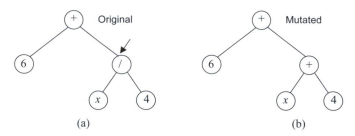

Figure 3.25: Mutation of an *S*-expression in GP. The mutation point is indicated by the arrow. (a) (+ 6 (/ x 4)). (b) (+ 6 (+ x 4)).

3.6.4. Selected Applications from the Literature: A Brief Description

This section illustrates the application of these three types of evolutionary algorithms (ES, EP, and GP) to three different problems. In the first example, a simple evolution strategy is applied to an engineering design problem. The second example illustrates the application of evolutionary programming to optimize the parameters of a voltage circuit, and the third example is a pattern classification task realized using a genetic programming technique. All descriptions are brief and aimed at illustrating the application potential of the approaches and the use of the three design principles: representation, objective, and function evaluation. Further details can be founded in the cited literature.

ES: Engineering Design

Engineering design is the process of developing a plan (e.g., a model, a system, a drawing) for a functional engineeing object (e.g., a building, a computer, a toaster). It often requires substantial research, knowledge, modelling and interactive adjustment and redesign until a useful and efficient plan is proposed. In (Hingston et al., 2002), the authors applied a simple evolution strategy to determine the geometry and operating settings for a crusher in a comminution circuit for ore processing. The task was to find combinations of design variables (including geometric shapes and machine settings) in order to maximize the capacity of a comminution circuit while minimizing the size of the crushed material.

The authors used a crusher with the shape of an upside down cone in which material is introduced from above and is crushed as it flows downward through the machine, as illustrated in Figure 3.26(a). Crushing is performed by an inner crushing surface, called mantle, mounted on the device and driven in an eccentric motion swivelling around the axis of the machine. The material is poured onto the top of the crushing chamber and is crushed between the mantle and the bowl liner when it passes through them. The area through which the material passes is called closed-side setting, and can be made wider or narrower depending on the desired size of the crushed material.

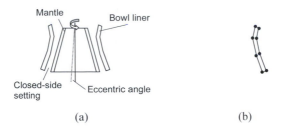

(a) (b)

Figure 3.26: (a) Simplified diagram of the crusher used in Hingston et al., 2002. (b) Shape representation of each liner.

Evolutionary algorithms are particularly suitable for solving such types of problems because they are too complex to be solved analytically; a well-defined evaluation function for the designs can be proposed; there is little information *a priori* to guide an engineer in the determination of near-optimal designs; the search space is too large to be searched exhaustively; and once a candidate design has been proposed, it is possible to use a simulation tool to evaluate it. These are some of the features that make specific problems suitable for natural computing approaches, as discussed in Section 1.5.

The authors proposed the following representation, objective, and fitness function.

Representation: The representation of the individuals of the population corresponds to real-valued vectors in specific domains representing the closed-side settings, eccentric angle and rotational speed. This is a straightforward representation, for all parameters assume real values. The geometric shapes of the bowl liners were represented as a series of line segments using a variable-length list of points, each represented by its *x-y* coordinates on the plane (Figure 3.26(b)). Mutation was performed using individual standard deviations for each attribute of the chromosome representation (Equation (3.7)) and with varying strategy parameters following Equation (3.8). The selection method was the $(\mu + \lambda)$-ES, where $\mu = \lambda = 1$. Note that not only the device parameters are being optimized, but also the shape (geometry) of the bowl liners.

Objective: The objective for this problem is to maximize a function *g*, which corresponds to maximizing the capacity of the comminution circuit while minimizing the size of the product.

Evaluation: The fitness function used is directly proportional to the crushing capacity *C* of the circuit in terms of tons per hour, and to the size measure of the crushed product *P*:

$$g(C,P) = 0.05 \times C + 0.95 \times P,$$

As one of the goals is to minimize the size of the crushed product, *P* refers to a size *s* in mm such that 80% of the product is smaller than *s* mm. Therefore, the fitness function *g(C,P)* to be maximized takes into account the capacity of the circuit and the percentage of material crushed within a given specification.

The constants multiplying each factor were chosen by the authors so as to equalize their variability, quantifying the importance of each component. Note that, in this case, after a plant has been designed by the evolutionary algorithm, a simulation tool is used to evaluate and compare alternative designs (candidate solutions), thus automating the process.

EP: Parameter Optimization

The process of engineering design involves mainly two levels: topology design and parameter optimization. In the previous example we discussed the use of an evolution strategy to perform both, topology design (bowl liner shape) and parameter design (combination of appropriate eccentric angle, closed-side setting, and rotation speed, which will influence C and P).

In (Nam et al., 2001), the authors applied an evolutionary programming approach to the problem of optimizing a set of parameters for a voltage reference circuit. Similarly to the engineering design problem mentioned above, in order to design an electronic circuit, several parameters of each part of the circuit have to be determined, but, in the present example, the topology of the circuit was already defined. The authors argued that such problems are highly multi-modal, making it difficult to apply more traditional techniques, such as gradient-based search and justifying the use of alternative approaches like EAs. Besides, the parameters of the circuit to be optimized are real-valued vectors, one reason why the authors chose to use evolutionary programming, though an evolution strategy could have been used instead or a genetic algorithm with float-point encoding.

The circuit whose parameters they wanted to optimize was an on-chip voltage reference circuit; that is, a circuit from semiconductor technology that generates a reference voltage independent of power fluctuation and temperature variation. Given the power voltage V_T and the temperature T, it is possible to write the reference voltage V_{ref} equation for the circuit as a function of several parameters, such as resistors, transistors, etc. The goal is to find appropriate parameter values that maintain the reference voltage stable when the temperature and power voltage vary within given ranges. A simplified equation for the reference voltage can be written as a function of only V_T and T:

$$V_{ref} = K_1 V_T + K_2 T \qquad (3.14)$$

where K_1 and K_2 are two constants, V_T is the power voltage and T is the temperature.

Representation: The representation used was real-valued vectors and each candidate solution corresponded to a combination of parameter values. The authors used a $(\mu + \lambda)$ selection scheme and a simplified mutation scheme. Equation (3.15) below describes how the authors evolved the attribute values of the vector to be optimized $\mathbf{x}$ and the strategy parameters as well.

$$x_i^{t+1} = x_i^t + \sigma_i . N_i(0,1)$$
$$\sigma_i^t = \beta_i . f(\mathbf{x}^t) \qquad (3.15)$$

where x_i^t is the i-th parameter of one chromosome at iteration t, σ_i its associated variance, $N_i(0,1)$ is a normal distribution of zero mean and standard deviation 1, β_i is a positive constant, and $f(\mathbf{x})$ is the fitness of chromosome $\mathbf{x}$. To determine $fit(\mathbf{x})$, a cost function was proposed taking into account a set of parameters to be optimized.

Objective: The objective is to minimize a function $f(\mathbf{x})$ that takes into account a reference voltage, temperature compensation and an active voltage constraint.

Evaluation: Based on this objective, the proposed fitness function was:

$$f(\mathbf{x}) = K_1|V_{25}(\mathbf{x}) - 2.5| + K_2(|V_0(\mathbf{x}) - V_{25}(\mathbf{x})| + |V_{25}(\mathbf{x}) - V_{100}(\mathbf{x})|) +$$
$$+ K_3|V_{act0}(\mathbf{x}) - 0.8|,$$

where V_0, V_{25}, and V_{100} represent reference voltages at 0°C, 25°C, and 100°C, respectively, and V_{act0} is an active bias voltage.

In this approach a simulation tool is also used to evaluate the performance of the evolved parameters.

GP: Pattern Classification

Adaptation problems often present themselves as problems of classification or function approximation. In classification problems, the objective is to discern a pattern from others and to develop a procedure capable of successfully performing the classification. The procedure is usually developed using a set of input data with known classification and new, previously unseen, data is used to evaluate the suitability of the adaptation (classification) process. The *generalization capability* of the classifier is assessed on these novel data. Learning relationships that successfully discriminate among examples associated with problem solving choices is a typical application of natural computing.

One possible application of genetic programming is in the design of classifiers (Koza, 1992, 1994b). Consider the input data set illustrated in Figure 3.27.

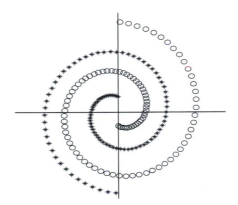

Figure 3.27: Input data set used to test the performance of many classifiers.

This is a typical data set used to test some classification techniques, including neural networks (Chapter 4), whose difficulty lies in the fact that data are presented using Cartesian coordinates making them nonlinearly separable.

This problem can be solved using a genetic programming approach (Koza, 1992). In this case, the GP will have to provide as output a computer program (*S*-expression) that determines to which spiral a given point belongs. Assume that the points belonging to one of the spirals correspond to class +1 and the points belonging to the other spiral correspond to class −1.

Representation: It is important to remember that the representation in GP corresponds to a function set F and a terminal set T believed to be sufficient for solving the problem. Due to the nature of the spirals problem, the terminal set T consists of the x and y coordinates of all given points (190 in the picture shown). As some numerical constants $c \in \Re$ may be needed to process all these data, some constants, $c \in [-1,+1]$, can be added to the terminal set. Thus, the terminal set is $T = \{X,Y,\Re\}$, where X and Y are the sets of coordinates for the x and y variables of the data set and $\Re$ is the set of real numbers belonging to the interval $[-1,+1]$. To determine the function set for this problem, we must have in mind that the resultant computer program has to determine to which of the spirals a given point belongs. Thus, the programs to be evolved may include the four arithmetic operations, a conditional comparative function for decision making, and the trigonometric *sine* and *cosine* functions. The function set for this problem is then $F = \{+,-,\times,/,\text{IFLTE},sin,cos\}$, in which these functions take 2, 2, 2, 2, 4, 1, and 1 arguments, respectively. Function IFLTE (If-Less-Than-Or-Equal-To) is a four argument conditional comparative operator that executes its third argument if its first argument is less than its second argument and, otherwise, executes the fourth (else) argument. The conditional IFLTE is used so as to allow the classification of points into one of the classes, and functions *sine* and *cosine* are included so as to increase nonlinearity. In order to have a binary classification in the output, a wrapper must be employed mapping any positive value returned by the evolved program when executed to class +1, and to class −1 otherwise.

Objective: The objective is to maximize the number of points correctly classified or, equivalently, minimize the number of points misclassified.

Evaluation: Assuming the objective is to maximize the correct classification rate of the *S*-expression, fitness can be calculated by taking the total number of correctly classified points.

3.7   FROM EVOLUTIONARY BIOLOGY TO COMPUTING

This chapter described the standard evolutionary algorithms with particular emphasis on the genetic algorithms. Table 3.2 suggests an interpretation of the biological terminology into the evolutionary algorithms.

Table 3.2: Interpretation of the biological terminology into the computational world for genetic algorithms.

Biology (Genetics)	Evolutionary Algorithms
Chromosome	Data structure
Genotype	Encoding of a potential candidate solution to a problem (equivalent to a chromosome)
Phenotype	Decoded value of one or more chromosomes
Gene	Element occupying a given position in a data structure
Locus	Position occupied by a gene in a data structure
Alleles	Variations of an element (gene) that can occupy a locus
Crossover	Exchange of portions between data structures
Mutation	Replacement of a given gene by a different one
Fitness	Value that indicates the quality of an individual in relation to the problem being solved
Selection	Process that allows the survival and reproduction of the fittest individuals in detriment of the less fit ones

3.8 SCOPE OF EVOLUTIONARY COMPUTING

Given a problem, how do we know if an evolutionary algorithm can solve it effectively? Brief discussions about it have been made throughout this chapter and Chapter 1, but there is no rigorous answer to this question, though some key intuitive aspects can be highlighted (Mitchell, 1998):

- If the search space is large, neither perfectly smooth nor unimodal, is unknown, or if the fitness function is noisy, then an EA will have a good chance of being a competitive approach.

- If the search space is smooth or unimodal, then *gradient* or *hill-climbing* methods perform much better than evolutionary algorithms.

- If the search space is well understood (such as in the *traveling salesman problem* – TSP), heuristics can be introduced in specific methods, including the EAs, such that they present good performances.

Beasley (2000) divided the possible application areas of evolutionary algorithms into five broad categories: planning (e.g., routing, scheduling and packing); design (e.g., signal processing); simulation, identification, control (general plant control); and classification (e.g., machine learning, pattern recognition and classification). More specifically, evolutionary algorithms have been applied in fields like art and music composition (Bentley, 1999; Bentley and Corne, 2001), electronics (Zebulum et al., 2001), language (O'Neill and Ryan, 2003), robotics (Nolfi and Floreano, 2000), engineering (Dasgupta, and Michalewicz, 1997), data mining and knowledge discovery (Freitas and Rozenberg, 2002), industry (Karr and Freeman, 1998), signal processing (Fogel, 2000b), and many others.

3.9 SUMMARY

This text introduced some basic concepts for problem solving, focusing optimization problems. Several versions of the hill-climbing procedure were presented for comparison with simulated annealing and evolutionary algorithms. The emphasis was always on how the standard algorithms were developed, their respective sources of inspiration, and how to interpret the natural motivation into the computational domain. It was discussed that simulated annealing was developed using ideas gleaned from statistical thermodynamics, while evolutionary algorithms were inspired by the neo-Darwinian theory of evolution. All strategies reviewed are conceptually very simple and constitute general-purpose search techniques.

Basic principles of genetics were reviewed so as to prepare the reader for an appropriate understanding of evolutionary algorithms, in particular genetic algorithms and their functioning. Some discussion about the philosophy of the theory of evolution and what types of outcomes can be expected from EAs was also provided. The focus of the chapter was on a description of evolutionary algorithms - those inspired by evolutionary biology - as a problem-solving approach. Any algorithm based upon a population of individuals (data structures) that reproduce, and suffer variation followed by selection, can be characterized as an EA. The standard evolutionary algorithm was presented in Algorithm 3.6, and a set of elementary genetic operators (crossover and mutation) was described for the standard genetic algorithm and the other traditional evolutionary algorithms.

Particular attention was also given to a perspective of evolution as an algorithmic process applied in problem solving. Under this perspective, the problem to be solved plays the role of the environment, and each individual of the population is associated with a candidate solution to the problem. Thus, an individual will be more adapted to the environment (will have a higher fitness value) whenever it corresponds to a 'better' solution to the problem. At each evolutionary generation, improved candidate solutions should be produced, though there is no guarantee that an optimal solution can be found. The evolutionary algorithm thus constitutes a useful iterative and parallel search procedure, suitable for solving search and optimization problems that involve a vast search space.

3.9.1. The Blind Watchmaker

To conclude this chapter on evolutionary biology and algorithms, it is important to say a few words about the type of search being performed by evolution and thus evolutionary algorithms.

Until Darwin's proposal, almost everyone embraced the idea that living systems were designed by some God-like entity. Even scientists were convinced by the so-called *watchmaker* argument; that is, the argument from design, proposed by theologian William Paley in his 1802 book *Natural Theology*. Paley noted that watches are very complex and precise objects. If you found a watch on the ground, you could not possibly believe that such a complex object had been created by random chance. Instead, you would naturally conclude that the watch must have had a maker; that there must have existed sometime, somewhere, an *artificer*, who designed and built it for the purpose which we find it actually to answer. For Paley, the same logic argument applies to living systems. Therefore, living systems, like watches, must have a maker, concluded Paley.

The point to be raised here is the one discussed by Richard Dawkins in his 1986 awarded book *The Blind Watchmaker*, namely, that evolution by means of natural selection might be seen as a *blind watchmaker* (see Chapter 8).

"In the case of living machinery, the 'designer' is unconscious natural selection, the blind watchmaker." (Dawkins, 1986, p. 45)

The philosophical discussions presented so far have already focused on the algorithmic (designer's) potentiality of evolution. What has to be emphasized now is that it is a blind (i.e., *unsupervised*) process.

While studying the example given in Section 3.5.3, one might be inclined to think that evolutionary algorithms are employed when the solution to a problem is known. However, this is not usually the case and deserves some comments. While in the first example, provided for the GA, the solution was indeed known, in the second example no assumption about the optimum of the problem had to be made so that the GA could be applied. The first example demonstrated that GAs are capable of generating an increasingly more fit population of individuals, thus being capable of evolving (adapting to the environment). The second example showed that GAs are capable of finding optimal solutions to problems. And both examples showed that though GAs might sometimes use information about the solution of a problem, they are never told how to find the solution, i.e., they perform a *blind search*. By simply initializing a population of chromosomes (represented using some type of data structure), creating copies of them (reproduction) and applying random genetic operators (performing variation and selection), it is possible to determine (unknown) solutions to complex problems. The same holds true for the other evolutionary algorithms presented: evolution strategies, evolutionary programming and genetic programming. All of them follow the same standard evolutionary procedure (Algorithm 3.6).

3.10 EXERCISES

3.10.1. Questions

1. This text introduced three versions of hill-climbing. The last version, namely stochastic hill-climbing, has many features in common with the simulated annealing algorithm. Do you find it necessary to know the inspiration taken from statistical thermodynamics in order to understand simulated annealing? Justify your answer.

2. Based upon the discussion presented concerning the evolution of species, provide an answer for the following classical question. What came first, the egg or the chicken? What arguments would you use to support your answer?

3. It is known that the eyes evolved more than once, in different times, and for different species. What do you think would be a good 'fitness measure' for a working 'eye'? Take the particular case of human eyes for example.

4. If evolution is considered to be a never-ending process, do humans present any sign of evolution from the last few thousand years? Justify your answer.

5. In addition to the classic example of the evolution of the moths in Great Britain (Section 3.4.5), can you provide any other example of an organism whose evolution can be (has been) clearly accompanied over the last years (or centuries, or thousands of years)?

6. Name at least three difficulties for the theory of evolution. Can you think of some arguments that could be used to refute these difficulties?

7. How would you explain the many similarities among living organisms? For instance, consider the case of whales and fishes. They belong to different species, but are similar in structure and habitat. Why would that be so?

8. If biology (e.g., evolution) can be seen as engineering, suggest another biological phenomenon or process that can be abstracted to an algorithmic level. Write down a procedure to implement it and propose two practical applications for this procedure.

3.10.2. Computational Exercises

1. Implement the various hill-climbing procedures and the simulated annealing algorithm to solve the problem exemplified in Section 3.3.3. Use a real-valued representation scheme for the candidate solutions (variable x).

    By comparing the performance of the algorithms, what can you conclude?

    For the simple hill-climbing, try different initial configurations as attempts at finding the global optimum. Was this algorithm successful?

    Discuss the sensitivity of all the algorithms in relation to their input parameters.

2. Implement and apply the hill-climbing, simulated annealing, and genetic algorithms to maximize function $g(x)$ used in the previous exercise assuming a bitstring representation.

    Tip: the perturbation to be introduced in the candidate solutions for the hill-climbing and simulated annealing algorithms may be implemented similarly to the point mutation in genetic algorithms. Note that in this case, no concern is required about the domain of x, because the binary representation already accounts for it.

    Discuss the performance of the algorithms and assess their sensitivity in relation to the input parameters.

3. Implement a standard genetic algorithm for the example of Section 3.5.3.

4. The algorithms presented in Section 3.3 were all devised assuming maximization problems. How would you modify the procedures in case minimization was desired? Rewrite the codes if necessary.

5. Apply an evolution strategy (ES) with no correlated mutation to the function maximization problem of Exercise 1.

6. Repeat the previous exercise with an evolutionary programming (EP) technique. Compare the results with those obtained in Exercises 1 and 5.

7. Determine, using genetic programming (GP), the computer program (S-expression) that produces exactly the outputs presented in Table 3.3 for each value of x. The following hypotheses are given:

    *   Use only functions with two arguments (binary trees).
    *   Largest depth allowed for each tree: 4.
    *   Function set: $F = \{+, *\}$.
    *   Terminal set: $T = \{0, 1, 2, 3, 4, 5, x\}$.

Table 3.3: Input data for the GP.

x	Program output
-10	153
-9	120
-8	91
-7	66
-6	45
-5	28
-4	15
-3	6
-2	1
-1	0
0	3
1	10
2	21
3	36
4	55
5	78
6	105
7	136
8	171
9	210
10	253

8. Solve the SPIR pattern recognition task of Section 3.6.4 using genetic programming. Assume the same hypotheses given in the example.

3.10.3. Thought Exercises

1. Evolutionary algorithms were introduced as general-purpose search algorithms developed with inspiration in evolutionary biology and applied for problem solving. Most of their applications are of an optimization type, but they can also be applied for design, arts, and so forth. For instance, one can use an EA to design tires for racing cars. In such a case, the tires would have to be encoded using some suitable representation, the evaluation function would have to take into account the endurance, shape, adherence to the ground in normal and wet conditions, etc., and the evaluation function would have to allow the distinction between candidate solutions (individual chromosomes) to your problem.

    Suggest a novel application for an evolutionary algorithm. This can be from your domain of expertise or any field of your interest. Provide a suitable representation, objective, and evaluation function. If you are using a binary representation, then the crossover and mutation operators described may be suitable for evolving a population of candidate solutions. However, if the representation chosen is not binary, suggest new crossover and mutation operators suitable for the proposed representation.

2. How would you implement a GA for a two variable numeric function? What would change: the representation, evaluation function, or genetic operators?

3. How would you hybridize a local search procedure, such as the standard hill-climbing, with a genetic algorithm? Provide a detailed discussion, including a discussion about the benefits of this hybridization.

3.10.4. Projects and Challenges

1. The *traveling salesman problem* (TSP) is a popular problem in combinatorial optimization with applications in various domains, from fast-food delivery to the design of VLSI circuits. In its simplest form, the traveling salesman must visit every city in a given territory exactly once, and then return to the starting city. Given the cost of travel between all cities (e.g., distance or cost in money), the question posed by the TSP is related to what should be the itinerary for the minimal cost of the whole tour.

    Implement an evolutionary algorithm to solve the pictorial TSP illustrated in Figure 3.28. The cities were placed on a regular grid to facilitate analysis and are labeled by a number on their top left corner. Three main aspects deserve careful examination: representation, evaluation function, and genetic operators.

    First, one has to define a representation for the chromosomes. Is a binary representation, such as the one used in standard genetic algorithms, suitable? If not, explain why, and suggest a new representation.

    Second, the definition of an evaluation function requires the knowledge of the objective. In this case, assume that the objective is simply to minimize the distance traveled by the salesman. The coordinates of each city on the x-y plane can be extracted from Figure 3.28.

Lastly, if the representation you chose is not binary, what should be the crossover and mutation operators employed? Can you see any relationship between these operators and the biological crossover and mutation?

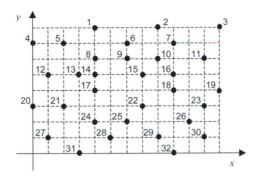

Figure 3.28: Simple TSP with 32 cities. The cities are placed on a regular grid on the x-y plane, in which each point (●) represents a city, the number next to each city corresponds to its index, and each square on the grid corresponds to one unit of distance (uod - e.g., Km).

2. In the evolutionary algorithm developed in the previous exercise, add a standard hill-climbing procedure as a means of performing local search in the individuals of the population.

3. Maximize function $g(x) = 2^{-2((x-0.1)/0.9)^2} (\sin(5\pi x))^6$, $x \in [0,1]$, using an evolution strategy with correlated mutation.

3.11 REFERENCES

[1] Anderson, R. L. (1953), "Recent Advances in Finding Best Operating Conditions", *Jour. of Am. Stat. Assoc.*, **48**, pp. 789–798.

[2] Bäck, T. and Schwefel, H.-P. (1993), "An Overview of Evolutionary Algorithms for Parameter Optimization", *Evolutionary Computation*, **1**(1), pp. 1–23.

[3] Bäck, T. (1996), *Evolutionary Algorithms in Theory and Practice*, Oxford University Press.

[4] Bäck, T., Fogel, D. B. and Michalewicz, Z. (eds.) (2000a), *Evolutionary Computation 1: Basic Algorithms and Operators*, Institut of Physics Publishing.

[5] Bäck, T., Fogel, D. B. and Michalewicz, Z. (eds.) (2000b), *Evolutionary Computation 2: Advanced Algorithms and Operators*, Institute of Physics Publishing.

[6] Banzhaf, W., Nordin, P., Keller, R. E. and Francone, F. D. (1998), *Genetic Programming – An Introduction: On the Automatic Evolution of Computer Programs and Its Applications*, Morgan Kaufmann Publishers.

[7] Beasley, D. (2000), "Possible Applications of Evolutionary Computation", In Bäck et al. (2000a), Chapter 2, pp. 4–19.

[8] Bentley, P. J. and Corne, D. W. (2001), *Creative Evolutionary Systems*, Morgan Kaufmann.

[9] Bentley, P. J. (1999), *Evolutionary Design by Computers*, Morgan Kaufmann.

[10] Bertsimas, D. and Tsitsiklis, J. (1993), "Simulated Annealing", *Stat. Sci.*, 8(1), pp. 10–15.

[11] Beyer, H.-G. (2001), *Theory of Evolution Strategies*, Springer-Verlag.

[12] Bremermann, H. J. (1962), "Optimization through Evolution and Recombination", in M. C. Yovits, G. T. Jacobi and D. G. Goldstein (eds.), *Self-Organizing Systems*, Spartan, Washington, D.C., pp. 93–106.

[13] Černý, V. (1985), "Thermodynamical Approach to the Travelling Salesman Problem: An Efficient Simulation Algorithm", *J. of Optim. Theory and App.*, 45(1), pp. 41–51.

[14] Colby, C. (1997), *Introduction to Evolutionary Biology*, TalkOrigins, [OnLine] (July, 11th, 2002) www.talkorigins.org/faqs/faq-intro-to-biology.html.

[15] Cramer, N. L. (1985), "A Representation for the Adaptive Generation of Simple Sequential Programs", In J. J. Grefenstette (ed.), *Proceedings of the First International Conference on Genetic Algorithms and Their Applications*, Erlbaum.

[16] Darwin, C. R. (1859), *The Origin of Species*, Wordsworth Editions Limited (1998).

[17] Dasgupta, D. and Michalewicz, Z. (1997), *Evolutionary Algorithms in Engineering Applications*, Springer-Verlag.

[18] Davis, L. (ed.) (1991), *Handbook of Genetic Algorithms*, Van Nostrand Reinhold.

[19] Dawkins, R. (1986), *The Blind Watchmaker*, Penguin Books.

[20] Dennett, D. C. (1995), *Darwin's Dangerous Idea: Evolution and the Meanings of Life*, Penguim Books.

[21] Eigen, M. (1992), *Steps Towards Life*, Oxford University Press.

[22] Elton, C. (1997), *Animal Ecology*, Sidgwick and Jackson, London.

[23] Fogel, D. B. (1992), *Evolving Artificial Intelligence*, Doctoral Dissertation, UCSD.

[24] Fogel, D. B. (2000a), *Evolutionary Computation: Toward a New Philosophy of Machine Intelligence*, 2nd Ed., The IEEE Press.

[25] Fogel, D. B. (2000b), *Evolutionary Computation: Principles and Practice for Signal Processing*, SPIE – The International Society for Optical Engineering.

[26] Fogel, D. B. (ed.) (1998), *Evolutionary Computation: The Fossil Record*, The IEEE Press.

[27] Fogel, L. J., Owens, A. J. and Walsh, M. J. (1966), *Artificial Intelligence through Simulated Evolution*. John Wiley.

[28] Fraser, A. S. (1959), "Simulation of Genetic Systems by Automatic Digital Computers", *Aust. Jour. Of Biol. Sci.*, **10**, pp. 489–499.

[29] Freitas, A. A. and Rozenberg, G. (2002), *Data Mining and Knowledge Discovery with Evolutionary Algorithms*, Springer-Verlag.

[30] Friedberg, R. M. (1958), "A Learning Machine: Part I", *IBM Jour. of Research and Development*, **2**, pp. 2–13.

[31] Futuyma, D. J. (1998), *Evolutionary Biology*, 3rd Ed., Sinauer Associates, Inc.

[32] Gardner, E. J., Simmons, M. J. and Snustad, D. P. (1991), *Principles of Genetics*, John Wiley & Sons, Inc.

[33] Goldberg, D. E. (1989), *Genetic Algorithms in Search, Optimization and Machine Learning*, Addisson-Wesley Reading, MA.

[34] Griffiths, A. J. F., Miller, J. H., Suzuki, D. T., Lewontin, R. C. and Gelbart, W. M. (1996), *An Introduction to Genetic Analysis*, W. H. Freeman and Company.

[35] Hingston, P., Barone, L. and While, L. (2002), "Evolving Crushers", *Proc. of the IEEE Congress on Evolutionary Computation*, **2**, pp. 1109–1114.

[36] Holland, J. J. (1975), *Adaptation in Natural and Artificial Systems*, MIT Press.

[37] Karr, C. L. and Freeman, L. M. (1998), *Industrial Applications of Genetic Algorithms*, CRC Press.

[38] Kinnear Jr., K. E. (ed.) (1994), *Advances in Genetic Programming*, MIT Press.

[39] Kirkpatrick, S., Gerlatt, C. D. Jr. and Vecchi, M. P. (1983), "Optimization by Simulated Annealing", *Science*, 220, pp. 671–680.

[40] Koza, J. R. (1992), *Genetic Programming: On the Programming of Computers by Means of Natural Selection*, MIT Press.

[41] Koza, J. R. (1994a), *Genetic Programming II: Automatic Discovery of Reusable Programs*, MIT Press.

[42] Koza, J. R. (1994b), "Genetic Programming as a Means for Programming Computers by Natural Selection", *Statistics and Computing*, **4**(2), pp. 87–112.

[43] Krebs, C. J. (1994), *Ecology: The Experimental Analysis of Distribution and Abundance*, 4th Ed., Harper Collins College Publishers.

[44] Man, K. F., Tang, K. S. and Kwong, S. (1999), *Genetic Algorithms: Concepts and Designs*, Springer Verlag.

[45] Mayr, E. (1988), *Toward a New Philosophy of Biology: Observations of an Evolutionist*, Belknap, Cambridge, MA.

[46] Mendel, G. (1865), "Versuche uber pflanzenhybriden", *J. Hered.*, **42**, pp. 1–47, English translation "Experiments in Plant Hybridization", Harvard University Press, Cambridge, MA.

[47] Metropolis, N., Rosenbluth, A.W., Rosenbluth, M. N., Teller, A.H. and Teller, E. (1953), "Equations of State Calculations by Fast Computing Machines", *J. Chem. Phys.*, 21, pp. 1087–1092.

[48] Michalewicz, Z. (1996), *Genetic Algorithms + Data Structures = Evolution Programs*, Springer-Verlag.

[49] Michalewicz, Z. and Fogel, D. B. (2000), *How To Solve It: Modern Heuristics*, Springer-Verlag, Berlin.

[50] Mitchell, M. (1996), *An Introduction to Genetic Algorithms*, The MIT Press.

[51] Nam, D. K., Seo, Y. D., Park, L. J., Park, C. H. and Kim, B. S. (2001), "Parameter Optimization of an On-Chip Voltage Reference Circuit Using Evolutionary Programming", *IEEE Trans. Evolutionary Computation*, **5**(4), pp 414–421.

[52] Nolfi, S. and Floreano, D. (2000), *Evolutionary Robotics: The Biology, Intelligence, and Technology of Self-Organizing Machines (Intelligent Robotics and Autonomous Agents)*, MIT Press.

[53] O'Neill, M. and Ryan, C. (2003), *Grammatical Evolution: Evolutionary Automatic Programming in an Arbitrary Language*, Kluwer Academic Publishers.

[54] Paley, W. (1802), *Natural Theology – or Evidences of the Existence and Attributes of the Deity Collected from the Appearances of Nature*, Oxford: J. Vincent.

[55] Pincus, M. (1970), "A Monte Carlo Method for the Approximate Solution of Certain Types of Constrained Optimization Problems", *Oper. Res.*, 18, pp. 1225–1228.

[56] Rechenberg, I. (1965), "Cybernetic Solution Path of an Experimental Problem", *Roy. Aircr. Establ. Libr. Transl.*, 1122, Farnborough, Hants, UK.

[57] Russel, P. J. (1996), *Genetics*, 4th Ed., Harper Collins College Publishers.

[58] Russell, S. J. and Norvig, P. (1995), *Artificial Intelligence A Modern Approach*, Prentice Hall, New Jersey, U.S.A.

[59] Sarle, W. (1993), *Kangoroos*, article posted on *comp.ai.neural-nets* on the 1st September.

[60] Schwefel, H.-P. (1965), *Kybernetische Evolution als Strategie der Experimentellen Forschung in der Strömungstechnik*, Diploma Thesis, Technical University of Berlin, March.

[61] Schwefel, H.-P. (1981), *Numerical Optimization of Computer Models*, Wiley, Chichester.

[62] Schwefel, H.-P. (1995), *Evolution and Optimum Seeking*, Wiley: New York.

[63] Wolpert, D. H. and Macready, W. G. (1997), "No Free Lunch Theorems for Optimization", *IEEE Trans. on Evolutionary Computation*, **1**(1), pp. 67–82.

[64] Wright, S. (1968–1978), *Evolution and the Genetics of Population*, 4 vols., University of Chicago Press, Chicago, IL.

[65] Zebulum, R. S., Pacheco, M. A., Vellasco, M. M. B. and Zebulum, R. S. (2001), *Evolutionary Electronics: Automatic Design of Electronic Circuits and Systems by Genetic Algorithms*, CRC Press.

CHAPTER 4

NEUROCOMPUTING

"Inside our heads is a magnificent structure that controls our actions and somehow evokes an awareness of the world around ... It is hard to see how an object of such unpromising appearance can achieve the miracles that we know it to be capable of."
(R. Penrose, The Emperor's New Mind, Vintage, 1990; p. 483)

"Of course, <u>something</u> about the tissue in the human brain is necessary for our intelligence, but the physical properties are not sufficient ... Something in the <u>patterning</u> of neural tissue is crucial."
(S. Pinker, How the Mind Works, The Softback Preview, 1998; p. 65)

4.1 INTRODUCTION

How does the brain process information? How is it organized? What are the biological mechanisms involved in brain functioning? These form just a sample of some of the most challenging questions in science. Brains are especially good at performing functions like pattern recognition, (motor) control, perception, flexible inference, intuition, and guessing. But brains are also slow, imprecise, make erroneous generalizations, are prejudiced, and are incapable of explaining their own actions.

Neurocomputing, sometimes called *brain-like computation* or *neurocomputation*, but most often referred to as *artificial neural networks* (ANN)[1], can be defined as information processing systems (computing devices) designed with inspiration taken from the nervous system, more specifically the brain, and with particular emphasis on problem solving. S. Haykin (1999) provides the following definition:

> "A[n artificial] neural network is a massively parallel distributed processor made up of simple processing units, which has a natural propensity for storing experiential knowledge and making it available for use." (Haykin, 1999; p. 2)

Many other definitions are available, such as

> "A[n artificial] neural network is a circuit composed of a very large number of simple processing elements that are neurally based." (Nigrin, 1993; p. 11)

> "... neurocomputing is the technological discipline concerned with parallel, distributed, adaptive information processing systems that develop infor-

---

[1] Although neurocomputing can be viewed as a field of research dedicated to the *design* of brain-like computers, this chapter uses the word neurocomputing as a synonym to artificial neural networks.

mation processing capabilities in response to exposure to an information environment. The primary information processing structures of interest in neurocomputing are neural networks..." (Hecht-Nielsen, 1990, p. 2)

Neurocomputing systems are distinct from what is now known as *computational neuroscience*, which is mainly concerned with the development of biologically-based computational models of the nervous system. Artificial neural networks, on the other hand, take a loose inspiration from the nervous system and emphasize the problem solving capability of the systems developed. However, most books on computational neuroscience not only acknowledge the existence of artificial neural networks, but also use several ideas from them in the proposal of more biologically plausible models. They also discuss the ANN suitability as models of real biological nervous systems.

Neurons are believed to be the basic units used for computation in the brain, and their simplified abstract models are the basic processing units of neurocomputing devices or artificial neural networks. Neurons are connected to other neurons by a small junction called synapse, whose capability of being modulated is believed to be the basis for most of our cognitive abilities, such as perception, thinking, and inferring. Therefore, some essential information about neurons, synapses, and their structural anatomy are relevant for the understanding of how ANNs are designed taking inspiration from biological neural networks.

The discussion to be presented here briefly introduces the main aspects of the nervous system used to devise neurocomputing systems, and then focuses on some of the most commonly used artificial neural networks, namely, single- and multi-layer perceptrons, self-organizing networks, and Hopfield networks. The description of the many algorithms uses a matrix notation particularly suitable for the software implementation of the algorithms. Appendix B.1 provides the necessary background on linear algebra. The biological plausibility of each model is also discussed.

4.2   THE NERVOUS SYSTEM

All multicellular organisms have some kind of nervous system, whose complexity and organization varies according to the animal type. Even relatively simple organisms, such as worms, slugs, and insects, have the ability to learn and store information in their nervous systems. The nervous system is responsible for informing the organism through sensory input with regards to the environment in which it lives and moves, processing the input information, relating it to previous experience, and transforming it into appropriate actions or memories.

The nervous system plays the important role of processing the incoming information (signals) and providing appropriate actions according to these signals. The elementary processing units of the nervous system are the *neurons*, also called *nerve cells*. *Neural networks* are formed by the interconnection of many neurons. Each neuron in the human brain has on the order of hundreds or thousands of connections to other neurons.

Anatomically, the nervous system has two main divisions: *central nervous system* (CNS) and *peripheral nervous system* (PNS), the distinction being their different locations. Vertebrate animals have a *bony spine* (*vertebral column*) and a *skull* (*cranium*) in which the central parts of the nervous system are housed. The peripheral part extends throughout the remainder of the body. The part of the (central) nervous system located in the skull is referred to as the *brain*, and the one found in the spine is called the *spinal cord*. The brain and the spinal cord are continuous through an opening in the base of the skull; both are in contact with other parts of the body through the nerves.

The brain can be further subdivided into three main structures: the *brainstem*, the *cerebellum*, and the *forebrain*, as illustrated in Figure 4.1. The *brainstem* is literally the stalk of the brain through which pass all the nerve fibers relaying input and output signals between the spinal cord and higher brain centers. It also contains the cell bodies of neurons whose axons go out to the periphery to innervate the muscles and glands of the head. The structures within the brainstem are the *midbrain*, *pons,* and the *medulla*. These areas contribute to functions such as breathing, heart rate and blood pressure, vision, and hearing. The cerebellum is located behind the brainstem and is chiefly involved with skeletal muscle functions and helps to maintain posture and balance and provides smooth, directed movements. The *forebrain* is the large part of the brain remaining when the brainstem and cerebellum have been excluded. It consists of a central core, the *diencephalon*, and right and left *cerebral hemispheres* (the *cerebrum*).

The outer portion of the cerebral hemispheres is called *cerebral cortex*. The cortex is involved in several important functions such as thinking, voluntary movements, language, reasoning, and perception. The *thalamus* part of the diencephalon is important for integrating all sensory input (except smell) before it is presented to the cortex. The *hypothalamus*, which lies below the thalamus, is a tiny region responsible for the integration of many basic behavioral patterns, which involve correlation of neural and endocrine functions. Indeed, the hypothalamus appears to be the most important area to regulate the internal environment (homeostasis). It is also one of the brain areas associated with emotions.

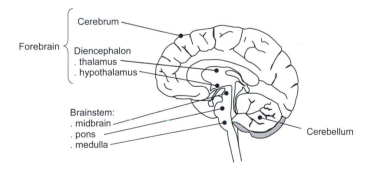

Figure 4.1: Structural divisions of the brain as seen in a midsagittal section.

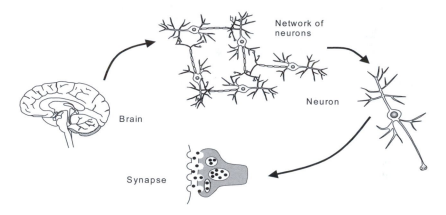

Figure 4.2: Some levels of organization in the nervous system.

4.2.1. Levels of Organization in the Nervous System

What structures really constitute a level of organization in the nervous system is an empirical not an *a priori* matter. We cannot tell, in advance of studying the nervous system, how many levels there are, nor what is the nature of the structural and functional features of any given level (Churchland and Sejnowski, 1992). Therefore, only the structures particularly interesting for the understanding, description, and implementation of artificial neural networks will be described here.

The nervous system can be organized in different levels: molecules, synapses, neurons, networks, layers, maps, and systems (Figure 4.2). An easily recognizable structure in the nervous system is the neuron, which is a cell specialized in signal processing. Depending on environmental conditions, the neurons are capable of generating a signal, more specifically an *electric potential*, that is used to transmit information to other cells to which it is connected. Some processes in the neuron utilize cascades of biochemical reactions that influence information processing in the nervous system. Many neuronal structures can be identified with specific functions. For instance, the *synapses* are important for the understanding of signal processing in the nervous system (Trappenberg, 2002).

Neurons and Synapses

Neurons use a variety of specialized biochemical mechanisms for information processing and transmission. These include *ion channels* that allow a controlled influx and outflux of currents, the generation and propagation of *action potentials*, and the release of *neurotransmitters*. Signal transmission between neurons is the core of the information processing capabilities of the brain. One of the most exciting discoveries in neuroscience was that the effectiveness of the signal transmission can be modulated in various ways, thus allowing the brain to adapt to different situations. It is believed to be the basis of associations, memories, and many other mental abilities (Trappenberg, 2002). *Synaptic plasticity*, that is the capability of synapses to be modified, is a key ingredient in most models described in this chapter.

Figure 4.3 shows a picture of a schematic generic neuron labeling its most important structural parts. The biological neuron is a single *cell*, thus containing a *cell body* with a *nucleus* (or *soma*) containing DNA, it is filled with fluid and cellular organelles, and is surrounded by a *cell membrane*, just like any other cell in the body. Neurons also have specialized extensions called *neurites*, that can be further distinguished into *dendrites* and *axons*. While the dendrites receive signals from other neurons, the axon propagates the output signal to other neurons.

One peculiarity about neurons is that they are specialized in signal processing utilizing special electrophysical and chemical processes. They can receive and send signals to many other neurons. The neurons that send signals, usually termed *sending* or *presynaptic neurons*, contact the *receiving* or *postsynaptic neurons* in specialized sites named *synapses* either at the cell body or at the dendrites. The synapse is thus the junction between the presynaptic neuron's axon and the postsynaptic neuron's dendrite or cell body.

The general information processing feature of synapses allow them to alter the state of a postsynaptic neuron, thus eventually triggering the generation of an electric pulse, called *action potential*, in the postsynaptic neuron. The action potential is usually initiated at the *axon hillock* and travels all along the axon, which can finally branch and send information to different regions of the nervous system. Therefore, a neuron can be viewed as a device capable of receiving diverse input stimuli from many other neurons and propagating its single output response to many other neurons.

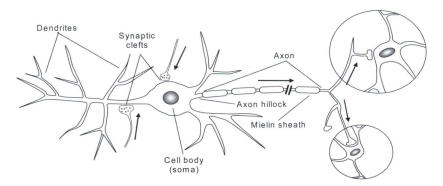

Figure 4.3: Schematic neuron similar in appearance to the pyramidal cells in the brain cortex. The parts outlined are the major structural components of most neurons. The direction of signal propagation between and within neurons is shown by the dark arrows.

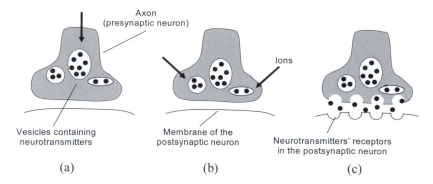

Figure 4.4: Diagram of a chemical synapse. The action potential arrives at the synapse (a) and causes ions to be mobilized in the axon terminal (b), thus causing vesicles to release neurotransmitters into the cleft, which in turn bind with the receptors on the postsynaptic neurons (c).

Various mechanisms exist to transfer information (signals) among neurons. As neurons are cells encapsulated in membranes, little openings in these membranes, called *channels*, allow the transfer of information among neurons. The basic mechanisms of information processing are based on the movement of charged atoms, *ions*, in and out of the channels and within the neuron itself. Neurons live in a liquid environment in the brain containing a certain concentration of ions, and the flow of ions in and out of a neuron through channels is controlled by several factors. A neuron is capable of altering the intrinsic electrical potential, the *membrane potential*, of other neurons, which is given by the difference between the electrical potential within and in the surroundings of the cell.

When an action potential reaches the terminal end of an axon, it mobilizes some ions by opening voltage-sensitive channels that allow the flow of ions into the terminal and possibly the release of some of these stored ions. These ions then promote the release of *neurotransmitters* (chemical substances) into the *synaptic cleft*, which finally diffuse across the cleft and binds with receptors in the postsynaptic neurons. As outcomes, several chemical processes might be initiated in the postsynaptic neuron, or ions may be allowed to flow into it. Figure 4.4 summarizes some of the mechanisms involved in synapse transmission.

As the electrical effects of these ions (action potential) propagate through the dendrites of the receiving neuron and up to the cell body, the process of information can begin again in the postsynaptic neuron. When ions propagate up to the cell body, these signals are *integrated* (summed), and the resulting membrane potential will determine if the neuron is going to *fire*, i.e., to send an output signal to a postsynaptic neuron. This only occurs if the membrane potential of the neuron is greater than the neuron *threshold*. This is because the channels are particularly sensitive to the membrane potential; they will only open when the potential is sufficiently large (O'Reilly and Munakata, 2000). The action of neural firing is also called *spiking, firing a spike*, or the *triggering of an action potential*. The spiking is a very important electrical response of a neuron; once generated it does not change its shape with increasing current. This phenomenon is called the *all-or-none* aspect of the action potential.

Different types of neurotransmitters and their associated ion channels have distinct effects on the state of the postsynaptic neuron. One class of neurotransmitters opens channels that will allow positively charged ions to enter the cell, thus triggering the increase of the membrane potential that drives the postsynaptic neurons towards their excited state. Other neurotransmitters initiate processes that drive the postsynaptic potential towards a resting state; a potential known as the *resting potential*. Therefore, neurotransmitters can promote the initiation of *excitatory* or *inhibitory* processes.

Networks, Layers, and Maps

Neurons can have *forward* and *feedback* connections to other neurons, meaning that they can have either one way or reciprocal connections with other neurons in the nervous system. These interconnected neurons give rise to what is known as *networks of neurons* or *neural networks*. For instance, within a cubic millimeter of cortical tissue, there are approximately 10^5 neurons and about 10^9 synapses, with the vast majority of these synapses arising from cells located within the cortex (Churchland and Sejnowski, 1992). Therefore, the degree of interconnectivity in the nervous system is quite high.

A small number of interconnected neurons (units) can exhibit complex behaviors and information processing capabilities that cannot be observed in single neurons. One important feature of neural networks is the representation of information (knowledge) in a *distributed* way, and the *parallel processing* of this information. No single neuron is responsible for storing "a whole piece" of knowledge; it has to be distributed over several neurons (connections) in the network. Networks with specific architectures and specialized information processing capabilities are incorporated into larger structures capable of performing even more complex information-processing tasks.

Many brain areas display not only networks of neurons but also *laminar organization*. Laminae are *layers of neurons* in register with other layers, and a given lamina conforms to a highly regular pattern of where it projects to and from where it receives projections. For instance, the *superior colliculus* (a particular layered midbrain structure; see Figure 4.1) receives visual inputs in superficial layers, and in deeper layers it receives tactile and auditory input. Neurons in an intermediate layer of the superior colliculus represent information about eye movements (Churchland and Sejnowski, 1992).

One of the most common arrangements of neurons in the vertebrate nervous systems is a layered two-dimensional structure organized with a *topographic* arrangement of unit responses. Perhaps the most well-known example of this is the mammalian cerebral cortex, though others such as the superior colliculus also use maps. The cerebral cortex is the outside surface of the brain. It is a two-dimensional structure extensively folded with fissures and hills in many larger or more intelligent animals. Two different kinds of cortex exist: an older form with three sub-layers called *paleocortex*, and a newer form that is most prominent in animals with more complex behavior, a structure with six or more sub-layers called *neocortex*.

Studies in human beings by neurosurgeons, neurologists, and neuropathologists have shown that different cortical areas have separate functions. Thus, it is possible to identify visible differences between different regions of the cerebral cortex, each presumably corresponding to a processing module. Figure 4.5 illustrates some specific functional areas in the cerebral cortex of humans. Note that regions associated with different parts of the body can have quite sharp boundaries.

In general, it is known that neocortical neurons are organized into six distinct layers, which can be subdivided into *input, hidden,* and *output* layer. The input layer usually receives the sensory input, the output layer sends commands and outputs to other portions of the brain, and the hidden layers receive inputs locally from other cortical layers. This means that hidden layers neither directly receive sensory stimuli nor produce motor and other outputs.

One major principle of organization within many sensory and motor systems is the *topographic map*. For instance, neurons in visual areas of the cortex (in the rear end of the cortex, opposite to the eyes; see Figure 4.5) are arranged topographically, in the sense that adjacent neurons have adjacent visual receptive fields and collectively they constitute a map of the retina. Because neighboring processing units are concerned with similar representations, topographic mapping is an important means whereby the brain manages to save on wiring and also to share wire.

Networking, topographic mapping, and layering are all special cases of a more general principle: the exploitation of geometric and structural properties in information processing design. Evolution has shaped our brains so that these structural organizations became efficient ways for biological systems to assemble in one place information needed to solve complex problems.

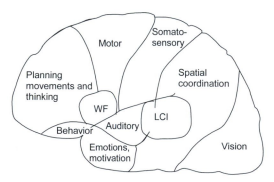

Figure 4.5: Map of specific functional areas in the cerebral cortex. WF: word formation; LCI: Language comprehension and intelligence.

4.2.2. Biological and Physical Basis of Learning and Memory

The nervous system is continuously modifying and updating itself. Virtually all its functions, including perception, motor control, thermoregulation, and reasoning, are modifiable by experience. The topography of the modifications appears not to be final and finished, but an ongoing process with a virtually interminable schedule. Behavioral observations indicate degrees of plasticity in the nervous system: there are fast and easy changes, slower and deeper modifiability, and more permanent but still modifiable changes.

In general, global learning is a function of local changes in neurons. There are many possible ways a neuron could change to embody adaptations. For instance, new dendrites might sprout out, or there might be extension of existing branches, or existing synapses could change, or new synapses might be created. In the other direction, pruning could decrease the dendrites or bits of dendrites, and thus decrease the number of synapses, or the synapses on the remaining branches could be shut down altogether. These are all postsynaptic changes in the dendrites. There could also be changes in the axons; for instance, there might be changes in the membrane, or new branches might be formed, and genes might be induced to produce new neurotransmitters or more of the old ones. Presynaptic changes could include changes in the number of vesicles released per spike and the number of transmitter molecules contained in each vesicle. Finally, the whole cell might die; taking with it all the synapses it formerly supported (Churchland and Sejnowsky, 1992).

This broad range of structural adaptability can be conveniently condensed in the present discussion by referring simply to synapses, since every modification either involves synaptic modification directly or indirectly, or can be reasonably so represented. Learning by means of setting synaptic efficiency is thus the most important mechanism in neural networks, biological and artificial. It depends on both individual neuron-level mechanisms and network-level principles to produce an overall network that behaves appropriately in a given environment.

Two of the primary mechanisms underlying learning in the nervous system are the *long-term potentiation* (LTP) and *long-term depression* (LTD), which refer to the strengthening and weakening of weights in a nontransient form. Potentiation corresponds to an increase in the measured depolarization or excitation delivered by a controlled stimulus onto a receiving neuron, and depression corresponds to a decrease in the measured depolarization. In both cases, the excitation or inhibition of a membrane potential may trigger a complex sequence of events that ultimately result in the modification of synaptic efficiency (strength).

Such as learning, *memory* is an outcome of an adaptive process in synaptic connections. It is caused by changes in the synaptic efficiency of neurons as a result of neural activity. These changes in turn cause new pathways or facilitated pathways to develop for transmission of signals through the neural circuits of the brain. The new or facilitated pathways are called *memory traces*; once established, they can be activated by the thinking mind to reproduce the memories. Actually, one of the outcomes of a learning process can be the creation of a more permanent synaptic modification scheme, thus resulting in the memorization of an experience.

Memories can be classified in a number of ways. One common classification being into:

- *Short-term memory*: lasts from a few seconds to a few minutes, e.g., one's memory of a phone number.

- *Intermediate long-term memory*: lasts from minutes to weeks, e.g., the name of a nice girl/boy you met in a party.

- *Long-term memory*: lasts for an indefinite period of time, e.g., your home address.

While the two first types of memories do not require many changes in the synapses, long-term memory is believed to require *structural changes* in synapses. These structural changes include an increase in number of vesicle release sites for secretion of neurotransmitter substances, an increase in the number of presynaptic terminals, an increase in the number of transmitter vesicles, and changes in the structures of the dendritic spines.

Therefore, the difference between learning and memory may be sharp and conceptual. Learning can be viewed as the adaptive process that results in the change of synaptic efficiency and structure, while memory is the (long-lasting) result of this adaptive process.

4.3   ARTIFICIAL NEURAL NETWORKS

Artificial neural networks (ANN) present a number of features and performance characteristics in common with the nervous system:

- The basic information processing occurs in many simple elements called (artificial) *neurons*, *nodes* or *units*.

- These neurons can *receive* and *send stimuli* from and to other neurons and the environment.

- Neurons can be connected to other neurons thus forming networks of neurons or *neural networks*.

- Information (signals) are transmitted between neurons via connection links called *synapses*.

- The efficiency of a synapse, represented by an associated *weight value* or *strength*, corresponds to the information stored in the neuron, thus in the network.

- Knowledge is acquired from the environment by a process known as *learning*, which is basically responsible for *adapting* the connection strengths (weight values) to the environmental stimuli.

One important feature of artificial neural networks is where the knowledge is stored. Basically, what is stored is the connection strengths (synaptic strengths) between units (artificial neurons) that allow patterns to be recreated. This feature has enormous implications, both for processing and learning. The knowledge representation is set up so that the knowledge necessarily influences the course of processing; it becomes a part of the processing itself. If the knowledge is incorporated into the strengths of the connections, then learning becomes a matter of finding the appropriate connection strengths so as to produce satisfactory patterns of activation under some circumstances.

This is an extremely important feature of ANNs, for it opens up the possibility that an information processing mechanism could *learn*, by tuning its connections strengths, to capture the interdependencies between activations presented to it in the course of processing. Another important implication of this type of representation is that *the knowledge is distributed* over the connections among a large number of units. There is no 'special' unit reserved for particular patterns.

An artificial neural network can be characterized by three main features: 1) a set of *artificial neurons*, also termed *nodes*, *units*, or simply *neurons*; 2) the pattern of connectivity among neurons, called the network *architecture* or *structure*; and 3) a method to determine the weight values, called its *training* or *learning* algorithm. Each one of these features will be discussed separately in the following sections.

4.3.1. Artificial Neurons

In the biological neuron, inputs come into the cell primarily through channels located in synapses, allowing ions to flow into and out of the neuron. A membrane potential appears as a result of the integration of the neural inputs, and will then determine whether a given neuron will produce a spike (action potential) or not. This spike causes neurotransmitters to be released at the end of the axon, which then forms synapses with the dendrites of other neurons. The action potential only occurs when the membrane potential is above a critical threshold level. Different inputs can provide different amounts of activation depending on how much neurotransmitter is released by the sender and how many channels in the postsynaptic neuron are opened. Therefore, there are important features of the biological synapses involved in the information processing of neurons.

The net effect of these biological processes is summarized in the computational models discussed here by a *weight* (also called *synaptic strength*, *synaptic efficiency*, *connection strength*, or *weight value*) between two neurons. Furthermore, the modification of one or more of these weight factors will have a major contribution for the neural network learning process. This section reviews three models of neuronal function. The first is the McCulloch-Pitts model in which the neuron is assumed to be computing a logic function. The second is a simple analog *integrate-and-fire* model. And the third is a generic connectionist neuron, which integrates its inputs and generates an output using one particular *activation function*. These neuronal models may be interchangeably called *nodes*, *units*, *artificial neurons*, or simply neurons. It is important to have in mind, though, that the nodes most commonly used in artificial neural networks bear a far resemblance with real neurons.

The McCulloch and Pitts Neuron

W. McCulloch and W. Pitts (1943) wrote a famous and influential paper based on the computations that could be performed by two-state neurons. They did one of the first attempts to understand nervous activity based upon elementary neural computing units, which were highly abstract models of the physiological properties of neurons and their connections. Five physical assumptions were made for their calculus (McCulloch and Pitts, 1943):

1. The behavior of the neuron is a binary process.
2. At any time a number of synapses must be excited in order to activate the neuron.
3. Synaptic delay is the only significant delay that affects the nervous system.
4. The excitation of a certain neuron at a given time can be inhibited by an inhibitory synapse.
5. The neural network has a static structure; that is, a structure that does not change with time.

McCulloch and Pitts conceived the neuronal response as being equivalent to a *proposition* adequate to the neuron's stimulation. Therefore, they studied the behavior of complicated neural networks using a notation of the symbolic *logic of propositions* (see Appendix B.4.2). The 'all-or-none' law of nervous activity was sufficient to ensure that the activity of any neuron could be represented as a proposition.

It is important to note that according to our current knowledge of how the neuron works - based upon electrical and chemical processes - neurons are not realizing any proposition of logic. However, the model of McCulloch and Pitts can be considered as a special case of the most general neuronal model to be discussed in the next sections, and is still sometimes used to study particular classes of nonlinear networks. Additionally, this model has caused a major impact mainly among computer scientists and engineers, encouraging the development of artificial neural networks. It has even been influential in the history of computing (cf. von Neumann, 1982).

The McCulloch and Pitts neuron is binary, i.e., it can assume only one of two states (either '0' or '1'). Each neuron has a fixed *threshold* θ and receives inputs from synapses of identical weight values. The neuronal mode of operation is simple. At each time step t, the neuron responds to its synaptic inputs, which reflect the state of the presynaptic neurons.

If no inhibitory synapse is active, the neuron integrates (sums up) its synaptic inputs, generating the *net input* to the neuron, u, and checks if this sum (u) is greater than or equal to the threshold θ. If it is, then the neuron becomes active, that is, responds with a '1' in its output ($y = 1$); otherwise it remains inactive, that is, responds with a '0' in its output ($y = 0$).

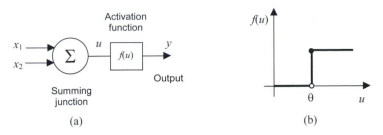

Figure 4.6: The McCulloch and Pitts neuron (a) and its threshold activation function (b).

a	b	a AND b	a	b	a OR b	a	NOT a
0	0	0	0	0	0	0	1
0	1	0	0	1	1	1	0
1	0	0	1	0	1		
1	1	1	1	1	1		

Figure 4.7: Truth tables for the connectives AND, OR, and NOT.

Although this neuron is quite simple, it already presents some important features in common with most neuron models, namely, the integration of the input stimuli to determine the *net input u* and the presence of an activation function (threshold). Figure 4.6 illustrates the simple McCulloch and Pitts neuron and its activation function.

To illustrate the behavior of this simple unit, assume two excitatory inputs x_1 and x_2 and a threshold $\theta = 1$. In this case, the neuron is going to fire; that is, to produce an output '1', every time x_1 or x_2 has a value '1', thus operating like the logical connective OR (see Figure 4.7). Assume now that the neuron threshold is increased to $\theta = 2$. In this new situation, the neuron is only going to be active (fire) if both x_1 and x_2 have value '1' simultaneously, thus operating like the logical connective AND (see Figure 4.7).

A Basic Integrate-and-Fire Neuron

Assume a noise-free neuron with the net input being a variable of time $u(t)$ corresponding to the membrane potential of the neuron. The main effects of some neuronal channels (in particular the sodium and leakage channels) can be captured by a simple equation of an integrator (Dayan and Abbot, 2001; Trappenberg, 2002):

$$\tau_m \frac{du(t)}{dt} = u_{res} - u(t) + R_m i(t) \tag{4.1}$$

where τ_m is the membrane time constant of the neuron determined by the average conductance of the channels (among other things); u_{res} is the resting potential of the neuron; i is the input current given by the sum of the synaptic currents generated by firings of presynaptic neurons; R_m is the resistance of the neuron to the flow of current (ions); and t is the time index.

Equation (4.1) can be very simply understood. The rate of variation of the membrane potential of the neuron is proportional to its current membrane potential, its resting potential, and the potential generated by the incoming signals to the neuron. Note that the last term on the right side of this equation is the Ohm's law ($u = R.i$) for the voltage generated by the incoming currents.

The input current $i(t)$ to the neuron is given by the sum of the incoming synaptic currents depending on the *efficiency* of individual synapses, described by the variable w_j for each synapse j. Therefore, the total input current to the neuron can be written as the sum of the individual synaptic currents multiplied by a weight value

$$i(t) = \sum_j \sum_{t_j^f} w_j f(t - t_j^f)$$ (4.2)

where the function $f(\cdot)$ parameterizes the form of the postsynaptic response. This function was termed *activation function* in the McCulloch and Pitts neuron discussed above and this nomenclature will be kept throughout this text. The variable t_j^f denotes the firing time of the presynaptic neuron of synapse j. The firing time of the postsynaptic neuron is defined by the time the membrane potential u reaches a threshold value θ,

$$u(t^f) = \theta$$ (4.3)

In contrast to the firing time of the presynaptic neurons, the firing time of the integrate-and-fire neuron has no index. To complete the model, the membrane potential has to be reset to the resting state after the neuron has fired. One form of doing this is by simply resetting the membrane potential to a fixed value u_{res} immediately after a spike.

The Generic Neurocomputing Neuron

The computing element employed in most neural networks is an integrator, such as the McCulloch and Pitts and the integrate-and-fire models of a neuron, and computes based on its connection strengths. Like in the brain, the artificial neuron is an information-processing element that is fundamental to the operation of the neural network. Figure 4.8 illustrates the typical artificial neuron, depicting its most important parts: the synapses, characterized by their *weight values* connecting each input to the neuron; the *summing junction* (*integrator*); and the *activation function*.

Specifically, an input signal x_j at the input of synapse j connected to neuron k is multiplied by the synaptic weight w_{kj}. In this representation, the first subscript of the synaptic weight refers to the neuron, and the second subscript refers to the synapse connected to it. The summing junction adds all input signals weighted by the synaptic weight values plus the neuron's bias b_k; this operation constitutes the dot (or inner) product (see Appendix B.1); that is, a linear combination of the inputs with the weight values, plus the bias b_k. Finally, an activation function is

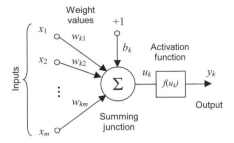

Figure 4.8: Nonlinear model of a neuron.

used to limit the amplitude of the output of the neuron. The activation function is also referred to as a *squashing function* (Rumelhart et al., 1986) for it limits the permissible amplitude range of the output signal to some finite value.

The bias has the effect of increasing or decreasing the net input to the activation function depending on whether it is positive or negative, respectively. In the generic neuron it is usually used in place of a fixed threshold θ for the activation function. For instance, the McCulloch and Pitts neuron (Figure 4.6) will fire when its net input is greater than θ:

$$y = f(u) = \begin{cases} 1 & \text{if } u \geq \theta \\ 0 & \text{otherwise} \end{cases}$$

where $u = x_1 + x_2$ for the neuron of Figure 4.6. It is possible to replace the threshold θ by a bias weight b that will be multiplied by a constant input of value '1':

$$y = f(u) = \begin{cases} 1 & \text{if } u \geq 0 \\ 0 & \text{otherwise} \end{cases}$$

where $u = x_1 + x_2 - b$. Note however, that while the bias is going to be adjusted during learning, the threshold assumes a fixed value.

It is important to note here that the output of this generic neuron is simply a number, and the presence of discrete action potentials is ignored. As real neurons are limited in dynamic range from zero-output firing rate to a maximum of a few hundred action potentials per second, the use of an activation function could be biologically justified.

Mathematically, the neuron k can be described by a simple equation:

$$y_k = f(u_k) = f\left(\sum_{j=1}^{m} w_{kj} x_j + b_k \right) \tag{4.4}$$

where x_j, $j = 1, \ldots, m$, are the input signals; w_{kj}, $j = 1, \ldots, m$, are the synaptic weights of neuron k; u_k is the net input to the activation function; b_k is the bias of neuron k; $f(\cdot)$ is the activation function; and y_k is the output signal of the neuron.

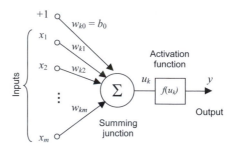

Figure 4.9: Reformulated model of a neuron.

It is possible to simplify the notation of Equation (4.4) so as to account for the presence of the bias by simply defining a constant input signal $x_0 = 1$ connected to the neuron k with associated weight value $w_{k0} = b_k$. The resulting equation becomes,

$$y_k = f(u_k) = f\left(\sum_{j=0}^{m} w_{kj}x_j\right) \tag{4.5}$$

Figure 4.9 illustrates the reformulated model of a neuron.

Types of Activation Function

The activation function, denoted by $f(u_k)$, determines the output of a neuron k in relation to its net input u_k. It can thus assume a number of forms, of which some of the most frequently used are summarized below (see Figure 4.10):

- *Linear function*: relates the net input directly with its output.

$$f(u_k) = u_k \tag{4.6}$$

- *Step or threshold function*: the output is '1' if the net input is greater than or equal to a given threshold θ; otherwise it is either '0' or '−1', depending on the use of a binary $\{0,1\}$ or a bipolar $\{-1,1\}$ function. The binary step function is the one originally used in the McCulloch and Pitts neuron:

$$f(u_k) = \begin{cases} 1 & \text{if } u_k \geq \theta \\ 0 & \text{otherwise} \end{cases} \tag{4.7}$$

- *Signum function*: it is similar to the step function, but has a value of '0' when the net input to the neuron is $u = 0$. It can also be either bipolar or binary; the bipolar case being represented in Equation (4.8).

$$f(u_k) = \begin{cases} 1 & \text{if } u_k > \theta \\ -1 & \text{if } u_k < \theta \end{cases} \tag{4.8}$$

- *Sigmoid function*: it is a strictly increasing function that presents saturation and a graceful balance between linear and nonlinear behavior. This is the most commonly used activation function in the ANN literature. It has an s-shaped form and can be obtained by several functions, such as the *logistic function*, the *arctangent function*, and the *hyperbolic tangent* function; the difference being that, as u_k ranges from $-\infty$ to $+\infty$, the logistic function ranges from 0 to 1, the arctangent ranges from $-\pi/2$ to $+\pi/2$, and the hyperbolic tangent ranges from -1 to $+1$:

$$Logistic: f(u_k) = \frac{1}{1 + \exp(-u_k)} \tag{4.9}$$

- *Radial basis function*: it is a nonmonotonic function that is symmetric around a base value. This is the main type of function for *radial basis function neural networks*. The expression shown below is that of the Gaussian bell curve:

$$f(u_k) = \exp(-u_k^2) \tag{4.10}$$

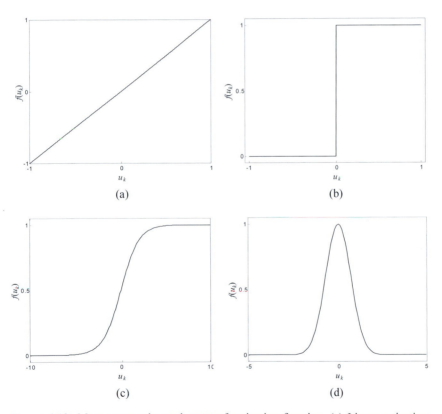

(a)   (b)

(c)   (d)

Figure 4.10: Most commonly used types of activation function. (a) Linear activation, $u_k \in [-1,+1]$. (b) Step function with $\theta = 0$, $u_k \in [-1,+1]$. (c) Logistic function, $u_k \in [-10,+10]$. (d) Gaussian bell curve, $u_k \in [-5,+5]$.

It is important to stress that the functions described and illustrated in Figure 4.10 depict only the general shape of those functions. However, it should be clear that these basic shapes can be modified. All these functions can include parameters with which some features of the functions can be changed, such as the *slope* and *offset* of a function. For instance, by multiplying u_k in the logistic function by a value β, it is possible to control the smoothness of the function:

$$f(u_k) = \frac{1}{1 + \exp(-\beta.u_k)}$$

In this case, for larger values of β, this function becomes more similar to the threshold function, while for smaller values of β, the logistic function becomes more similar to a linear function.

4.3.2. Network Architectures

One of the basic prerequisites for the emergence of complex behaviors is the interaction of a number of individual agents. In the nervous system, it is known that individual neurons can affect the behavior (firing rate) of others, but, as a single entity, a neuron is meaningless. This might be one reason why evolution drove our brains to a system with such an amazingly complex network of interconnected neurons.

In the human brain not much is known of how neurons are interconnected. Some particular knowledge is available for specific brain portions, but little can be said about the brain as a whole. For instance, it is known that the cortex can be divided into a number of different cortical areas specialized in different kinds of processing; some areas perform pattern recognition, others process spatial information, language processing, and so forth. In addition, it is possible to define anatomic neuronal organizations in the cortex in terms of *layers of neurons*, which are very important for the understanding of the detailed biology of the cortex (Figure 4.5).

In the domain of artificial neural networks, a layer of neurons will refer to functional layers of nodes. As not much is known about the biological layers of neurons, most neurocomputing networks employ some standardized architectures, specially designed for the engineering purposes of solving problems. There are basically three possible layers in a network: an *input layer*, one or more *intermediate* (or *hidden*) *layers*, and an *output layer*. These are so-called because the input layer receives the input stimuli directly from the environment, the output layer places the network output(s) to the environment, and the hidden layer(s) is not in direct contact with the environment.

The way in which neurons are structured (interconnected) in an artificial neural network is intimately related with the learning algorithm that is used for training the network. In addition, this interconnectivity affects the network storage and learning capabilities. In general, it is possible to distinguish three main types of network architectures (Haykin, 1999): *single-layer feedforward networks, multi-layer feedforward networks,* and *recurrent networks*.

Single-Layer Feedforward Networks

The simplest case of layered networks consists of an input layer of nodes whose output feeds the output layer. Usually, the input nodes are linear, i.e., they simply propagate the input signals to the output layer of the network. In this case, the input nodes are also called sensory units because they only sense (receive information from) the environment and propagate the received information to the next layer. In contrast, the output units are usually processing elements, such as the neuron depicted in Figure 4.9, with a nonlinear type of activation function. The signal propagation in this network is purely positive or *feedforward*; that is, signals are propagated from the network input to its outputs and never the opposite way (*backward*). This architecture is illustrated in Figure 4.11(a) and the direction of signal propagation in Figure 4.11(b). In order not to overload the picture, very few connection strengths were depicted in Figure 4.11(a), but these might be sufficient to give a general idea of how the weights are assigned in the network.

The neurons presented in Figure 4.11 are of the generic neuron type illustrated in Figure 4.9. (Note that input x_0 is assumed to be fixed in '1' and the weight vector that connects it to all output units w_{i0}, $i = 1, \ldots, o$, corresponds to the bias of each output node $b_i = w_{i0}$, $i = 1, \ldots, o$.) The weight values between the inputs and the output nodes can be written using matrix notation (Appendix B.1) as

$$\mathbf{W} = \begin{bmatrix} w_{10} & w_{11} & \cdots & w_{1m} \\ \vdots & \vdots & \ddots & \vdots \\ w_{o0} & w_{o1} & \cdots & w_{om} \end{bmatrix} \tag{4.11}$$

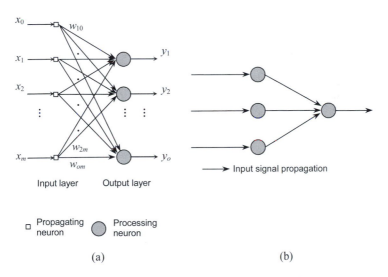

Input layer   Output layer

□ Propagating neuron   ● Processing neuron

(a)   (b)

Figure 4.11: *Single-layer feedforward* neural network. (a) Network architecture. (b) Direction of propagation of the input signal.

where the first index i, $i = 1, \dots, o$, of each element corresponds to the postsynaptic node and the second index j, $j = 0, \dots, m$, corresponds to the presynaptic node (just remember 'to-from' while reading from left to right). Therefore, each row of matrix $\mathbf{W}$ corresponds to the weight vector of an output unit, and each element of a given column j, $j = 0, \dots, m$, corresponds to the strength with which an input j is connected to each output unit i. Remember that $j = 0$ corresponds to the unitary input that will multiply the neuron's bias.

The output of each postsynaptic neuron i in Figure 4.11(a), $i = 1, \dots, o$, is computed by applying an activation function $f(\cdot)$ to its net input, which is given by the linear combination of the network inputs and the weight vector,

$$y_i = f(\mathbf{w}_i.\mathbf{x}) = f(\Sigma_j\, w_{ij}.x_j)\, , j = 0, \dots, m \qquad (4.12)$$

The output vector of the whole network $\mathbf{y}$ is given by the activation function applied to the product of the weight matrix $\mathbf{W}$ by the input vector $\mathbf{x}$,

$$\mathbf{y} = f(\mathbf{W}.\mathbf{x}) \qquad (4.13)$$

Assuming the general case where the weight and input vectors can take any real value, the dimension of the weight matrix and each vector is $\mathbf{W} \in \mathfrak{R}^{o \times (m+1)}$, $\mathbf{w}_i \in \mathfrak{R}^{1 \times (m+1)}$, $i = 1, \dots, o$, $\mathbf{x} \in \mathfrak{R}^{(m+1) \times 1}$, and $\mathbf{y} \in \mathfrak{R}^{o \times 1}$.

Multi-Layer Feedforward Networks

The second class of feedforward neural networks is known as multi-layer networks. These are distinguished from the single-layered networks by the presence of one or more *intermediate* or *hidden* layers. By adding one or more hidden layers, the nonlinear computational processing and storage capability of the network is increased, for reasons that will become clearer further in the text. The output of each network layer is used as input to the following layer. A multi-layer feedforward network can be thought of as an assembly line: some basic material is introduced into the production line and passed on, the second stage components are assembled, and then the third stage, up to the last or output stage, which delivers the final product.

The learning algorithm typically used to train this type of network requires the *backpropagation* of an error signal calculated between the network output and a desired output. This architecture is illustrated in Figure 4.12(a) and the direction of signal propagation is depicted in Figure 4.12(b).

In such networks, there is one weight matrix for each layer, and these are going to be denoted by the letter $\mathbf{W}^k$ with a superscript k indicating the layer. Layers are counted from left to right, and the subscripts of this matrix notation remain the same as previously. Therefore, w_{ij}^k corresponds to the weight value connecting the postsynaptic neuron i to the presynaptic neuron j at layer k.

In the network of Figure 4.12(a), $\mathbf{W}^1$ indicates the matrix of connections between the input layer and the first hidden layer; matrix $\mathbf{W}^2$ contains the connections between the first hidden layer and the second hidden layer; and matrix $\mathbf{W}^3$ contains the connections between the second hidden layer and the output layer.

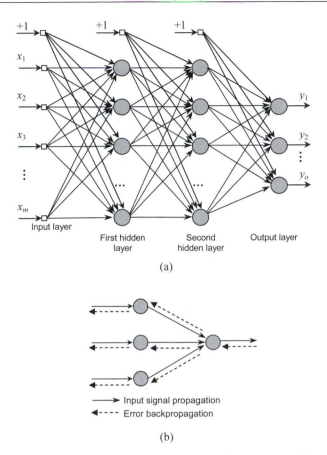

(a)

(b)

Figure 4.12: *Multi-layer feedforward* neural network. (a) Network architecture (legend as depicted in Figure 4.11). (b) Direction of propagation of the functional and error signals. (The weight values were suppressed from the picture not to overload it.)

In feedforward networks the signals are propagated from the inputs to the network output in a layer-by-layer form (from left to right). Therefore, the network output is given by (in matrix notation):

$$\mathbf{y} = \mathbf{f}^3(\mathbf{W}^3 \, \mathbf{f}^2(\mathbf{W}^2 \, \mathbf{f}^1(\mathbf{W}^1 \mathbf{x}))) \tag{4.14}$$

where $\mathbf{f}^k$ is the vector of activation functions of layer k (note that nodes in a given layer may have different activation functions); $\mathbf{W}^k$ is the weight matrix of layer k and $\mathbf{x}$ is the input vector.

It can be observed from Equation (4.14) that the network output is computed by recursively multiplying the weight matrix of a given layer by the output produced by each previous layer. The expression to calculate the output of each layer is given by Equation (4.12) using the appropriate weight matrix and inputs.

It is important to note that if the intermediate nodes have linear activation functions there is no point in adding more layers to this network, because $f(x) = x$ for linear functions, and thus the network output would be given by

$$\mathbf{y} = \mathbf{f}^3(\mathbf{W}^3\,\mathbf{W}^2\,\mathbf{W}^1\mathbf{x}).$$

and this expression can be reduced to,

$$\mathbf{y} = \mathbf{f}^3(\mathbf{Wx}),$$

where $\mathbf{W} = \mathbf{W}^3\,\mathbf{W}^2\,\mathbf{W}^1$.

Therefore, if one wants to increase the computational capabilities of a multi-layer neural network by adding more layers to the network, nonlinear activation functions have to be used in the hidden layers.

Recurrent Networks

The third class of networks is known as recurrent networks, distinguished from feedforward networks for they have at least one *recurrent* (or *feedback*) loop. For instance, a recurrent network may consist of a single layer of neurons with each neuron feeding its output signal back to the input of other neurons, as illustrated in Figure 4.13. In a feedback arrangement, there is communication between neurons; there can be cross talk and plan revision; intermediate decisions can be taken; and a mutually agreeable solution can be found.

Note that the type of feedback loop illustrated in Figure 4.13 is distinguished from the backpropagation of error signals briefly mentioned in the previous section. The recurrent loop has an impact on the network learning capability and performance because it involves the use of particular branches composed of retard units (Z^{-1}) resulting in a nonlinear dynamic behavior (assuming that the network has nonlinear units).

The network illustrated in Figure 4.13 is fully recurrent, in the sense that all network units have feedback connections linking them to all other units in the network. This network also assumes an input vector $\mathbf{x}$ weighted by the weight vector $\mathbf{w}$ (not shown).

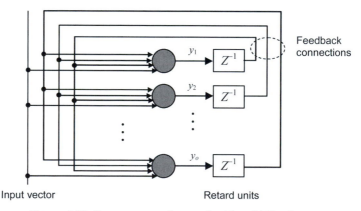

Figure 4.13: Recurrent neural network with no hidden layer.

The network output y_i, $i = 1, \ldots, o$, is a composed function of the weighted inputs at iteration t, plus the weighted outputs in the previous iteration $(t - 1)$:

$$y_i(t) = f(\mathbf{w}_i.\mathbf{x}(t) + \mathbf{v}_i.\mathbf{y}(t-1)) = f(\Sigma_j \ w_{ij}.x_j(t) + \Sigma_k \ v_{ik}.y_k(t-1)) \ ,$$
$$j = 1, \ldots, m; \ k = 1, \ldots, o$$

(4.15)

where $\mathbf{v}_i$, $i = 1, \ldots, o$ is the vector weighting the retarded outputs fed back into the network ($\mathbf{W} \in \mathfrak{R}^{o \times m}$, $\mathbf{w}_i \in \mathfrak{R}^{1 \times m}$, $i = 1, \ldots, o$, $\mathbf{x} \in \mathfrak{R}^{m \times 1}$, $\mathbf{y} \in \mathfrak{R}^{o \times 1}$, $\mathbf{V} \in \mathfrak{R}^{o \times o}$, and $\mathbf{v}_i \in \mathfrak{R}^{1 \times o}$).

4.3.3. Learning Approaches

It has been discussed that one of the most exciting discoveries in neuroscience was that synaptic efficiency could be modulated to the input stimuli. Also, this is believed to be the basis for learning and memory in the brain. Learning thus involves the adaptation of synaptic strengths to environmental stimuli. Biological neural networks are known not to have their architectures much altered throughout life. New neurons cannot routinely be created to store new knowledge, and the number of neurons is roughly fixed at birth, at least in mammals. The alteration of synaptic strengths may thus be the most relevant factor for learning. This change could be the simple modification in strength of a given synapse, the formation of new synaptic connections or the elimination of pre-existing synapses. However, the way learning is accomplished in the biology of the brain is still not much clear.

In the neurocomputing context, *learning* (or *training*) corresponds to the process by which the network's "free parameters" are adapted (adjusted) through a mechanism of presentation of environmental (or input) stimuli. In the standard neural network learning algorithms to be discussed here, these free parameters correspond basically to the connection strengths (weights) of individual neurons. It is important to have in mind though, that more sophisticated learning algorithms are capable of dynamically adjusting several other parameters of an artificial neural network, such as the network architecture and the activation function of individual neurons. The environmental or input stimuli correspond to a set of *input data* (or *patterns*) that is used to *train* the network.

Neural network learning basically implies the following sequence of events:

- Presentation of the input patterns to the network.
- Adaptation of the network free parameters so as to produce an altered pattern of response for the input data.

In most neural network applications the network weights are first adjusted according to a given learning rule or algorithm, and then the network is applied to a new set of input data. In this case there are two steps involved in the use of a neural network: 1) network training, and 2) network application.

With standard learning algorithms a neural network learns through an iterative process of weight adjustment. The type of learning is defined by the way in which the weights are adjusted. The three main learning approaches are: 1) *supervised learning*, 2) *unsupervised learning*, and 3) *reinforcement learning*.

Supervised Learning

This learning strategy embodies the concept of a *supervisor* or *teacher*, who has the knowledge about the environment in which the network is operating. This knowledge is represented in the form of a set of *input-output samples* or *patterns*. Supervised learning is typically used when the class of data is known *a priori* and this can be used as the supervisory mechanism, as illustrated in Figure 4.14. The network free parameters are adjusted through the combination of the input and error signals, where the error signal is the difference between the desired output and the current network output.

To provide the intuition behind supervised learning, consider the following example. Assume you own an industry that produces patterned tiles. In the quality control (QC) part of the industry, an inspection has to be made in order to guarantee the quality in the patterns printed on the tiles. There are a number of tiles whose patterns are assumed to be of good quality, and thus pass the QC. These tiles can be used as input samples to train the artificial neural network to classify good quality tiles. The known or desired outputs are the respective classification of the tiles as good or bad quality tiles.

In the artificial neural network implementation, let neuron j be the only output unit of a feedforward network. Neuron j is stimulated by a signal vector $\mathbf{x}(t)$ produced by one or more hidden layers, that are also stimulated by an input vector. Index t is the discrete time index or, more precisely, the time interval of an iterative process that will be responsible for adjusting the weights of neuron j. The only output signal $y_j(t)$, from neuron j, is compared with a *desired output*, $d_j(t)$. An error signal $e_j(t)$ is produced:

$$e_j(t) = d_j(t) - y_j(t) \tag{4.16}$$

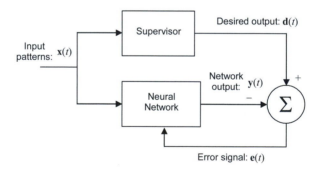

Error signal: $\mathbf{e}(t)$

Figure 4.14: In supervised learning, the environment provides input patterns to train the network and a supervisor has the knowledge about the environment in which the network is operating. At each time step t, the network output is compared with the desired output (the error is used to adjust the network response).

The error signal acts as a control mechanism responsible for the application of corrective adjustments in the weights of neuron j. The goal of supervised learning is to make the network output more similar to the desired output at each time step; that is, to *correct the error* between the network output and the desired output. This objective can be achieved by minimizing a *cost function* (also called *performance index*, *error function*, or *objective function*), $\Im(t)$, which represents the instant value of the error measure:

$$\Im(t) = \frac{1}{2} e_j^2(t) \qquad (4.17)$$

Generalization Capability

Once the supervised learning process is complete, it is possible to present an *unseen* pattern to the network, which will then classify that pattern, with a certain degree of accuracy, into one of the classes used in the training process. This is an important aspect of an artificial neural network, its *generalization capability*. In the present context, *generalization* refers to the performance of the network on (new) patterns that were not used in the network learning process.

This can be illustrated with the help of the following example. Assume we want to use a multi-layer feedforward neural network to approximate the function $sin(x).cos(2x)$ depicted by a solid line in Figure 4.15(a). It is only necessary a few training samples to adjust the weight vectors of a multi-layer feedforward neural network with a single hidden layer composed of five sigmoidal units. The network contains a single input and a single output unit, but these details are not relevant for the present discussion.

It is known that most real-world data contains noise, i.e., irrelevant or meaningless data, or data with disturbances. To investigate the relationship between noisy data and the network capability of approximating these data, some noise was added to the input data by simply adding a uniform distribution of zero mean and variance 0.15 to each input datum. Figure 4.15(a) shows the input data and the curve representing the original noise-free function $sin(x).cos(2x)$. Note that the noisy data generated deviates a little from the function to be approximated.

In this case, if the neural network is not trained sufficiently or if the network architecture is not appropriate, the approximation it will provide for the input data will not be satisfactory; an *underfitting* will occur (Figure 4.15(b)). In the opposite case, when the network is trained until it perfectly approximates the input data, some *overfitting* will occur, meaning that the approximation is too accurate for this noisy data set (Figure 4.15(c)). A better training is the one that establishes a compromise between the approximation accuracy and a good generalization for unseen data (Figure 4.15(d)).

From a biological perspective, generalization is very important for our creation of models of the world. Think of generalization according to the following example. If you just memorize some specific facts about the world instead of trying to extract some simple essential regularity underlying these facts, then you would be in trouble when dealing with novel situations where none of the specifics appear. For instance, if you memorize a situation where you almost

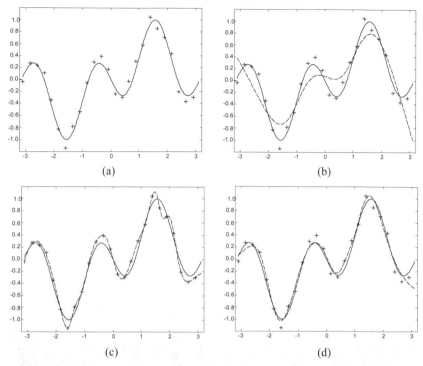

Figure 4.15: Function $sin(x) \times cos(2x)$ with uniformly distributed noise over the interval [-0.15, 0.15]. Legend: + input patterns; — desired output; - - - network output. (a) Training patterns with noise and the original function to be approximated (solid line). (b) Training process interrupted too early, *underfitting* (dashed line). (c) Too much training, *overfitting* (dashed line). (d) Better trade-off between quality of approximation and generalization capability.

drowned in seawater (but you survived) as 'stay away from seawater', then you may be in trouble if you decide to swim in a pool or a river. Of course, if the problem is that you don't know how to swim, then, knowing that seawater in particular is dangerous might not be a sufficiently good 'model' of the world to prevent you from drowning in other types of water. Your internal model has to be general enough so as to suggest 'stay away from water'.

Unsupervised Learning

In the *unsupervised* or *self-organized* learning approach, there is no supervisor to evaluate the network performance in relation to the input data set. The intuitive idea embodied in unsupervised learning is quite simple: given a set of input data, what can you do with it? For instance, if you are given a set of balloons, you could group them by colors, or by shape, or by any other attribute you can qualify.

Note that in unsupervised learning there is no error information being fed back into the network; the classes of the data are *unknown* or *unlabelled* and the presence of the supervisor no longer exists. The network adapts itself to statistical regularities in the input data, developing an ability to create internal representations that encode the features of the input data and thus, generate new classes automatically. Usually, self-organizing algorithms employ a *competitive learning* scheme.

In *competitive learning*, the network output neurons compete with one another to become activated, with a single output neuron being activated at each iteration. This property makes the algorithm appropriate to discover salient statistical features within the data that can then be used to classify a set of input patterns.

Individual neurons learn to specialize on groups (*clusters*) of similar patterns; in effect they become *feature extractors* or *feature detectors* for different classes of input patterns. In its simplest form, a *competitive neural network*, i.e., a neural network trained using a competitive learning scheme, has a single layer of output neurons that is fully connected. There are also lateral connections among neurons, as illustrated in Figure 4.16, capable of inhibiting or stimulating neighbor neurons.

For a neuron i to be the winner, the distance between its corresponding weight vector $\mathbf{w}_i$ and a certain input pattern $\mathbf{x}$ must be the smallest measure (among all the network output units), given a certain metric $\| \cdot \|$, usually taken to be the Euclidean distance. Therefore, the idea is to find the output neuron whose weight vector is most similar (has the shortest distance) to the input pattern presented.

$$i = \arg \min_i \|\mathbf{x} - \mathbf{w}_i\|, \quad \forall i \qquad (4.18)$$

If a neuron does not respond to a determined input pattern (i.e., is not the winner), no learning takes place for this neuron. However, if a neuron i wins the competition, then an adjustment $\Delta \mathbf{w}_i$ is applied to the weight vector $\mathbf{w}_i$ associated with the winning neuron i,

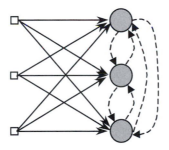

Figure 4.16: Simple competitive network architecture with direct excitatory (feedforward) connections (solid arrows) from the network inputs to the network outputs and inhibitory lateral connections among the output neurons (dashed arrows).

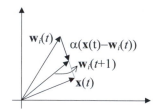

Figure 4.17: Geometric interpretation of the adjustment performed in unsupervised learning. The weight vector of the winning neuron i is moved toward the input pattern $\mathbf{x}$.

$$\Delta \mathbf{w}_i = \begin{cases} \alpha(\mathbf{x} - \mathbf{w}_i) & \text{if } i \text{ wins the competition} \\ 0 & \text{if } i \text{ loses the competition} \end{cases} \tag{4.19}$$

where α is a *learning rate* that controls the step size given by $\mathbf{w}_i$ in the direction of the input vector $\mathbf{x}$. Note that this learning rule works by simply moving the weight vector of the winning neuron in the direction of the input pattern presented, as illustrated in Figure 4.17.

Reinforcement Learning

Reinforcement learning (RL) is distinguished from the other approaches as it relies on learning from direct interaction with the environment, but does not rely on explicit supervision or complete models of the environment. Often, the only information available is a scalar evaluation that indicates how well the artificial neural network is performing. This is based on a framework that defines the interaction between the neural network and its environment in terms of the current values of the network's free parameters (weights), network response (actions), and rewards. Situations are mapped into actions so as to maximize a numerical reward signal (Sutton and Barto, 1998). Figure 4.18 illustrates the network-environment interaction in a reinforcement learning system, highlighting that the network output is fed into the environment that provides it with a reward signal according to how well the network is performing.

In reinforcement learning, the artificial neural network is given a goal to achieve. During learning, the neural network *tries* some actions (i.e., output values) on its environment, then it is *reinforced* or *penalized* by receiving a scalar evaluation (the *reward* or *penalty value*) for its actions. The reinforcement learning algorithm selectively retains the outputs that maximize the received reward over time. The network learns how to achieve its goal by trial-and-error interactions with its environment. At each time step t, the learning system receives some representation of the *state* of the environment $\mathbf{x}(t)$, it provides an output $y(t)$, and one step later it receives a scalar reward $r(t + 1)$, and finds itself in a new state $\mathbf{x}(t + 1)$. Thus, the two basic concepts behind reinforcement learning are *trial and error search* and *delayed reward*.

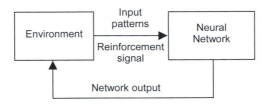

Figure 4.18: In reinforcement learning the network output provides the environment with information about how well the neural network is performing.

The intuitive idea behind reinforcement learning can also be illustrated with a simple example. Assume that you are training in a simulator to become a pilot without any supervisor, and your goal in today's lesson is to land the plane smoothly. Given a certain state of the aircraft, you perform an action that results in the plane crashing on the floor. You thus receive a negative reinforcement (or *punishment*) to that action. If, with the same state of the aircraft as before, you had performed an action resulting in a smooth landing, then you would have received a positive reinforcement (or *reward*). By interacting with the environment and maximizing the amount of reward, you can learn how to smoothly land the plane. We have long been using reinforcement learning techniques to teach animals to play tricks for us, mainly in circuses. For example, consider teaching a dolphin a new trick: you cannot tell it what to do, but you can reward or punish it if it does the right or wrong thing. It has to figure out what it did that made it get the reward or the punishment. This process is also known as the *credit assignment problem*.

4.4   TYPICAL ANNS AND LEARNING ALGORITHMS

The previous section introduced the fundamentals of neurocomputing. It was argued that an artificial neural network can be designed by 1) choosing some abstract models of neurons; 2) defining a network architecture; and 3) choosing an appropriate learning algorithm. In the present section, some of the most commonly used neural networks will be described, focusing on the main three aspects above: type of neuron, network architecture, and learning algorithm. As will be explained throughout this section, the type of learning algorithm is intimately related with the type of application the neural network is going to be used on. In particular, supervised learning algorithms are used for function approximation, pattern classification, control, identification, and other related tasks. On the other hand, unsupervised learning algorithms are employed in data analysis, including clustering and knowledge discovery.

4.4.1. Hebbian Learning

After the 1943 McCulloch and Pitts' paper describing a logical calculus of neuronal activity, N. Wiener published a famous book named *Cybernetics* in 1948, followed by the publication of Hebb's book *The Organization of Behavior*. These were some landmarks in the history of neurocomputing.

In Hebb's book, an explicit statement of a physiological learning rule for synaptic modification was presented for the first time. Hebb proposed that the connectivity of the brain is continually changing as an organism learns distinct functional tasks, and that neural assemblies are created by such changes. Hebb stated that the effectiveness of a variable synapse between two neurons is increased by the repeated activation of one neuron by the other across that synapse. Quoting from Hebb's book *The Organization of Behavior*:

"When an axon of cell A is near enough to excite a cell B and repeatedly or persistently takes part in firing it, some growth process or metabolic change takes place in one or both cells such that A's efficiency as one of the cells firing B, is increased." (Hebb, 1949; p. 62)

This postulate requires that change occur in the synaptic strength between cells when the presynaptic and postsynaptic cells are active at the same time. Hebb suggested that this change is the basis for *associative learning* that would result in a lasting modification in the activity pattern of a spatially distributed assemble of neurons.

This rule is often approximated in artificial neural networks by the *generalized Hebb rule*, where changes in connection strengths are given by the product of presynaptic and postsynaptic activity. The main difference being that change is now a result of both stimulatory and inhibitory synapses, not only excitatory synapses. In mathematical terms,

$$\Delta w_{ij}(t) = \alpha \, y_i(t) \, x_j(t) \qquad (4.20)$$

where Δw_{ij} is the change to be applied to neuron i, α is a *learning rate* parameter that determines how much change should be applied, y_i is the output of neuron i, x_j is the input to neuron i (output of neuron j), and t is the time index. Equation (4.20) can be expressed generically as

$$\Delta w_{ij}(t) = g(y_i(t), x_j(t)) \qquad (4.21)$$

where $g(\cdot, \cdot)$ is a function of both pre- and postsynaptic signals. The weight of a given neuron i is thus updated according to the following rule:

$$w_{ij}(t+1) = w_{ij}(t) + \Delta w_{ij}(t) \qquad (4.22)$$

Equation (4.20) clearly emphasizes the correlational or associative nature of Hebb's updating rule. It is known that much of human memory is *associative*, just as the mechanism suggested by Hebb. In an associative memory, an event is linked to another event, so that the presentation of the first event gives rise to the linked event. In the most simplistic version of association, a *stimulus* is linked to a *response*, so that a later presentation of the stimulus evokes the response.

Two important aspects of the Hebbian learning given by Equation (4.20) must be emphasized. First, it is a general updating rule to be used with different types of neurons. In the most standard case, the Hebbian network is but a single neuron with linear output, but more complex structures could be used. Second, this learning rule is unsupervised; that is, no information about the desired behavior is accounted for. However, the Hebb rule can also be employed when the target output for each input pattern is known. This means that the Hebb rule can also be used in supervised learning. In such cases, the modified Hebb rule has the current output substituted by the desired output:

$$\Delta w_{ij}(t) = \alpha \, d_i(t) \, x_j(t) \tag{4.23}$$

where, d_i is the desired output of neuron i.

Other variations of the Hebb learning rule take into account, for example, the difference between the neuron's output and its desired output. This leads to the Widrow-Hoff and perceptron learning rules to be described later.

Biological Basis of Hebbian Synaptic Modification

Physiological evidence suggests the existence of Hebb synapses at various locations in the brain. Advances in modern neurophysiological techniques have allowed us to see what appears to be Hebbian modification in several parts of the mammalian brain (Anderson, 1995). A part of the cerebral cortex, the *hippocampus*, shows an effect called *long-term potentiation*, in which its neurons can be induced to display long-term, apparently permanent, increases in activity with particular patterns of stimulation.

Kelso et al. (1986) have presented an experimental demonstration of Hebbian modification using a slice of rat hippocampus maintained outside the animal. They showed that when a presynaptic cell is excited and the postsynaptic cells inhibited, little or no change was seen in the efficiency of a synapse. If the postsynaptic cell was excited by raising the membrane potential at the same time the presynaptic cell was active, the excitatory postsynaptic potential from the presynaptic cell was significantly increased in a stable and long-lasting form.

4.4.2. Single-Layer Perceptron

Rosenblatt (1958, 1962) introduced the *perceptron* as the simplest form of a neural network used for the classification of *linearly separable* patterns. Perceptrons constituted the first model of supervised learning, though some perceptrons were self-organized. Before describing the perceptron learning algorithm, the next section discusses in some more detail the problem of linear separability.

Linear Separability

Linear separability can be easily understood with a simple example. Without loss of generality, assume there is a set of input patterns to be classified into a single class. Assume also there is a classifier system that has to respond TRUE if a given input pattern is a member of a certain class, and FALSE if it is not. A TRUE response is represented by an output '1' and a FALSE response by an output '0' of the classifier.

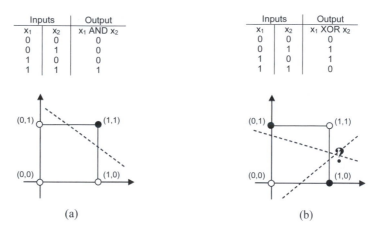

Inputs		Output
x_1	x_2	x_1 AND x_2
0	0	0
0	1	0
1	0	0
1	1	1

Inputs		Output
x_1	x_2	x_1 XOR x_2
0	0	0
0	1	1
1	0	1
1	1	0

(a)   (b)

Figure 4.19: Examples of linear and nonlinear separability. (a) The logic function AND is linearly separable. (b) The logic function XOR is nonlinearly separable.

As one of two responses is required, there is a *decision boundary* that separates one response from the other. Depending on the number m of variables of the input patterns, the decision boundary can have any shape in the space of dimension m. If $m = 2$ and the classifier can be set up so that there is a line dividing all input patterns that produce output '1' from those patterns that produce output '0', then the problem is *linearly separable*; else it is *nonlinearly separable*. This holds true for a space of any dimension, but for higher dimensions, a plane or hyper-plane should exist so as to divide the space into a '1'-class and a '0'-class region.

Figure 4.19 illustrates one linearly separable function and one nonlinearly separable function. Note that, in Figure 4.19(a), several lines could be drawn so as to separate the data into class '1' or '0'. The regions that separate the classes to which each of the input patterns belong to are often called *decision regions*, and the (hyper-)surface that defines these regions are called *decision surfaces*. Therefore, a classification problem can be viewed as the problem of finding decision surfaces that correctly classify the input data. The difficulty lying is that it is not always trivial to automatically define such surfaces.

Simple Perceptron for Pattern Classification

Basically, the perceptron consists of a single layer of neurons with adjustable synaptic weights and biases. Under suitable conditions, in particular if the training patterns belong to linearly separable classes, the iterative procedure of adaptation for the perceptron can be proved to converge to the correct weight set. These weights are such that the perceptron algorithm converges and positions the decision surfaces in the form of (hyper-)planes between the classes. This proof of convergence is known as the *perceptron convergence theorem*.

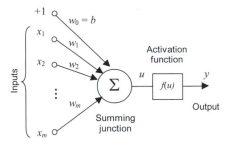

Figure 4.20: The simplest perceptron to perform pattern classification.

The simple perceptron has a single layer of feedforward neurons as illustrated in Figure 4.11(a). The neurons in this network are similar to those of McCulloch and Pitts with a signum or threshold activation function, but include a bias. The key contribution of Rosenblatt and his collaborators was to introduce a learning rule to train the perceptrons to solve pattern recognition and classification problems.

Their algorithm works as follows. For each training pattern $\mathbf{x}_i$, the network output response y_i is calculated. Then, the network determines if an error e_i occurred for this pattern by comparing the calculated output y_i with a *desired value* d_i for that pattern, $e_i = d_i - y_i$. Note that the desired output value for each training pattern is known (supervised learning). The weight vector connecting the inputs (presynaptic neurons) to each output (postsynaptic neuron) and the neuron's bias are updated according to the following rules:

$$\mathbf{w}(t+1) = \mathbf{w}(t) + \alpha\, e_i\, \mathbf{x} \qquad (4.24)$$

$$b(t+1) = b(t) + \alpha\, e \qquad (4.25)$$

where $\mathbf{w} \in \Re^{1 \times m}$, $\mathbf{x} \in \Re^{1 \times m}$, and $b \in \Re^{1 \times 1}$.

Consider now the simplest perceptron case with a single neuron, as illustrated in Figure 4.20. The goal of this network, more specifically neuron, is to classify a number of input patterns as belonging or not belonging to a particular class. Assume that the input data set is given by the N pairs of samples $\{\mathbf{x}_1,d_1\}$, $\{\mathbf{x}_2,d_2\}$, ... , $\{\mathbf{x}_N,d_N\}$, where $\mathbf{x}_j$ is the input vector j and d_j its corresponding *target* or *desired* output. In this case, the target value is '1' if the pattern belongs to the class, and '−1' (or '0') otherwise.

Let $\mathbf{X} \in \Re^{m \times N}$, be the matrix of N input patterns of dimension m each (the patterns are placed column wise in matrix $\mathbf{X}$), and $\mathbf{d} \in \Re^{1 \times N}$ be the vector of desired outputs. The learning algorithm for this simple perceptron network is presented in Algorithm 4.1, where $f(\cdot)$ is the signum or the threshold function. The stopping criterion for this algorithm is either a fixed number of iteration steps (max_it) or the sum E of the squared errors e_i^2, $i = 1, ... , N$, for each input pattern being equal to zero. The algorithm returns the weight vector $\mathbf{w}$ as output.

```
procedure [w] = perceptron(max_it,α,X,d)
    initialize w       //set it to zero or small random values
    initialize b       //set it to zero or small random value
    t ← 1; E ← 1
    while t < max_it & E > 0 do,
        E ← 0
        for i from 1 to N do,  //for each training pattern
            yᵢ ← f(wxᵢ + b)              //network output for xᵢ
            eᵢ ← dᵢ – yᵢ                  //determine the error for xᵢ
            w ← w + α eᵢ xᵢ              //update the weight vector
            b ← b + α eᵢ                 //update the bias term
            E ← E + eᵢ²                  //accumulate the error
        end for
        t ← t + 1
    end while
end procedure
```

Algorithm 4.1: Simple perceptron learning algorithm. The function $f(\cdot)$ is the signum (or the threshold) function, and the desired output is '1' if a pattern belongs to the class and '−1' (or '0') if it does not belong to the class.

Assuming the input patterns are linearly separable, the perceptron will be capable of solving the problem after a finite number of iteration steps and, thus, the error should be used as the stopping criterion. The parameter α is a *learning rate* that determines the step size of the adaptation in the weight values and bias term. Note, from Algorithm 4.1, that the perceptron learning rule only updates a given weight when the desired response is different from the actual response of the neuron.

Multiple Output Perceptron for Pattern Classification

Note that the perceptron updating rule uses the error-correction learning of most supervised learning techniques, as discussed in Section 4.3.3. This learning rule, proposed to update the weight vector of a single neuron, can be easily extended to deal with networks with more than one output neuron, such as the network presented in Figure 4.11(a). In this case, for each input pattern $\mathbf{x}_i$ the vector of network outputs is given by Equation (4.13) explicitly including the bias vector $\mathbf{b}$ as follows:

$$\mathbf{y} = f(\mathbf{Wx}_i + \mathbf{b}) \tag{4.26}$$

where $\mathbf{W} \in \mathfrak{R}^{o \times m}$, $\mathbf{x}_i \in \mathfrak{R}^{m \times 1}$, $i = 1, \ldots, N$, $\mathbf{y} \in \mathfrak{R}^{o \times 1}$, and $\mathbf{b} \in \mathfrak{R}^{o \times 1}$. The function $f(\cdot)$ is the signum or the threshold function.

Let the matrix of desired outputs be $\mathbf{D} \in \mathfrak{R}^{o \times N}$, where each column of $\mathbf{D}$ corresponds to the desired output for one of the input patterns. Therefore, the error signal for each input pattern $\mathbf{x}_i$ is calculated by simply subtracting the vectors $\mathbf{d}_i$ and $\mathbf{y}_i$ that are of same dimension: $\mathbf{e}_i = \mathbf{d}_i - \mathbf{y}_i$, where $\mathbf{e}_i, \mathbf{d}_i, \mathbf{y}_i \in \mathfrak{R}^{o \times 1}$. The algorithm to implement the perceptron with multiple output neurons is presented in

```
procedure [W] = perceptron(max_it,α,X,D)
    initialize W      //set it to zero or small random values
    initialize b      //set it to zero or small random values
    t ← 1; E ← 1
    while t < max_it & E > 0 do,
        E ← 0
        for i from 1 to N do,    //for each training pattern
            yᵢ ← f(Wxᵢ + b)        //network outputs for xᵢ
            eᵢ ← dᵢ - yᵢ           //determine the error for xᵢ
            W ← W + α eᵢ xᵢᵀ       //update the weight matrix
            b ← b + α eᵢ           //update the bias vector
            E ← E + sum(e²ⱼᵢ)      //j = 1,...,o

            veᵢ ← eᵢ
        end for
        t ← t + 1
    end while
end procedure
```

Algorithm 4.2: Learning algorithm for the perceptron with multiple outputs. The function $f(\cdot)$ is the signum (or the threshold) function, and the desired output is '1' if a pattern belongs to the class and '−1' (or '0') if it does not belong to the class. Note that e_i, d_i, y_i, and **b** are now vectors and **W** is a matrix. e_{ji} corresponds to the error of neuron j when presented with input pattern i.

Algorithm 4.2. As with the simple perceptron, in the perceptron with multiple output neurons the network will converge if the training patterns are linearly separable. In this case, the error for all patterns and all outputs will be zero at the end of the learning phase.

Examples of Application

To illustrate the applicability of the perceptron network, consider the two problems below. The first example - simple classification problem - illustrates the potentiality of the perceptron to represent Boolean functions, and the second example - character recognition - illustrates its capability to recognize a simple set of binary characters.

Simple Classification Problem

Consider the problem of using the simple perceptron with a single neuron to represent the AND function. The training data and its graphical interpretation are presented in Figure 4.21. The training patters to be used as inputs to Algorithm 4.1 are, in matrix notation:

$$\mathbf{X} = \begin{bmatrix} 0 & 0 & 1 & 1 \\ 0 & 1 & 0 & 1 \end{bmatrix} \quad \mathbf{d} = \begin{bmatrix} 0 & 0 & 0 & 1 \end{bmatrix}$$

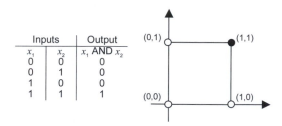

Inputs		Output
x_1	x_2	x_1 AND x_2
0	0	0
0	1	0
1	0	0
1	1	1

Figure 4.21: The AND function and its graphical representation.

The boundary between the values of x_1 and x_2 for which the network provides a response '0' (not belonging to the class) and the values for which the network responds '1' (belonging to the class) is the separating line given by

$$w_1 x_1 + w_2 x_2 + b = 0 \qquad (4.27)$$

Assuming the activation function of the network is the threshold with $\theta = 0$, the requirement for a positive response from the output unit is that the net input it receives be greater than or equal to zero, that is, $w_1 x_1 + w_2 x_2 + b \geq 0$. During training, the values of w_1, w_2, and b are determined so that the network presents the correct response for all training data.

For simplicity, the learning rate is set to 1, $\alpha = 1$, and the initial values for the weights and biases are taken to be zero, $\mathbf{w} = [0\ 0]$ and $b = 0$. By running Algorithm 4.1 with $\mathbf{X}$, $\mathbf{d}$, and the other parameters given above, the following values are obtained for the perceptron network with a single output.

$$w_1 = 2;\ w_2 = 1;\ b = -3 \qquad (4.28)$$

By replacing these values in the line equation for this neuron we obtain

$$2x_1 + 1x_2 - 3 = 0 \qquad (4.29)$$

For each pair of training data, the following net input u_i, $i = 1, \dots, N$, is calculated: $\mathbf{u} = [-3, -2, -1, 0]$. Therefore, the network response, $\mathbf{y} = f(\mathbf{u})$, for each input pattern is $\mathbf{y} = [0, 0, 0, 1]$, realizing the AND function as desired.

The decision line between the two classes can be determined by isolating the variable x_2 as a function of the other variable in Equation (4.27)

$$x_2 = -\frac{w_1}{w_2} x_1 - \frac{b}{w_2} = -2x_1 + 3 \qquad (4.30)$$

Note that this line passes exactly on top of class '1', as illustrated in Figure 4.22.

An alternative form of initializing the weights for the perceptron network is by choosing random values within a given domain. For instance, a uniform or normal distribution with zero mean and predefined variance (usually less than or equal to one) could be used. Assume now that the following initial values for $\mathbf{w}$ and b were chosen using a uniform distribution of zero mean and variance one:

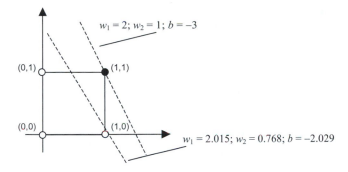

$w_1 = 2; w_2 = 1; b = -3$

(0,1) (1,1)

(0,0) (1,0)

$w_1 = 2.015; w_2 = 0.768; b = -2.029$

Figure 4.22: Decision lines for the simple perceptron network realizing the logic function AND.

$$w_1 = 0.015; w_2 = 0.768; b = 0.971$$

After training the network, the following values for the weights were determined:

$$w_1 = 2.015; w_2 = 0.768; b = -2.029$$

These values define the decision line below, as depicted in Figure 4.22.

$$x_2 = -\frac{w_1}{w_2}x_1 - \frac{b}{w_2} = -2.624\,x_1 + 2.642$$

Two important issues can be noticed here. First, the network response is dependent upon the initial values chosen for its weights and biases. Second, there is not a single correct response for a given training data set. These and other important issues on neural network training will be discussed later.

Character Recognition

As a second example of how to use the perceptron for pattern classification, consider the eight input patterns presented in Figure 4.23. Each input vector is a 120-tuple representing a letter expressed as a pattern on a 12×10 grid of pixels. Each character is assumed to have its own class. One output unit is assigned to each character/class, thus there are eight categories to which each character could be assigned. Matrix $\mathbf{X} \in \Re^{120 \times 8}$, where each column of $\mathbf{X}$ corresponds to an input pattern, and matrix $\mathbf{D} \in \Re^{8 \times 8}$ is a diagonal matrix, meaning that input pattern number 1 (character '0') should activate only output 1, input pattern number 2 (character '1') should activate only output 2, and so on.

Figure 4.23: Training patterns for the perceptron.

```
procedure [y] = run_perceptron(W,b,Z)
    for i from 1 to N do,   //for each training pattern
        yᵢ ← f(Wzᵢ + b)         //network outputs for zᵢ
    end for
end procedure
```

Algorithm 4.3: Algorithm used to run the trained single-layer perceptron network.

Figure 4.24: Noisy patterns used to test the generalization capability of the single-layer perceptron network trained to classify the patterns presented in Figure 4.23.

After training, the network correctly classifies each of the training patterns. Algorithm 4.3 can be used to run the network in order to evaluate if it is correctly classifying the training patterns. The weight matrix **W** and the bias vector **b** are those obtained after training. Matrix **Z** contains the patterns to be recognized; these can be the original training data **X**, the training data added with noise, or completely new unseen data. Note that, in this case, the network is first trained and then applied to classify novel data; that is, its generalization capability for unseen data can be evaluated.

To test how this network generalizes, some random *noise* can be inserted into the training patterns by simply changing a value '0' by '1' or a value '1' by '0', respectively, with a given probability. Figure 4.24 illustrates the training patterns with 5% noise.

4.4.3. ADALINE, the LMS Algorithm, and Error Surfaces

Almost at the same time Rosenblatt introduced the perceptron learning rule, B. Widrow and his student M. Hoff (Widrow and Hoff, 1960) developed the Widrow-Hoff learning rule, also called the *least mean squared (LMS) algorithm* or *delta rule*. They introduced the ADALINE (ADAptive LInear NEuron) network that is very similar to the perceptron, except that its activation function is linear instead of threshold. Although both networks, ADALINE and perceptron, suffer from only being capable of solving linearly separable problems, the LMS algorithm is more powerful than the perceptron learning rule and has found many more practical uses than the perceptron.

As the ADALINE employs neurons with a linear activation function, the neuron's output is equal to its net input. The goal of the learning algorithm is to minimize the *error* between the network output and the desired output. This allows the network to perform a continuous learning even after a given input pattern has been learnt. One advantage of the LMS algorithm over the perceptron learning rule is that the resultant network after training is not too sensitive to noise. It could be observed in Figure 4.22, that the decision surfaces generated by the perceptron algorithm usually lie very close to some input data. This is very much the case for the perceptron algorithm, but it is not the case for the LMS algorithm, making the latter more robust to noise. In addition, the LMS algorithm has been broadly used in signal processing applications.

LMS Algorithm (Delta Rule)

The LMS algorithm or delta rule is another example of supervised learning in which the learning rule is provided with a set of inputs and corresponding desired outputs $\{\mathbf{x}_1, \mathbf{d}_1\}$, $\{\mathbf{x}_2, \mathbf{d}_2\}$, ..., $\{\mathbf{x}_N, \mathbf{d}_N\}$, where $\mathbf{x}_j$ is the input vector j and $\mathbf{d}_j$ its corresponding *target* or *desired* output vector.

The delta rule adjusts the neural network weights and biases so as to minimize the difference (error) between the network output and the desired output over all training patterns. This is accomplished by reducing the error for each pattern, one at a time, though weight corrections can also be accumulated over a number of training patterns.

The *sum-squared error* (SSE) for a particular training pattern is given by

$$\Im = \sum_{i=1}^{o} e_i^2 = \sum_{i=1}^{o} (d_i - y_i)^2 \tag{4.31}$$

The gradient of $\Im$, also called the *performance index* or *cost function*, is a vector consisting of the partial derivatives of $\Im$ with respect to each of the weights. This vector gives the direction of most rapid increase of $\Im$; thus, its opposite direction is the one of most rapid decrease in the error value. Therefore, the error can be reduced most rapidly by adjusting the weight w_{IJ} in the following manner

$$w_{IJ} = w_{IJ} - \alpha \frac{\partial \Im}{\partial w_{IJ}} \tag{4.32}$$

where w_{IJ} is the weight from the J-th presynaptic neuron to the I-th postsynaptic neuron, and α is a learning rate.

It is necessary now to explicitly determine the gradient of the error with respect to the arbitrary weight w_{IJ}. As the weight w_{IJ} only influences the output (postsynaptic) unit I, the gradient of the error is

$$\frac{\partial \Im}{\partial w_{IJ}} = \frac{\partial}{\partial w_{IJ}} \sum_{i=1}^{o} (d_i - y_i)^2 = \frac{\partial}{\partial w_{IJ}} (d_I - y_I)^2$$

It is known that

$$y_I = f(\mathbf{w}_I.\mathbf{x}) = f(\Sigma_j w_{Ij}.x_j)$$

Then, we obtain

$$\frac{\partial \Im}{\partial w_{IJ}} = -2(d_I - y_I) \frac{\partial y_I}{\partial w_{IJ}} = -2(d_I - y_I)x_J$$

Therefore, similarly to the updating rule of Hebb's network, the LMS also assumes an updating value Δw_{IJ} to be added to the weight w_{IJ}. The delta rule for updating the weight from the J-th presynaptic neuron to the I-th postsynaptic neuron is given by

$$\Delta w_{IJ} = \alpha \, (d_I - y_I) \, x_J \qquad (4.33)$$

$$w_{IJ} = w_{IJ} + 2\alpha \, (d_I - y_I) \, x_J \qquad (4.34)$$

The following equation for updating the bias b_I can be obtained by calculating the gradient of the error in relation to the bias b_I

$$b_I = b_I + 2\alpha \, (d_I - y_I) \qquad (4.35)$$

For each input pattern x_i, the LMS algorithm can be conveniently written in matrix notation as

$$\mathbf{W} = \mathbf{W} + \alpha \, \mathbf{e}_i \, \mathbf{x}_i^T$$

$$\mathbf{b} = \mathbf{b} + \alpha \, \mathbf{e}_i$$

where $\mathbf{W} \in \Re^{o \times m}$, $\mathbf{x}_i \in \Re^{m \times 1}$, $i = 1, \ldots, N$, $\mathbf{e}_i \in \Re^{o \times 1}$, and $\mathbf{b} \in \Re^{o \times 1}$. (Note that 2α was replaced by α.)

The beauty of this algorithm is that at each iteration it calculates an approximation to the gradient vector by simply multiplying the error by the input vector. And this approximation to the gradient can be used in a steepest descent-like algorithm with fixed learning rate.

Error Surfaces

Assume a neural network with n weights. This set of weights can be viewed as a point in an n-dimensional space, called *weight space*. If the neural network is used to classify a set of patterns, for each of these patterns the network will generate an error signal. This means that every set of weights (and biases) has an associated scalar error value; if the weights are changed, a new error value is determined. The error values for every set of weights define a surface in the weight space, called the *error surface* (Anderson, 1995; Hagan et al., 1996).

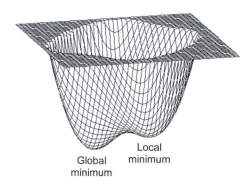

Figure 4.25: An error surface in weight space. The total error is a function of the values of the weights, so every set of weights has an associated error.

The question that arises now is what is the function of learning? The way the LMS algorithm, and the backpropagation algorithm to be explained in the next section, view learning is as the minimization of the error to its smallest value possible. This corresponds to finding an appropriate weight set that leads to the smallest error in the error surface. Figure 4.25 depicts an error surface in weight space. Note that this surface potentially has many *local minima* and one or more *global minimum*. In the example depicted the error surface contains a single local optimum and a single global optimum. Any gradient-like method, such as the LMS and the backpropagation algorithm, can only converge to local optima solutions. Therefore, the choice of an appropriate initial weight set for the network is crucial.

4.4.4. Multi-Layer Perceptron

The multi-layer perceptron is a kind of multi-layer feedforward network such as the one illustrated in Figure 4.12. Typically, the network consists of a set of input units that constitute the input layer, one or more hidden layers, and an output layer. The input signal propagates through the network in a forward direction, layer by layer. This network is a generalization of the single-layer perceptron discussed previously.

In the late 1960's, M. Minsky and S. Papert (1969) released a book called *Perceptrons* demonstrating the limitations of single-layered feedforward networks, namely, the incapability of solving nonlinearly separable problems. This caused a significant impact on the interest in neural network research during the 1970s. Both, Rosenblatt and Widrow were aware of the limitations of the perceptron network and proposed multi-layer networks to overcome this limitation. However, they were unable to generalize their algorithms to train these more powerful networks.

It has already been discussed that a multi-layer network with linear nodes is equivalent to a single-layered network with linear units. Therefore, one necessary condition for the multi-layer network to be more powerful than the single-layer networks is the use of nonlinear activation functions. More specifically, multi-layer networks typically employ a sigmoidal activation function (Figure 4.10).

Multi-layer perceptrons (MLP) have been applied successfully to a number of problems by training them with a supervised learning algorithm known as the *error backpropagation* or simply *backpropagation*. This algorithm is also based on an error-correction learning rule and can be viewed as a generalization of the LMS algorithm or delta rule.

The backpropagation algorithm became very popular with the publication of the *Parallel Distributed Processing* volumes by Rumelhart and collaborators (Rumelhart et al., 1986; McClelland et al., 1986). These books were also crucial for the re-emergence of interest for the research on neural networks. Basically, the error backpropagation learning consists of two passes of computation: a *forward* and a *backward* pass (see Figure 4.12(b)). In the forward pass, an input pattern is presented to the network and its effects are propagated through the network until they produce the network output(s). In the backward pass, the synaptic weights, so far kept fixed, are updated in accordance with an error correction rule.

The Backpropagation Learning Algorithm

To derive the backpropagation learning algorithm, consider the following notation:

i, j	Indices referring to neurons in the network
t	Iteration counter
N	Number of training patterns
M	Number of layers in the network
$y_j(t)$	Output signal of neuron j at iteration t
$e_j(t)$	Error signal of output unit j at iteration t
$w_{ji}(t)$	Synaptic weight connecting the output of unit j to the input of unit i at iteration t
$u_j(t)$	Net input of unit j at iteration t
$f_j(\cdot)$	Activation function of unit j
$\mathbf{X}$	Matrix of input (training) patterns
$\mathbf{D}$	Matrix of desired outputs
$x_i(t)$	i-th element of the input vector at iteration t
$d_j(t)$	j-th element of the desired output vector at iteration t
α	Learning rate

The description to be presented here follows that of Hagan et al. (1996) and de Castro (1998). For multi-layer networks, the output of a given layer is the input to the next layer,

$$\mathbf{y}^{m+1} = \mathbf{f}^{m+1} (\mathbf{W}^{m+1}\mathbf{y}^m + \mathbf{b}^{m+1}), \quad m = 0, 1, \dots , M-1 \qquad (4.36)$$

where M is the number of layers in the network, and the superindex m refers to the layer taken into account (e.g., $m = 0$: input layer, $m = 1$: first hidden layer, ... , $m = M-1$: output layer). The nodes in the input layer ($m = 0$) receive the input patterns

$$\mathbf{y}^0 = \mathbf{x} \qquad (4.37)$$

that represent the initial condition for Equation (4.36). The outputs in the output layer are the network outputs:

$$\mathbf{y} = \mathbf{y}^M \qquad (4.38)$$

Equation (4.14) demonstrates that the network output can be a function of only the input vector $\mathbf{x}$ and the weight matrices $\mathbf{W}^m$. If we explicitly consider the bias terms $\mathbf{b}^m$, Equation (4.13) becomes Equation (4.39) for the network of Figure 4.26:

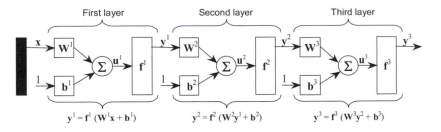

First layer   Second layer   Third layer

$$\mathbf{y}^1 = \mathbf{f}^1(\mathbf{W}^1\mathbf{x} + \mathbf{b}^1) \qquad \mathbf{y}^2 = \mathbf{f}^2(\mathbf{W}^2\mathbf{y}^1 + \mathbf{b}^2) \qquad \mathbf{y}^3 = \mathbf{f}^3(\mathbf{W}^3\mathbf{y}^2 + \mathbf{b}^3)$$

Figure 4.26: An MLP network with three layers. (Matrix notation.)

$$\mathbf{y} = \mathbf{y}^3 = \mathbf{f}^3(\mathbf{W}^3\mathbf{f}^2(\mathbf{W}^2\mathbf{f}^1(\mathbf{W}^1\mathbf{x} + \mathbf{b}^1) + \mathbf{b}^2) + \mathbf{b}^3) \tag{4.39}$$

The backpropagation algorithm to MLP networks is a generalization of the LMS algorithm that uses as a performance index the *mean squared error* (MSE). A set of input patterns and desired outputs is provided, as follows:

$$\{(\mathbf{x}_1, \mathbf{d}_1), (\mathbf{x}_2, \mathbf{d}_2), \ldots, (\mathbf{x}_N, \mathbf{d}_N)\}$$

where $\mathbf{x}_i$ is the i-th input to the network and $\mathbf{d}_i$ is the corresponding desired output, $i = 1, \ldots, N$. From Equation (4.39) it is possible to observe that if ω is the vector of network parameters (weights and biases), then the network output can be given as a function of ω and x:

$$y^M = f(\omega, x)$$

After each training pattern is presented to the network, the network output is compared with the desired output. The algorithm must adjust the vector of parameters so as to minimize the mathematical expectation of the mean squared error:

$$\Im(\omega) = E(e(\omega)^2) = E((d - y(\omega))^2) \tag{4.40}$$

If the network has multiple outputs, Equation (4.40) becomes

$$\Im(\omega) = E(\mathbf{e}(\omega)^T \mathbf{e}(\omega)) = E(\,(\mathbf{d} - \mathbf{y}(\omega))^T\,(\mathbf{d} - \mathbf{y}(\omega))\,)$$

Similarly to the LMS algorithm, the mean squared error can be approximated by the following expression:

$$\hat{\Im}(\omega) = \mathbf{e}(t)^T\mathbf{e}(t) = (\mathbf{d}(t) - \mathbf{y}(t))^T\,(\mathbf{d}(t) - \mathbf{y}(t))$$

where the expectation of the mean squared error was substituted by the error at iteration t. In order not to overload the notation, assume $\hat{\Im}(\omega) = \Im(\omega)$.

The updating rule known as steepest descent to minimize the squared error is given by

$$w_{ij}^m(t+1) = w_{ij}^m(t) - \alpha\frac{\partial\Im(t)}{\partial w_{ij}^m} \tag{4.41}$$

$$b_i^m(t+1) = b_i^m(t) - \alpha\frac{\partial\Im(t)}{\partial b_i^m} \tag{4.42}$$

where α is the learning rate.

The most elaborate part is the determination of the partial derivatives that will produce the components of the gradient vector. To determine these derivatives it will be necessary to apply the chain rule a number of times.

The Chain Rule

For a multi-layer network, the error is not a direct function of the weights in the hidden layers, reason why the calculus of these derivatives is not straightforward.

As the error is an indirect function of the weights in the hidden layers, the chain rule must be used to determine the derivatives. The chain rule will be used to determine the derivatives in Equation (4.41) and Equation (4.42).

$$\frac{\partial \Im}{\partial w_{ij}^m} = \frac{\partial \Im}{\partial u_i^m} \times \frac{\partial u_i^m}{\partial w_{ij}^m} \tag{4.43}$$

$$\frac{\partial \Im}{\partial b_i^m} = \frac{\partial \Im}{\partial u_i^m} \times \frac{\partial u_i^m}{\partial b_i^m} \tag{4.44}$$

The second term of both equations above can be easily determined for the net input of layer m as an explicit function of the weights and biases in this layer:

$$u_i^m = \sum_{j=1}^{S^{m-1}} w_{ij}^m y_j^{m-1} + b_i^m \tag{4.45}$$

where S^m is the number of neurons in layer m.

Therefore,

$$\frac{\partial u_i^m}{\partial w_{ij}^m} = y_j^{m-1}, \frac{\partial u_i^m}{\partial b_i^m} = 1 \tag{4.46}$$

Now define the *sensitivity* (δ) of $\Im$ to changes in the i-th element of the net input in layer m as

$$\delta_i^m \equiv \frac{\partial \Im}{\partial u_i^m} \tag{4.47}$$

Equation (4.43) and Equation (4.44) can now be simplified by

$$\frac{\partial \Im}{\partial w_{ij}^m} = \delta_i^m y_j^{m-1} \tag{4.48}$$

$$\frac{\partial \Im}{\partial b_i^m} = \delta_i^m \tag{4.49}$$

Thus, Equation (4.41) and Equation (4.42) become

$$w_{ij}^m(t+1) = w_{ij}^m(t) - \alpha \delta_i^m y_j^{m-1} \tag{4.50}$$

$$b_i^m (t+1) = b_i^m (n) - \alpha \delta_i^m \tag{4.51}$$

In matrix notation these equations become

$$\mathbf{W}^m (t+1) = \mathbf{W}^m (t) - \alpha \boldsymbol{\delta}^m (\mathbf{y}^{m-1})^T \tag{4.52}$$

$$\mathbf{b}^m (t+1) = \mathbf{b}^m (t) - \alpha \boldsymbol{\delta}^m \tag{4.53}$$

where

$$\boldsymbol{\delta}^m \equiv \frac{\partial \Im}{\partial \mathbf{u}^m} = \begin{bmatrix} \dfrac{\partial \Im}{\partial u_1^m} \\[2mm] \dfrac{\partial \Im}{\partial u_2^m} \\[1mm] \vdots \\[1mm] \dfrac{\partial \Im}{\partial u_{S^m}^m} \end{bmatrix} \tag{4.54}$$

Backpropagating the Sensitivities

It is now necessary to calculate the sensitivity $\boldsymbol{\delta}^m$ that requires another application of the chain rule. This process gives rise to the term *backpropagation* because it describes a recurrence relationship in which the sensitivity of layer m is determined from the sensitivity of layer $m + 1$.

To derive the recurrence relationship for the sensitivities, we will use the following Jacobian matrix

$$\frac{\partial \mathbf{u}^{m+1}}{\partial \mathbf{u}^m} = \begin{bmatrix} \dfrac{\partial u_1^{m+1}}{\partial u_1^m} & \dfrac{\partial u_1^{m+1}}{\partial u_2^m} & \cdots & \dfrac{\partial u_1^{m+1}}{\partial u_{S^m}^m} \\[3mm] \dfrac{\partial u_2^{m+1}}{\partial u_1^m} & \dfrac{\partial u_2^{m+1}}{\partial u_2^m} & \cdots & \dfrac{\partial u_2^{m+1}}{\partial u_{S^m}^m} \\[3mm] \vdots & \vdots & \ddots & \vdots \\[3mm] \dfrac{\partial u_{S^{m+1}}^{m+1}}{\partial u_1^m} & \dfrac{\partial u_{S^{m+1}}^{m+1}}{\partial u_2^m} & \cdots & \dfrac{\partial u_{S^{m+1}}^{m+1}}{\partial u_{S^m}^m} \end{bmatrix} \tag{4.55}$$

Next, we want to find an expression for this matrix. Consider the element i,j of this matrix

$$\frac{\partial u_i^{m+1}}{\partial u_j^m} = w_{ij}^{m+1} \frac{\partial y_j^m}{\partial u_j^m} = w_{ij}^{m+1} \frac{\partial f^m (u_j^m)}{\partial u_j^m} = w_{ij}^{m+1} \dot{f}^m (u_j^m) \tag{4.56}$$

where

$$\dot{f}^m (u_j^m) = \frac{\partial f^m (u_j^m)}{\partial u_j^m} \tag{4.57}$$

Therefore, the Jacobian matrix can be written as

$$\frac{\partial \mathbf{u}^{m+1}}{\partial \mathbf{u}^m} = \mathbf{W}^{m+1}\dot{\mathbf{F}}^m(\mathbf{u}^m) \tag{4.58}$$

where

$$\dot{\mathbf{F}}^m(\mathbf{u}^m) = \begin{bmatrix} \dot{f}^m(u_1^m) & 0 & \cdots & 0 \\ 0 & \dot{f}^m(u_2^m) & \cdots & 0 \\ \vdots & \vdots & \ddots & \vdots \\ 0 & 0 & \cdots & \dot{f}^m(u_{S^m}^m) \end{bmatrix} \tag{4.59}$$

It is now possible to write the recurrence relationship for the sensitivity using the chain rule in matrix form

$$\delta^m = \frac{\partial \Im}{\partial \mathbf{u}^m} = \left(\frac{\partial \mathbf{u}^{m+1}}{\partial \mathbf{u}^m}\right)^T \frac{\partial \Im}{\partial \mathbf{u}^{m+1}} = \dot{\mathbf{F}}^m(\mathbf{u}^m)(\mathbf{W}^{m+1})^T \frac{\partial \Im}{\partial \mathbf{u}^{m+1}}$$

$$= \dot{\mathbf{F}}^m(\mathbf{u}^m)(\mathbf{W}^{m+1})^T \delta^{m+1} \tag{4.60}$$

Note that the sensitivities are (back)propagated from the last to the first layer

$$\delta^M \rightarrow \delta^{M-1} \rightarrow \dots \rightarrow \delta^2 \rightarrow \delta^1 \tag{4.61}$$

There is a last step to be executed to complete the backpropagation algorithm. We need the starting point, δ^M, for the recurrence relation of Equation (4.60).

$$\delta_i^M = \frac{\partial \Im}{\partial u_i^M} = \frac{\partial (\mathbf{d} - \mathbf{y})^T (\mathbf{d} - \mathbf{y})}{\partial u_i^M} = \frac{\partial \sum_{j=1}^{S^M} (d_j - y_j)^2}{\partial u_i^M} = -2(d_i - y_i)\frac{\partial y_i}{\partial u_i^M} \tag{4.62}$$

Now, since

$$\frac{\partial y_i}{\partial u_i^M} = \frac{\partial y_i^M}{\partial u_i^M} = \frac{\partial f^M(u_j^M)}{\partial u_i^M} = \dot{f}^M(u_j^M) \tag{4.63}$$

It is possible to write

$$\delta_i^M = -2(d_i - y_i)\dot{f}^M(u_j^M) \tag{4.64}$$

Or equivalently in matrix notation

$$\delta^M = -2\dot{\mathbf{F}}^M(\mathbf{u}^M)(\mathbf{d} - \mathbf{y}) \tag{4.65}$$

Figure 4.27 provides an adaptation, using matrix notation, of the result presented by Narendra and Parthasarathy (1990) and describes a flow graph for the backpropagation algorithm.

Algorithm 4.4 presents the pseudocode for the standard backpropagation algorithm. The value of α should be small, $\alpha \in (0,1]$. As this algorithm updates the weight matrices of the network after the presentation of each input pattern, this could cause a certain bias toward the order in which the training patterns are presented. In order to avoid this, the patterns are presented to the network in a random order, as described in Algorithm 4.4.

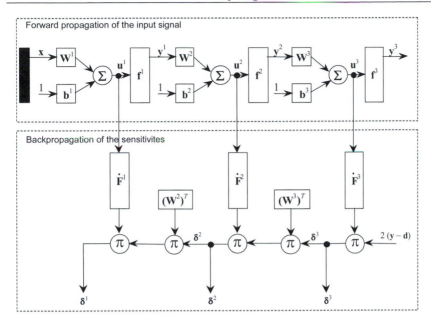

Figure 4.27: Architectural graph of an MLP network with three layers representing the forward propagation of the functional signals and the backward propagation of the sensitivities.

```
procedure [W] = backprop(max_it,min_err,α,X,D)
    for m from 1 to M do,
        initialize Wᵐ //small random values
        initialize bᵐ //small random values
    end for
    t ← 1
    while t < max_it & MSE > min_err do,
        vet_permut ← randperm(N)   //permutations of N
        for j from 1 to N do,      //for all input patterns
            //select the index i of pattern xᵢ to be presented
            i ← vet_permut(j) //present patterns randomly
            //forward propagation of the functional signal
```
$$\mathbf{y}^0 \leftarrow \mathbf{x}_i \qquad\qquad //\text{Equation (4.37)}$$
```
            for m from 0 to M - 1 do,
```
$$\mathbf{y}_i^{m+1} \leftarrow \mathbf{f}^{m+1}(\mathbf{W}^{m+1}\mathbf{y}_i^m + \mathbf{b}^{m+1}) \qquad //\text{Equation (4.36)}$$
```
            end for
            //backpropagation of sensitivities
```
$$\boldsymbol{\delta}_i^M \leftarrow -2\dot{\mathbf{F}}^M(\mathbf{u}_i^M)(\mathbf{d}_i - \mathbf{y}_i) \qquad //\text{Equation (4.65)}$$
```
            for m from M - 1 down to 1 do,
```
$$\boldsymbol{\delta}_i^m \leftarrow \dot{\mathbf{F}}^m(\mathbf{u}_i^m)(\mathbf{W}^{m+1})^T\boldsymbol{\delta}_i^{m+1}, \qquad //\text{Equation (4.60)}$$
```
            end for
            //update weights and biases
            for m from 1 to M do,
```
$$\mathbf{W}^m \leftarrow \mathbf{W}^m - \alpha\boldsymbol{\delta}_i^m(\mathbf{y}_i^{m-1})^T, \qquad //\text{Equation (4.52)}$$
$$\mathbf{b}^m \leftarrow \mathbf{b}^m - \alpha\boldsymbol{\delta}_i^m, \qquad //\text{Equation (4.53)}$$
```
            end for
            //calculate the error for pattern i
```
$$E_i \leftarrow \mathbf{e}_i^T\mathbf{e}_i = (\mathbf{d}_i - \mathbf{y}_i)^T(\mathbf{d}_i - \mathbf{y}_i)$$
```
        end for
        MSE ← 1/N.sum(Eᵢ)                //Mean Square Error
        t ← t + 1
    end while
end procedure
```

Algorithm 4.4: Learning algorithm for the MLP network trained via the backpropagation algorithm.

Universal Function Approximation

An MLP network can be seen as a generic tool to perform a *nonlinear input-output mappings*. More specifically, let *m* be the number of inputs to the network and *o* the number of outputs. The input-output relationship of the network defines a mapping from an input *m*-dimensional Euclidean space into an output *o*-dimensional Euclidean space, which is infinitely continuously differentiable.

Cybenko (1989) was the first researcher to rigorously demonstrate that an MLP neural network with a single hidden layer is sufficient to uniformly approximate any continuous function that fits a unit hypercube.

The *universal function approximation theorem* is as follows:

Theorem: Let $f(\cdot)$ be a nonconstant continuous, limited, and monotonically increasing function. Let I_m be a unit hypercube of dimension m, $(0,1)^m$. The space of continuous functions in I_m is denoted by $C(I_m)$. Thus, given any function $g \in C(I_m)$ and $\varepsilon > 0$, there is an integer M and sets of real-valued constants α_i and w_{ij}, where $i = 1, \dots , o$ and $j = 1, \dots , m$, it is possible to define

$$F(x_1, x_2, \dots, x_m) = \sum_{i=1}^{o} \alpha_i f\left(\sum_{j=1}^{m} w_{ij} x_j - w_{i0} \right) \tag{4.66}$$

as an approximation to the function $g(.)$ such that

$|F(x_1, \dots , x_m) - g(x_1, \dots , x_m)| < \varepsilon$ for all $\{x_1, \dots , x_m\} \in I_m$.

Proof: see Cybenko (1989).

This theorem is directly applicable to the MLP networks. First, note that the sigmoidal function used in the MLP networks is continuous, nonconstant, limited, and monotonically increasing; satisfying the constraints imposed to $f(\cdot)$. Then, note that Equation (4.66) represents the outputs of an MLP network, as illustrated in Figure 4.28.

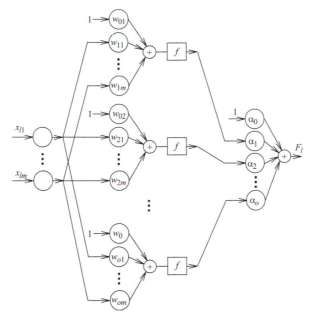

Figure 4.28: An MLP network as a universal function approximator. The network parameters compose Equation (4.66). (Courtesy of © Fernando J. Von Zuben.)

In summary, the theorem states that an MLP network with a single hidden layer is capable of uniform approximation, given an appropriate training data set to represent the function. However, the theorem does not say anything regarding the number of units in the hidden layer necessary to perform an approximation with the precision ε.

Some Practical Aspects

There are a several aspects of the MLP network training that deserve some comments. Some of these aspects are also valid for the other networks presented and to be presented in this chapter.

- *Network architecture*: the number of network inputs is usually defined by the training data, while the number of outputs and hidden units are design issues. For instance, if a data set to be classified has a single class, one or two output units can be used. In the case of a single sigmoidal output in the range $(-1,1)$, a network output less than '0' can be considered as not belonging to the class, and an output greater than or equal to zero can be considered as belonging to the class. If two output units are used, then one of them corresponds to belonging to the class, while the other corresponds to not belonging to the class. Finally, the number of hidden units is chosen so that the network can appropriately classify or approximate the training set. By increasing the number of hidden units, one increases the network mapping capability. It is important to have in mind that an excessive number of hidden units may result in overfitting, while a small number may lead to underfitting.

- *Generalization*: we have already discussed that too much training may result in overfitting, while too little training may result in underfitting. Overfitting is only possible if there are enough hidden units to promote an increasingly better representation of the data set. On the other side, if the network is not big enough to perform an adequate input-output mapping, underfitting may occur.

- *Convergence*: the MLP network is usually trained until a minimum value for the error is achieved; for instance, stop training if the MSE < 0.001. Another choice for the *stopping criterion* is to stop training if the estimated gradient vector at iteration t has a small norm; for instance, stop training if $\|\partial\mathfrak{J}/\partial w\| < 0.001$. A small value for the norm of the gradient value indicates that the network is close to a local minimum.

- *Epoch*: after all the training patterns have been presented to the network and the weights adjusted, an *epoch* is said to have been completed. This terminology is used in all artificial neural network learning algorithms, including the perceptron and self-organizing maps.

- *Data normalization*: when the activation functions of the neurons have a well-defined range, such as all sigmoidal functions, it is important to normalize the data such that they belong to a range equal to or smaller than the range of the activation functions. This helps to avoid the saturation of the neurons in the network. Normalizing the training data is also important when their attributes range over different scales. For example, a data set about cars may include variables such as color (symbolic or discrete attribute), cost (real-valued attribute), type (off road, van, etc.), and so on. Each of these attributes assumes a value on a different range and, thus, influences the network with different degrees. By normalizing all attributes to the same range, we reduce the importance of the differences in scale.

- *Initialization of the weight vectors and biases*: most network models, mainly multi-layer feedforward networks, are sensitive to the initial values chosen for the network weights. This sensitivity can be expressed in several ways, such as guarantee of convergence, convergence speed, and convergence to local optima solutions. Usually, small random values are chosen to initialize the network weights.

- *Learning rate*: the learning rate α plays a very important role in neural network training because it establishes the size of the adaptation step to be performed on a given weight. Too small learning rates may result in a very long learning time, while too large learning rates may result in instability and nonconvergence.

Biological Plausibility of Backpropagation

The backpropagation of error signals is probably the most problematic feature in biological terms. Despite the apparent simplicity and elegance of the backpropagation algorithm, it seems quite implausible that something like the equations described above are computed in the brain (Anderson, 1995; O'Reilly and Munakata, 2000; Trappenberg, 2002).

In order to accomplish this, some form of information exchange between postsynaptic and presynaptic neurons should be possible. For instance, there would be the requirement, in biological terms, that the sensitivity values were propagated backwards from the dendrite of the postsynaptic neuron, across the synapse, into the axon terminal of the presynaptic neuron, down the axon of this neuron, and then integrated and multiplied by both the strength of that synapse and some kind of derivative, and then propagated back out its dendrites, and so on. If a neuron were to gather the backpropagated errors from all the other nodes to which it projects, some synchronization issues would arise, and it would also affect the true parallel processing in the brain.

However, it is possible to rewrite the backpropagation equations so that the error backpropagation between neurons takes place using standard activation signals. This approach has the advantage that it ties into the psychological interpretation of the teaching signal (desired output) as an actual state of experience that reflects something like an outcome or corrected response. The result is a very powerful learning algorithm that need not ignore issues of biological plausibility (Hinton and McClelland, 1987; O'Reilly, 1996).

Examples of Application

Two simple applications for the MLP network will be presented here. The first problem - universal function approximation - aims at demonstrating that the MLP network with nonlinear hidden units is capable of universal approximation, and how this is accomplished. The second example - design of a controller - is a classic example from the literature that demonstrates the capability of using an MLP network trained with error backpropagation to implement a self-learning controller.

Universal Function Approximation

To illustrate the universal function approximation of the MLP neural networks, consider the following example: approximate one period of the function $sin(x).cos(2x)$ using an additive composition of basic functions ($f(\cdot)$ is the hyperbolic tangent function) under the form of Equation (4.66), as illustrated in Figure 4.29.

The MLP network depicted in Figure 4.30 is used to approximate this function. The architecture is: $m = 1$ input, $S^1 = 5$ hidden units in a single hidden layer, and $o = 1$ output unit. The training algorithm is the backpropagation introduced above. By looking at this picture, and assuming the network has a linear output, y is expressed as

$$y = w_{10} + w_{11}z_1 + w_{12}z_2 + w_{13}z_3 + w_{14}z_4 + w_{15}z_5 \qquad (4.67)$$

where

$$z_l = f\left(\sum_{j=1}^{m} v_{lj}x_j + v_{l0}\right), \quad l = 1, \dots, 5 \qquad (4.68)$$

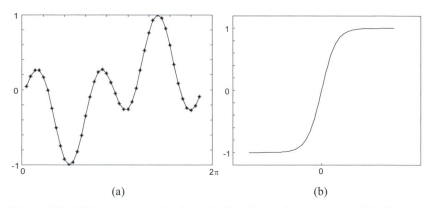

(a)   (b)

Figure 4.29: Universal approximation. (a) Function to be approximated (information available: 42 training data equally sampled in the domain of the function. (b) Basic function to be used in the approximation: hyperbolic tangent, satisfying the constraints of the universal function approximation theorem.

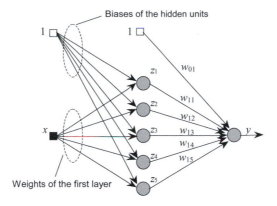

Figure 4.30: MLP network trained via the error backpropagation algorithm used to approximate the function $sin(x).cos(2x)$. The activation functions are hyperbolic tangents.

where $f(\cdot)$ is the hyperbolic tangent, m is the number of network inputs, v_{lj} is the weight of the synapse connecting input j to the hidden unit l. (Compare Equation (4.67) and Equation (4.68) with Equation (4.66).)

After training the network so as to define appropriate weight sets, let us analyze the components of Equation (4.67) that determine the network output. Figure 4.31 depicts the activation functions (basic functions) of the hidden units after network training. Note that these curves are hyperbolic tangents translated or scaled along the axes. Figure 4.32 presents the linear combination of the outputs of the hidden units, where the coefficients of the combination are the weights of the second layer in the network. Finally, the resultant approximation is presented in Figure 4.33.

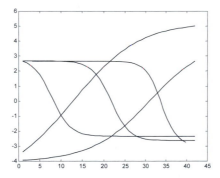

Figure 4.31: Activation functions of the hidden units, after the network training, multiplied by the corresponding weights of the output layer. Note that all functions are hyperbolic tangents (basic functions).

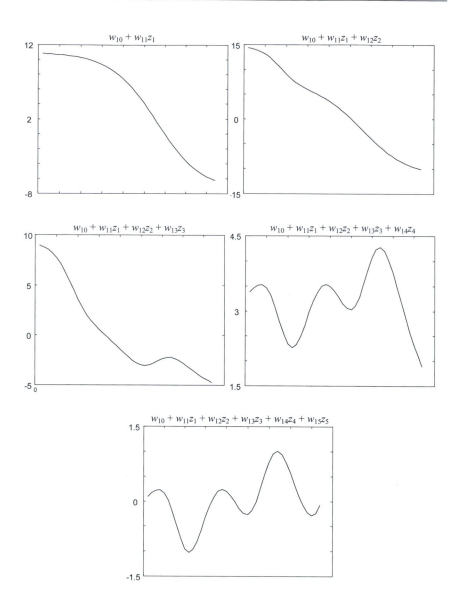

Figure 4.32: Linear combination of the outputs of the hidden units. The coefficients of the linear combination are the weights of the second layer. This figure presents the sum of the curves that compose Figure 4.31, corresponding to the additive composition (linear combination) of the basic functions of the network's neurons.

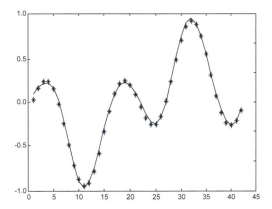

Figure 4.33: Approximation obtained by the MLP network for the function $sin(x).cos(2x)$. The '*' correspond to the training samples, and the line (—) is the network approximation.

Design of a Controller

In a classic paper from the ANN literature, Nguyen and Widrow (1990) applied a simple multi-layer perceptron network trained with the standard error back-propagation algorithm to design a neuro-controller to control the steering of a trailer truck while backing up to a loading dock from any initial position. The authors wanted to demonstrate how a simple neural network architecture could be used to implement a self-learning nonlinear controller for the truck when only backing up was allowed. The network architecture was a standard MLP with a single hidden layer, 25 hyperbolic tangents in the hidden layer and a single sigmoid (tanh) output unit.

The critical state variables representing the position of the truck and that of the loading dock are the angle of the cab θ_{cab}, the angle of the trailer $\theta_{trailer}$, and the Cartesian position of the rear of the center of the trailer $(x_{trailer}, y_{trailer})$. The truck is placed at some initial position and is backed up while being steered by the controller, and the run ends when the truck gets to the dock, as illustrated in Figure 4.34. The goal is to cause the back of the trailer to be parallel to the loading dock; that is, $\theta_{trailer} = 0$ and to have $(x_{trailer}, y_{trailer})$ as close as possible to (x_{dock}, y_{dock}). The final cab angle is not important. An objective function that allows the controller to learn to achieve these objectives by adapting the network weights and biases is:

$$\Im = E[\alpha_1(x_{dock} - x_{trailler})^2 + \alpha_2(y_{dock} - y_{trailler})^2 + \alpha_3(0 - \theta_{trailler})^2]$$

where the constants α_1, α_2 and α_3 are chosen by the user so as to weigh the importance of each error component, and $E[\cdot]$ corresponds to the average (expectation) over all training runs.

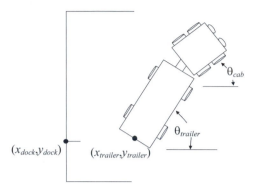

Figure 4.34: Illustration of the truck, trailer, loading dock, and the parameters involved. (Adapted with permission from [Nguyen and Widrow, 1990], © IEEE Press.)

The training of the controller was divided into several 'lessons'. In the beginning, the controller was trained with the truck initially very close to the dock and the trailer pointing at the dock. Once the controller was proficient at working with these initial conditions, the problem was made harder by placing the truck farther away from the dock and with increasingly more difficult angles. This way, the controller learned how to deal with easy situations first, and then was adapted to deal with harder scenarios.

The MLP neural network trained with the standard backpropagation of errors learning algorithm demonstrated to be capable of solving the problem by controlling the truck very well. The truck and trailer were placed in various different initial positions, and backing up was performed successfully in each case. The trained controller was capable of controlling the truck from initial positions it had never seen before, demonstrating the generalization capability of the neural network studied.

4.4.5. Self-Organizing Maps

The *self-organizing map* (SOM) discussed in this section is an unsupervised (self-organized or competitive; Section 4.3.3) network introduced by T. Kohonen (1982). The network architecture is a single layer feedforward network with the postsynaptic neurons placed at the nodes of a *lattice* or *grid* that is usually uni- or bi-dimensional, as illustrated in Figure 4.35. This network is trained via a competitive learning scheme, thus there are lateral connections between the output units which are not shown in the picture. (See Appendix B.4.5 for a brief introduction to data clustering.)

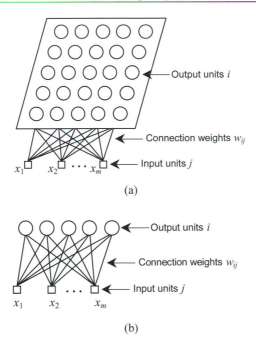

Figure 4.35: Typical architectures (single layer feedforward) of a self-organized map. (a) Bi-dimensional output grid. (b) Uni-dimensional output grid.

The output units become selectively tuned to various input signal patterns or clusters of patterns through an unsupervised learning process. In the standard version of the SOM to be presented here, only one neuron or local group of neurons at a time gives the active response to the current input. The locations of the responses tend to become ordered as if some meaningful coordinate system for different input features were being created over the network (Kohonen, 1990). A self-organizing map is thus characterized by the formation of a topographic map of the input patterns in which the spatial location or coordinates of a neuron in the network are indicative of intrinsic statistical features contained in the input patterns.

For instance, Figure 4.36 depicts a possible synaptic weight distribution for a uni-dimensional self-organizing map with two inputs, $m = 2$. Note that, although the output array allows that the neurons can be freely positioned in the space of the input data, its dimension is still one. Thus, the interpretation of this uni-dimensional output array, after self-organization, occurs in a uni-dimensional space independently of the original data dimension.

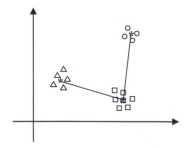

Figure 4.36: Possible distribution of weight vectors for two-input patterns and a uni-dimensional array of neurons in a SOM. The black stars correspond to the position of the SOM weight vectors and the lines connecting them represent their neighborhood relationships. Triangles, rectangles, and circles correspond to different clusters of data that are presented to the network without their labels; that is, the network is not informed that they belong to different groups.

Self-Organizing Map Learning Algorithm

The main objective of the self-organizing map is to transform an input pattern of dimension *m* into a uni- or bi-dimensional discrete map, and to perform this transformation adaptively in a topologically ordered form. Each unit in the grid presented in Figure 4.35 is fully connected to all the presynaptic units in the input layer. The uni-dimensional array of Figure 4.35(b) is a particular case of the bi-dimensional grid of Figure 4.35(a).

The self-organizing learning algorithm proceeds by first initializing the synaptic strengths of the neurons in the network, what usually follows the same rule of the other networks discussed; that is, by assigning them small values picked from a random or uniform distribution. This stage is important so that no *a priori* order is imposed into the map. After initialization, there are three essential processes involved in the formation of the self-organized map (Haykin, 1999):

- *Competition*: for each input pattern the neurons in the network will compete with one another in order to determine the winner.

- *Cooperation*: the winning neuron determines the spatial location (neighborhood) around which other neighboring neurons will also be stimulated. It is crucial to the formation of ordered maps that the cells doing the learning are not affected independently of one another, but as topologically related subsets, on each of which a similar kind of correction is imposed.

- *Adaptation*: the winning neuron and its neighbors will have their associated weight vectors updated. Usually the degree of adaptation of the weight vectors is proportional to their distance, in the topological neighborhood, to the winner: the closer to the winner, the higher the degree of adaptation, and vice-versa.

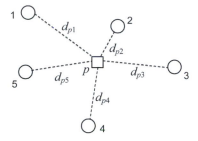

Figure 4.37: Competitive phase.

Let $\mathbf{x} = [x_1, x_2, \ldots, x_m]^T \in \mathfrak{R}^m$ be an input vector (training pattern) that is assumed to be presented in parallel to all the neurons $i = 1, \ldots, o$, in the network. The whole data set $\mathbf{X}$ is composed on N patterns, $\mathbf{X} \in \mathfrak{R}^{m \times N}$. The weight vector of unit i is $\mathbf{w}_i = [w_{i1}, w_{i2}, \ldots, w_{im}] \in \mathfrak{R}^{1 \times m}$, and thus the weight matrix of the network is $\mathbf{W} \in \mathfrak{R}^{o \times m}$, where o is the number of output (postsynaptic) units in the network.

Competition

For each input pattern, the o neurons in the network will compete to become activated. One simple way of performing the competition among the output units is by applying the competitive learning scheme discussed in Section 4.3.3. That is, for a neuron i to be the winner, the distance between its corresponding weight vector $\mathbf{w}_i$ and a certain input pattern $\mathbf{x}$ must be the smallest measure (among all the network output units), given a certain metric $\| \cdot \|$, usually taken to be the Euclidean distance. Let $i(\mathbf{x})$ indicate the winner neuron for input pattern $\mathbf{x}$. It can thus be determined by

$$i(\mathbf{x}) = \arg \min_j \|\mathbf{x} - \mathbf{w}_j\|, \; j = 1, \ldots, o \tag{4.69}$$

Neuron $i(\mathbf{x})$ that satisfies Equation (4.69) is known as the best-matching, winner, or winning unit for the input pattern $\mathbf{x}$.

Figure 4.37 illustrates the competition in a SOM. Assume a map with 5 neurons and an input pattern p presented to the network. Each value $d_{pj}, j = 1, \ldots, 5$, corresponds to the Euclidean distance between the neuron weight vector and the input pattern p. In this example, the winner neuron is number 2, because it has the smallest Euclidean distance to the input pattern and is, thus, the neuron most activated by the input pattern p.

Cooperation

The winning neuron locates the center of a topological neighborhood of cooperating neurons. There are neurobiological evidences for the lateral interaction among a set of excited neurons. In particular, a neuron that is firing tends to excite the neurons in its immediate neighborhood more than those far away from it. One form of simulating this is by defining topological neighborhoods around a winning neuron, in which a similar kind of adaptation to that of the winner will be imposed.

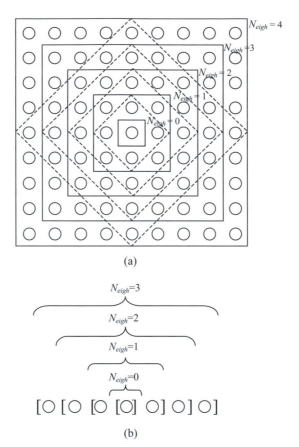

(a)

(b)

Figure 4.38: Examples of topological neighborhoods for bi-dimensional (a) and uni-dimensional grids (b).

In biophysically inspired neural networks, there are a number of ways of defining the topological neighborhood by using various kinds of lateral feedback connections and other lateral interactions. In the present discussion, lateral interaction is enforced directly by defining a neighborhood set N_{eigh} around a winning unit u. At each iteration, all the units within N_{eigh} are updated, whereas units outside N_{eigh} are left intact. The width or radius of N_{eigh} is usually time variant; N_{eigh} is initialized very wide in the beginning of the learning process and shrinks monotonically with time. Figure 4.38 illustrates some typical topological neighborhoods for bi-dimensional and uni-dimensional output grids. In Figure 4.38(a), two different forms of implementing a squared neighborhood for a bi-dimensional grid are depicted. Note that other neighborhoods are still possible, such as a hexagonal (or circular) neighborhood.

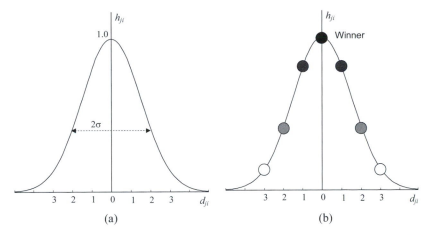

(a)   (b)

Figure 4.39: Neighborhood. (a) Gaussian neighborhood. (b) Neighborhood relationships: gray levels indicate the neighborhood degree of each unit. The winner neuron has a higher value for h, $h_{ji(\mathbf{x})}$, while h_{ji} of the other neurons decreases for larger distances from the winner. (Courtesy of © Lalinka C. T. Gomes)

Another form of simulating the influence of a neuron in its immediate neighborhood is to define a general kernel function $h_{ji(\mathbf{x})}$ that allows a decay of the neurons updating smoothly with lateral distance (Figure 4.39). One form of implementing this is by giving function $h_{ji(\mathbf{x})}$ a "bell curve" shape. More specifically, let $h_{ji(\mathbf{x})}$ denote the topological neighborhood centered on the winning neuron $i(\mathbf{x})$ and encompassing a set of neighbor neurons $j \in N_{eigh}$, and let $d_{i(\mathbf{x})j}$ denote the lateral distance between the winning neuron $i(\mathbf{x})$ and its neighbor j:

$$h_{ji(\mathbf{x})} = \exp(-\|\mathbf{r}_j - \mathbf{r}_{i(\mathbf{x})}\|^2/2\sigma^2) \tag{4.70}$$

where $\mathbf{r}_j$ and $\mathbf{r}_{i(\mathbf{x})}$ are discrete vectors that indicate the coordinates of cells j and $i(\mathbf{x})$, respectively, on the bi-dimensional grid of the SOM. In this case, the value of $\sigma = \sigma(t)$ will be responsible for controlling the neighborhood influence along the iterative procedure of adaptation. N_{eigh} can be kept constant and $\sigma(t)$ iteratively reduced until the network weights stabilize.

Adaptation

The updating process for the weight vectors is similar to the one presented in Equation (4.19) for competitive learning schemes, but takes into account the topological neighborhood of the winning unit (Figure 4.40).

If the neighborhood is defined topologically as presented in Figure 4.38, then the updating rule presented in Equation (4.71) can be used to adjust the weight vectors.

$$\mathbf{w}_i(t+1) = \begin{cases} \mathbf{w}_i(t) + \alpha(t)[\mathbf{x}(t) - \mathbf{w}_i(t)] & \text{if } i \in N_{eigh} \\ \mathbf{w}_i(t) & \text{if } i \notin N_{eigh} \end{cases} \tag{4.71}$$

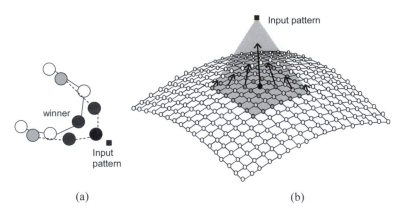

(a)   (b)

Figure 4.40: Adaptive procedure. The winner and its neighbors are moved in the direction of the input pattern. The winner unit performs a greater updating in the direction of the input pattern, followed by its immediate neighbors. (a) Uni-dimensional neighborhood. (b) Bi-dimensional neighborhood. (Courtesy of © Lalinka C. T. Gomes)

where the learning rate $\alpha(t)$ decreases with time, $0 < \alpha(t) < 1$.

However, if the neighborhood is defined using a general kernel function, an alternative notation is to introduce the scalar kernel function $h_{ji(\mathbf{x})} = h_{ji(\mathbf{x})}(t)$, described previously, directly in the weight adjusting equation

$$\mathbf{w}_i(t+1) = \mathbf{w}_i(t) + \alpha(t)\, h_{ji(\mathbf{x})}(t)\, [\mathbf{x}(t) - \mathbf{w}_i(t)] \qquad (4.72)$$

where $h_{ji(\mathbf{x})}(t) = 1$ within N_{eigh}, and $h_{ji(\mathbf{x})}(t) = 0$ outside N_{eigh}.

It is sometimes useful to keep the initial values for the parameters α and N_{eigh} (or σ) fixed for a number of iterations, usually 1,000, before starting to cool (reduce) them. Following this approach may result in an initial period corresponding to a self-organizing or *ordering phase* of the map, followed by a *convergence phase* in which, as the values of α and N_{eigh} (or σ) are monotonically reduced, the network will later suffer very little adaptation. Algorithm 4.5 summarizes the learning algorithm for the Kohonen self-organizing map. One simple form of cooling α and σ is by adopting a geometric decrease, i.e., by multiplying α and σ by a constant value smaller than one. Another form is by adopting the equations below (Haykin, 1999):

$$\alpha(t) = \alpha_0 \exp(-t/\tau_1) \qquad (4.73)$$

$$\sigma(t) = \sigma_0 \exp(-t/\tau_2) \qquad (4.74)$$

where $\alpha_0 \approx 0.1$, $\tau_1 \approx 1,000$, $\sigma_0 =$ radius of the grid (all neurons are assumed to be neighbors of the winning neuron in the beginning of learning), and $\tau_2 = 1,000/\log(\sigma_0)$.

```
procedure [W] = som(max_it,α₀,σ₀,τ₁,τ₂,X)
    initialize W        //small random values (wᵢ≠wⱼ for i≠j)
    t ← 1
    while t < max_it do,
        vet_permut ← randperm(N)   //permutations of N
        for j from 1 to N do,        //for all input patterns
            //select the index i of pattern xᵢ to be presented
            i ← vet_permut(j)  //present patterns randomly
            //competition and determination of the winner i(x)
            i(x) ← arg minⱼ ||xᵢ – wⱼ(t)||, j = 1,...,o
            //calculate the neighborhood function
            hⱼᵢ₍ₓ₎ = exp(–||rⱼ – rᵢ₍ₓ₎||²/2σ²)
            //cooperation and adaptation
            wᵢ(t+1) = wᵢ(t) + α(t) hⱼᵢ₍ₓ₎(t) [x(t) – wᵢ(t)]
        end   for
        //update α and the neighborhood function
        α ← reduce(α)          //until α reaches a minimal value
        σ ← reduce(σ)          //until only the winner is updated
        t ← t + 1
    end while
end procedure
```

Algorithm 4.5: Learning algorithm for the Kohonen self-organized map (SOM).

Biological Basis and Inspiration for the Self-Organizing Map

It has been long since a topographical organization of the brain, in particular of the cerebral cortex, has been suggested. These could be deduced from functional deficits and behavioral impairments produced by various kinds of lesions, hemorrhages, tumors, and malformations. Different regions in the brain thereby seemed to be dedicated to specific tasks (see Figure 4.5).

Nowadays, more advanced techniques, such as *positron emission topography* (PET), *magneto encephalography* (MEG), and *electroencephalogram* (EEG), can be used to measure nervous activity, thus allowing the study of the behavior of living brains. After a large number of experimentation and observation, a fairly detailed organization view of the brain has been proposed. Especially in higher animals, such as the human beings, the various cortices in the cell mass seem to contain many kinds of topographic maps, such that a particular location of the neural response in the map often directly corresponds to a specific modality and quality of sensory signal.

The development of the self-organizing maps as neural models is motivated by this distinctive feature of local brain organization into topologically ordered maps. The computational map constitutes a basic building block in the information-processing infrastructure of the nervous system. A computational map is defined by an array of neurons representing slightly different tuned processors or filters, which operate on the sensory information-bearing signals in parallel (Haykin, 1999).

The use of these computational maps offers some interesting properties:

- At each stage of representation, each incoming piece of information is kept in its proper context.

- Neurons dealing with closely related pieces of information are close together so they can interact via short synaptic connections.

The self-organizing map, or perhaps more accurately named *artificial topographic map*, was designed to be a system capable of learning through self-organization in a neurobiologically inspired manner. Under this perspective, the most relevant aspect of the nervous system to be taken into account is its capability of topographical map formation. As stated by T. Kohonen:

> "The spatial location or coordinates of a cell in the network then correspond to a particular domain of input signal patterns." (Kohonen, 1990; p. 1464)

Examples of Applications

To illustrate the behavior and applicability of the SOM, it will first be applied to a structured random distribution of data. This will demonstrate the self-organizing properties of the network. The second problem - animals' data set - demonstrates the SOM capability of identifying groups of data, and allowing for the visualization of useful relationships within a high-dimensional data set in a low dimensional grid, usually of one or two dimensions.

Structured Random Distribution

To illustrate the behavior of the SOM, consider an input data set composed of 500 bi-dimensional samples uniformly distributed over the interval $[-1,1]$, as depicted in Figure 4.41(a). The input patterns were normalized to the unit interval so that their distribution lies on the surface of a unit circle. A bi-dimensional grid with six rows and six columns (6×6) of neurons is used. Figure 4.41(b) to (d) shows three stages of the adaptive process as the network learns to represent the input distribution.

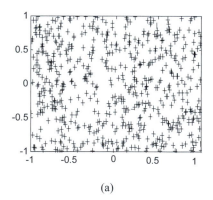

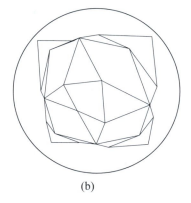

(a)   (b)

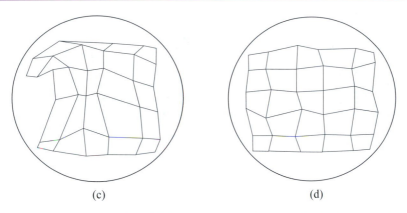

(c)   (d)

Figure 4.41: Application of the SOM network with a bi-dimensional grid to a randomly distributed input data set. (a) Input data set for a bi-dimensional SOM. (b) Condition of the grid after 1,000 iterations. (c) Condition of the grid after 5,000 iterations. (d) Grid after 10,000 iterations.

Semantic Maps

This section illustrates the SOM behavior by applying it to the Animals data set. This high-dimensional data set was originally proposed by Ritter and Kohonen (1989) to verify the SOM capability of creating a topographic map based on a symbolic data set. The data set is composed of 16 input vectors each representing an animal with the binary feature attributes as shown in Table 4.1.

Table 4.1: Animals data set: animals' names and their attributes.

		Dove	Hen	Duck	Goose	Owl	Hawk	Eagle	Fox	Dog	Wolf	Cat	Tiger	Lion	Horse	Zebra	Cow
	Small	1	1	1	1	1	1	0	0	0	0	1	0	0	0	0	0
Is	Medium	0	0	0	0	0	0	1	1	1	1	0	0	0	0	0	0
	Big	0	0	0	0	0	0	0	0	0	0	0	1	1	1	1	1
	Two legs	1	1	1	1	1	1	1	0	0	0	0	0	0	0	0	0
	Four legs	0	0	0	0	0	0	0	1	1	1	1	1	1	1	1	1
Has	Hair	0	0	0	0	0	0	0	1	1	1	1	1	1	1	1	1
	Hooves	0	0	0	0	0	0	0	0	0	0	0	0	0	1	1	1
	Mane	0	0	0	0	0	0	0	0	0	1	0	0	1	1	1	0
	Feathers	1	1	1	1	1	1	1	0	0	0	0	0	0	0	0	0
	Hunt	0	0	0	0	1	1	1	1	0	1	1	1	1	0	0	0
Likes to	Run	0	0	0	0	0	0	0	0	1	1	0	1	1	1	1	0
	Fly	1	0	0	1	1	1	1	0	0	0	0	0	0	0	0	0
	Swim	0	0	1	1	0	0	0	0	0	0	0	0	0	0	0	0

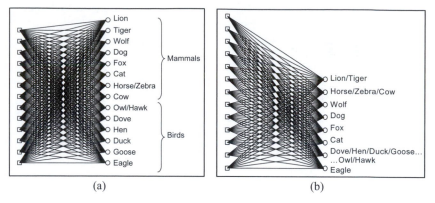

(a)   (b)

Figure 4.42: Uni-dimensional map containing labeled units with strongest responses to their respective inputs. (a) Output grid with 14 units. (b) Output grid with 8 units.

A value of '1' in Table 4.1 corresponds to the presence of an attribute, while a value of '0' corresponds to the lack of this attribute. The authors suggested that the interestingness of this data set lies in the fact that the relationship between the different symbols may not be directly detectable from their encodings, thus not presuming any metric relations even when the symbols represent similar items.

Figure 4.42 illustrates the performance of a SOM with a uni-dimensional output grid applied to the Animals data set. In Figure 4.42(a) an output grid with 14 units was used and in Figure 4.42(b) a grid with 8 units was used. Note the interesting capability of the SOM to cluster data which are semantically similar without using any labeled information about the input patterns. By observing Table 4.1 it is possible to note that the horse and zebra, as well as the owl and the hawk, have the same set of attributes. Note that animals with similar features are mapped onto the same or neighboring neurons.

4.4.6. Discrete Hopfield Network

J. Hopfield (1982, 1984) realized that in physical systems composed of a large number of simple elements, interactions among large numbers of elementary components yield complex collective phenomena, such as the stable magnetic orientations and domains in a magnetic system or the vortex patterns in fluid flow. Particularly, Hopfield noticed that there are classes of physical systems whose spontaneous behavior can be used as a form of general and error-correcting *content-addressable memory* (CAM). The primary function of a content-addressable memory, also called an *associative memory*, is to retrieve a pattern stored in the memory as a response to the presentation of an incomplete or noisy version of that pattern, as illustrated in Figure 4.43. In a CAM, the memory is reached not by knowing its address, but rather by supplying some subpart of the memory.

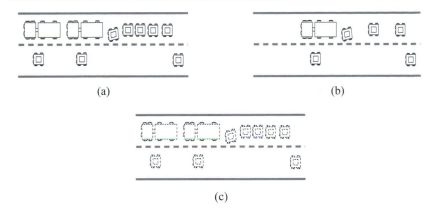

Figure 4.43: Content-addressable memory (CAM) or associative memory. Given a memorized pattern, the role of the CAM is to restore the stored pattern given an incomplete or noisy version of it. (a) Memorized patterns. (b) Incomplete patterns. (c) Noisy patterns.

By assuming that the time evolution of a physical system can be described by a set of general coordinates, then a point in a space, called state space, represents the instantaneous condition of the system. Therefore, the equations of motion of the system describe a flow in state space. Assuming also systems whose flow is performed toward locally stable points from anywhere within given neighborhoods of these stable points, these systems become particularly useful as content-addressable memories.

The original neuron model used by Hopfield (1982) was that of McCulloch and Pitts (1943), whose output could be either '0' or '1', and the inputs were external and from other neurons as well. Thus, his analysis involved networks with full back-coupling; that is, recurrent networks (such as the one illustrated in Figure 4.13, but with no external inputs). Other important features of the Hopfield network were the storage (instead of iterative learning) of the memories of the system, and the use of an asynchronous procedure of memory retrieval, as will be further discussed.

Recurrent Neural Networks as Nonlinear Dynamical Systems

The Hopfield neural networks were proposed by J. Hopfield in the early 1980s (Hopfield, 1982, 1984) as neural models derived from statistical physics. They incorporate a fundamental principle from physics: the storage of information in a dynamically stable configuration. Each pattern is stored in a valley of an energy surface. The nonlinear dynamics of the network is such that the energy is minimized and valleys represent points of stable equilibrium, each one with its own basin of attraction. Therefore, a network memory corresponds to a point of stable equilibrium, thus allowing the system to behave as an associative memory. Appendix B.4.3 brings an elementary discussion about nonlinear dynamical systems.

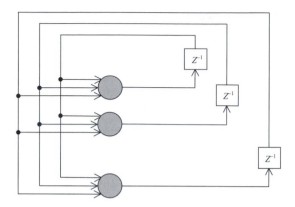

Figure 4.44: Recurrent neural network with no hidden layer.

In this case, convergent flow to stable states becomes the essential feature of the CAM operation. There is a simple mathematical condition that guarantees that the state space flow algorithm converges to stable states. Any symmetric matrix **W** with zero-diagonal elements (i.e., $w_{ij} = w_{ji}$, $w_{ii} = 0$) will produce such a flow. The proof of this property follows from the construction of an appropriate energy function that is always decreased by any state change produced by the algorithm.

Assume a single-layer neural network composed of N neurons and a corresponding set of unit delays, forming a multiple-loop feedback system, as illustrated Figure 4.44. Note that this network is a particular case of the network presented in Figure 4.13.

Assume that the neurons in this network have symmetric coupling described by $w_{ij} = w_{ji}$ ($i,j = 1, \ldots, N$), where w_{ij} is the synaptic strength connecting the output of unit i to the input of unit j. Assume also that no self-feedback is allowed, i.e., $w_{ii} = 0$ ($i = 1, \ldots, N$). Let $u_j(t)$ be the net input to neuron j and $y_j(t)$ its corresponding output $y_j(t) = f(u_j(t))$, where $f(\cdot)$ is the activation function of unit j.

Let now the vector **u**(t) be the state of the network at time t. Then, from the basic integrate and fire model of a neuron given by Equation (4.1) it is possible to define an additive model of a neuron as given by Equation (4.75).

$$\tau_m \frac{du(t)}{dt} = u_{res} - u(t) + R_m i(t) \qquad (4.1)$$

where τ_m is the membrane time constant of the neuron determined by the average conductance of the channels (among other things); u_{res} is the resting potential of the neuron; $i(t)$ is the input current given by the sum of the synaptic currents generated by firings of presynaptic neurons; and R_m is the resistance of the neuron to the flow of current (ions).

$$C_j \frac{du_j}{dt} = I_j - \frac{u_j}{R_j} + \sum_{\substack{i=1 \\ i \neq j}}^{N} w_{ji} f_i(u_i), j = 1, \dots, N \qquad (4.75)$$

where R_j and C_j are the leakage resistance and capacitance of the neuron respectively, I_j corresponds to an externally applied bias, u_j is the net input to unit j, and $f_i(\cdot)$ is the activation function of unit i.

To the dynamics provided by Equation (4.75) it is possible to derive an energy function as given by Equation (4.76) (for details see Hopfield, 1984).

$$E = -\frac{1}{2} \sum_{i=1}^{N} \sum_{\substack{j=1 \\ j \neq i}}^{N} w_{ji} y_i y_j + \sum_{j=1}^{N} \frac{1}{R_j} \int_0^{y_j} f_j^{-1}(z) dz + \sum_{j=1}^{N} b_j y_j \qquad (4.76)$$

This energy function is a monotonically decreasing function of the state of the network. Therefore, the stationary states of this network are such that they correspond to equilibrium points (Hopfield, 1984). These equilibrium points are termed attractors because there is a neighborhood (basin of attraction) over which these points exert a dominant influence.

Discrete Hopfield Network

The discrete Hopfield networks constitute a particular case of recurrent networks for which the state space is discrete and the neuron models are the ones proposed by McCulloch and Pitts (Section 4.3.1). This allows it to be seen as a content-addressable memory or nonlinear associative memory. In this case, the equilibrium points of the network are known to correspond to the patterns to be memorized. However, the network weights that produce the desired equilibrium points are not known, and thus must be determined according to some rule.

Suppose the goal is to memorize a set $\mathbf{X}$ of m-dimensional vectors (bipolar strings built from the set $\{-1,1\}$), denoted by $\{\mathbf{x}_i \mid i = 1, \dots, N\}$ in a network with m neurons. These vectors are called the *fundamental memories* of the system and represent the patterns to be memorized. According to the generalized Hebb's rule, the weight matrix $\mathbf{W} \in \Re^{m \times m}$ of the network is determined by, in vector notation:

$$\mathbf{W} = \frac{1}{m} \sum_{i=1}^{N} \mathbf{x}_i (\mathbf{x}_i)^T \qquad (4.77)$$

As discussed previously, to guarantee the convergence of the algorithm to stable states, it is necessary to set $w_{ii} = 0$. Therefore, Equation (4.77) becomes

$$\mathbf{W} = \frac{1}{m} \sum_{i=1}^{N} \mathbf{x}_i (\mathbf{x}_i)^T - \frac{N}{m} \mathbf{I} \qquad (4.78)$$

where $\mathbf{I}$ is the identity matrix.

Equation (4.78) satisfies all the conditions required for the convergence of this discrete network to stable states: it is symmetric and with zero-valued diagonal. A monotonically decreasing energy function can thus be defined (see Hopfield 1982).

```
procedure [W,Y] = hopfield(X,Z)
    //storage phase
    W ← (1/m)X.X^T − (N/m)I
    //retrieval phase
    vet_permut ← randperm(m)        //permutations of n
    for j from 1 to m do,           //for all input patterns
        i ← vet_permut(j)           //randomly select a neuron
        while y_i is changing do,
            y_i ← sign(W.z_i)
        end while
    end for
end procedure
```

Algorithm 4.6: Discrete Hopfield network.

After storing the memories in the network (*storage phase*), its retrieval capability can now be assessed by presenting new or noisy patterns to the network (*retrieval phase*). Assume an m-dimensional vector $\mathbf{z} \in \Re^{m \times 1}$ is taken and presented to the network as its initial state. The vector $\mathbf{z}$, called *probe*, can be an incomplete or noisy version of one of the fundamental memories of the network.

The memory retrieval process obeys a dynamical rule termed *asynchronous updating*. Randomly select a neuron j of the network as follows:

$$s_j(0) = z_j; \quad \forall j$$

where $s_j(0)$ is the state of neuron j at iteration $t = 0$, and z_j is the j-th element of vector $\mathbf{z}$. At each iteration, update one randomly selected element of the state vector $\mathbf{s}$ according to the rule:

$$s_j(t+1) = \text{sign}\left[\sum_{i=1}^{m} w_{ji} s_i(t)\right], \quad j = 1, \dots, m \tag{4.79}$$

where the function $\text{sign}(u_j)$ is the signum function: it returns 1 if $u_j > 0$, −1 if $u_j < 0$, and remains unchanged if $u_j = 0$.

Repeat the iteration until the state vector $\mathbf{s}$ remains unchanged. The fixed point, stable state, computed ($\mathbf{s}_{\text{fixed}}$) corresponds to the output vector returned by the network: $\mathbf{y} = \mathbf{s}_{\text{fixed}}$.

Let $\mathbf{X} \in \Re^{m \times N}$ be the matrix of N input patterns of dimension m each (the patterns are placed columnwise in matrix $\mathbf{X}$), and $\mathbf{Z} \in \Re^{m \times n}$ be the matrix of n probe patterns of dimension m each. Algorithm 4.6 presents the pseudocode for the discrete Hopfield network, where $\mathbf{Y} \in \Re^{m \times N}$ is the matrix of restored patterns.

Spurious Attractors

If the Hopfield network stores N fundamental memories using the generalized Hebb's rule, the stable states present in the energy surface are not restricted to the fundamental memories stored. Any stable state that is not associated with the fundamental memories are named *spurious attractors*. Spurious attractors are due to several factors. The energy function is symmetrical; any linear combination of an odd number of stable states is also a stable state; and for a large number of fundamental memories the energy function generates stable states that are not correlated with any of the fundamental memories.

Example of Application

To illustrate the error-correcting capability of the Hopfield network, consider the same character recognition problem discussed in Section 4.4.2 and illustrated in Figure 4.45(a). In the present case, the patterns are assumed to be bipolar, that is, composed of a bit −1 (dark gray) or +1 (light gray). As each pattern has a dimension $m = 120$, the network used in the experiment has $m = 120$ neurons, and therefore $m^2 - m = 12,280$ connections.

To test the storage capability of the network, the same patterns (fundamental memories) can be presented to the network. To demonstrate its error-correcting capability, introduce random noise in the patterns by independently reversing some pixels from +1 to −1, or vice-versa, with a given probability. The corrupted (noisy) patterns are used as probes to the network (Figure 4.45(b)).

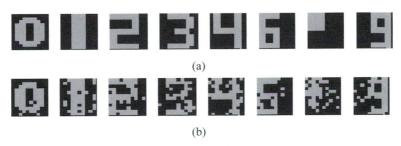

(a)

(b)

Figure 4.45: (a) Fundamental memories for the Hopfield network. (b) Noisy patterns.

4.5 FROM NATURAL TO ARTIFICIAL NEURAL NETWORKS

Table 4.2 summarizes the interpretation of biological neural networks as artificial neural networks. The study of any mathematical model involves the mapping of the mathematical domain into some observable part of the natural world. The nervous system studied here includes neurons, synaptic strengths, neural firings, networks of neurons, and several other features of the brain viewed at a specific level of description. In Section 4.4, where the typical neural networks were described, the models introduced were invariably placed in the biological context. Even in the case of the Hopfield network, where this description was not so explicit, it was discussed that a continuous model of such network uses a simple integrate-and-fire neuron, and thus the network can also be viewed as searching for minima in an energy surface.

Table 4.2: Interpretation of the biological terminology into the computational domain.

Biology (Nervous System)	Artificial Neural Networks
Neuron	Unit, node, or (artificial) neuron
Synaptic contact	Connection
Synaptic efficacy (such as the number of vesicles released with each action potential, the amount of neurotransmitter within each vesicle, and the total number of channel receptors exposed to neurotransmitters)	Weight, connection strength, or synaptic strength
Neural integration over the input signals	Net input to the neuron
Triggering of an action potential (spike)	Output of the neuron given by an activation function applied to its net input
Limited range firing rate	Limited output of the processing strength by an activation function
Threshold of activation	Bias
Laminae (layers) of neurons	Layers of neurons
Network of neurons	Network of neurons
Excitation/inhibition	Positive/negative weight

4.6   SCOPE OF NEUROCOMPUTING

Artificial neural networks can, in principle, compute any computable function, i.e., they can do everything a standard digital computer can do, and can thus be classified as a type of *neurocomputer*. Artificial neural networks are particularly useful for clustering, classification, and function approximation (mapping) problems to which hard and fast rules cannot be easily applied. As discussed in Section 4.4.4, multi-layer perceptrons are universal function approximators, thus any finite-dimensional vector function on a compact set can be approximated to an arbitrary precision by these networks. Self-organizing feature maps, like the one described in this chapter, find applications in diverse fields, such as unsupervised categorization of data (clustering), pattern recognition, remote sensing, robotics, processing of semantic information, and knowledge discovery in databases (data mining). Recurrent networks have been broadly applied to time series prediction, pattern recognition, system identification and control, and natural language processing. There are also many other types of neural networks and learning algorithms which find applications in different domains.

Vasts lists of practical and commercial applications of ANN, including links to other sources, can be found at the Frequently Asked Questions (FAQ) site for neural networks:

ftp://ftp.sas.com/pub/neural/FAQ7.html#A_applications.

Similarly to the case of evolutionary algorithms, the applications range from industry to arts to games.

4.7 SUMMARY

The nervous system is one of the major control systems of living organisms. It is responsible, among other things, for receiving stimuli from the environment, processing them, and then promoting an output response. It has been very influential in the design of computational tools for problem solving that use neuron-like elements as the main processing units.

In a simplified form, the nervous system is composed of nerve cells, mainly neurons, structured in networks of neurons. Through synaptic modulation the nervous system is capable of generating internal representations of environmental stimuli, to the point that these stimuli become unnecessary for the emergence of ideas and creativity. The internal models themselves are capable of self-stimulation.

Artificial neural networks were designed as highly abstract models based upon simple mathematical models of biological neurons. The assembly of artificial neurons in network structures allows the creation of powerful neurocomputing devices, with applications to several domains, from pattern recognition to control.

This chapter only scratched the surface of the mature and vast field of neural networks. It presented a framework to design ANN (neuron models, network structures, and learning algorithms), and then proceeded with a review of the standard most well-known types of neural networks developed with inspiration in the nervous systems: Hebb net, Perceptron network, ADALINE, MLP, SOM, and discrete Hopfield network. There are several other types of neural networks that receive great attention and have applications in various domains, but that we could not talk about here due to space constraints. Among these, it is possible to remark the *support vector machines* (Cortes and Vapnik, 1995; Vapnik, 1998) and the *radial basis function* (Broomhead and Lowe, 1988; Freeman and Saad, 1995) neural networks. Appendix C provides a quick guide to the literature with particular emphasis on textbooks dedicated to neural networks. All readers interested in knowing more about the many types of neural networks may consult the literature cited.

4.8 EXERCISES

4.8.1. Questions

1. It has been discussed, in Section 4.2, that a neuron can fire by receiving input signals from other neurons. In most cases it is assumed that these signals come from the environment through sensory cells in the skin, eyes, and so forth. However, it is also known that the brain is capable of presenting thought processes from stimuli generated within the brain itself. Discuss how these "self-stimuli" can be generated.

2. Artificial neural networks have been extensively used in pattern recognition tasks, such as character and face recognition. Independently of the potentiality of the network implemented, would you expect an ANN to outperform human beings, for instance in noisy character recognition? Justify your answer.

3. Classify the list of topics below as fact or fiction (argument in favor of your classification and cite works from the literature as support):

    a) We only use 10% of the potential of our brains.

    b) The number of neurons we are born with is the number of neurons we die with.

    c) Neurons cannot reproduce.

    d) Men have more neurons than women.

    e) Women have a broader peripherical vision than men.

4.8.2. Computational Exercises

1. Section 4.3.1 introduced the McCulloch and Pitts neuron and presented one form of implementing it so as to simulate the logical AND and OR functions. Assuming there is a NOT neuron, as illustrated in Figure 4.46, connect some AND, OR, and NOT neurons such that the resulting artificial neural network performs the XOR function with inputs a and b. Draw the resultant network and describe the required threshold for each neuron in the network.

a	b	a XOR b
0	0	0
0	1	1
1	0	1
1	1	0

Figure 4.46: A NOT neuron and the truth-table for the XOR function.

2. In Section 4.4.1, the extended Hebbian updating rule for the weight vector was presented. Write a pseudocode for the Hebb network assuming that the network is composed of a single neuron with linear activation and is going to be used for supervised learning:

    Input patterns: $\mathbf{X} \in \Re^{m \times N}$.

    Desired outputs: $\mathbf{d} \in \Re^{1 \times N}$.

    Network outputs: $\mathbf{y} \in \Re^{1 \times N}$.

    The procedure should return as outputs the network output y for each training pattern, the weight vector $\mathbf{w}$ and the bias term b.

Inputs		Desired output	Inputs		Desired output
x_1	x_2	x_1 AND x_2	x_1	x_2	x_1 OR x_2
0	0	0	0	0	0
0	1	0	0	1	1
1	0	0	1	0	1
1	1	1	1	1	1

Figure 4.47: Boolean functions AND and OR with binary outputs.

Inputs		Desired output	Inputs		Desired output
x_1	x_2	x_1 AND x_2	x_1	x_2	x_1 OR x_2
0	0	−1	0	0	−1
0	1	−1	0	1	+1
1	0	−1	1	0	+1
1	1	+1	1	1	+1

Figure 4.48: Boolean functions AND and OR with bipolar outputs.

3. Apply the extended Hebb network procedure, written in the previous exercise to realize the Boolean functions AND and OR, as illustrated in Figure 4.47.

    What happens when the first, second, and third patterns are presented to the network in the AND function? The same happens with the first pattern of function OR. Explain why the learning rule is not capable of generating a correct decision line?

    Change the binary vector of desired outputs by a bipolar vector, that is, every value '0' in vector **d** should correspond to a value '−1', as illustrated in Figure 4.48, and reapply the Hebb network to this problem. What happens? Is the network capable of solving the problem? Justify your answer.

    Finally, change the input patterns into the bipolar form and try to solve the problem using the simple Hebb network.

4. Apply the simple perceptron algorithm (Algorithm 4.1) to the AND and OR functions with binary inputs and desired outputs. Initialize the weight vector and bias all with zero, and with small random numbers as well. Compare the results of the different initialization methods and draw the decision lines.

5. Apply the simple perceptron algorithm (Algorithm 4.1) to the XOR function. Is it capable of finding a decision line to correctly classify the inputs into '1' or '0'? Why using bipolar inputs or desired outputs still does not solve the problem in this case?

6. Implement the perceptron algorithm for multiple output neurons (Algorithm 4.2) and train the network to classify the characters presented in Figure 4.23.

    Using the weight matrix and bias vector trained with the noise-free characters, insert some random noise in these patterns and test the network sensitivity in relation to the noise level; that is, test how the network performance degrades in relation to the level of noise introduced in the test set. Try various levels of noise, from 5% to 50%.

7. Determine the gradient of an arbitrary bias for an ADALINE network with linear output nodes. That is,

    Determine $\dfrac{\partial \mathfrak{I}}{\partial b_l}$, where $\mathfrak{I} = \sum_{i=1}^{o} e_i^2 = \sum_{i=1}^{o} (d_i - y_i)^2$, and $y_i = \Sigma_j w_{ij}.x_j + b_i$,

    $j = 1, \dots, m$.

8. Write a pseudocode for the ADALINE network and compare it with the pseudocode for the Perceptron network.

9. Repeat Exercise 7 above using an Adaline network. Compare the results with those obtained using the perceptron.

10. Algorithm 4.3 presents the algorithm to run the single-layer perceptron network. Write a pseudocode for an algorithm to run the multi-layer perceptron (MLP) network. Assume generic activation functions and a number of hidden layers to be entered by the user. (The trained weight matrices and bias vectors must be entered as inputs to the algorithm.)

11. Apply an MLP network with two sigmoid hidden units and a single linear output unit trained with the standard backpropagation learning algorithm to solve the XOR problem.

    Test the network sensitivity to various values of α; that is, test the convergence time of the algorithm for different values of α. Tip: start with $\alpha = 0.001$ and vary it until $\alpha = 0.5$.

    For a fixed value of α draw the decision boundaries constructed for each hidden unit, and the decision boundary constructed by the complete network.

    Can you solve this problem with a single hidden unit? Justify your answer.

12. Apply an MLP network with threshold (McCulloch and Pitts) units, two hidden and a single output unit, trained with the backpropagation learning algorithm to solve the XOR problem.

    Determine the weights of the network and draw the decision boundaries constructed for each hidden unit, and the decision boundary constructed by the complete network.

13. Repeat the previous exercise assuming the neural network has sigmoidal activation functions in the hidden and output layers.

14. It is known that the network presented in Figure 4.49 is capable of solving the XOR problem with a single hidden unit.

    Can you apply the backpropagation algorithm to train this network? If yes, how? If not, why?

    Find an appropriate set of weights so that the network with sigmoid hidden units and linear (or sigmoid) output unit solves the XOR problem.

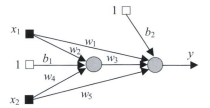

Figure 4.49: Network with connections from the input units directly to the output unit.

15. Train an MLP network to approximate one period of the function $sin(x).cos(2x)$, $x \in [0,2\pi]$, with only 36 training samples uniformly distributed over the domain of x. Test the network sensitivity with relation to:

a) The number of hidden units.

b) The learning rate α.

16. Insert some noise into the data set of the previous exercise and retrain the MLP network with this new data set. Study the generalization capability of this network (Section 4.4.4).

17. Apply the MLP network trained with the backpropagation learning algorithm to solve the character recognition task of Section 4.4.2.

Determine a suitable network architecture and then test the resultant network sensitivity to noise in the test data. Test different noise levels, from 5% to 50%.

18. Apply an MLP network to solve the two-spirals problems (Figure 4.50) using in the previous chapter to illustrate the application of genetic programming. This data set is composed of 190 labeled samples (95 samples in each class), generated using polar coordinates.

Determine a suitable number of neurons in the MLP network to be trained using the standard backpropagation algorithm. Assume the algorithm has converged when $MSE \leq 10^{-3}$, and plot the decision surface generated separating both classes.

Test the sensitivity of the trained network to various levels of noise added to the input data.

19. The *encoding/decoding* (ENC/DEC) problem is a sort of *parity* problem; that is, the output is required to be one if the input pattern contains an odd number of ones, and zero otherwise. Assuming a data set with 10 samples ($N = 10$) of dimension 10 each ($m = 10$), and a network with 10 outputs ($o = 10$), the input vector is composed of N bits, from which only one has a value of '1', and the others have values '0'. The desired output vector is identical to the input vector so that the network has to make an auto-association between the input and output values. The goal is to learn a map-

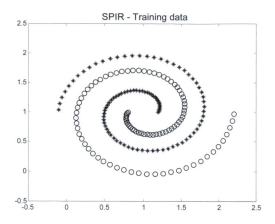

Figure 4.50: Input data set for training a classifier.

ping from the m inputs to the h hidden units (*encoding*) and then from the h hidden units to the o outputs (*decoding*). Usually, in a network m-h-o (m inputs, h hidden units, and o outputs) $h < N$, where N is the number of input patterns. If $h \leq \log_2 N$ it is said to be a tight encoder.

Train an MLP network with $h = 10$ to solve an ENC/DEC problem of dimension 10.

Is it possible to train an MLP network so that it becomes a tight encoder?

20. Apply the self-organizing map (SOM) learning algorithm to the clustering problem presented in Section 5.6.2, Exercise 3 of the Swarm Intelligence chapter (Chapter 5). Try first a uni-dimensional output grid, and then a bi-dimensional output grid. Compare and discuss the results of each method.

21. Apply the self-organized map to the Animals data set presented in Section 4.4.5. Consider a bi-dimensional output grid with 10 rows by 10 columns of neurons (10 × 10).

22. Apply a bi-dimensional SOM to the two-spirals problem considering unlabeled data. Depict (plot) the final configuration of the neurons in the network.

23. Implement the discrete Hopfield network and apply it to the character recognition problem, as described in Section 4.4.2. Test the network sensitivity to noise, that is, introduce noise in the patterns with increasing probability (5%, 10%, 15%, 20%, 25%, ... , 50%) and study the error-correcting capability of the network.

24. A more efficient rule, than the generalized Hebb rule, to define the weight matrix of a discrete Hopfield network is the *projection rule* or *pseudo-inverse rule*. Although this rule does not present the biological motivation of the Hebb rule, it explores some algebraic features of the equilibrium points.

A vector $\mathbf{x}_i$ is going to be an equilibrium point if $\mathbf{W}\mathbf{x}_i = \mathbf{x}_i$, $i = 1, \dots, n$. Let $\mathbf{X}$ be the matrix of all input patterns, then $\mathbf{W}\mathbf{X} = \mathbf{X}$. A solution to this equation can be given by the following expression:

$$\mathbf{W} = \mathbf{X}(\mathbf{X}^T\mathbf{X})^{-1}\mathbf{X}^T$$

The necessary condition for $(\mathbf{X}^T\mathbf{X})^{-1}$ to exist is that the vectors $\mathbf{x}_i$, $i = 1, \dots, n$ be linearly independent, and thus, matrix $\mathbf{X}$ has full rank.

Test if the input patterns used in the previous exercise are linearly independent (see Appendix B.1.2) and use the projection rule to store the patterns in the discrete Hopfield network.

Test the error-correcting capability of the network trained with this rule and compare with the results obtained in the previous exercise.

4.8.3. Thought Exercises

1. Discuss the similarities and differences between the perceptron and the ADALINE updating rules. Why is the ADALINE more powerful than the perceptron algorithm?

2. In Appendix B.4.4 some elements of graph theory are introduced. Provide a graphical description of a neural network, more specifically, propose a signal-flow graph of the artificial neuron depicted in Figure 4.9.

3. Chapter 3 introduced the concept of a fitness or adaptive landscape whose contours represent different levels of adaptation to the environment. Populations of individuals at high levels (peaks) of this landscape are more adapted to the environment, while populations at lower levels (valleys) are less adapted. Compare the concept of a fitness landscape to the error surface discussed in Section 4.3.3.

4.8.4. Projects and Challenges

1. In Section 4.3.1 a basic integrate-and-fire neuron model was presented according to Equation (4.1) and Equation (4.3). Equation (4.1) corresponds to an inhomogeneous linear differential equation that can be solved numerically, and also analytically in some cases. Assuming that the sum of the input currents to the neuron $i(t)$ is constant; that is, $i(t) = i$, determine the analytical solution to Equation (4.1).

   Assume now that the activation threshold is set to $\theta = 10$. Plot a graph with the variation of the solution to Equation (4.1), $u(t)$, as a function of time t for $Ri = 8$.

   Finally, plot a graph with the variation of the solution to Equation (4.1), $u(t)$, as a function of time t for $Ri = 10$. If the neuron fires a spike, reset its membrane potential to a fixed value $u_{res} = 1$.

   Can you observe spikes in your graphs? If not, explain why.

2. In Section 4.4.5 the SOM learning algorithm was presented. A competitive learning rule based upon the minimization of the Euclidean distance bet-

ween the input pattern and the neurons' weights was presented. Another criterion to determine the winning neuron, i.e., the best matching neuron, is based on the maximization of the inner product between the weight vectors and the input pattern ($\mathbf{w}_j^T\mathbf{x}$).

Show that minimizing the Euclidean distance between a given input pattern and the neurons' weights is mathematically equivalent to maximizing the inner product $\mathbf{w}_j^T\mathbf{x}$.

Implement this new competitive rule and compare qualitatively the results with those obtained with the Euclidean distance.

Use the same initialization values for the network weights and the same parameters for the algorithm and compare quantitatively and qualitatively both competitive strategies.

3. Explain why a multi-layer feedforward neural network with sigmoidal output units cannot learn the simple function $y = 1/x$, with $x \in (0,1)$. How can you modify this network so that it can learn this function.

4. If an evolutionary algorithm is a general-purpose search and optimization technique, it can be applied to determine suitable network architectures and weight values for a neural network. Implement an evolutionary algorithm to determine a suitable MLP network architecture trained with the backpropagation algorithm to solve the IRIS problem described below.

There are a vast number of data sets available for testing and benchmarking neural networks and learning algorithms. For instance, the University of California Irvine (UCI) provides many collections of data sets accessible via anonymous FTP at ftp.ics.uci.edu or their website at http://www.ics.uci.edu/~mlearn/MLRepository.html.

The IRIS data set is a very well-known classification problem containing 150 samples divided into three differences classes, each of which with 50 patterns. The samples have a dimension of 4 ($m = 4$) and refer to a type of iris of plants. One of the three classes is linearly separable from the others, and two of them are not. Further documentation about this problem can be downloaded from the UCI repository together with the data set.

Tip: define a suitable representation for the chromosomes, a fitness function, and appropriate genetic operators. Each chromosome will correspond to a neural network that will have to be trained via backpropagation in order to be evaluated. The objective is to find a network with a classification error less than 5%.

4.9 REFERENCES

[1]   Anderson, J. A. (1995), *An Introduction to Neural Networks*, The MIT Press.

[2]   Broomhead, D. S. and Lowe D. (1988), "Multivariable Functional Interpolation and Adaptive Networks", *Complex Systems*, **2**, pp. 321–355.

[3]   Churchland, P. and Sejnowski, T. J. (1992), *The Computational Brain*, MIT Press.

[4]   Cortes, C. and Vapnik, V. N. (1995), "Support Vector Networks", *Machine Learning*, **20**, pp. 273–297.

[5]   Cybenko G. (1989), "Approximation by Superpositions of a Sigmoidal Function", *Mathematics of Control Signals and Systems*, **2**, pp. 303–314.

[6]   Dayan, P. and Abbot, L. F. (2001), *Theoretical Neuroscience: Computational and Mathematical Modeling of Neural Systems*, The MIT Press.

[7]   de Castro, L. N. (1998), *Analysis and Synthesis of Learning Strategies for Artificial Neural Networks* (in Portuguese), M.Sc. Dissertation, School of Computer and Electrical Engineering, State University of Campinas, Brazil, 246 p.

[8]   Fausett, L. (1994), *Fundamentals of Neural Networks: Architectures, Algorithms, and Applications*, Prentice Hall.

[9]   Freeman, J. A. S. and Saad, D. (1995), "Learning and Generalization in Radial Basis Function Networks", *Neural Computation*, **7**, pp. 1000–1020.

[10]   Hagan, M. T., Demuth, H. B. and Beale, M. H. (1996), *Neural Network Design*, PWS Publishing Company.

[11]   Haykin, S. (1999), *Neural Networks: A Comprehensive Foundation*, 2nd Ed., Prentice Hall.

[12]   Hebb, D. O. (1949), *The Organization of Behavior: A Neuropsychological Theory*, New York: Wiley.

[13]   Hecht-Nielsen, R. (1990), *Neurocomputing*, Addison-Wesley Publishing Company.

[14]   Hinton, G. E. and McClelland, J. L. (1987), "Learning Representation by Recirculation", In D. Z. Anderson (ed.), *Neural Information Processing Systems*, New York: American Institute of Physics, pp. 358–366.

[15]   Hopfield, J. J. (1982), "Neural Networks and Physical Systems with Emergent Collective Computational Capabilities", *Proc. of the Nat. Acad. of Sci. USA*, **79**, pp. 2554–2558.

[16]   Hopfield, J. J. (1984), "Neurons with Graded Response Have Collective Computational Properties like those of Two-State Neurons", *Proc. of the Nat. Acad. of Sci. USA*, **81**, pp. 3088–3092.

[17]   Kelso, S. R., Ganong, A. H. and Brown, T. H. (1986), "Hebbian Synapses in Hippocampus", *Proc. of the Nat. Ac. of Sciences U.S.A.*, **83**, pp. 5326–5330.

[18]   Kohonen, T (1990), "The Self-Organizing Map", *Proceedings of the IEEE*, **78**(9), pp. 1464–1480.

[19]   Kohonen, T. (1982), "Self-Organized Formation of Topologically Correct Feature Maps", *Biological Cybernetics*, **43**, pp. 59–69.

[20]   Kohonen, T. (2000), *Self-Organizing Maps*, 3rd Ed., Springer-Verlag.

[21]   McClelland, J. L., Rumelhart, D. E. and the PDP Research Group (1986), *Parallel Distributed Processing: Explorations in the Microstructure of Cognition, Volume 2: Psychological and Biological Models*, The MIT Press.

[22]   McCulloch, W. and Pitts, W. H. (1943), "A Logical Calculus of the Ideas Immanent in Nervous Activity", *Bulletin of Mathematical Biophysics*, **5**, pp. 115–133.

[23]   Minsky, M. L. and Papert, S. A. (1969), *Perceptrons*, MIT Press.

[24]   Narendra, K. S. and Pathasarathy, K., (1990), "Identification and Control of Dynamical Systems Using Neural Networks", *IEEE Trans. on Neural Networks*, **1**(1), pp. 4–27.

[25]   Nguyen, D. H. and Widrow, B. (1990), "Neural Networks for Self-Learning Control Systems", *IEEE Control Systems Magazine*, April, pp. 18–23.

[26]   Nigrin, A. (1993), *Neural Networks for Pattern Recognition,* Cambridge, MA: The MIT Press.

[27]   O'Reilly, R. C. (1996), "Biologically Plausible Error-Driven Learning Using Local Activation Differences: The Generalized Recirculation Algorithm", *Neural Computation*, **8**(5), pp. 985–938.

[28]   O'Reilly, R. C. and Munakata, Y. (2000), *Computational Explorations in Cognitive Neuroscience: Understanding the Mind by Simulating the Brain*, The MIT Press.

[29]   Ritter, H. and Kohonen, T. (1989), "Self-Organizing Semantic Maps", *Biological Cybernetics*, **61**, pp. 241–254.

[30]   Rosenblatt, F. (1958), "The Perceptron: A Probabilistic Model for Information Storage and Organization in the Brain", *Psychological Review*, **65**, pp. 386–408.

[31]   Rosenblatt, F. (1962), *Principles of Neurodynamics*, Spartan Books.

[32]   Rumelhart, D. E., McClelland, J. L. and the PDP Research Group (1986), *Parallel Distributed Processing: Explorations in the Microstructure of Cognition, Volume 1: Foundations*, The MIT Press.

[33]   Sutton, R. S. and Barto, A. G. (1998), *Reinforcement Learning: An Introduction*, A Bradford Book.

[34]   Trappenberg, T. (2002), *Fundamentals of Computational Neuroscience*, Oxford University Press.

[35]   Vapnik, V. N. (1998), *Statistical Learning Theory*, New York: Wiley.

[36]   von Neumman, J. (1982), "First Draft of a Report on the EDVAC", In B. Randall (ed.), *The Origins of Digital Computers: Selected Papers*, 3rd Ed., Springer, Berlin.

[37]   Widrow, B. and Hoff, M. E. Jr. (1960), "Adaptive Switching Circuits", *WESCON Convention Record*, **4**, pp. 96–104.

[38]   Wiener, N. (1948), *Cybernetics*, New York: Wiley.

CHAPTER 5

SWARM INTELLIGENCE

"There is some degree of communication among the ants, just enough to keep them from wandering off completely at random. By this minimal communication they can remind each other that they are not alone but are cooperating with teammates. It takes a large number of ants, all reinforcing each other this way, to sustain any activity - such as trail-building - for any length of time."

(D. R. Hofstadter, Gödel, Escher, Bach: An Eternal Golden Braid, Penguin Books, 20th Anniversary Edition, 2000; p. 316)

"Ants are everywhere, but only occasionally noticed. They run much of the terrestrial world as the premier soil turners, channelers of energy, dominatrices of the insect fauna ... They employ the most complex forms of chemical communication of any animals and their organization provides an illuminating contrast to that of human beings. ... [Ants] represent the culmination of insect evolution, in the same sense that human beings represent the summit of vertebrate evolution."

(B. Hölldobler and E. O. Wilson, The Ants, Belknap Press, 1990, p. 1)

5.1. INTRODUCTION

Several species benefit from sociality in various ways, usually resulting in greater survival advantages. For instance, life in social groups may increase the mating chances, facilitate the retrieval of food, reduce the probability of attack by predators, allow for a division of labor, and facilitate hunting. Social behaviors have also inspired the development of several computational tools for problem-solving and coordination strategies for collective robotics. This chapter explores the approach known as *swarm intelligence* demonstrating how several ideas from natural systems have been used in the development and implementation of computational swarm systems.

The term *swarm intelligence* was coined in the late 1980s to refer to cellular robotic systems in which a collection of simple agents in an environment interact according to local rules (Beni, 1988; Beni and Wang, 1989). Some definitions of swarm intelligence can be found in the literature:

"Swarm intelligence is a property of systems of unintelligent agents of limited individual capabilities exhibiting collectively intelligent behavior." (White and Pagurek, 1998; p. 333)

"[Swarm Intelligence] include[s] any attempt to design algorithms or distributed problem-solving devices inspired by the collective behavior of social insects and other animal societies." (Bonabeau et al., 1999; p. 7)

Kennedy et al. (2001) quote a passage of a FAQ (Frequently Asked Questions) document from the Santa Fe Institute. They agree with their definition, though they add the possibility of escaping from physical space, allowing swarms to occur in cognitive space as well:

"We use the term 'swarm' in a general sense to refer to any such loosely structured collection of interacting agents. The classic example of a swarm is a swarm of bees, but the metaphor of a swarm can be extended to other systems with a similar architecture. An ant colony can be thought of as a swarm whose individual agents are ants, a flock of birds is a swarm whose agents are birds, traffic is a swarm of cars, a crowd is a swarm of people, an immune system is a swarm of cells and molecules, and an economy is a swarm of economic agents. Although the notion of a swarm suggests an aspect of collective motion in space, as in the swarm of a flock of birds, we are interested in all types of collective behavior, not just spatial motion." (quoted by Kennedy et al., 2001; p. 102)

In these definitions, an agent can be simply understood as an entity capable of sensing the environment and acting on it. The actions may include the modification of the environment or interactions with other agents.

Millonas (1994) suggests five basic principles of swarm intelligence systems:

- *Proximity*: individuals should be able to interact so as to form social links.

- *Quality*: individuals should be able to evaluate their interactions with the environment and one another.

- *Diversity*: diversity is fundamental in most natural computing approaches, for it improves the capability of the system reacting to unknown and unexpected situations.

- *Stability*: individuals should not shift their behaviors from one mode to another in response to all environmental fluctuation.

- *Adaptability*: the capability of adapting to environmental and populational changes is also very important for swarm systems.

Therefore, a *swarm system* is the one composed of a set of individuals capable of interacting with one another and the environment. And *swarm intelligence* is an emergent property of the swarm system as a result of its principles of proximity, quality, diversity, stability, and adaptability.

Two main lines of research can be identified in swarm intelligence: (i) the works based on social insects, and (ii) the works based on the ability of human societies to process knowledge. Although the resultant approaches are quite different in sequence of steps and sources of inspiration, they present some commonalities. In general terms, both of them rely upon a population (colony or swarm) of individuals (social insects or particles) capable of interacting (directly or indirectly) with the environment and one another. As a result of these interactions there might be a change in the environment or in the individuals of the populations, what may lead to useful emergent phenomena.

This chapter starts with a discussion of social insects with particular emphasis on ants. The foraging behavior, clustering of dead bodies and larval sorting, and the collective prey retrieval capabilities of ants will be discussed, always referring to the computational tools developed with inspiration in these phenomena. Algorithms for the solution of combinatorial optimization problems, clustering and robotics coordination are just a sample of what the collective behavior of ants can teach us. Insect societies have inspired much more research in computer science and engineering than only ant algorithms, and some of these other natural systems will be discussed in Chapter 8 under the umbrella of artificial life.

The last topic of this chapter, namely the Particle Swarm optimization (PS) approach, has its roots on artificial life, social psychology, engineering, and computer science. It is based on a swarm of particles that move around a search-space and that is capable of locating portions of this space corresponding to peaks of a quality (goodness) function. The swarm of particles self-organizes through an updating mechanism that takes into account the present position of the particle, its best position so far, and the best positions of a number of neighbors. The core idea of the PS approach is that individuals have some knowledge of their own past experience and the past experience of some of their neighbors, and they use this knowledge in order to drive their lives (minds and behaviors).

5.2. ANT COLONIES

Interactions among members of an animal group or society, particularly chemical and visual communication, may be important for the organization of a collective behavior. Furthermore, the interactions among individuals and the environment allow different collective patterns and decisions to appear under different conditions, with the same individual behavior. Though most clearly demonstrable in social insects, these principles are fundamental to schools of fish, flocks of birds, groups of mammals, and many other social aggregates (Deneubourg and Goss, 1989).

The analysis of collective behavior implies a detailed observation of both individual and collective behavior, sometimes combined with mathematical modeling linking the two. There have been a number of computational tools developed with the inspiration taken from the collective behavior of social insects, in particular ant colonies. The resulting models from observed behaviors have also played an important role in the development of computational tools for problem solving.

This section overviews some of the main works inspired by social insects, focusing on ant colonies. It starts with a discussion about *ant colony optimization*, composed of algorithms for discrete optimization developed with inspiration from observations of how some ant species forage for food. Field and experimental observations have led to the conclusion that some ant species organize dead bodies and the brood into clusters of items. One theoretical result derived to model such behaviors gave rise to *ant clustering algorithms*. Ant colonies have also provided good insights into how a set of non-intelligent agents is capable of performing complex tasks. Grouping a number of objects, foraging for food, and prey retrieval are just a sample of tasks performed by ants that inspired the development of collective robotic teams, named *swarm robotics*. Further examples of the collective behavior of other insect societies, such as bees and wasps, will be discussed in the context of Artificial Life in Chapter 8.

5.2.1. Ants at Work: How an Insect Society Is Organized

Ants are the most well studied social insects. People with the most diverse background and interests are fascinated by ants. Biologists and sociobiologists are interested in studying, among other things, how the many events occurring in different levels of an ant colony are related; how such simple organisms can perform fairly complex tasks like collective prey retrieval, foraging, and dead body clustering. Many of these scientists and even philosophers believe that a better understanding of social insects, in particular ant colonies, may lead to a better understanding of how nature works. They believe ant colonies have similar patterns of behavior to other systems, such as the nervous system.

The movie "Antz" by Dreamworks Pictures, directed by Eric Darnell, has an important intellectual appeal and provides an interesting, though not realistic, perspective on how an ant colony is organized. It starts with a worker ant named Z on the couch of a therapist office, moaning about his insignificance and the social order of the colony. "I feel insignificant", he complains. "You've made a big breakthrough", says the therapist. "I have?" replies Z. "Yes", answers the doctor, "You ARE insignificant".

One interesting aspect about Antz is that the whole film is presented under the perspective of an ant colony, with particular emphasis on individual ants. Z falls in love with the queen's daughter, princess Bala, who is engaged to the colony's General Mandible. But the princess does not love the general and foresees a boring and sad future life for herself. Eager to meet princess Bala, Z asks for the help of his best friend Weaver, a soldier ant. But the adventure of being a soldier leads Z into a terrifying battle with the dreaded termites, a species much larger and superior in weaponry. Z is the only survivor of the battle and builds a philosophical and social revolution. Meanwhile, the ambitious General Mandible wants to divide the colony into a superior and stronger race formed by soldiers, and a weak race formed by workers.

This popular view of an ant colony as a Stalinist or Marxist stereotype makes it clear the hierarchical view most of us have in relation to social insects. Most of us believe queen ants rule the colony, and we also usually believe in a hierarchical structure much like human societies. However, while ants are capable of remarkable coordinated activities, such as foraging for food, task allocation, nest building, and grouping of dead bodies, there is no centralized behavior in an ant colony. It is important to stress that the absence of centralized behaviors, so common in human societies, is a fundamental property of natural computing. Actually, the ant queen's main function is to lay eggs, not to rule the colony. Of course, by arguing that the perspective presented in the movie is somewhat misleading, I am not directly criticizing the movie itself, which I personally recommend (Section 5.6.1). Instead, it is a call for attention to the subject matter that is one of the central themes of natural computing: *adaptation and 'intelligence' emerging out of a decentralized system composed of 'unintelligent' agents.*

In a remarkably interesting, brief, and accessible description of how an ant colony is organized, D. Gordon (1999) discourses about her almost two decades of research with ants, particularly harvester ants. Her description ranges from the individual to the colony level and the implications and motivation for studying ants. The first pages of Chapter 7 of her book provide a very good summary of the complexity involved in a colony of ants and the many tasks they perform; reasons why I quote from it at length:

"[An ant colony] must collect and distribute food, build a nest, and care for the eggs, larvae and pupae. It lives in a changing world to which it must respond. When there is a windfall of food, more foragers are needed. When the nest is damaged, extra effort is required for quick repairs. Task allocation is the process that results in certain workers engaged in specific tasks, in numbers appropriate to the current situation.

Task allocation is a solution to a dynamic problem and thus it is a process of continual adjustment. It operates without any central or hierarchical control to direct individual ants in particular tasks. Although 'queen' is a term that reminds us of human political systems, the queen is not an authority figure. She lays eggs and is fed and cared for by the workers. She does not decide which worker does what. In a harvester ant colony, many feet of intricate tunnels and chambers and thousands of ants separate the queen, surrounded by interior workers, from the ants working outside the nest and using only the chambers near the surface. It would be physically impossible for the queen to direct every worker's decision about which task to perform and when. The absence of central control may seem counterintuitive, because we are accustomed to hierarchically organized social groups in many aspects of human societies, including universities, business, governments, orchestras, and armies. This mystery underlies the ancient and pervading fascination of social insect societies … .

No ant is able to assess the global needs of the colony, or to count how many workers are engaged in each task and decide how many should be allocated differently. The capacities of individuals are limited. Each worker need make only fairly simple decisions. There is abundant evidence, throughout physics, the social sciences, and biology, that such simple behavior by individuals can lead to predictable patterns in the behavior of the groups. It should be possible to explain task allocation in a similar way, as the consequence of simple decisions by individuals.

Investigating task allocation requires different lines of research. One is to find out what colonies do. The other is to find out how individuals generate the appropriate colony state. An ant is not very smart. It can't make complicated assessments. It probably can't remember anything for very long. Its behavior is based on what it perceives in its immediate environment." (Gordon, 1999; p. 117-119)

The many ant collective tasks to be described in this chapter gave rise to useful tools for problem solving: an algorithm for solving combinatorial optimization problems, and an algorithm for solving clustering problems. All these behaviors served also as inspiration for the development of what is to date known as *swarm robotics*. Swarm robotics is based on the use of a 'colony' of autonomous robots, or agents, with limited capabilities, in order to perform a particular task. The next three sections describe these patterns of behavior in some ant species and the respective systems developed inspired by them.

5.2.2. Ant Foraging Behavior

There is a large variety of ant species over the world. However, a number of them present similar foraging behaviors. Field and laboratory experiments with many species of ants have demonstrated an interesting capability of exploiting rich food sources without loosing the capability of exploring the environment, as well as finding the shortest path to a food source in relation to the nest (exploration × exploitation). Some of these observations have led to the development of models of ant behavior and further to the proposal of computational algorithms for the solution of complex problems. This section reviews one such algorithm.

Although most ants live in colonies, and thus present some form of social behavior, for few species of ants there is evidence of the presence of leaders, templates, recipes, or blueprints playing a role in the development of foraging patterns. Instead, the process appears to be based on local competition among information (in the form of varying concentrations of trail *pheromone*), and is used by individual ants to generate observable collective foraging decisions (Camazine et al., 2001).

The process of trail formation in ant colonies can be easily observable with the simple experiment illustrated in Figure 5.1. Place a dish of sugar solution (food source) in the vicinity of an ants' nest - Figure 5.1(a). After some time, *forager ants* will discover the sugar and shortly after, through a *recruitment* process, a number of foragers will appear at the food source - Figure 5.1(b). Observation will reveal ants trafficking between the nest and the food source as if they were following a *trail* on the ground - Figure 5.1(c). Note the presence of a few foragers not following the trail; an important behavior for the location of alternative food sources; that is, exploration of the environment.

Generally speaking, *recruitment* is a behavioral mechanism that enables an ant colony to assemble rapidly a large number of foragers at a desirable food source and to perform efficient decision making, such as the selection of the most profitable food source or the choice of the shortest path between the nest and the food source. Different recruitment mechanisms exist (Deneubourg et al., 1986). In *mass recruitment*, a scout (*recruiter*) discovers the food source and returns to the nest, laying a chemical (pheromone) trail. In the nest, some of its nestmates (the recruited) detect the trail and follow it to the food source. There they ingest food and return to the nest *reinforcing* the trail. In *tandem recruitment*, the scout invites ants at the nest to accompany her back to the food. One recruit succeeds in following the leader, the two ants being in close contact. In *group recruitment*, the recruiter leads a group of recruits to the food source by means of a short-range chemical attractant. After food ingestion, in each recruitment type, the recruited become recruiters. This typical self-reinforcing (positive feedback) process is the basis of many activities of ants' societies. The recruitment process slows down when there are fewer ants left to be recruited or when there are other forces, such as alternative food sources, competing for the ants' attention.

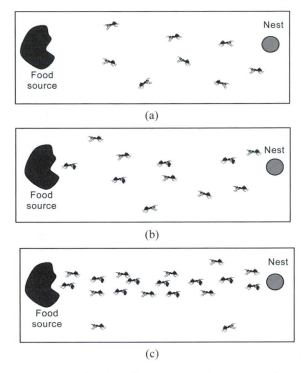

Figure 5.1: In the collective foraging of some ant species, ants recruit nestmates by releasing pheromone on the path from the food source to the nest; a pheromone trail is thus established. (a) Foraging ants. (b) A few ants find the food source and start recruiting nestmates by releasing pheromone. (c) A pheromone trail is formed.

The process of trail formation may be divided in two steps: *trail-laying* and *trail-following* (Camazine et al., 2001). Consider the particular case of mass recruitment and some of the experimental set ups used to observe it and propose mathematical models. The behavior of individual ants during trail recruitment starts when a forager ant, on discovering a food source, returns to the nest and lays a chemical trail all the way home. These chemicals are low molecular weight substances, called *pheromone*, produced in special glands or in the guts. Other ants, either foraging or waiting in the nest, are then stimulated under the influence of the pheromone to start exploiting the food source encountered. In addition to the pheromone trail, recruiter ants may provide other stimuli. The ants recruited follow the trail to the food source and load up with food. On their return journey to the nest, they add their own pheromone to the trail and may even provide further stimulation to other foragers in the nest and on the way. The pheromone thus has two important functions: 1) primarily to define the trail, and also 2) to serve as an orientation signal for ants traveling outside the nest. It is important to remark, however, that the pheromone evaporates, usually very slowly, with time. It means that, for example, if the food source is completely depleted, the trail will disappear after some time due to the lack of reinforcement.

Goss et al. (1989) and Deneubourg et al. (1990) performed a number of experiments and demonstrated that the trail-laying trail-following behavior of some ant species enables them to find the shortest path between a food source and the nest. They used a simple yet elegant experimental set up with the Argentine ant *Iridomyrmex humilis*. Basically, their experiment can be summarized as follows. Laboratory colonies of *I. humilis* were given access to a food source in an arena linked to the nest by a bridge consisting of two branches of different lengths, arranged so that a forager going in either direction (from the nest to the food source or vice-versa) must choose between one or the other branch, as illustrated in Figure 5.2.

One experimental observation is that, after a transitory phase that can last a few minutes, most of the ants choose the shortest path. It is also observed that an ant's probability of selecting the shortest path increases with the difference in length between the two branches. The explanation for this capability of selecting the shortest route from the nest to the food source comes from the use of pheromone as an indirect form of communication, known as *stigmergy*, mediated by local modifications of the environment. Before starting a discussion about stigmergy in the process of pheromone trail laying and following, it must be asked what would happen if the shorter branch were presented to the colony after the longer one. In this case, the shorter path would not be selected because the longer branch is already marked with a pheromone trail. However, if the pheromone evaporates quickly enough, longer paths would have trouble to maintain stable pheromone trails. This is not the case for most ant species, but it is the approach usually taken by engineers and computer scientists in their computational implementations of *ant colony optimization* (ACO) algorithms.

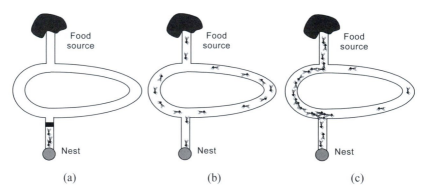

(a)   (b)   (c)

Figure 5.2: An experimental set up that can be used to demonstrate that the ant *I. Humilis* is capable of finding the shortest path between the nest and a food source. (b) The bridge is initially closed. (b) Initial distribution of ants after the bridge is open. (b) Distribution of ants after some time has passed since they were allowed to exploit the food source.

Another observation is that randomness plays an important role in ant forag-ing. Ants do not follow pheromone trails perfectly. Instead, they have a prob-abilistic chance of loosing their way as they follow trails. Deneubourg et al. (1986) argue that this 'ant randomness' is not a defective stage on an evolution-ary path 'towards an idealistic deterministic system of communication.' Rather, this randomness is an evolutionarily adaptive behavior. In some of their labora-tory experiments, they describe one case with two food sources near an ant nest: a rich food source far from the nest, and an inferior source close to the nest. Ini-tially, the ants discover the inferior food source and form a robust trail to that source. But some ants wander off the trail. These lost ants discover the richer source and form a trail to it. Since an ant's pheromone emission is related to the richness of the food source, the trail to the richer source becomes stronger than the original trail. Eventually, most ants shift to the richer source. Therefore, the randomness of the ants provides a way for the colony to explore multiple food sources in parallel; that is, randomness allows for exploration. While positive feedback through pheromone trails encourages exploitation of particular sources, randomness encourages exploration of multiple sources.

Stigmergy

The case of foraging for food in the Argentine ants shows how stigmergy can be used to coordinate the ant's foraging activities by means of self-organization. Self-organized trail-laying by individual ants is a way of modifying the envi-ronment to communicate to other nestmates to follow that trail. In the experi-mental set up of Figure 5.2, when the ants arrive at the branching point, they have to make a probabilistic choice between one of the two branches to follow. Such choice is biased by the amount of pheromone they smell on the branches. This behavior has a self-reinforcing (positive feedback) effect, because choosing a branch will automatically increase the probability that this branch will be cho-sen again by other ants. At the beginning of the experiment there is no phero-mone on the two branches and thus, the choice of a branch will be made with the same probability for both branches. Due to the different branch lengths, the ants traveling on the shortest branch will be the first ones to reach the food source. In their journey back to the nest, these ants will smell some pheromone trail (re-leased by themselves) on the shorter path and will thus follow it. More phero-mone will then be released on the shorter path, making it even more attractive for the other ants. Therefore, ants more and more frequently select the shorter path.

5.2.3. Ant Colony Optimization (ACO)

The choice of the shortest path enables ants to minimize the time spent traveling between nest and food source, thus taking less time to complete the route. This also allows ants to collect food more quickly and minimize the risk that this food source is found and monopolized by a stronger competitor, such as a larger col-ony. Shorter paths also mean lower transportation costs (Camazine et al., 2001).

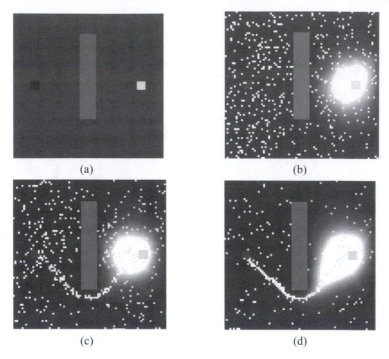

(a)   (b)

(c)   (d)

Figure 5.3: Artificial life simulation of pheromone trail laying and following by ants. (a) Environmental setup. The square on the left corresponds to the ants' nest, and the one on the right is the food source. In between the two there is an obstacle whose top part is slightly longer than the bottom part. (b) 500 ants leave the nest in search for food, and release pheromone (depicted in white color) while carrying food back to the nest. (c) The deposition of pheromone on the environment serves as a reinforcement signal to recruit other ants to gather food. (d) A strong pheromone trail in the shorter route is established.

To illustrate this, consider the artificial life simulation (Chapter 8) of pheromone trail-laying, trail-following illustrated in Figure 5.3. In this simulation, a food source is available (big square on the right hand side of the picture) together with an ant nest (big square on the left hand side of the picture) and an obstacle between them (rectangle centered). At first, the virtual ants explore the environment randomly. After they find the food source and start carrying it back to the nest, those ants that choose the shorter path to the nest return first, thus reinforcing the shorter path. Note that the top part of the obstacle is slightly longer than its bottom part, indicating that the shorter route from the nest to the food source is the bottom route. Next, the ants maintain almost only the shortest trail from the food source to the nest, leading to the exploitation of this trail.

The problem of finding the shortest route from the nest to the food source is akin to the well-known traveling salesman problem (TSP), discussed in Chapter 3. The salesman has to find the shortest route by which to visit a given number of cities, each exactly once. The goal is to minimize the cost (distance) of travel.

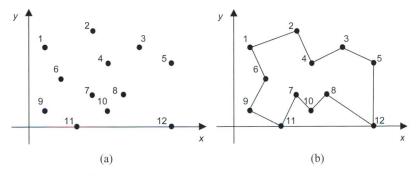

Figure 5.4: A simple instance of the TSP. (a) A set with 12 cities. (b) A minimal route connecting the cities.

Inspired by the experiments of Goss et al. (1989, 1990), Dorigo and collaborators (Dorigo et al., 1996) extended this ant model to solve the traveling salesman problem. Their approach relies on *artificial ants* laying and following *artificial pheromone trails*. Assume the simple TSP instance of Figure 5.4(a). A colony of artificial ants, each independently going from one city to another, favoring nearby cities but otherwise traveling randomly. While traversing a link, i.e., a path from one city to another, an ant deposits some pheromone on it. The amount of pheromone being inversely proportional to the overall length of the tour: the shorter the tour length, the more pheromone released; and vice-versa. After all the artificial ants have completed their tours and released pheromone, the links belonging to the highest number of shorter tours will have more pheromone deposited. Because the pheromone evaporates with time, links in longer routes will eventually contain much less pheromone than links in shorter tours. As the colony of artificial ants is allowed to travel through the cities a number of times, those tours less reinforced (by pheromone) will attract fewer ants in their next travel (Bonabeau and Théraulaz, 2000). Dorigo and collaborators (Dorigo et al., 1996) have found that by repeating this process a number of times, the artificial ants are able to determine progressively shorter routes, such as the one illustrated in Figure 5.4(b).

The Simple Ant Colony Optimization Algorithm (S-ACO)

Ant algorithms were first proposed by Dorigo et al. (1991) and Dorigo (1992) as a multi-agent approach to solve discrete optimization problems, such as the traveling salesman problem and the quadratic assignment problem (QAP). Similarly to all the other fields of investigation discussed in this book, there is a lot of ongoing research. Therefore, there are various versions and extensions of ant algorithms. This section reviews the simplest ACO algorithm, and the next section presents a general-purpose ACO algorithm along with its most relevant features.

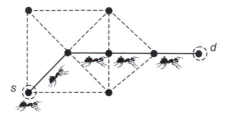

Figure 5.5: An ant travels through the graph from the source node *s* to the destination node *d* thus building a solution (path). (Note that this travel resembles the one performed by the traveling salesperson, who has a number of cities to visit and has to decide which path to take.)

Assuming a connected graph $G = (V, E)$, the simple ACO algorithm (S-ACO) can be used to find a solution to the shortest path problem defined on the graph *G*. (Appendix B.4.4 brings the necessary fundamentals from *graph theory*.) A solution is a path on the graph connecting a source node *s* to a destination node *d* and the path length is given by the number of edges traversed (Dorigo and Di Caro, 1999), as illustrated in Figure 5.5. Associated with each edge (i,j) of the graph there is a variable τ_{ij} termed *artificial pheromone trail*, or simply phero-mone. Every *artificial ant* is capable of "marking" an edge with pheromone and "smelling" (reading) the pheromone on the trail.

Each ant traverses one node per iteration step *t* and, at each node, the local in-formation about its pheromone level, τ_{ij}, is used by the ant such that it can prob-abilistically decide the next node to move to, according to the following rule:

$$p_{ij}^{k}(t) = \begin{cases} \dfrac{\tau_{ij}(t)}{\sum_{j \in N_i} \tau_{ij}(t)} & \text{if } j \in N_i \\ 0 & \text{otherwise} \end{cases} \qquad (5.1)$$

where $p_{ij}^{k}(t)$ is the probability that ant *k* located at node *i* moves to node *j*, $\tau_{ij}(t)$ is the pheromone level of edge (i,j), all taken at iteration *t*, and N_i is the set of one step neighbors of node *i*.

While traversing an edge (i,j), the ant deposits some pheromone on it, and the pheromone level of edge (i,j) is updated according to the following rule:

$$\tau_{ij}(t) \leftarrow \tau_{ij}(t) + \Delta\tau \qquad (5.2)$$

where *t* is the iteration counter and $\Delta\tau$ is the constant amount of pheromone de-posited by the ant.

The analysis of Equation (5.1) and Equation (5.2) lead to the conclusion that when an ant deposits some pheromone on a given edge (i,j), it increases the probability that this edge is selected by another ant, thus reinforcing the trail that passes through this edge. The presence of positive feedback favoring the selec-tion of short paths is clear in this simple model.

Preliminary experiments with the S-ACO algorithm demonstrated that it is capable of determining the shortest route between a simulated nest and food source in a computer simulation similar to the laboratory experiment illustrated in Figure 5.2. However, the behavior of the algorithm becomes unstable when the complexity of the searched graph increases (Dorigo and Di Caro, 1999). In order to overcome this limitation and to avoid a quick convergence of all ants towards sub-optimal routes, an evaporation of the pheromone trails was allowed, thus changing Equation (5.2) into:

$$\tau_{ij}(t) \leftarrow (1-\rho)\tau_{ij}(t) + \Delta\tau \qquad (5.3)$$

where $\rho \in (0,1]$ is the pheromone decay rate.

General-Purpose Ant Colony Optimization Algorithm

Ant algorithms constitute all algorithms for discrete optimization that took inspiration from the observation of the foraging behavior of ant colonies (Dorigo et al., 1999). Although some authors have already developed versions of ant algorithms for continuous optimization (e.g., Bilchev and Parmee, 1995), not much research has been conducted in this direction and, thus, the focus of the discussion to be presented here is on ACO for discrete optimization.

An ACO algorithm alternates, for a maximum number of iterations `max_it`, the application of two basic procedures (Bonabeau et al., 1999). Assume a search is being performed on a connected graph $G = (V,E)$ with V nodes and E edges:

1. A parallel solution construction/modification procedure in which a set of N ants builds/modifies in parallel N solutions to the problem under study.

2. A pheromone trail updating procedure by which the amount of pheromone trail on the problem graph edges is changed.

The process of building or modifying a solution is made in a probabilistic fashion, and the probability of a new edge to be added to the solution being built is a function of the edge *heuristic desirability* η, and of the *pheromone trail* τ deposited by previous ants. The heuristic desirability η expresses the likelihood of an ant moving to a given edge. For instance, in cases the minimal path is being sought among a number of edges, the desirability η is usually chosen to be the inverse of the distance between a pair of nodes. The modification (updating) of pheromone trails is a function of both the evaporation rate ρ (see Equation (5.3)) and the quality of the solutions produced. Algorithm 5.1 depicts the standard, or general-purpose, ACO algorithm to perform discrete optimization (Bonabeau et al., 1999). The parameter e is the number of edges on the graph, and `best` is the best solution (path) found so far; the pheromone level of each edge is only updated after all ants have given their contributions (laid some pheromone). The generic algorithm assumes that the heuristic desirability function of an edge and how to update the pheromone trail have already been defined.

```
procedure [best] = ACO(max_it, N, τ₀)
    initialize τᵢⱼ    //usually initialized with the same τ₀
    initialize best
    place each ant k on a randomly selected edge
    t ← 1
    while t < max_it do,
        for i = 1 to N do,   //for each ant
            build a solution by applying a probabilistic
            transition rule (e − 1) times.
            //The rule is a function of τ and η
            //e is the number of edges on the graph G
        end for
        eval the cost of every solution built
        if an improved solution is found,
            then update the best solution found
        end if
        update pheromone trails
        t ← t + 1
    end while
end procedure
```

Algorithm 5.1: Standard ACO for discrete optimization.

Selected Applications from the Literature: A Brief Description

Algorithm 5.1 presents the general-purpose ACO algorithm. It is important to note the basic features of an ACO algorithm:

- A colony of ants that will be used to build a solution in the graph.

- A probabilistic transition rule responsible for determining the next edge of the graph to which an ant will move.

- A heuristic desirability that will influence the probability of an ant moving to a given edge.

- The pheromone level of each edge, which indicates how 'good' it is.

To illustrate how to use this simple algorithm in practical problems, consider the following two examples. The first example - traveling salesman problem - was chosen because it was one of the first tasks used to evaluate an ACO algorithm. In addition, the TSP is a shortest path problem to which the ant colony metaphor is easily adapted, it is an NP-hard problem, it is didactic, and it has been broadly studied (Lawer et al., 1985). The second example - vehicle routing - illustrates how a small variation in the ACO for the traveling salesman problem allows it to be applied to the vehicle routing problem.

Traveling Salesman

In the TSP the goal is to find a closed tour of minimal length connecting e given cities, where each city must be visited once and only once. Let d_{ij} be the Euclidean distance between cities i and j given by:

$$d_{ij} = [(x_i - x_j)^2 + (y_i - y_j)^2]^{1/2} \tag{5.4}$$

where x_m and y_m are the coordinates of city m, on the x-y plane. This problem can be more generally defined on a graph $G = (V,E)$, where the cities correspond to the nodes V and the connections between them are the edges E.

In (Dorigo et al., 1996; Dorigo and Gambardella, 1997a,b), the authors introduced the Ant System (AS) as an ACO algorithm to solve the TSP problem. In AS ants build solutions to the TSP by moving on the problem graph from one city to another until they complete a tour. During each iteration of the AS, an ant k, $k = 1, \ldots, N$ (a colony of N ants is assumed), builds a tour by applying a probabilistic transition rule $(e - 1)$ times, as described in Algorithm 5.1. The iterations are indexed by the iteration counter t and a maximum number of iteration steps, max_it, is defined for the iterative procedure of adaptation, i.e., $1 \leq t \leq$ max_it. The description to be presented here follows those of (Dorigo et al., 1996; Dorigo and Gambardella, 1997a,b; Bonabeau et al., 1999).

For each ant, the transition from city i to city j at iteration t depends on: 1) the fact that the city has already been visited or not; 2) the inverse of the distance d_{ij} between cities i and j, termed *visibility* $\eta_{ij} = 1/d_{ij}$; and 3) the amount of pheromone τ_{ij} on the edge connecting cities i and j. As in the TSP problem each ant has to visit a city only once, some knowledge of the edges (cities) already visited has to be kept by the ants, thus they must possess some sort of memory, also called a *tabu list*. The memory is used to define the set J_i^k of cities that the ant k still has to visit while in city i.

The probability of an ant k going from city i to city j at iteration t is given by the following transition rule:

$$p_{ij}^k(t) = \begin{cases} \dfrac{[\tau_{ij}(t)]^\alpha . [\eta_{ij}]^\beta}{\sum_{l \in J_i^k} [\tau_{il}(t)]^\alpha . [\eta_{il}]^\beta} & \text{if } j \in J_i^k \\ 0 & \text{otherwise} \end{cases} \tag{5.5}$$

where $\tau_{ij}(t)$ is the pheromone level of edge (i,j), η_{ij} is the visibility of city j when in city i, and J_i^k is the tabu list of cities still to be visited by ant k from node i. The parameters α and β are user-defined and control the relative weight of trail intensity $\tau_{ij}(t)$ and visibility η_{ij}. For instance, if $\alpha = 0$, the closest cities are more likely to be chosen, and if $\beta = 0$, by contrast, only pheromone amplification is considered.

Similarly to the simple ACO algorithm (S-ACO), while traversing an edge (city), an ant lays some pheromone on that edge. In the AS, the quantity $\Delta\tau_{ij}^k(t)$ of pheromone released on each edge (i,j) by ant k depends on how well it has performed according to the following rule:

$$\Delta\tau_{ij}^k(t) = \begin{cases} Q/L^k(t) & \text{if } (i,j) \in T^k(t) \\ 0 & \text{otherwise} \end{cases} \tag{5.6}$$

where $L^k(t)$ is the length of the tour $T^k(t)$ performed by ant k at iteration t, and Q is another user-defined parameter.

The pheromone updating rule is the same as the one used for the simple ACO algorithm (Equation (5.3)) taking into account the differential amount of pheromone released by each ant in each iteration:

$$\tau_{ij}(t) \leftarrow (1-\rho)\tau_{ij}(t) + \Delta\tau_{ij}(t) \tag{5.7}$$

where $\rho \in (0,1]$ is the pheromone decay rate, $\Delta\tau_{ij}(t) = \sum_k \Delta\tau^k_{ij}(t)$, and $k = 1, \ldots , N$ is the index of ants.

According to Dorigo et al. (1996), the number N of ants is an important parameter of the algorithm. Too many ants reinforce sub-optimal trails leading to bad solutions, whereas too few ants would not result in a sufficient cooperative behavior due to pheromone decay. They suggested to use $N = e$; that is, a number of ants equals to the number of cities.

```
procedure [best] = AS-TSP(max_it,α,β,ρ,N,e,Q,τ₀,b)
    initialize τij   //usually initialized with the same τ₀
    place each ant k on a randomly selected city
    Let best be the best tour found from the beginning and
    Lbest its length
    t ← 1
    while t < max_it do,
        for i = 1 to N do,              //for every ant
            //e is the number of cities on the graph
            build tour Tᵏ(t) by applying (e-1) times the fol-
            lowing step:
            At city i, choose the next city j with probabil-
            ity given by Equation (5.5)
        end for
        eval the length of the tour performed by each ant
        if a shorter tour is found,
            then update best and Lbest
        end if
        for every city e do,
            Update pheromone trails by applying the rule:
            τij(t+1) ← (1-ρ)τij(t) + Δτij(t) + b.Δτᵇij(t), where
            Δτij(t) = Σk Δτᵏij(t), k = 1,...N;
```
$$\Delta\tau^k_{ij}(t) = \begin{cases} Q/L^k(t) & \text{if } (i,j) \in T^k(t) \\ 0 & \text{otherwise} \end{cases}, \text{ and}$$
$$\Delta\tau^b_{ij}(t) = \begin{cases} Q/L_{best} & \text{if } (i,j) \in best \\ 0 & \text{otherwise} \end{cases}.$$
```
        end for
        t ← t + 1
    end while
end procedure
```

Algorithm 5.2: Ant system for the traveling salesman problem (AS-TSP).

In an effort to improve the AS performance when applied to the TSP problem, the authors also introduced the idea of "elitist ants", a term borrowed from evolutionary algorithms. An elitist ant is the one that reinforces the edges belonging to the best route found so far, best. Such ants would reinforce the trail (cities or edges) of the best tour by adding $b.Q/L_{best}$ to the pheromone level of these edges, where b is the number of elitist ants chosen, and L_{best} is the length of the best tour found from the beginning of the trial. Algorithm 5.2 depicts the ant system for the traveling salesman problem (AS-TSP). The authors employed the following values for the parameters in their experiments: $\alpha = 1$, $\beta = 5$, $\rho = 0.5$, $N = e$, $Q = 100$, $\tau_0 = 10^{-6}$, and $b = 5$.

It is important to remark that the results presented in Dorigo et al. (1996) were interesting but disappointing. The AS-TSP could outperform other approaches, such as a GA, when applied to small instances of the TSP, but its performance was poor when applied to TSP problems with a large number of cities. In order to enhance even further the performance of the ACO approach, the authors altered the transition rule (Equation (5.5)), employed local updating rules of the pheromone trail (Equation (5.6)), and used a candidate list to restrict the choice of the next city to visit (Dorigo and Gambardella, 1997a,b). These modifications made the algorithm competitive in relation to other approaches from the literature. Further improvements, variations, and applications of ant colony optimization algorithms can be found in Corne et al. (1999), Dorigo et al. (2002), Dorigo and Stützle (2004), and de Castro and Von Zuben (2004).

Vehicle Routing

The simplest *vehicle routing problem* (VRP) can be described as follows: given a fleet of vehicles with uniform capacity, a common depot, and several costumer demands (represented as a collection of geographical scattered points), find the set of routes with overall minimum route cost which service all the demands. All the itineraries start and end at the depot and they must be designed in such a way that each costumer is served only once and just by one vehicle. Note that the VRP is closely related to the TSP as presented here; the VRP consists of the solution of many TSPs considering a single common start and end city for each TSP, which now has a limited capacity.

Bullnheimer et al. (1999a,b) and Bullnheimer (1999) modified the AS-TSP algorithm to solve the vehicle routing problem with one central depot and identical vehicles. The VRP problem was represented by a complete *weighted directed graph* $G = (V,E,w)$, where $V = \{v_0,v_1, \ldots , v_N\}$ is the set of nodes of G, $E = \{(v_i,v_j) : i \neq j\}$ its set of edges, and $w_{ij} \geq 0$ is a weight connecting nodes i and j. The node v_0 denotes the depot and the other vertices are the customers or cities. The weights w_{ij}, associated with edge (v_i,v_j), represent the distance (time or cost) between v_i and v_j. A demand $d_i \geq 0$ and a service time $\delta_i \geq 0$ are associated with each customer v_i. The respective values for the depot are $v_0 = d_0 = 0$.

The objective is to find the minimum cost vehicle routes where:

- Every customer is visited only once by only one vehicle.

- All vehicle routes begin and end at the depot.
- For every vehicle route the total demand does not exceed the vehicle capacity D.
- For every vehicle route the total route length, including service times, does not exceed a given threshold L.

Similarly to the AS-TSP, to solve the vehicle routing problem one ant is placed on each node of the graph and the ants construct solutions by sequentially visiting a city, until all cities have been visited. Whenever the choice of another city would lead to an unfeasible solution; that is, vehicle capacity greater than D or total route length greater than the threshold L, the depot is chosen and a new tour is started.

The VRP problem was solved by applying the procedure presented in Algorithm 5.1 with the probabilistic transition rule presented in Equation (5.5), which takes into account the pheromone trail τ_{ij} and the heuristic desirability η_{ij}. The heuristic desirability of a city follows a *parametrical saving function* (Paessens, 1988):

$$\eta_{ij} = w_{i0} + w_{0j} - g.w_{ij} + f.\,|w_{i0} - w_{0j}| \qquad (5.8)$$

where g and f are two user-defined parameters.

In the standard ant system algorithm, after an ant has constructed a solution, the pheromone trails are laid by all ants depending on the objective value of the solution. It was discussed in the previous example that by using σ elitist ants, the performance of the algorithm was improved for the TSP problem. In (Bullnheimer et al., 1999b), the authors suggested that by ranking the ants according to the solution quality and using only the best ranked ants, also called elitist ants, to update the pheromone trails would improve even further the performance of the algorithm for the VRP problem. The new updating rule is as follows:

$$\tau_{ij}(t) \leftarrow (1 - \rho)\tau_{ij}(t) + \Delta\tau^r_{ij}(t) + \sigma.\Delta\tau_{ij}^+(t) \qquad (5.9)$$

where $\rho \in (0,1]$ is the pheromone decay rate. Only if an edge (v_i, v_j) was used by the μ-th best ant, the pheromone trail is increased by a quantity

$$\Delta\tau^r_{ij}(t) = \sum_{\mu=1}^{\sigma-1} \Delta\tau^\mu_{ij}(t) \qquad (5.10)$$

where $\Delta\tau^\mu_{ij}(t) = (\sigma - \mu)/L_\mu(t)$ if ant μ uses the edge (v_i, v_j), and $\Delta\tau^\mu_{ij}(t) = 0$ otherwise. $L_\mu(t)$ is the length of the tour ant μ performs at iteration t. All edges belonging to the best tour determined so far are emphasized as if σ elitist ants had used them. Therefore, each elitist ant increases the trail intensity by an amount $\Delta\tau_{ij}^+(t)$ that is equal to $1/L^+$ if edge (v_i, v_j) belongs to the so far best solution, and zero otherwise.

Scope of ACO Algorithms

Although the traveling salesman problem is an intuitive application of ACO, many other discrete (combinatorial) optimization problems can be solved with

ACO. Bonabeau et al. (2000) claim that ACO is currently the best available heuristic for the sequential ordering problem, for real-world instances of the quadratic assignment problem, and is among the best alternatives for the vehicle and network routing problems. Good surveys of applications of ACO algorithms, including a number of main references can be found in Bonabeau et al. (1999, 2000), Dorigo et al. (1999), Dorigo and Di Caro (1999), and Dorigo and Stützle (2004). The main problems tackled by ACO algorithms are: TSP, network routing, graph coloring, shortest common super sequence, quadratic assignment, machine scheduling, vehicle routing, multiple knapsack, frequency assignment, and sequential ordering.

ACO algorithms are suitable for solving problems that involve graph searching, mainly minimal cost problems. Similarly to all the other approaches discussed in this text, ACO algorithms may be applied only when more classical approaches, such as dynamic programming or other methods cannot be (efficiently) applied. Dorigo and Di Caro (1999) suggest a list of particular cases of interest for ACO algorithms:

- *NP problems*: a problem is said to be solvable in polynomial time if there is an algorithm to solve it in a time that is a polynomial function of the size of the input. NP stands for *nondeterministic polynomial*. It is a large class of problems that have the property that if any solution exists, there is at least one solution which may be verified as correct in polynomial time. Less rigorously, a problem is NP if we can check a proposed solution in polynomial time. A more detailed description of problem complexity is provided in Appendix B.3.3.

- *Static and dynamic combinatorial optimization problems*: a combinatorial problem is the one in which there is a large discrete set of possible solutions. The static problems are those whose characteristics do not change over time, e.g., the classic traveling salesman problem. The shortest path problems in which the properties of the graph representation of the problem change over time constitute the dynamic problems (e.g., network routing).

- *Distributed problems*: those in which the computational architecture is spatially distributed (e.g., parallel or network processing), and may require a set of agents to find suitable solutions. The inherent distribution of agents' resources, such as knowledge and capability, promotes the need of a cooperative work.

From Natural to Artificial Ants

Most of the ideas of ACO were inspired by the foraging behavior of trail-laying trail-following ants. In particular, the main features of ACO algorithms are the use of: 1) a colony of cooperating individuals, 2) a pheromone trail for local indirect (stigmergic) communication, 3) a sequence of local moves to find the shortest paths, and 4) a probabilistic decision rule using local information (Dorigo et al., 1999). Table 5.1 clarifies how the ACO approach uses the biological terminology borrowed from the natural world of ants.

Table 5.1: Interpretation of the biological terminology into the computational Ant Colony Optimization (ACO) algorithms.

Biology (Ant Foraging)	ACO Algorithms
Ant	Individual (agent) used to build (construct) a solution to the problem
Ant colony	Population (colony) of cooperating individuals known as artificial ants
Pheromone trail	Modification of the environment caused by the artificial ants in order to provide an indirect mean of communication with other ants of the colony This allows the assessment of the quality of a given edge on a graph
Pheromone evaporation	Reduction in the pheromone level of a given path (edge) due to aging

5.2.4. Clustering of Dead Bodies and Larval Sorting in Ant Colonies

Chrétien (1996) has performed a number of experiments to study the organization of cemeteries of the ant *Lasius niger*. It has been observed, in several species of ants, workers grouping corpses of ants, or parts of dead ants, to clean up their nests, thus forming cemeteries. Further research on different ant species (Deneubourg et al., 1991) has confirmed the cemetery organization capability of ants.

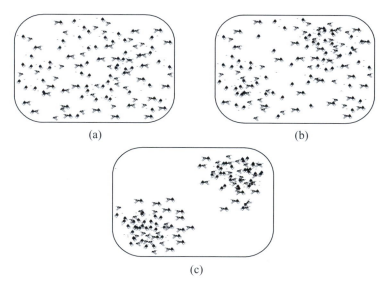

(a)   (b)

(c)

Figure 5.6: A qualitative example of cemetery formation in ants. (a) The corpses are initially randomly distributed over the arena. (b) After some time (e.g., a few hours) the workers start piling the corpses. (c) Clusters of corpses are formed by the worker ants.

The phenomenon observed in these experiments is the aggregation of dead bodies by workers. If dead bodies, or more precisely, items belonging to dead bodies, are randomly distributed in space at the beginning of the experiment, the workers will form clusters with these items within a few hours (Bonabeau, 1997). If the experimental arena is not sufficiently large or if it contains spatial heterogeneities, the clusters will be formed along the borders of the arena or along the heterogeneities. Figure 5.6 presents a qualitative illustration of how ants form clusters of corpses in a given arena. It is assumed that a number of corpses are spread over the arena and worker ants are allowed to pick them up and drop them somewhere within the arena.

The basic mechanism behind this type of clustering (aggregation) phenomenon is an attraction between dead items (corpses) mediated by ant workers. Small clusters of items grow by attracting more workers to deposit more dead items. It is this positive feedback mechanism that leads to the formation of larger and larger clusters of dead items. However, the exact individual behavior that leads to this positive feedback mechanism and the adaptive significance of corpse clustering are still unclear. Cleaning the nest by getting rid of dead items is an important task, though (Bonabeau, 1991; Bonabeau et al., 1999).

Another related phenomenon is larval sorting by the ant *Leptothorax unifasciatus*, a phenomenon widespread and broadly studied for this ant species (Franks and Sendova-Franks, 1992). Workers of this species gather the larvae according to their size. All larvae tend to be clustered, with small larvae located in the center and larger larvae in the periphery of the brood.

Stigmergy

The stigmergic principle is also clear in this corpse-gathering behavior of ant colonies. The observations show that ants tend to put the corpses of dead ants together in cemeteries in certain places far from the nest and that grow in size with time. If a large number of corpses are scattered outside the nest, the ants from the nest will pick them up, carry them about for a while and drop them. Within a short period of time, it can be observed that corpses are being placed in small clusters. As time goes by, the number of clusters decrease and the size of the remaining clusters increase, until eventually all the corpses are piled in one or two large clusters. Therefore, the increase in cluster size reinforces the enlargement of a cluster, and the reverse is also true.

5.2.5. Ant Clustering Algorithm (ACA)

In addition to path optimization to food sources, *data clustering* is another problem solved by real ants. Clustering problems have already been explored in the previous chapter while describing the self-organizing networks and some of their applications. Before proceeding with the presentation of the clustering algorithm derived from models of cemetery organization and brood sorting in ant colonies, the reader should refer to Appendix B.4.5 for further comments on data clustering.

The Standard Ant Clustering Algorithm (ACA)

In order to model the two phenomena of dead body clustering and larval sorting in some ant species, Deneubourg et al. (1991) have proposed two closely related models. Both models rely on the same principles, but the first one (cemetery organization) is a special case of the second one (brood sorting). While the clustering model reproduces experimental observations more faithfully, the second one has given rise to more practical applications. In their study, a colony of ant-like agents, randomly moving in a bi-dimensional grid, was allowed to move basic objects so as to cluster them.

The general idea is that isolated items should be picked up and dropped at some other location where more items of that type are present. Assume that there is a single type of item in the environment, and a number of "artificial ants" whose role will be to carry an item and drop it somewhere in the environment. The probability p_p that a randomly moving ant, who is not currently carrying an item, picks up an item is given by

$$p_p = \left(\frac{k_1}{k_1 + f} \right)^2 \qquad\qquad (5.11)$$

where f is the *perceived fraction of items* in the neighborhood of the ant, and k_1 is a threshold constant. For $f \ll k_1$, $p_p \approx 1$, thus the probability that the ant picks up an item is high when there are not many items in its neighborhood. In the case $f \gg k_1$, $p_p \approx 0$, thus the probability that an item is removed from a dense cluster is small; that is, when there are many items in the neighborhood of an item it tends to remain there.

The probability p_d that a randomly moving ant, who is currently carrying an item, deposits that item in a certain position is given by

$$p_d = \left(\frac{f}{k_2 + f} \right)^2 \qquad\qquad (5.12)$$

where k_2 is another threshold constant. In this case, if $f \ll k_2$, then $p_d \approx 0$, thus the probability that the ant deposits the item is low when there are not many items in its neighborhood. In the case $f \gg k_2$, $p_d \approx 1$, thus the probability that the item is deposited in a dense cluster is high; that is, items tend to be deposited in regions of the arena where there are many items.

In order to use this theoretical model to generate an *ant clustering algorithm* (ACA), two main aspects have yet to be discussed. First, in which kind of environment are the ants going to move? That is, how to project the space of attributes so that it is possible to visualize clusters? Second, how is the function f (the perceived fraction of items) going to be defined?

Projecting the Space of Attributes into a Bi-Dimensional Grid

In the standard ant clustering algorithm (Lumer and Faieta, 1994), ants are assumed to move on a discrete 2-D board. This board may be considered as a bi-

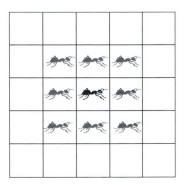

Figure 5.7: A bi-dimensional grid with 5 × 5 cells. An ant (dark ant in the center cell) is placed on a cell with its eight neighbors depicted (gray ants surrounding the center ant).

dimensional grid B of $m \times m$ cells. An ant is allowed to travel from one side of the board to another, meaning that the board is toroidal, and the ant perceives a surrounding region area, i.e., a squared neighborhood $\text{Neigh}_{(s \times s)}$ of $s \times s$ cells surrounding the current site (cell) r. In the example illustrated in Figure 5.7 the reference ant, the one in the middle, perceives a neighborhood of 3×3 cells.

This representation of the arena by a bi-dimensional board resembles the bi-dimensional grid of Kohonen's self-organizing map (Section 4.4.5), but is more akin to a bi-dimensional grid in a *cellular automaton* (Section 7.3). Therefore, each ant is modeled by an automaton able to move on the grid and displace objects according to probabilistic rules which are based only upon information of the local environment (neighborhood). The combined actions of a set (colony) of these automata (ants) will lead to the clustering in the same spatial region of the grid of data items belonging to the same classes.

In the algorithm proposed by Lumer and Faieta (1994), the bi-dimensional grid is used as a low dimensional space in which to project the space of attributes of the input data. In this case, the dimension is $z = 2$ so that, after the clustering by ants, clusters present the following property: intra-cluster distances should be small in relation to inter-cluster distances. Therefore, for $z = 2$, the input data are mapped onto a board composed of m^2 cells. Such a mapping, after the ant clustering process, must preserve the neighborhood inherent in the input data without creating too many new neighbors in z-dimensions that do not exist in the L-dimensional space corresponding to the original space of the input data set.

The idea of the ant clustering algorithm is thus to initially project the input data items in the bi-dimensional grid and scatter them randomly, as illustrated in Figure 5.8. The artificial ants (agents) are also spread all over the bi-dimensional grid, and allowed to move, pick up, and deposit items according to some similarity measure and given probabilities.

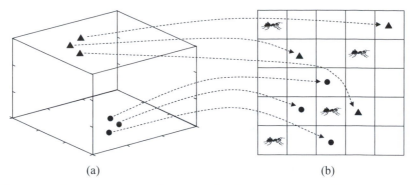

(a)   (b)

Figure 5.8: Projecting the input data into the bi-dimensional grid. (a) The input data lies in a three-dimensional space. (b) Before the ant clustering process starts, each input datum is projected into a random position of the bi-dimensional grid.

Defining the Perceived Fraction of Items: f

The second aspect that has to be accounted for is the definition of how f, the perceived fraction of items, is evaluated. Similarly to evolutionary algorithms (Chapter 3), f will be defined by the problem under study. For instance, in the context of robotic systems, Deneubourg et al. (1991) defined f as the quotient between the number M of items encountered during the last T time units and the largest possible number of items that can be encountered during these T time units.

Assuming that ants move in a bi-dimensional grid, the standard ACA can be summarized as in Algorithm 5.3. Let 'unloaded ant' be an ant not carrying an item, assume f is problem dependent, p_p and p_d are the probabilities of picking up and dropping off an item, respectively, and N is the number of ants. In this standard algorithm a cell can only be occupied by a single item.

Selected Applications from the Literature: A Brief Description

The standard ant clustering algorithm is described in Algorithm 5.3. Its applications in clustering range from exploratory data analysis to graph partitioning. To illustrate how to use the standard algorithm to different types of data, including to real-world data, it will be first described its application to simple numeric data and then to bioinformatics data. Both examples project the data into a bi-dimensional grid, but use different local density functions and picking up and dropping off probability functions. The examples focus on how to use the ACA standard algorithm to tackle these problems. Further details can be found in the cited literature.

Exploratory Data Analysis

Lumer and Faieta (1994) applied the standard ACA to exploratory data analysis. Their aim was to identify clusters in an unlabelled data set. To accomplish this

```
procedure [] = ACA(max_it, N, k₁, k₂)
    project every item i on a random cell of the grid
    place every ant on a random cell of the grid unoccupied
    by ants
    t ← 1
    while t < max_it do,
        for i = 1 to N do,    //for every ant
            compute f(xᵢ)
            if unloaded ant AND cell occupied by item xᵢ, then
                compute pₚ(xᵢ)        //pick up probability
                pick up item xᵢ with probability pₚ(xᵢ)
            else if ant carrying item xᵢ AND cell empty, then
                compute p_d(xᵢ)        //drop off probability
                deposit (drop) item xᵢ with probability p_d(xᵢ)
            end if
            move randomly to an unoccupied neighbor
        end for
        t ← t + 1
    end while
    print location of items
end procedure
```

Algorithm 5.3: Standard ant clustering algorithm (ACA).

task, the input data items belonging to an L-dimensional Euclidean space, $\mathbf{X} \in \mathfrak{R}^L$, were projected into a bi-dimensional grid Z^2, which can be considered as a discretization of a real space. The artificial ants were allowed to move about in this grid (discrete space) and perceive a surrounding region of area s^2 that corresponds to a square $\text{Neigh}_{(s \times s)}$ of $s \times s$ cells surrounding the current cell r, as illustrated in Figure 5.7.

The authors suggested the following function f to calculate the local density of items in the neighborhood of object $\mathbf{x}_i$ situated at site r:

$$f(\mathbf{x}_i) = \begin{cases} \dfrac{1}{s^2} \displaystyle\sum_{x_j \in \text{Neigh}_{(s \times s)}(r)} \left[1 - \dfrac{d(\mathbf{x}_i, \mathbf{x}_j)}{\alpha} \right] & \text{if } f > 0 \\ 0 & \text{otherwise} \end{cases}$$

(5.13)

where $f(\mathbf{x}_i)$ is a measure of the average similarity of object $\mathbf{x}_i$ with another object $\mathbf{x}_j$ in the neighborhood of $\mathbf{x}_i$, α is a factor that defines the scale of dissimilarity, and $d(\mathbf{x}_i, \mathbf{x}_j)$ is the Euclidean distance between two data items in their original L-dimensional space. The parameter α determines when two items should or should not be located next to each other. For instance, if α is too large, there is not much discrimination between two items, leading to the formation of clusters composed of items that should not belong to the same cluster; and vice-versa.

Note: the distance $d(\cdot, \cdot)$ between items is calculated in their original space, not in the projected space.

Equation (5.13) takes into account the neighborhood of a cell ($1/s^2$), and the relative distance between items in this neighborhood. It proposes that the smaller the distance between the item an ant is carrying and the items in its neighborhood, the higher the local density of items perceived and, thus, the higher the probability of this item being dropped in that position (cell). Inversely, the more different the item being carried is from the items on a given neighborhood, the smaller the likelihood that this item is dropped in that vicinity.

The probabilities of picking up or depositing an item are given by the following equations:

$$p_p(\mathbf{x}_i) = \left(\frac{k_1}{k_1 + f(\mathbf{x}_i)} \right)^2 \tag{5.14}$$

$$p_d(\mathbf{x}_i) = \begin{cases} 2f(\mathbf{x}_i) & \text{if } f(\mathbf{x}_i) < k_2 \\ 1 & \text{otherwise} \end{cases} \tag{5.15}$$

where k_1 and k_2 are two constants playing similar roles to the constants in Equation (5.11) and Equation (5.12) respectively.

To illustrate the performance of this algorithm, the authors applied it, with p_p and p_d determined by Equation (5.14) and Equation (5.15), to an artificially generated set of points in $\Re^2$ (see Computational Exercise 3, Section 5.6.2).

Although the results obtained by Lumer and Faieta (1994) with the standard ant clustering algorithm when applied to the simple benchmark problem of Computational Exercise 3 were good, the authors observed the presence of more clusters in the projected space than in the original distribution of data. In order to alleviate this problem, they proposed a number of variations in their initially proposed algorithm: ants with different speeds, ants with memory, and behavioral switches:

1) *The use of ants with different moving speeds*: each ant in the colony is allowed to have its own speed. The speed of the swarm is distributed uniformly in the interval [1,v_{max}], where $v_{max} = 6$ is the maximal speed of an ant, and v corresponds to the number of grid units (cells) walked per time unit by an ant along a given grid axis. This pace influences the probability an ant picks up or drops a given item according to the following equation:

$$f(\mathbf{x}_i) = \begin{cases} \dfrac{1}{s^2} \sum_{\mathbf{x}_j \in \text{Neigh}_{(s \times s)}(r)} \left[1 - \dfrac{d(\mathbf{x}_i, \mathbf{x}_j)}{\alpha + \alpha(v-1)/v_{max}} \right] & \text{if } f > 0 \\ 0 & \text{otherwise} \end{cases} \tag{5.16}$$

Therefore, fast moving ants are not as selective as slow ants in their estimation of the average similarity of an object to its neighbors.

2) *A short-term memory for the ants*: ants can remember the last m items they have dropped along with their locations. Each time an item is picked up, it is compared to the elements in the ant's memory, and the ant goes toward the location of the memorized element most similar to the one just collected. This behavior leads to the formation of a smaller number of sta-

tistically equivalent clusters, because similar items have a lower probability of starting a new cluster.

3) *Behavioral switches*: ants are endowed with the capability of destroying clusters that have not been reinforced by the piling up of more items during some specific time interval. This is necessary because items are less likely to be removed as clusters of similar objects are formed.

Bioinformatics Data

As discussed previously, ant clustering algorithms often generate a number of clusters that is much larger than the natural number of clusters in the data set. Besides, the standard ACA does not stabilize in a particular clustering solution; it constantly constructs and deconstructs clusters during adaptation. To overcome these difficulties, Vizine et al. (2005) proposed three modifications in the standard ACA: 1) a cooling schedule for k_1; 2) a progressive vision field that allows ants to 'see' over a wider area; and 3) the use of a pheromone function added to the grid as a way to promote reinforcement for the dropping of objects at denser regions of the grid.

Concerning the cooling schedule for k_1, the adopted scheme is simple: after one cycle has passed (10,000 ant steps), the value of the parameter k_1 starts being geometrically decreased, at each cycle, until it reaches a minimal allowed value, k_{1min}, which corresponds to the stopping criterion for the algorithm:

$$k_1 \leftarrow 0.98 \times k_1$$
$$k_{1min} = 0.001$$

(5.17)

In the standard ACA, the value of the perceived fraction of items, $f(\mathbf{x}_i)$, depends on the vision field, s^2, of each ant. The definition of a fixed value for s^2 may sometimes cause inappropriate behaviors, because a fixed perceptual area does not allow the distinction of clusters with different sizes. To overcome this difficulty, a progressive vision scheme was proposed (Vizine et al., 2005). When an ant i perceives a 'large' cluster, it increments its perception field (s_i^2) up to a maximal size. A large cluster is detected using function $f(\mathbf{x}_i)$; when $f(\mathbf{x}_i)$ is greater than a given threshold θ, then the vision s^2 of an ant is incremented by n_s units until it reaches its maximum value:

$$\text{If } f(\mathbf{x}_i) > \theta \quad \text{and} \quad s^2 \leq s^2_{max}$$
$$\text{then } s^2 \leftarrow s^2 + n_s$$

(5.18)

where $s^2_{max} = 7 \times 7$ and $\theta = 0.6$ in their implementation.

Inspired by the way termites use pheromone to build their nests (Section 2.2.3), Vizine et al. (2005) proposed adding pheromone to the objects the ants carry and allowing the transference of this pheromone to the grid when an object is dropped.

The pheromone added to the carried items lead to the introduction of a pheromone level function $\phi(i)$ in the grid, which corresponds to how much pheromone is present at each grid cell i. During each iteration, the artificial pheromone $\phi(i)$

pheromone $\phi(i)$ at each cell of the grid evaporates at a fixed rate. This phero-mone influences the picking and dropping probabilities as follows:

$$P_p(i) = \frac{1}{f(i)\phi(i)}\left(\frac{k_1}{k_1 + f(i)}\right)^2 \tag{5.19}$$

$$P_d(i) = f(i)\phi(i)\left(\frac{f(i)}{k_2 + f(i)}\right)^2 \tag{5.20}$$

Note that according to Equation (5.19) the probability that an ant picks up an item from the grid is inversely proportional to the amount of pheromone at that position and also to the density of objects around i. The reverse holds for Equation (5.20); the probability of an ant dropping an item on the grid is directly proportional to the amount of pheromone at that position and to the density of objects around i.

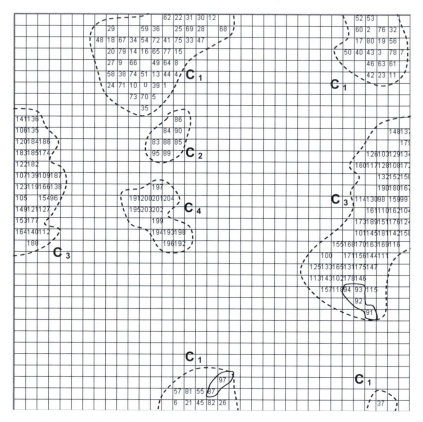

Figure 5.9: Final grid configuration for the bioinformatics data after the algorithm converged.

One of the problems the authors used to evaluate their proposal was in the field of bioinformatics, more specifically, the *yeast galactose* data set composed of 20 attributes (Yeung et al., 2003). Note that although each datum has dimension 20 all of them will be projected onto the bi-dimensional grid of the ACA. The authors used a subset of 205 data items, whose expression patterns or attributes reflect four functional categories (clusters) of genes formed by 83, 15, 93 and 14 genes (data). The authors used 10 ants to run the algorithm, $n_{ants} = 10$, a grid of size 35×35, $\alpha = 1.05$, $\theta = 0.6$, $k_1 = 0.20$, and $k_2 = 0.05$.

Figure 5.9 illustrates one grid configuration after the algorithm converged. Note the presence of four clusters of data (within dotted lines) and some misclassifications made by the algorithm (within solid lines). The numbers indicate the label of each datum in the data set. Note also the toroidal nature of the grid; that is, some clusters cross from one side of the picture to another, both from top to bottom and from left to right.

Scope of Ant Clustering Algorithms

The examples of application and the nature of the ant clustering algorithm suggest its potentiality for clustering problems. In more general terms, ACAs are suitable for exploratory data analysis, in which an unlabelled set of data is available and some information must be extracted from it. For instance, useful information is the degree of similarity between items and how to infer the cluster a new item (unknown to the system) belongs to.

It was discussed that the standard ACA fails in determining the appropriate number of clusters. However, some modifications can be introduced in the standard algorithm so as to make it suitable for determining a more accurate number of clusters in relation to the actual clusters of the input data set.

The ACA performs a dimensionality reduction in the space of attributes of the input data set into a bi-dimensional grid. This may be a useful approach when dealing with spaces of very high dimension, for it allows a visualization of neighborhood (similarity) relationships between the data items in the data set. Nevertheless, it may also cause some loss of information as a consequence of dimensionality reduction.

From Natural to Artificial Ants

Ant clustering algorithms were developed as a generalization of models of dead body clustering and brood sorting in ant colonies. The main features of ant clustering algorithms are a colony of ants that move within a bi-dimensional grid used to project the items (e.g., data and graph vertices) to be clustered. Ants are allowed to move items throughout this grid in a probabilistic manner, which is proportional to the degree of similarity between items. Table 5.2 presents one interpretation of the biological terminology to the computational ant clustering algorithm.

Table 5.2: Interpretation of the biological terminology into the computational ant clustering algorithm (ACA).

Biology (Ant Clustering)	Ant Clustering Algorithm
Environment (Arena)	Bi-dimensional grid in which items are projected and ants are allowed to move
Ant	Agent-like entity capable of moving about, picking up, and depositing items on the grid
Ant colony	Population (colony or swarm) of cooperating individuals known as artificial ants
Dead bodies or larvae	Items or objects (e.g., data patterns and vertices of a graph)
Pile (heap) of dead bodies	Clusters of items
Visibility of an ant	Perceived fraction of items f

5.2.6. Summary of Swarm Systems Based on Social Insects

This section discussed the behavior of social insects, focusing the case of ant colonies. It was observed that several different patterns of behavior can be seen in nature, from foraging for food to cemetery organization. These different behaviors lead to several computational tools and systems with various applications, such as combinatorial optimization and data analysis.

The majority of ant colony optimization algorithms are used to perform some sort of graph search associated with a combinatorial optimization problem. During the operation of an ACO algorithm, a colony of artificial ants is initialized in randomly chosen nodes of the graph G, and is allowed to move through adjacent nodes of this graph. The move of an ant from one node to another follows a probabilistic transition rule that is proportional to the pheromone level of a neighbor edge and the visibility (heuristic desirability) of the ant. By moving, ants incrementally build solutions to the problem and deposit pheromone on the edges traversed. This pheromone trail information will direct the search process of other ants, indicating that a certain route has been taken by (an)other ant(s). Pheromone trails are allowed to evaporate; that is, the level of pheromone of an edge is decreased over time, thus avoiding a rapid convergence towards sub-optimal solutions.

The cemetery organization and larval sorting behavior of ants have led to the proposal of a theoretical model capable of reproducing a similar behavior. This model then led to the development of the ant clustering algorithm with potential applicability to exploratory data analysis, among other clustering tasks. The algorithm works by projecting the objects to be clustered into a bi-dimensional grid and probabilistically placing these objects onto the grid according to the density of similar items in a given region of the grid. The main aspects to be defined in this algorithm are the probability of placing or moving an item at or from a given position (site or cell) of the grid, and how to define the density function of items in this area of the grid.

5.3. SWARM ROBOTICS

In *autonomous robotics*, the so-called *swarm-based*, or *collective robotics*, relies on the use of metaphors and inspiration from biological systems, in particular social insects, to the design of distributed control systems for groups of robots (Kube and Bonabeau, 2000). Biological analogies and inspiration are thus pervasive in swarm robotics (Kube et al., 2004).

Well-known collective behaviors of social insects provide striking evidence that systems composed of simple *agents* (ants, bees, wasps, and termites) can accomplish sophisticated tasks in the real world. It is widely held that the cognitive capabilities of these insects are very limited, and that complex behaviors are an emergent property of interactions among the agents, which are themselves obeying simple rules of behavior (Cao et al., 1997). Thus, rather than following the AI tradition of modeling robots as rational, deliberative agents, researchers in swarm robotics have chosen a more bottom-up approach in which individual agents are more like *artificial insects*, in particular *artificial ants* - they follow simple rules and are reactive to environmental and group changes. Therefore, the concept of *self-organization* forms the core of swarm robotics.

Groups of mobile robots are built aimed basically at studying issues such as group architecture, conflict resolution, origin of cooperation, learning, and geometric problems. There has been an increasing interest in systems composed of multiple autonomous mobile robots exhibiting cooperative behavior. The current interest for swarm-based robotics is due to several factors (Cao et al., 1997; Kube and Bonabeau, 2000):

- Some tasks may be inherently too complex (or impossible) for a single robot to accomplish. For instance, moving an item (e.g., a load) that is much heavier than the robot itself.

- Performance benefits can be gained from the use of multiple robots. For example, a task such as garbage collection may be performed faster by a team of robots.

- Building and using simple robots is usually easier, cheaper, more flexible and more fault tolerant than having a single robot with sophisticated sensors, actuators and reasoning capabilities.

- The decrease in price of simple commercial robots, such as the Khepera® (cf. K-Team Web Site, the current enterprise responsible for commercializing the Khepera robots), make them more accessible for research experiments.

- The progress of mobile robotics over the last few years facilitated the implementation and testing of groups (swarms) of robots.

- The field of *Artificial Life* (Chapter 8), where the concept of emergent behavior is emphasized as essential to the understanding of fundamental properties of life, has done much to propagate ideas about collective behavior in biological systems, particularly social insects.

- The constructive, synthetic approach of swarm robotics can yield insights into fundamental problems in social and life sciences (e.g., cognitive psychology and theoretical biology).

Up to the early 1990s, the majority of research projects in robotics were concentrated in designing a single autonomous robot capable of functioning, and usually performing a given task, in a dynamic environment. However, by organizing multiple robots into collections of task achieving populations, useful tasks may be accomplished with fairly simple behavior-based control mechanisms. This is the approach adopted by the swarm-based robotics, in which a swarm of robots cooperate to accomplish a given task.

In swarm robotics, a set of (simple) individual behavior rules is defined as a means to determine how the robots are going to interact with one another and the environment. Task accomplishment is a result of the emergent group level behavior. Therefore, if the behavior of the swarm is an emergent property, how can one evaluate the rules set for the robots as appropriate or not? There is no answer to such question yet, though some insight can be gained from studies in the field of artificial life and research with the swarms of robots. Another difficulty in swarm robotics is that, due to the lack of global knowledge, a group of robots may stagnate or find itself in a deadlock where it can make no progress. Despite these difficulties and the youth of the field, it is still very promising, though some authors believe many potential applications of swarm robotics may require miniaturization (Bonabeau et al., 1999; Kube and Bonabeau, 2000).

There are a large number of researchers studying swarm robotics and a vast literature on the topic. Thus, instead of trying to make a broad survey, the aim of this section is to describe a selected sample of real-world implementations of swarm robotics. The focus will be on the type of collective behavior modeled, and on the main behavioral rules and environmental set ups used. No detail will be provided regarding technical, implementation or architectural issues. The examples of collective behavior to be described here were inspired basically by the following natural behaviors of ants:

- Foraging for food.
- Clustering of dead bodies.
- Clustering around food (recruitment).
- Collective prey retrieval.

5.3.1. Foraging for Food

Krieger et al. (2000) and Krieger and Billeter (2000) showed that an ant-inspired system for task allocation based on variation in individual stimulus thresholds provides a simple, robust and efficient way to regulate activities of a team of robots. In this case, the task was to collect food items when robots have no initial information about the environment and the total number of robots in the colony.

The essential components of social organization governing ant colonies were used to program swarms of robots foraging from a central nest. The authors chose foraging as the task assigned to the robots due to several reasons: 1) foraging is the task for which mechanisms of task allocation are best understood; 2) foraging efficiency is a key factor influencing colony productivity; and 3) foraging is the task that has received the greatest attention in cooperative robotics.

Given a colony of robots with one central nest and a set of "food items", all within a closed arena, the task the robots have to accomplish corresponds to collecting food so as to maintain a minimal energy level for the colony. The experimental setup and behavior rules for the robots can be summarized as follows:

1. Colony-level information about the energy of the colony is assessed and updated by each individual robot.

2. While in the nest, the robots receive information about the colony energy level via radio messages from a control station.

3. Upon return to the nest, each robot renews its energy reserves, resulting in a decrease in colony energy, and unload (if loaded) the food item collected in a basket, thereby increasing the energy of the colony.

4. The robots were programmed so as to avoid one another in order to minimize collisions and negative interactions.

Individual variation in the tendency to perform a task was also implemented in the robots, thus modeling the different individual stimulus thresholds for task allocation observed in ants and bees.

5. Therefore, robots do not leave the nest simultaneously, but only when the colony energy drops below a pre-defined threshold of a particular robot.

6. Finally, for some trials, robots were programmed to recruit another robot when they identify an area rich in food items, thus mimicking the tandem recruitment behavior observed in many ant species (Section 5.2.2).

In summary, the robots have generalized information about the energy level of the colony, ways to avoid interfering with another robot in space and time, and, in some experiments, the ability to recruit other robots to collect food. Their objective was to maintain the energy of the colony above a minimum threshold level.

The authors used Khepera miniature robots, incremented with a gripper turret, a custom made detection module and a radio turret, designed as a research tool at the *Laboratoire de Micro-Informatique* of the Swiss Federal Institute of Technology at Lausanne, Switzerland. A video of their experiment can be found at http://www.nature.com/nature/journal/v406/n6799/extref/406992ai1.mov.
Further information is available at the K-Team and Nature web sites, and at the Collective Robotics Group web site: http://diwww.epfl.ch/w3lami/collective/. Figure 5.10 presents some screen shots from their movie.

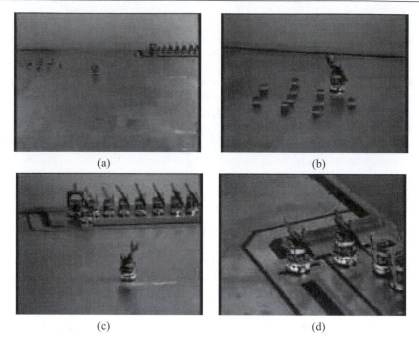

Figure 5.10: Screen shots from the movie by Krieger et al. (2000). (a) A robot leaves the nest at the top right corner and moves towards the food items. (b) The robot collects a food item and returns to the nest (c). The robot recruits another robot to collect food (tandem recruitment). (Reproduced with permission from [Krieger et al., 2000], © M. B. Krieger, J.-B. Billeter and L. Keller.)

The experiments performed showed that giving different but individually fixed activation levels to the robots results in an efficient dynamical task allocation, with great flexibility with regard to team (colony) size. Furthermore, relatively complex tasks can be performed by relatively simple and autonomous robots regulated in a decentralized way. It was shown that this strategy allows a swarm of robots to continue to work efficiently in case the number of robots is reduced by individual failures. Finally, it was shown that tandem recruitment of robots in an ant-like manner is an efficient strategy for exploiting clustered resources.

Note that there is a slight difference between tandem recruitment and the mass recruitment that gave rise to the ant colony optimization algorithm. In the former the recruitment is a result of direct contact between ants, while in the latter the interaction occurs indirectly via pheromone trails.

5.3.2. Clustering of Objects

Beckers et al. (1994) developed a system using multiple robots to gather together a dispersed set of objects (pucks) into a single cluster within an arena, much like the corpse-gathering behavior of ants discussed in Section 5.2.4. A robot was designed so that it could move small numbers of objects, and such that the probability of an object to be left (deposited) in a given position of the arena was directly proportional to the number of objects in that region. This was accomplished by effectively sensing a very local density of objects via a simple threshold mechanism. The plan was to evaluate the performance of the robots with this mechanism and to develop the strategy and the behaviors required until the task could be performed reliably.

In their simulations, the robots were equipped with a C-shaped gripper to gather objects instead of a clipper as used in the works of Krieger et al. (2000). The gripper was equipped with a microswitch activated by the presence of three or more pucks within the gripper. Figure 5.11 illustrates the type of robot with a gripper used in the experiments performed by Beckers et al. (1994).

In their experiments, the robots have only three behaviors, and only one is active at any time:

1. When no sensor is activated, a robot executes the default behavior of moving in a straight line until an obstacle is detected or until a microswitch is activated.

2. On detecting an obstacle, the robot executes an obstacle avoidance behavior by turning on the spot away from the obstacle with a random angle. The default behavior takes over again and the robot moves in a straight line in the new direction. If the robot is pushing some of the pucks (spread over the arena) when it encounters an obstacle, the pucks will be retained by the gripper during the turn.

3. When the gripper pushes three or more pucks, the microswitch is activated triggering the puck dropping behavior, which consists of backing up by reversing both motors releasing the pucks from the gripper. It then executes a turn with a random angle and returns to the default behavior.

As the robots behave autonomously, with all sensors, motors, and control circuits independent of one another, there is no explicit communication between robots. Therefore, the resulting behavior of the swarm of robots is an emergent property achieved through stigmergic self-organization based upon the reaction of the robots to the local configuration of the environment (arena).

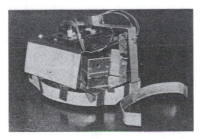

Figure 5.11: Robot equipped with a gripper to gather objects (pucks). (Reproduced with permission from [Beckers et al., 1994], © MIT Press.)

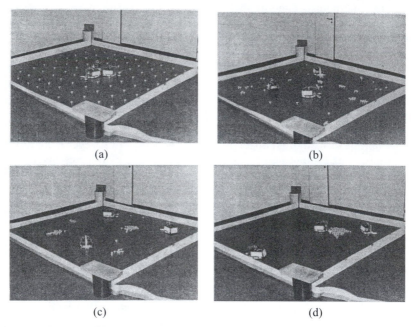

(a)   (b)

(c)   (d)

Figure 5.12: Type of behavior obtained in one of the experiments performed by Beckers et al. (1994). (a) Initial configuration of the environment. (b), (c) Evolution of the experiment after some time. (d) Final clustering of the pucks. (Reproduced with permission from [Beckers et al., 1994], © MIT Press.)

Figure 5.12 summarizes the type of behavior obtained in their experiments. Initially the pucks are distributed all over the arena (Figure 5.12(a)). Qualitatively, the robots initially move forward gathering pucks into the gripper one at a time, and dropping them in case three or more have been gathered. After a short period of time, a number of small clusters will have been formed (Figure 5.12(b)). By striking the clusters with a small angle, the robots remove some pucks from the clusters, resulting in the formation of larger clusters (Figure 5.12(c)). Finally, one single large cluster remains (Figure 5.12(d)).

The experimental results presented demonstrate that the simple behavioral rules allow the control and coordination of a number of robots without direct communication. This is a real-world evidence that stigmergic behaviors can be implemented in a swarm of robots so that they carry out a predefined task. The authors also suggested that this approach to multiple robotic systems is robust to individual robots' failures. This is mainly because each robot is seen as an individual agent whose rules of behavior are independent of the others' rules. The authors pointed out that such swarm robotic system has as advantage a gain in the speed of accomplishing some tasks because the addition of more robots does not require any reconfiguration of the system.

5.3.3. Collective Prey Retrieval

Several species of ants are capable of collectively retrieving preys so large that they could not possibly be retrieved by a single ant. If a single ant finds a prey and is capable of moving it alone, it takes the prey to the nest. Otherwise, the ant recruits nestmates via tandem recruitment (direct contact) or pheromone laying. When even a group of ants is not capable of moving a large prey item, specialized workers with large mandibles may be recruited in some species to cut the prey into smaller pieces. Such phenomena of large prey retrieval by ant colonies, though broadly studied, have no formal description yet (Bonabeau et al., 1999; Kube and Bonabeau, 2000).

There are several interesting aspects in the collective prey retrieval of ants. Some of these aspects have been explored for the development of behavioral rules for swarm robotic systems. One interesting aspect, agreed upon by most researchers (e.g., Moffett, 1988; Franks, 1989), is that collective transport is more efficient than solitary transport for large prey; i.e., the total weight that can be carried by a group of ants increases nonlinearly with the group size.

Another interesting aspect is that resistance to transport seems to be the reason why a food item is carried or dragged, and also if there will be more than one ant involved in the prey retrieval process. If a single ant tries to carry the item and fails, more ants may be recruited to help. The ant spends some time testing the resistance of the item to dragging before *realigning* the orientation of its body without releasing the item. In case realignment is not sufficient, the ant releases the item and finds another position to grasp it. If several of these *repositioning* attempts are not sufficient to result in a movement of the item, then the ant eventually recruits other nestmates (Sudd, 1960, 1965).

It is believed that the number of ants involved in prey retrieval is a function of the difficulty encountered in first moving the prey, not only of prey weight. A prey item that resists motion stimulates the recruitment of more ants. Success in carrying a prey item in one direction is followed by another attempt in the same direction. Finally, recruitment ceases as soon as a group of ants is capable of carrying the prey in a well-defined direction. In such situation the group size is adapted to the prey size.

Therefore, if a certain prey item cannot be moved, ants exhibit realigning and repositioning behaviors, and ultimately more ants are recruited to help. Only when realignment is not possible, repositioning is performed (Sudd, 1960, 1965). Coordination in collective transport thus seem to occur through the item being transported: a movement of one ant engaged in group transport is likely to modify the stimuli perceived by the other group members, possibly producing, in turn, orientational or positional changes in these ants (Bonabeau et al., 1999). This is another clear example of stigmergy.

Cooperative Box Pushing

Inspired by these (and other) observations of how ants cooperate in collective prey transport, Kube and Zhang (1992, 1994a,b), Kube and Bonabeau (2000), and Kube et al. (2004) have consistently studied the problem of collective prey retrieval in ants with a view of implementing a robotic system to perform a task known as *cooperative box-pushing*. This is equivalent to cooperative prey retrieval in ants. They have initially introduced a simulation model for the robots, and lately implemented a physical system with a group of homogeneous (identical) robots.

The initial task under study was undirected box-pushing, in which a group of robots found a box and pushed it in a direction that was dependent upon the initial configuration. The next task studied was a directed box-pushing, in which the robots were supposed to push the box toward a predefined goal position. Finally, the transport task, which is a variation of the directed box-pushing with multiple sequenced goals, was studied and the robots had to push the box from one location to the next (Kube and Bonabeau, 2000).

Five hierarchically organized behaviors were defined for the robotic experiments of (Kube and Zhang, 1992, 1994a,b). Behavior 1 has a higher priority than behavior 2, and so forth:

1. An <u>avoid</u> behavior to prevent collisions.
2. A <u>goal</u> behavior to direct the robot towards the goal.
3. A <u>slow</u> behavior to reduce the robots' speed.
4. A <u>follow</u> behavior to direct the robot towards its nearest neighbor.
5. A <u>find</u> behavior that allows for exploration.

These hierarchical behaviors were used to simulate the simple three behavioral rules for the robots:

1. A single robot (ant) tries to drag or carry the object (food item) with numerous realignment and repositioning strategies.
2. If unsuccessful, the robot (ant) will recruit other robots (ants) to assist in the task.
3. The group of robots (ants) will collectively and cooperatively try to move the object (food item) by realigning and repositioning themselves until a satisfactory configuration of the robots leads to a motion in the object.

The robotic implementation presented by Kube and Zhang (1992, 1994, 1997) demonstrated that a group of simple robots is capable of cooperatively performing a task without any direct communication if the robots have a common goal and try to minimize collisions. In the undirected box-pushing task, a lit object was placed on the arena and had to be pushed towards any of its ends (see Figure 5.13). The group of robots eventually moves the lit box in a direction dependent on the initial configuration of the system. Both execution time and average success rates increased with group size up to a point, when interference among robots starts to play a role.

One problem identified with their initial approach was *stagnation*, when no improvement in task accomplishing can be made. To tackle this problem, Kube and Zhang (1994b) implemented the mechanisms of realignment and repositioning, known as stagnation recovery behaviors, inspired by the ants. In this case,

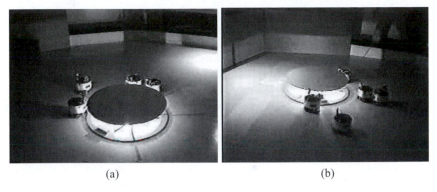

(a)   (b)

Figure 5.13: Transport experiments performed by Kube and Zhang (1997). (a) Robots grouping around the box and pushing it toward the right corner of the arena (b). Robots pushing the box to the left side of the arena. (Reproduced with permission from [Kube and Zhang, 1997], © C. R. Kube and H. Zhang.)

each robot exerts a force on the box side at a certain angle, thus producing a resultant force and torque applied to the center of the box. If the total resultant force is greater than a threshold, the box translates in the plane. Else, if the total torque exceeds a torque threshold, the box rotates. Stagnation here refers to any situation where a robot is in contact with the box and it is not moving. The experimental results demonstrated that the application of random pushing motions by either realigning the pushing angle or repositioning the pushing force is an effective technique against stagnation.

Some of their experiments were recorded and can be downloaded from Kube's web site at http://www.cs.ualberta.ca/~kube. In the videos, the realignment and repositioning of the robots can be clearly observed. The path used to take the lit object from its original position to the lit corner of the arena (corresponding to the nest or goal) is not optimal and may vary from one experiment to another. However, the robotic system described gives some clues about cooperative transport in ants and task accomplishment without direct communication. At a fundamental level, as their model reproduces some of the collective features of cooperative prey retrieval in ants, it suggests that these assumptions serve the purpose of explaining similar behaviors in ants. Actually, it is the first formalized model of cooperative transport in ants (Bonabeau et al., 1999; Kube and Bonabeau, 2000). Figure 5.13 presents two screen shots of their video recordings depicting the experimental environment and task achievement (box pushed towards the lit corner on the left hand side of the arena).

Recruitment of Nestmates

One of the problems involved in large prey retrieval is the recruitment of ants around the prey. In the context of collective prey retrieval, Hölldobler and Wilson (1978) and Höldobler (1983) have suggested two categories of recruitment via chemical cues: short-range and long-range recruitment. In the former catego-

ry - short-range recruitment - when a large prey item is found by an ant it secretes some chemical substance in its vicinity to attract other ants. This is
equivalent to the group recruitment described previously. In the later category -
long-range recruitment - the ant that finds a large prey item secretes some
pheromone making a trail connecting the food item to the nest. This is equivalent to the mass recruitment discussed in Section 5.2.2.

Inspired by the collective behaviors of ants, researchers at the MIT Artificial
Intelligence Lab were interested in developing a community of cubic-inch microrobots to simulate what they termed AntFarm. There were two main goals for
this project. The first was to try to integrate several sensors and actuators into a
small package for the microrobots. The second was to form a structured robotic
community from the interactions of many simple individuals whose communication was performed using infrared (IR) emitters.

One of the behaviors of ants they wanted to include in the AntFarm project
was that of recruitment of ants around a food item. The robots are initially scattered within a given arena and a food item is placed on it. Then, the task to be
performed in this part of their system was that of finding the item and recruiting
other ants for retrieval. Figure 5.14 presents two screen shots from the movie
available for their experiments (from http://www.ai.mit.edu/projects/ants/). In
Figure 5.14(a) the food item is initially placed in a given position of the arena.
The ants are recruited and cluster around the food item (Figure 5.14(b)).

Three basic rules were used for recruitment:

1.   Once a robot detects food, it emits an "I found food" IR signal. Any robot
     within about 12 inches of it can detect the signal and head towards it.

2.   When a robot receives the "I found food" signal, it heads towards the ro-
     bot with the food while transmitting "I see an Ant with food".

3.   Any robot within range of the second robot receives the "I see an Ant
     with food" signal, heads towards the second robot, and transmits "I see an

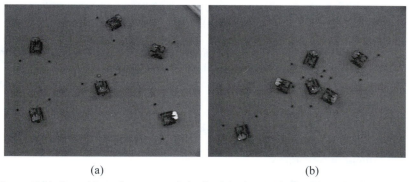

(a)   (b)

Figure 5.14: Recruiting of ants around the food item. (a) The food item is placed in the
center of the arena. (b) Ants are recruited to the site around the food item. (Reproduced
with permission from [http://www.ai.mit.edu/projects/ants/social-behavior.html], © MIT
Computer Science and Artificial Intelligence Laboratory.)

Ant that sees an Ant with food", and so on.

It can be observed by the rules above that this system embodies the idea of short-range recruitment proposed by Höldobler and Wilson (1978). The communication among ants, in this case, is mediated by an IR signal that can be likened to the chemical released in the air when an ant finds a large food item it cannot carry by itself.

5.3.4. Scope of Swarm Robotics

As the production cost of integrated circuits and robots decreases, it is more and more tempting to consider applications involving teams of simple robots rather than a single complex robot. Several authors in swarm robotics have suggested that the potential applications of swarm-based robotics require miniaturization (Bonabeau et al., 1999; Kube and Bonabeau, 2000). Very small robots, micro- and nano-robots, which will, by construction, have limited sensing and computing capabilities, will have to operate in large groups or swarms in order to perform useful tasks.

Défago (2001) identified four broad application areas in which swarm-robots, more specifically micro- and nano-robotics, is already being used or could potentially be used:

- *Exploration*: the unmanned exploration of inaccessible environments, such as the space, deep sea, plumbing, etc., is one application for cooperative and nano-robotics.

- *Manufacturing*: the manufacturing of human-scale objects by nanofactories, also called assemblers, is unlikely to be achieved by a single assembler. This will almost certainly require very large teams of assemblers to cooperate.

- *Medicine and surgery*: there are several medical applications using colonies of nano-robots, such as micro-surgery and the control of the precise location and the exact amount of drugs delivered in the body. Some researchers are already developing nano-robots made of a combination of organic and inorganic components (e.g., Bachand and Montemagno, 2000).

- *Smart materials*: these are materials made of a multitude of tiny modules linked to one another forming a large structure capable of autonomously changing its shape and physical properties.

5.3.5. Summary of Swarm Robotics

This section has briefly described some works from the literature that drew inspiration from the behavior of ants in order to design cooperative teams of robots capable of performing particular tasks. The ant-inspired tasks discussed include foraging for food, clustering of dead bodies and larval sorting, and collective prey retrieval.

From an architectural point of view, there are three main advantages of the swarm-robotics approach (Martinoli, 2001). First, the resulting collective systems are scalable, because the control architecture is exactly the same from a few to thousands of robots. Second, such systems are flexible, because individual robots can be dynamically added or removed without explicit reorganization by the operator. Third, these systems are robust, not only due to robot redundancy but also due to the minimalist robot design.

The approaches derived or inspired by social insects demonstrated to be feasible and interesting, though they are not the single alternative for the control and coordination of groups of small robots. The reader may be disappointed by the great simplicity of the tasks reported here as the state of the art in swarm robotics. Besides, collective robotics, including swarm-robotics, still lacks a rigorous theoretical foundation. In spite of these apparent drawbacks, it must be kept in mind that swarm robotics may still be one of the most useful strategies for collective robotics in the light of miniaturization; that is, micro- and nano-robots, though much more has to be done than has actually already been accomplished.

5.4. SOCIAL ADAPTATION OF KNOWLEDGE

The two previously described sections focused on how the social behavior of insect societies led to the development of computational algorithms for problem solving and strategies for the coordination of collective robotics. The *particle swarm* (PS) algorithm, to be discussed in this section, has as one of its motivations to create a simulation of human social behavior; that is, the ability of human societies to process knowledge (Kennedy and Eberhart, 1995; Kennedy, 1997; Kennedy, 2004). As is characteristic of all swarm intelligence approaches (actually, most biologically inspired algorithms), PS also takes into account a population of individuals capable of interacting with the environment and one another, in particular some of its neighbors. Thus, population level behaviors will emerge from individual interactions. Although the original approach has also been inspired by particle systems (Chapter 7) and the collective behavior of some animal societies, the main focus of the algorithm is on its social adaptation of knowledge; the same focus taken here.

A very simple sociocognitive theory underlies the particle swarm approach (Kennedy et al., 2001; Kennedy, 2004). The authors theorize that the process of cultural adaptation comprises a low-level component corresponding to the actual behavior of individuals, and a high-level component in the formation of patterns across individuals. Therefore, each individual within a population has its own experience and they know how good it is and, as social beings, they also have some knowledge of how other neighboring individuals have performed.

These two types of information correspond to *individual learning* and *cultural or social transmission*, respectively. The probability that a given individual takes a certain decision is a function of how successful this decision was to him/her in the past. The decision is also affected by social influences, though the exact rules in human societies are not very clear (Kennedy, 2004; Kennedy et al., 2001). In the particle swarm proposal, individuals tend to be influenced by the best successes of anyone they are connected to, i.e., the members of their social (sociometric) neighborhood that have had the most success so far.

Kennedy (1998) and Kennedy et al. (2001) use three principles to summarize the process of cultural adaptation. These principles may be combined so as to enable individuals to adapt to complex environmental challenges:

- *Evaluate*: the capability of individuals to sense the environment allows them to quantify their degree of *goodness* (quality or suitability) in relation to some parameter(s) or task(s), such as height when one wants to become a basketball player.

- *Compare*: people usually use others as standards for assessing themselves, what may serve as a kind of motivation to learn and change. For instance, we look at other people and evaluate our wealth, look, humor, beauty, exam mark, whom to vote, and so forth.

- *Imitate*: human imitation comprises taking the perspective of someone else, not only by imitating a behavior but by realizing its purpose and executing the behavior of others when it is appropriate. Imitation is central to human sociality and the acquisition and maintenance of mental abilities.

5.4.1. Particle Swarm

In the particle swarm (PS) algorithm, individuals searching for solutions to a given problem learn from their own past experience and from the experiences of others. Individuals evaluate themselves, compare to their neighbors and imitate only those neighbors who are superior to themselves. Therefore, individuals are able to evaluate, compare and imitate a number of possible situations the environment offers them.

Although there are two basic versions of the PS algorithm, binary and real-valued, this section focuses only on the latter one, namely, the real-valued or continuous PS. This is mainly because this version has a much broader applicability and it has been more extensively studied as well. Under this perspective, the PS algorithm can be viewed as a numeric optimization procedure inspired by the social adaptation of knowledge discussed above. It searches for optima in an L-dimensional real-valued space, $\Re^L$. Therefore, the discussion regarding problem solving as a search in a search space presented in Section 3.2 is also appropriate for the purposes of this section.

In real-valued spaces, the variables of a function to be optimized can be conceptualized as a vector that corresponds to a point in a multidimensional search space. Multiple individuals can thus be plotted within a single set of coordinates, where a number of individuals will correspond to a set of points or particles in the space. Individuals similar to one another in relevant features appear closer to one another in the space, as illustrated in Figure 5.15.

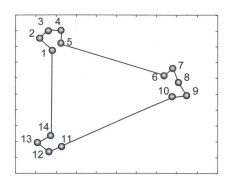

Figure 5.15: Particles in a bi-dimensional real-valued search-space. Particles are connected to their topological neighbors, and neighbors tend to cluster in the same regions of the space.

The individuals are viewed as points in a search-space and their change over time is represented as movements. In this case, individuals truly correspond to *particles*. Forgetting and learning are seen as a decrease or increase on some dimensions, attitude changes are seen as movements between the negative and positive ends of an axis, and emotion and mood changes of numerous individuals can be plotted conceptually in a coordinate system. The swarm of particles corresponds to the multiple individuals in the population.

One insight from social psychology is that these particles will tend to move toward one another and to influence one another as individuals seek agreement with their neighbors (Kennedy et al., 2001; Kennedy, 2004). Another insight is that the space in which the particles move is heterogeneous with respect to evaluation; that is, some regions are better than others, meaning that they have a higher goodness value.

In mathematical terms, the PS algorithm can be implemented as follows. The position of a particle i is given by $\mathbf{x}_i$, which is an L-dimensional vector in $\Re^L$. The change of position of a particle is denoted by $\Delta\mathbf{x}_i$, which is a vector that is added to the position coordinates in order to move the particle from one iteration t to the other $t + 1$:

$$\mathbf{x}_i(t + 1) = \mathbf{x}_i(t) + \Delta\mathbf{x}_i(t + 1) \tag{5.21}$$

The vector $\Delta\mathbf{x}_i$ is commonly referred to as the velocity $\mathbf{v}_i$ of the particle. The particle swarm algorithm samples the search-space by modifying the velocity of each particle. The question that remains is how to define the velocity of the particle so that it moves in the appropriate directions and with the appropriate "step size" in the search-space. In addition, the neighborhood of each particle (individual) has to be defined.

The inspiration taken from the social-psychological sciences suggests that individuals (particles) should be influenced by their own previous experience and the experience of its neighbors. Neighborhood here does not necessarily mean similarity in the parameter space; instead, it refers to topological similarity in a given structure of the population. In a social network, a given individual particle is capable of influencing the ones to which it is socially connected. Neighborhood is thus defined for each individual particle based on its position in a topological array, usually implemented as a ring structure (the last member is a neighbor of the first one).

There are a number of different schemes to connect the individuals of the population. Most particle swarm implementations use one of two simple sociometric principles. The first, called *gbest* (*g* for global), conceptually connects all members of the population to one another (Figure 5.16(a)). The effect of this is that each particle is influenced by the very best performance of any member of the entire population. The second, termed *lbest* (*l* for local), creates a neighborhood for each individual comprising itself and its *k*-nearest neighbors in the population (Figure 5.16(a)). To illustrate, consider the particles in Figure 5.15: each individual particle is connected with its 2-nearest neighbors in an *lbest* scheme ($k = 2$). Thus, particle $\mathbf{x}_3$ has neighbors $\mathbf{x}_2$ and $\mathbf{x}_4$, particle $\mathbf{x}_8$ has neighbors $\mathbf{x}_7$ and $\mathbf{x}_9$, and so on.

A particle will move in a certain direction as a function of its current position $\mathbf{x}_i(t)$, its change in position $\Delta\mathbf{x}_i(t)$, the location of the particle's best success so far $\mathbf{p}_i$, and the best position found by any member of its neighborhood $\mathbf{p}_g$:

$$\mathbf{x}_i(t+1) = f(\mathbf{x}_i(t), \Delta\mathbf{x}_i(t), \mathbf{p}_i, \mathbf{p}_g) \tag{5.22}$$

The influence of the terms $\Delta\mathbf{x}_i(t)$, $\mathbf{p}_i$, and $\mathbf{p}_g$ can be summarized by a change $\Delta\mathbf{x}_i(t+1)$ to be applied at iteration $t+1$:

$$\Delta\mathbf{x}_i(t+1) = \Delta\mathbf{x}_i(t) + \varphi_1 \otimes (\mathbf{p}_i - \mathbf{x}_i(t)) + \varphi_2 \otimes (\mathbf{p}_g - \mathbf{x}_i(t)) \tag{5.23}$$

where φ_1 and φ_2 represent positive random vectors composed of numbers drawn from uniform distributions with a predefined upper limit: $\varphi_1 = U(0, AC_1)$ and $\varphi_2 = U(0, AC_2)$; $U(0, AC)$ is a vector composed of uniformly distributed random numbers, and AC is called the *acceleration constant*.

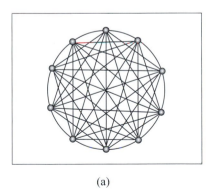
(a)

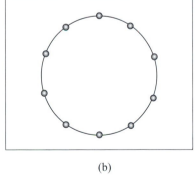
(b)

Figure 5.16: Illustration of *gbest* and *lbest* neighborhoods. (a) *gbest*. (b) *lbest* for $k = 2$.

According to Kennedy (2004), the sum of the two acceleration constants should equal 4.1, $AC_1 + AC_2 = 4.1$ (usually both are 2.05). The second term on the right hand side of Equation (5.23) is proportional to the difference between the particle's previous best and its current position, and the last term on the right hand side of Equation (5.23) is proportional to the difference between the neighborhood's best and the current position of the particle. The symbol $\otimes$ represents the elementwise vector multiplication.

In (Kennedy, 1997), the author clearly identifies the meaning of each term in Equation (5.23) and performs a number of experiments, with a single application problem, in order to evaluate the importance of "social" and "cognitive" interactions in the PS algorithm. He defines Equation (5.23) without the last term as the "cognition-only" model, and without the second term, but with all the others, as the "social-only" model.

In order to limit the change in position of a particle so that the system does not "explode", two values v_{min} and v_{max} are defined for the change Δx, thus guaranteeing that the particles oscillate within some predefined boundaries:

If $\Delta x_{id} > v_{max}$, then $\Delta x_{id} = v_{max}$,

Else if $\Delta x_{id} < v_{min}$ then $\Delta x_{id} = v_{min}$

```
procedure [X] = PS(max_it,AC₁,AC₂,vₘₐₓ,vₘᵢₙ)
    initialize X   //usually xᵢ, ∀i, is initialized at random
    initialize Δxᵢ //at random, Δxᵢ ∈ [vₘᵢₙ,vₘₐₓ]
    t ← 1
    while t < max_it do,
        for i = 1 to N do,             //for each particle
            if g(xᵢ) > g(pᵢ),
                then pᵢ = xᵢ,           //best indiv. performance
            end if
            g = i                      //arbitrary
            //for all neighbors
            for j = indexes of neighbors
                if g(pⱼ) > g(pg),
                    then g = j,    //index of best neighbor
                end if
            end for
            Δxᵢ ← Δxᵢ + φ₁⊗(pᵢ - xᵢ) + φ₂⊗(pg - xᵢ)
            Δxᵢ ∈ [vₘᵢₙ,vₘₐₓ]
            xᵢ ← xᵢ + Δxᵢ
        end for
        t ← t + 1
    end while
end procedure
```

Algorithm 5.4: Standard particle swarm optimization (PS) algorithm.

Let $\mathbf{X} = \{\mathbf{x}_1, \ldots, \mathbf{x}_N\}$ be the swarm of particles, $g(\mathbf{x}_i)$ the goodness of particle $\mathbf{x}_i$, $\Delta\mathbf{x}_i$ its change in position, $v_{\min}$ and $v_{\max}$ the lower and upper limit for the change in position (velocity), respectively, and L the dimension of the particles. Assume a maximal number of iterations max_it to be run and a swarm of a given size N (according to the authors, the swarm size is usually $N \in [10,50]$). The original PS algorithm is described in Algorithm 5.4. In the section on applications (Section 5.4.2), two simple modifications of the algorithm that considerably improve performance will be introduced.

Note that solving a problem using the PS algorithm involves the same three aspects suggested in the previous chapter: the choice of a representation (in this case it is fixed: real-valued vectors), the choice of an objective function, and the choice of a goodness function according to the desired objective.

5.4.2. Selected Applications from the Literature: A Brief Description

Algorithm 5.4 presents the original particle swarm optimization algorithm developed taking inspiration from the social adaptation of knowledge and bird flocking behavior. The description presented here emphasized the social behavior of human beings as the primary motivation for the algorithm because this is the emphasis given in Kennedy et al. (2001) and Kennedy (2004), two of the most important fundamental works of the field.

To illustrate in what types of problem the algorithm has been applied to, two applications will be presented. The first example - optimization of neural network weights - illustrates the importance of the PS algorithm as a tool to be hybridized with other strategies. The second example - numerical function optimization - presents two slight variations of the original algorithm and discusses its potentiality to solve multi-dimensional function optimization problems in general.

Optimization of Neural Network Weights

One of the first applications of the particle swarm algorithm was to the problem of defining appropriate weights for artificial neural networks (Kennedy and Eberhart, 1995). It has been discussed in the previous chapter that a neural network is comprised of a structured set of nodes partially or fully connected. The neural network performs a mapping from a set of input data into one or more output node(s). Each network node is a processing unit that receives stimuli from the environment or (an)other node(s), processes these inputs and produces an output signal. Assuming that the nodes in the network are homogeneous; that is, of same (predefined) type, and the network structure is of a fixed given size, the resultant network learning task is reduced to the problem of determining an appropriate weight set that will lead to a satisfactory network behavior when given input stimuli are presented. Figure 5.17 illustrates a simple network with two input nodes, one output node, and two intermediary nodes between the input and output nodes.

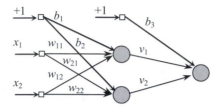

Figure 5.17: An example of a simple artificial neural network with two input nodes, two intermediate or hidden nodes, and one output node.

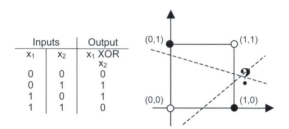

Figure 5.18: The XOR function and its graphical representation. The black dot corresponds to 1, while the circles are equivalent to 0.

Each connection between the nodes has an assigned weight w_{ij}, v_i, and b_i, where w and v correspond to weights of different layers in the network, and b_i corresponds to the bias of a node. Usually, the weight vectors of this type of network are determined according to a learning rule or algorithm, responsible for updating the network weights along an iterative procedure of adaptation. In the present example, a PS algorithm will be used to determine the network weights instead of a standard learning algorithm.

Consider the problem of using the neural network of Figure 5.17 to determine a mapping capable of simulating the exclusive-or (XOR) logic function depicted in Figure 5.18. The difficulty in solving this problem, seen as a classification problem, resides in the fact that it is not possible to separate the input patterns that lead to an output 1 from the input patterns that result in an output 0 by simply drawing a line or a plane on the graph. That means that the two classes available (0 or 1) are not linearly separable. This is a classic benchmark problem for the network structure presented in Figure 5.17, though nowadays there are various algorithms that solve this problem very easily.

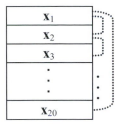

Figure 5.19: The structured neighborhood used in the PS implementation to define neural network weights. Particle $\mathbf{x}_1$ is connected to $\mathbf{x}_2$ and $\mathbf{x}_{20}$, particle $\mathbf{x}_2$ is connected to $\mathbf{x}_1$ and $\mathbf{x}_3$, and so forth. This is a circular neighborhood such as the one depicted in Figure 5.16(b).

In the experimental results reported in (Kennedy and Eberhart, 1995; Kennedy, 1997; Kennedy et al., 2001), the authors used a swarm composed of 20 particles ($N = 20$) initialized at random. Each particle has a dimension $L = 9$ corresponding to the nine weights of the network ($w_{11}, w_{12}, w_{21}, w_{22}, b_1, b_2, b_3, v_1, v_2$) to be determined. The swarm of particles can be defined by using a matrix $\mathbf{X}$ with 20 rows and 9 columns, $\mathbf{X} \in \Re^{20 \times 9}$, where each row corresponds to a particle of dimension $L = 9$.

The neighborhood is of the 2-nearest neighbor type, as illustrated in Figure 5.19. The goodness of each particle is defined as the error (e.g., the mean squared error, MSE) between the current network output and the desired outputs given in Figure 5.18. The goal is thus to determine an appropriate weight set that minimizes the mean squared error between the network output for all input patterns and the desired output.

As each particle corresponds to one of the parameters to be adjusted in the neural network, the movements of the particles in the search space allow for a selection of particles that provide the desired network performance. In their simulations the desired criterion was $MSE < 0.02$. Experimental results were very satisfactory, and presented for different values of v_{max}.

Numerical Function Optimization

Similarly to the application of evolutionary algorithms, Angeline (1998), and Shi and Eberhart (1999) applied the PS algorithm to perform numeric function optimization. The authors used some standard test functions (e.g., Sphere, Rosenbrock, Rastrigin, Griewank) taken from the literature of evolutionary algorithms to assess the performance of the PS algorithm:

Sphere
$$f_0(x) = \sum_{i=1}^{L} x_i^2$$

Rosenbrock $$f_1(x) = \sum_{i=1}^{L} \left(100(x_{i+1} - x_i^2)^2 + (x_i - 1)^2\right)$$

Rastrigin $$f_2(x) = \sum_{i=1}^{L} \left(x_i^2 - 10.\cos(2\pi x_i) + 10\right)$$

Griewank $$f_3(x) = \frac{1}{4000} \sum_{i=1}^{L} x_i^2 - \prod_{i=1}^{L} \cos\left(\frac{x_i}{\sqrt{i}}\right) + 1$$

In (Shi and Eberhart, 1999), all functions were implemented using $L = 10, 20$ and 30 dimensions, and the PS algorithm was run for 1000, 1500, and 2000 iterations, respectively. Also, the authors studied different population sizes for the PS, $N = 20, 40, 80$ and 160. In this application, each particle corresponds to one candidate solution to the problem, in a form similar to that used in genetic algorithms, but with real-valued particles instead of a binary representation. They suggested that by using a decreasing factor termed *inertia weight* w the PS algorithm was able to better balance exploration with exploitation:

$$\Delta x_i(t + 1) = w. \Delta x_i(t) + \varphi_1 \otimes (p_i - x_i(t)) + \varphi_2 \otimes (p_g - x_i(t)) \qquad (5.24)$$

where φ_1 and φ_2 were defined as previously and w was assigned an initial value but was allowed to be reduced (or cooled, such as the temperature in simulated annealing) during the iterative procedure of adaptation. The decreasing inertia weight started at 0.9 and was linearly decreased until $w = 0.4$. The limiting values for Δx were set equal to v_{max} according to a pre-defined table.

Another variant of the original algorithm that is currently known as the standard particle swarm algorithm (Kennedy, 2004) involves the use of a global constriction coefficient χ (Clerc and Kennedy, 2002) as follows:

$$\Delta x_i(t + 1) = \chi \left(\Delta x_i(t) + \varphi_1 \otimes (p_i - x_i(t)) + \varphi_2 \otimes (p_g - x_i(t))\right) \qquad (5.25)$$

where the parameter $\chi \approx 0.729$.

There are some java applets that show the movement of particles when applied to some of these functions. Most of them use the decreasing inertia weight. Some of the codes and demos available about the PSO algorithm can be found at the PSO website: http://www.swarmintelligence.org.

5.4.3. Scope of Particle Swarm Optimization

The original applications of PS focused primarily on neural network parameter adjustment and model selection. This corresponds to the problem of designing neural networks, including the network weights adjustment and structure definition processes. Model selection has been one of the most important application areas of evolutionary algorithms as well. Actually, some similarities can be observed between these approaches, but their distinctions are also undeniable; for instance, there can be no evolution if there is no selection, and PS does not account for selection, individuals move but never die. Kennedy et al. (2001) and Kennedy (2004) provide a good discussion about the similarities and differences between PS and other approaches.

PS algorithms have been applied to a number of numeric and parameter optimization problems, including real-world tasks. For instance, works can be found in the literature applying the PS approach to tasks like human tremor analysis, milling optimization, ingredient mix optimization, reactive power and voltage control, battery pack state-of-charge estimation, and improvised music composition.

5.4.4. From Social Systems to Particle Swarm

Section 5.4 discussed the importance of the social influence for the acquisition and improvement of knowledge. Some of these concepts led to the proposal of the particle swarm algorithm. Table 5.3 summarizes how to interpret the social-psychological views presented as a particle swarm approach.

Table 5.3: Interpretation of the sociocognitive domain into the particle swarm approach.

Social-Psychology	PS Algorithms
Individual (minds)	Particles in space
Population of individuals	Swarm of particles
Forgetting and learning	Increase or decrease in some attribute values of the particle
Individual own experience	Each particle has some knowledge of how it performed in the past and uses it to determine where it is going to move to
Social interactions	Each particle also has some knowledge of how other particles around itself performed and uses it to determine where it is going to move to

5.4.5. Summary of Particle Swarm Optimization

Although not explicitly discussed in the chapter that introduces the PS algorithm in Kennedy's book (Kennedy et al., 2001; Chapter 7), the algorithm was also motivated by the social behavior of organisms such as bird flocking and fish schooling (Kennedy and Eberhart, 1995). The algorithm is not only a tool for optimization, but also a tool for representing sociocognition of human and artificial agents, based on principles of social psychology. Some scientists suggest that knowledge is optimized by social interaction and thinking is not only private but also a result of interactions with other people. There are reports in the literature of feral humans, human children who grew up in the wild are commonly uneducable, unable to learn or to speak, unable to adapt to social life, and never able to show much evidence of cognition or thinking (Kennedy et al., 2001; Kennedy, 2004).

The PS algorithm as an optimization tool provides a population-based search procedure in which individuals, called particles, change their position with time. In real-valued PS, the particles fly around in a multidimensional search-space. During flight, each particle adjusts its position according to its own experience and according to the experience of a number of neighboring particles, making use of its best position so far and the best position of its neighbors as well. Therefore, the PS algorithm combines local search with global search, attempting to balance exploration with exploitation.

A particle swarm is thus another self-organizing system whose global dynamics emerge from local interactions. As each individual trajectory is adjusted toward its success and the success of neighbors, the swarm converges or clusters in optimal regions of the search-space.

5.5. SUMMARY

This chapter has discussed four main approaches in swarm intelligence: ant colony optimization, ant clustering, swarm robotics, and particle swarm. Ant colony optimization algorithms are suitable for finding minimal cost paths on a graph, mainly in those cases in which other more classical algorithms cannot be efficiently applied. Although each ant of the colony builds a solution to the problem at hand, good quality solutions can only emerge from the cooperative behavior of the colony. The communication among ants is performed indirectly via pheromone trails. Note also that the ants themselves are not adaptive individuals. Instead, they adapt the environment by releasing pheromone, thus, changing the way the problem is represented and perceived by other ants. This is one example of a stigmergic algorithm.

The clustering of dead bodies and larval sorting in ants has led to the development of the ant clustering algorithm. ACA is suitable for exploratory data analysis, in particular for the clustering of unlabelled data sets. It works by projecting the data into a bi-dimensional grid in which ants are allowed to move about carrying items to and from specific cells. Regions containing a large number of similar data reinforce the release of similar data, and vice-versa.

Most collective robotic systems were inspired by the collective behavior of ant colonies. Foraging, clustering of dead bodies, clustering around food, and collective prey retrieval have motivated the development of robotic systems to perform equivalent tasks. These systems are believed to be much promising when miniaturization is pursued. In such cases, they can aid in the unmanned exploration of inaccessible environments, in the large scale manufacturing of products, and in the medical sciences and surgery, among other things. The swarm robotics systems described here are other examples of stigmergic systems.

The particle swarm approach is inspired by the human social adaptation of knowledge and has demonstrated good performances in a number of applications, such as numeric function optimization and optimal design of other natural computing techniques (e.g., neural networks). The algorithm is conceptually simple and easily implementable. Although this chapter has only presented its most well-known version, real-valued PS, binary and discrete versions of the PS algorithm are also available.

5.6. EXERCISES

5.6.1. Questions

1. Describe how the ant colony of the movie "Antz" is organized. Contrast the perspective of ant colonies presented in the movie with that of real ant colonies.

2. How many queens an ant or a termite colony may have? Discuss.

3. It is known that scientists do not have a detailed knowledge of the inner workings of insect swarms. The identification of the rules that govern individual behaviors is a major challenge, and without such knowledge it is hard to develop appropriate models and thus to predict their behavior.

    Can this (apparent) unpredictability of behavior be a drawback of the swarm intelligence approaches inspired by social insects?

    Would it be possible that a swarm of robots goes out of control due to this lack of knowledge?

4. The Ant Farm project of the MIT Artificial Intelligence Lab had as one of its final goals to design a structured community of micro-robots inspired by the social behavior of ants. By running the video recording of some of their experiments (http://www.ai.mit.edu/projects/ants/social-behavior.html), it is possible to observe a recruitment behavior of robots. Based upon the discussion presented in Section 5.3.3 and the information available at their website, identify which type of communication is being performed by the robots: direct or indirect communication. Comment on its efficiency.

    Can this recruitment strategy be used in conjunction with the work presented in Kube's videos (Section 5.3.3)? Discuss.

5. Section 5.3 was dedicated to swarm robotics. Suggest two real-world applications (different from those presented here) for each of these systems and discuss how these could be accomplished by swarm robotics.

6. In Section 2.2.3, it was described the basic behavior of termites' nest building. It was seen that, during nest building, termites are stimulated to work according to the current configuration of the local environment and other termites. In this case, the accumulation of pellets reinforces the dropping of more pellets in a given portion of the space. The intermediate results of their building process are pillars, while the final result is a mound. Is the model presented in Section 5.2.4 of the present chapter for the clustering of dead bodies and larval sorting appropriate to model termite nest building as well? Discuss.

7. The particle swarm approach was developed with an inspiration in the collective behavior of animals, such as flocks of birds, and can also be viewed as a sociocognitive theory of knowledge. In this case sociocognition refers to those qualities of thought about social beings and objects with their own features and in relation to others. In the particular case of human beings, it suggests that in order for us to become intelligent we need to interact with other humans. Using the authors own words:

    "In order for a brain to become mental, it needs to have interaction with other minds. A fetal brain grown to adulthood in a vat containing nutritious fluids would never acquire the qualities of mind. If it were transplanted into a Frankesteinian body, it would be unable to reason, remember, categorize, communicate, or do any of the other things that a minimal mind should be able to. Feral humans, that is, humans who have grown up in the wild, have never been known to show signs of having anything like a mind - it is not sufficient to have perceptual stimuli; social ones are necessary. The relation between mind and society is unquestionable ... ." (Kennedy et al., 2001; p. 118)

    According to the authors, for a healthy human brain to develop into a mind, it has to become conscious and learn to think. These are some minimum requirements for something to be called a mind. To what extent do you believe that the interactions with other beings are important for the development of our intelligence and mind? There are known examples of human beings who were born in the wild and raised by wolves or monkeys and lately brought into towns. What happened to these human beings? Were they ever capable of adapting to social life and develop several basic skills such as eating with knifes? Could they ever fall in love?

8. Similarly to the Kohonen's self-organizing map (SOM) presented in the previous chapter, the ant clustering algorithm (ACA) is applicable to the unsupervised classification of data. Both algorithms make a projection of the input data set into an output grid. In the case of the SOM, the grid can be of any dimension, while for the ACA it is assumed a bi-dimensional grid. Given their main features, compare both strategies?

9. The particle swarm algorithm presents two features in common with the Kohonen self-organizing map studied in the previous chapter: (i) the use of a topological neighborhood among the components of the system; and (ii) the fact that particles are moved in the space taking into account their neighborhood.

    Discuss the similarities and differences between the neighborhood in SOM and in PS.

    Provide a geometric interpretation, in a 2D space, of the movement of a particle in the space taking into account Equations (5.21) and (5.23).

5.6.2. Computational Exercises

1. Write a pseudocode for the simple ACO (S-ACO) algorithm considering pheromone evaporation, implement it computationally, and apply it to solve the TSP instance presented in Section 3.10.4. Discuss the results obtained.

    Remove the pheromone evaporation term (Equation (5.3)), apply the algorithm to the same problem, and discuss the results obtained.

2. Apply the AS-TSP algorithm described in Algorithm 5.2 to solve the simple TSP instance described in Section 3.10.4.

    Compare the performance of the AS-TSP with that of the genetic algorithm applied to this same problem.

    Study and discuss the influence of parameters α and β in the performance of the algorithm.

3. In the first experiment performed by Lumer and Faieta (1994), the authors applied the ant clustering algorithm to a simple data set generated by four Gaussian distributions of mean μ and standard deviation σ, $G(\mu,\sigma)$. The authors used the distributions below to generate four different clusters of 200 data points each ($M = 800$, where M is the total number of data points), and the following simulation parameters $k_1 = 0.1$, $k_2 = 0.15$, $\alpha = 0.5$, $s^2 = 9$, and max_it $= 10^6$ to run their algorithm:

    $x \propto G(0.2,0.1)$, $y \propto G(0.2,0.1)$;

    $x \propto G(0.2,0.1)$, $y \propto G(0.8,0.1)$.

    $x \propto G(0.8,0.1)$, $y \propto G(0.2,0.1)$;

    $x \propto G(0.8,0.1)$, $y \propto G(0.8,0.1)$;

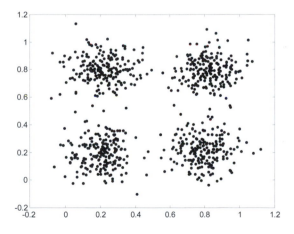

Figure 5.20: Distribution of points in $\Re^2$.

Each set of 200 points generated by each of the distributions above correspond to a cluster in $\Re^2$, as illustrated in Figure 5.20.

Implement the ant clustering algorithm (ACA) presented in Section 5.2.5 and test its performance when applied to the problem described above. Discuss the results obtained.

Was the algorithm capable of clustering similar data items into the same clusters in the bi-dimensional grid?

How many clusters appeared on the grid?

4. In Exercise 3 above, instead of using Equation (5.15) to determine the probability of depositing an item, use Equation (5.12).

   What happens with the performance of the algorithm?

5. Repeat the previous exercise introducing the three modifications proposed by Lumer and Faieta (1994) for the standard ACA: ants with different speeds, short term memory, and behavioral switches. Compare the performance of the modified algorithm with its standard version.

6. Repeat Exercise 3 introducing the three modifications proposed by Vizine et al. (2005): a cooling schedule for k_1; a progressive vision scheme; and the addition of pheromone to the environment. Compare the performance of this version of the algorithm with that of Exercises 3 and 4.

7. The Animals data set was originally proposed by Ritter and Kohonen (1989) to verify the self-organizing capability of one artificial neural network in creating a topographic map based on a symbolic data set (Section 4.4.5). The data set is composed of 16 input vectors each representing an animal with the binary feature attributes as shown in Table 5.4. A value of 1 in this table corresponds to the presence of an attribute, while a value of 0 corresponds to the lack of this attribute.

   Apply the ant clustering algorithm (ACA) to the animals data set presented above and study its performance. The local density f of items in the neighborhood of an object $\mathbf{x}_i$ situated at site r can be computed by a variation of Equation (15) with the distance between two items $\mathbf{x}_i$ and $\mathbf{x}_j$, $d(\mathbf{x}_i,\mathbf{x}_j)$, given by their Hamming distance in their original attribute space:

$$f(\mathbf{x}_i) = \begin{cases} \dfrac{1}{s^2} \displaystyle\sum_{x_j \in \text{Neigh}_{(s \times s)}(r)} \left[L - \dfrac{d(\mathbf{x}_i, \mathbf{x}_j)}{\alpha} \right] & \text{if } f > 0 \\ 0 & \text{otherwise} \end{cases}$$

   where L is the length (number of attributes) of $\mathbf{x}_i$.

8. Apply the PS algorithm described in Section 5.4.1 to the maximization problem of Example 3.3.3. Compare the relative performance of the PS algorithm with that obtained using a standard genetic algorithm.

9. Modify the standard PS algorithm in order to accommodate the inertia weight given in Equation (5.24) and apply it to the same problem as above. Compare the results obtained.

Table 5.4: Animals data set: animals' names and their attributes.

		Dove	Hen	Duck	Goose	Owl	Hawk	Eagle	Fox	Dog	Wolf	Cat	Tiger	Lion	Horse	Zebra	Cow
Is	Small	1	1	1	1	1	1	0	0	0	0	1	0	0	0	0	0
	Medium	0	0	0	0	0	0	1	1	1	1	0	0	0	0	0	0
	Big	0	0	0	0	0	0	0	0	0	0	0	1	1	1	1	1
Has	Two legs	1	1	1	1	1	1	1	0	0	0	0	0	0	0	0	0
	Four legs	0	0	0	0	0	0	0	1	1	1	1	1	1	1	1	1
	Hair	0	0	0	0	0	0	0	1	1	1	1	1	1	1	1	1
	Hooves	0	0	0	0	0	0	0	0	0	0	0	0	0	1	1	1
	Mane	0	0	0	0	0	0	0	0	0	1	0	0	1	1	1	0
	Feathers	1	1	1	1	1	1	1	0	0	0	0	0	0	0	0	0
Likes to	Hunt	0	0	0	0	1	1	1	1	0	1	1	1	1	0	0	0
	Run	0	0	0	0	0	0	0	0	1	1	0	1	1	1	1	0
	Fly	1	0	0	1	1	1	1	0	0	0	0	0	0	0	0	0
	Swim	0	0	1	1	0	0	0	0	0	0	0	0	0	0	0	0

10. Repeat the previous exercise for the constriction factor χ.

5.6.3. Thought Exercises

1. The ant clustering algorithm (ACA) is based upon the projection of an input data set into a bi-dimensional grid. Is it possible to apply this algorithm without projecting the input data into a lower dimensional grid? What are the advantages and disadvantages of this dimensionality reduction process?

2. Equation (5.6) corresponds to the amount of pheromone each ant k is going to release on the edge (i,j) when this is traversed. As it stands, the amount of pheromone is absolute, in the sense that it is only proportional to the quality of the tour performed by this particular ant. Modify this equation so that the relative performance of ant k is taken into account. Discuss.

   Modify your implementation of ACO and investigate the effects of your proposed relative pheromone deposition rate.

5.6.4. Projects and Challenges

1. In Bullnheimer et al. (1999a) the authors applied the ant colony optimization algorithm in several instances of the vehicle routing problem (VRP) extracted from Christofides et al. (1979). All these instances are benchmark for combinatorial optimization algorithms and can be found in the literature.

   Implement the ACO algorithm and apply it to two instances of the VRP and compare its performance with other results from the literature.

2. The three main problem solving techniques discussed in this chapter are the ant colony optimization algorithms (ACO), the ant clustering algorithms (ACA), and the particle swarm (PS) algorithm. All these approaches involve a number of user-defined parameters, such as α, β, ρ, N, Q, and b for the ACO algorithm, k_p and k_d for the ACA, and w or χ, the number of particles, etc., for the PS algorithm.

    Choose one benchmark problem from the literature for each of these algorithms and study their sensitivity to these user-defined parameters. For example, by varying α, what happens to the performance of the ACO algorithm, and so on.

3. The ACO algorithm was originally developed to solve combinatorial optimization problems. However, it is known that some colonies of real ants are capable of finding the minimal path between the nest and the food source in the real world, i.e., in a continuous environment. Can you develop a modified (or new) ACO algorithm to solve continuous optimization problems? What would be the requirements for the artificial ants, visibility (if it is to be included), and artificial pheromone trails? Provide a number of problems in which this new ACO algorithm could be applied.

4. In Section 5.4.2 we described the use of a particle swarm algorithm for determining an appropriate weight set for a multi-layer perceptron neural network. Repeat Project 4 of Section 4.8.4 using a PS algorithm to train the neural network instead of using an evolutionary algorithm.

5.7. REFERENCES

[1] Angeline, P. A. (1998), "Evolutionary Optimization Versus Particle Swarm Optimization: Philosophy and Performance Differences", In. V. W. Porto, N. Saravanan, D. Waagen and A. E. Eiben (eds.), *Lecture Notes in Computer Science*, **1447**, Springer-Verlag, pp. 601–610.

[2] Bachand, G. D. and Montemagno, C. D. (2000), "Constructing Organic/Inorganic Nems Devices Powered by Biomolecular Motors", *Biomedical Microdevices*, **2**, pp. 179–184.

[3] Beckers, R., Holland, O. E. and Deneubourg, J. L. (1994), "From Local Actions to Global Tasks: Stigmergy and Collective Robotics", In R. A. Brooks and P. Maes (eds.) *Proc. of the 4th Int. Workshop on the Synthesis and Simulation of Life, Artificial Life IV*, MIT Press, pp. 181–189.

[4] Beni, G. (1988), "The Concept of Cellular Robotic Systems", *Proc. of the IEEE Int. Symp. on Intelligent Control*, pp. 57–62.

[5] Beni, G. and Wang, J. (1989), "Swarm Intelligence", *Proc. of the 7th Annual Meeting of the Robotics Society of Japan*, pp. 425–428.

[6] Bilchev, G. and Parmee, I. C., (1995), "The Ant Colony Metaphor for Searching Continuous Design Spaces", *Evolutionary Computing, AISB Workshop*, pp. 25–39

[7] Bonabeau, E. (1997), "From Classical Models of Morphogenesis to Agent-Based Models of Pattern Formation", *Artificial Life*, **3**, pp. 191–211.

[8] Bonabeau, E. and Théraulaz, G. (2000), "Swarm Smarts", *Scientific American*, March, pp. 72–79.

[9] Bonabeau, E., Dorigo, M. and Théraulaz, G. (1999), *Swarm Intelligence from Natural to Artificial Systems*, Oxford University Press.

[10] Bonabeau, E., Dorigo, M. and Théraulaz, G. (2000), "Inspiration for Optimization from Social Insect Behavior", *Nature*, **406**, 6 July 2000, pp. 39–42.

[11] Bullnheimer, B. (1999), *Ant Colony Optimization in Vehicle Routing*, Ph.D. Thesis, University of Vienna, January.

[12] Bullnheimer, B., Hartl, R. F. and Strauss, C. (1999a), "An Improved Ant System Algorithm for the Vehicle Routing Problem", *Annals of Operations Research*, **89**, pp. 319–328.

[13] Bullnheimer, B., Hartl, R. F. and Strauss, C. (1999b), "Applying the Ant System to the Vehicle Routing Problem", S.Voss, S. Martello, I. H. Osman and C. Roucairol (eds.), In *Metaheuristics: Advances and Trends in Local Search for Optimization*, Kluwer Academic Publishers, Boston, pp. 285–296.

[14] Camazine, S., Deneubourg, J.-L., Franks, N. R., Sneyd, J., Theraulaz, G. and Bonabeau, E. (2001), *Self-Organization in Biological Systems*, Princeton University Press.

[15] Cao, Y. U., Fukunaga, A. S. and Kahng, A. B. (1997), "Cooperative Mobile Robotics: Antecedents and Directions", *Autonomous Robots*, **4**, pp. 1–23.

[16] Chrétien, L. (1996), *Organisation Spatiale du Materiel Provenant de l'Excavation du Nid Chez* Messor Barbarus *et Agregation des Cadaures d'Ouvrieres Chez* Lasius Niger *(Hymenoptera: Formicidae)*, Ph.D. Thesis, Department of Animal Biology, Université Libre de Bruxelles, Belgium.

[17] Christofides, N., Mingozzi, A. and Toth, P. (1979), "The Vehicle Routing Problem", In N. Christofides, A. Mingozzi, P. Toth, and C. Sandi (eds.), *Combinatorial Optimization*, Wiley, Chichester.

[18] Clerc, M. and Kennedy, J. (2002), "The Particle Swarm Explosion, Stability, and Convergence in a Multidimensional Complex Space", *IEEE Trans. on Evolutionary Computation*, **6**(1), pp. 58–73.

[19] Corne, D., Dorigo, M. and Glover, F. (eds.) (1999), *New Ideas in Optimization*, McGraw-Hill.

[20] de Castro, L. N. and Von Zuben, F. J. (2004), *Recent Developments in Biologically Inspired Computing*, Idea Group Publishing.

[21] Défago, X. (2001), "Distributed Computing on the Move: From Mobile Computing to Cooperative Robotics and Nanorobotics", In *Proc. of the 1st ACM Int. Workshop on Principles of Mobile Computing (POMC'01)*, pp. 49–55.

[22] Deneubourg, J.-L., Aron, S., Goss, S. and Pasteels, J. M. (1990), "The Self-Organizing Exploratory Pattern of the Argentine Ants", *Jour. of Insect Behavior*, **3**, pp. 159–168.

[23] Deneubourg, J.-L., Goss, S., Franks, N., Sendova-Franks, A., Detrain, C. and Chrétien, L.. (1991), "The Dynamics of Collective Sorting: Robot-Like Ant and Ant-Like Robot", In J. A. Meyer and S. W. Wilson (eds.) *Simulation of Adaptive Behavior: From Animals to Animats*, pp. 356–365, MIT Press/Bradford Books, Cambridge, MA.

[24] Deneubourg, J.-L. and Goss, S. (1989), "Collective Patterns and Decision Making", *Ethology, Ecology & Evolution*, **1**, pp. 295–311.

[25] Deneubourg, J.-L., Aron, S., Goss, S., Pasteels, J. M. and Duerinck, D. (1986), "Random Behaviour, Amplification Processes and Number of Participants: How They Contribute to the Foraging Properties of Ants", *Physica 22D*, pp. 176–186.

[26] Dorigo, M., Di Caro, G. and Sampels, M. (2002), *Ant Algorithms: Third International Workshop (ANTS 2002)*, Lecture Notes in Computer Science, 2463, Springer-Verlag.

[27] Dorigo M. and Gambardella, L. M. (1997b), "Ant Colonies for the Traveling Salesman Problem", *BioSystems*, **43**, pp. 73–81.

[28] Dorigo, M. (1992), *Optimization, Learning and Natural Algorithms*, (in Italian), Ph.D. Thesis, Dipartimento di Elettronica, Politecnico di Milano, IT.

[29] Dorigo, M. and Gambardella, L. M. (1997a). "Ant Colony System: A Cooperative Learning Approach to the Traveling Salesman Problem", *IEEE Trans. on Evolutionary Computation*, **1**(1), pp. 53–66.

[30] Dorigo, M. and Di Caro, G. (1999), "The Ant Colony Optimization Meta-Heuristic", In D. Corne, M. Dorigo, and F. Glover (eds.), *New Ideas in Optimization*, McGraw-Hill, pp. 13–49.

[31] Dorigo, M., Di Caro, G. and Gambardella, L. M. (1999), "Ant Algorithms for Discrete Optimization", *Artificial Life*, **5**(3), pp. 137–172.

[32] Dorigo, M., Maniezzo, V. and Colorni, A. (1991), "Positive Feedback as a Search Strategy", *Tech. Report 91-016*, Dipartimento di Elettronica, Politecnico di Milano, IT.

[33] Dorigo, M., Maniezzo, V. and Colorni, A. (1996), "The Ant System: Optimization by a Colony of Cooperating Agents", *IEEE Trans. on Systems, Man, and Cybernetics – Part B*, **26**(1), pp. 29–41.

[34] Dorigo, M. and Stützle, T. (2004), *Ant Colony Optimization*, MIT Press.

[35] Franks, N. R. (1986), "Teams in Social Insects: Group Retrieval of Prey by Army Ants (*Eciton burcheli*, Hymenoptera: Formicidae)", *Behav. Ecol. Sociobiol.*, **18**, pp. 425–429.

[36] Franks, N. R. and Sendova-Franks, A. B. (1992), "Brood Sorting by Ants: Distributing the Workload Over the Work Surface", *Behav. Ecol. Sociobiol.*, **30**, pp. 109-123.

[37] Gordon, D. (1999), *Ants at Work: How an Insect Society is Organized*, W. W. Norton, New York.

[38] Goss, S., Aron, S., Deneubourg, J. L. and Pasteels, J. M. (1989), "Self Organized Shortcuts in the Argentine Ants", *Naturwissenchaften*, **76**, pp. 579–581.

[39] Goss, S., Beckers, R., Deneubourg, J. L., Aron, S. and Pasteels, J. M. (1989), "How Trail Laying and Trail Following Can Solve Foraging Problems for Ant Colonies", In R. N. Hughes (ed.), *Behavioural Mechanisms of Food Selection*, NATO-ASI Series, **G20**, Springer-Verlag, pp. 661–678.

[40] Höldobler, B. (1983), "Territorial Behavior in the Green Tree Ant (*Oecophylla smaragdina*)", *Biotropica*, **15**, pp. 241–250.

[41] Höldobler, B. and Wilson, E. O. (1978), "The Multiple Recruitment Systems of the African Weaver Ant *Oecophylla longinola* (Latreille)", *Behav. Ecol. Sociobiol.*, **3**, pp. 19–60.

[42] Kennedy, J. (2004), "Particle Swarms: Optimization Based on Sociocognition", In L. N. de Castro and F. J. Von Zuben, *Recent Developments in Biologically Inspired Computing*, Idea Group Publishing, Chapter X, pp. 235–269.

[43] Kennedy, J. (1997a), "The Particle Swarm: Social Adaptation of Knowledge", *Proc. of the IEEE Int. Conf. on Evolutionary Computation*, pp. 303–308.

[44] Kennedy, J. (1997b), "Thinking is Social: Experiments with the Adaptive Culture Model", *Journal of Conflict Resolution*, **42**, pp. 56–76.

[45] Kennedy, J. and Eberhart, R. (1995), "Particle Swarm Optimization", *Proc. of the IEEE Int. Conf. on Neural Networks*, Perth, Australia, **4**, pp. 1942–1948.

[46] Kennedy, J., Eberhart, R. and Shi. Y. (2001), *Swarm Intelligence*, Morgan Kaufmann Publishers.

[47] Krieger M. B. and Billeter J.-B. (2000), "The Call of Duty: Self-Organised Task Allocation in a Population of Up to Twelve Mobile Robots. *Robotics and Autonomous Systems*, **30**(1–2), pp. 65–84.

[48] Krieger, M. J. B., Billeter, J.-B. and Keller, L. (2000), "Ant-Task Allocation and Recruitment in Cooperative Robots", *Nature*, **406**, 31 August, pp. 992–995.

[49] K-Team Web Site: http://www.k-team.com/robots/khepera/

[50] Kube, C. R., Parker, C. A. C., Wang, T. and Zhang, H. (2004), "Biologically Inspired Collective Robotics", In L. N. de Castro and F. J. Von Zuben, *Recent Developments in Biologically Inspired Computing*, Idea Group Publishing, Chapter XV, pp. 367–397.

[51] Kube, C. R. and Bonabeau, E. (2000), "Cooperative Transport by Ants and Robots", *Robotics and Autonomous Systems*, **30**(1–2), pp. 85–101.

[52] Kube, C. R. and Zhang, H. (1997), "Multirobot Box-Pushing", *IEEE International Conference on Robotics and Automation*, Video Proceedings, 4 minutes.

[53] Kube, C. R. and Zhang, H. (1992), "Collective Robotic Intelligence", In *From Animals to Animats: Int. Conf. on the Simulation of Adaptive Behavior*, pp. 460–468.

[54] Kube, C. R. and Zhang, H. (1994a), "Collective Robotics: From Social Insects to Robots", *Adaptive Behavior*, **2**, pp. 189–218.

[55] Kube, C. R. and Zhang, H. (1994b), "Stagnation Recovery Behaviours for Collective Robotics", In *1994 IEEE/RSJ/GI International Conference on Intelligent Robots and Systems*, pp. 1893–1890.

[56] Lawer, E. L., Lenstra, J. K., Rinnooy Kan, A. H. G. and Shmoys, D. B. (1985), *The Traveling Salesman Problem: A Guided Tour of Combinatorial Optimization*, Wiley-Interscience.

[57] Lumer, E. D. and Faieta, B. (1994), "Diversity and Adaptation in Populations of Clustering Ants", In D. Cliff, P. Husbands, J. A. Meyer, S.W. Wilson (eds.), *Proc. of the 3rd Int. Conf. on the Simulation of Adaptive Behavior: From Animals to Animats*, **3**, MIT Press, pp. 499–508.

[58] Martinoli, A (2001), "Collective Complexity out of Individual Simplicity", *Artificial Life*, **7**(3), pp. 315–319.

[59] Millonas, M. M. (1994), "Swarm, Phase Transitions, and Collective Intelligence", In C. Langton (ed.), *Proc. of Artificial Life III*, Addison-Wesley, pp. 417–445.

[60] Moffett, M. W. (1988), "Cooperative Food Transport by an Asiatic Ant", *National Geographic Res.*, **4**, pp. 386–394.

[61] Nature Web Site: http://www.nature.com

[62] Paessens, H. (1988), "The Savings Algorithm for the Vehicle Routing Problem", *European Journal of Operations Research*, **34**, pp. 336–344.

[63] Shi, Y. and Eberhart, R. C. (1999), "Empirical Study of Particle Swarm Optimization", *Proc. of the IEEE Congress on Evolutionary Computation*, pp. 1945–1950.

[64] Sudd, J. H. (1960), "The Transport of Prey by an Ant *Pheidole Crassinoda*", *Behavior*, **16**, pp. 295–308.

[65] Sudd, J. H. (1965), "The Transport of Prey by Ant", *Behavior*, **25**, pp. 234–271.

[66] Vizine, A. L., de Castro, L. N., Hruschka, E. R., and Gudwin, R. R. (2005), "Towards Improving Clustering Ants: An Adaptive Ant Clustering Algorithm", *Informatica*, **29**(2005), pp. 143–154.

[67] White, T. and Pagurek, B. (1998), "Towards Multi-Swarm Problem Solving in Networks", *Proc. of the 3rd Int. Conf. on Multi-Agent Systems* (ICMAS'98), pp. 333–340.

[68] Yeung, K. Y., Medvedovic, M. and Bumgarner, R. E. (2003), "Clustering Gene-Expression Data with Repeated Measurements", *Genome Biology*, **4**(5), article R34.

CHAPTER 6

IMMUNOCOMPUTING

"Like the central nervous systems, the immune system is individualized. Identical twins, born with identical DNA, develop different immune systems, just as they develop different brains. Each person's immune system records a unique history of individual life, because ... the immune system, like the brain, organizes itself through experience. Thus the brain and the immune system establish individuality at two levels: they help us adapt to life and so preserve us, and they make a record of what has happened."

(I. R. Cohen, Tending Adam's Garden: Evolving the Cognitive Immune Self, Academic Press, 2004; p. 5)

"There is probably no system, living or manmade, that can protect itself as effectively, on so many levels, and from so many different diseases and infections as the human immune system."

(N. Forbes, Imitation of Life: How Biology is Inspiring Computing, MIT Press, 2004; p. 98)

6.1 INTRODUCTION

The immune system of vertebrates is an intricate collection of distributed cells, molecules, and organs that altogether play the important role of maintaining a dynamic internal state of equilibrium in our bodies. Its complexity has been compared to that of the brain's in many respects: immune systems are capable of recognizing foreign and internal signals; controlling the action of immune components; influencing the behavior of other systems, such as the nervous and the endocrine systems; and, most importantly, learning how to fight against disease causing agents and extracting information from them.

Immunocomputing is a terminology introduced to describe a new computational approach that aims at implementing information processing principles from proteins and immune networks in new kinds of computer algorithms and software, leading to the concept of an immunocomputer (Tarakanov et al., 2003). *Artificial immune systems* (AIS)[1], by contrast, is a terminology that refers to adaptive systems inspired by theoretical and experimental immunology with the goal of solving problems (de Castro and Timmis, 2002). They encompass any system or computational tool that extracts ideas and metaphors from the biological immune system in order to solve problems.

Together with swarm intelligence, AIS constitute one of the youngest fields of investigation in nature-inspired computing. Despite its youth, the bibliography on AIS is becoming vast. Added to the success of many applications, the creati-

---

[1] Although AIS and immunocomputing mean slightly different approaches, in this chapter the terminology immunocomputing will be used as a synonym for artificial immune systems.

on of the series of International Conferences on Artificial Immune Systems (ICARIS) in 2002 (Timmis and Bentley, 2002; Timmis et al., 2003; Nicosia et al., 2004; Jacob et al., 2005), in addition to the many special sessions and tracks on the major evolutionary computation conferences, helped to promote the field and attract more researchers.

This chapter provides a basic introduction to artificial immune systems. It starts with a review of some immunological background necessary for the development and understanding of AIS, and then follows with the description of a framework to design artificial immune systems. The focus is on four important immunological principles and theories: pattern recognition within the immune system; adaptive immune responses; self/nonself discrimination; and the immune network theory. The framework to design an AIS introduces a generic structure to create abstract models of immune cells and molecules, presents some measures to quantify the interactions of these elements with each other and the environment, and describes some general-purpose immune algorithms. Although most sections describe portions of the immune system, Section 6.2.7 on the mammalian immune system presents a slightly broader view of the immune system to illustrate part of its complexity and the various types of cells and molecules involved in an immune response.

6.2   THE IMMUNE SYSTEM

All living beings have the ability to present resistance to disease-causing agents, known as *pathogens*. These include viruses, bacteria, fungi, and parasites. The nature of this resistance varies from one species to the other, and is a function of the complexity of the organism. Mammals, in particular human beings, have developed a highly sophisticated immune system that acts together with several other bodily systems (such as the nervous and the endocrine system) to maintain life. The primary role of the immune system is to protect our bodies against infections caused by pathogens (Janeway et al., 1999; Tizard, 1995).

The defense must occur in many levels and has to cover the whole body. Therefore, various levels of defense mechanisms and barriers have evolved in order to result in sufficient protection. For instance, physical barriers such as the skin, and biochemical barriers such as the pH levels and the saliva, are sometimes considered as parts of the immune system. Apart from these physical and biochemical barriers, the immune system can be divided into *innate immune system* and *adaptive immune system*, composed of diverse sets of cells, molecules and organs that work in concert to protect the organism.

The innate immune system is very important as a first line of defense against several types of pathogens and is also crucial for the regulation of the adaptive immune system. Cells belonging to the innate immune system are capable of recognizing generic molecular patterns (a type of molecular signature) that are only present in pathogens, and can never be found in the cells of the host. Once a pathogen has been recognized by a cell of the innate immune system, this cell signals (through chemical messengers) other immune cells, including those of the adaptive immune system, to start fighting against the pathogen. Therefore, the innate immune system plays a major role in providing *co-stimulatory* signals for the adaptive immune system. Co-stimulatory signals are usually provided by the innate immune system when the organism is being damaged in some way, such as when cells are being killed by viruses. For the most types of pathogens, the adaptive immune system cannot act without the co-stimulatory signals provided by the innate immune system.

Nevertheless, not all pathogens can be recognized by the innate system. Some specific pathogens are only recognized by cells and molecules of the adaptive immune system, also called specific immune system. The adaptive immune system possesses some particular features that are important from a biological and computational perspective. For instance, it can adapt to those molecular patterns previously seen and it generates and maintains a stable memory of known patterns.

After a certain pathogen has been eliminated by the adaptive system, the innate system plays a role in signaling the adaptive system that the foreign agent has been defeated. Another way the innate immunity is important for the adaptive immunity is in that the latter usually requires some time before it starts acting. Thus, the innate immune system tries to get the pathogen at bay until the adaptive immune system can act, but the innate system by itself is usually not capable of removing the infection.

Once the adaptive immune system is prepared to act, it can adapt to the invading pathogen and create specific molecular patterns to fight against the same or a similar future infection of this type. This adaptive (evolutionary and learning) capability has been explored by immunologists and medics for over two centuries in order to protect us against known pathogens. The vaccination principle is very much embodied in this learning capability of the immune system. By inoculating a healthy individual with weakened or dead samples of some specific disease-causing agents (pathogens), the immune system is allowed to generate sets of molecular structures specific in recognizing and fighting against that pathogen, without subjecting the organism to the unpleasant symptoms of the disease. The mechanisms underlying this adaptability of the immune system have also been broadly explored in immunocomputing, and these will be discussed later.

Last, but not least, there are theories that suggest the immune system is a dynamic system whose cells and molecules are capable of interacting with each other. This viewpoint establishes the idea that pathogens are responsible for modifying the structure of a dynamic immune system, while the other more traditional perspectives suggest that the immune system is composed of discrete sets of cells and molecules that are only activated by pathogens. This section reviews some basic immune theories and principles that have been used for the design and application of artificial immune systems, including the more controversial *immune network theory*. The adaptive immune system is emphasized due to its adaptation, learning, and memory capabilities.

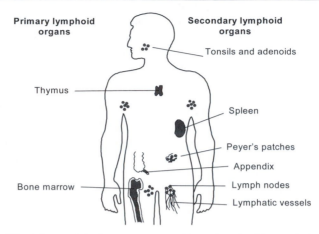

Figure 6.1: Physiology of the immune system. (Reproduced with permission from [de Castro and Timmis, 2002], © L. N. de Castro and J. Timmis.)

6.2.1. Physiology and Main Components

One remarkable feature of the immune system is its distributivity all over the body. There are cells, molecules, and organs in all parts of the organism, as illustrated in Figure 6.1. The organs composing the immune system, sometimes referred to as the *lymphoid system*, are called *lymphoid organs*. The *primary lymphoid organs* are those responsible for the production, growth, development, and maturation of *lymphocytes*, which are specialized white blood cells that bear specific receptors to recognize pathogens. The *secondary lymphoid organs* are the sites where the lymphocytes interact with pathogens.

The two main types of lymphocytes are the *B-cells* and the *T-cells*, or *B-lymphocytes* and *T-lymphocytes*. The B-cells are so named because they mature (i.e., become able of fighting against diseases) in the bone marrow, and the T-cells mature in the thymus. Both present several important features from the biological and computational perspectives. They have surface receptor molecules capable of recognizing specific molecular patterns present on pathogens. They are also capable of multiplying themselves through a *cloning* process, and the B-cells are capable of suffering genetic variation during reproduction. Therefore, these cells have their number varied with time and B-cells can have their molecular structure changed as well.

The two primary lymphoid organs, bone marrow and thymus, have important functions in the production and maturation of immune cells. All blood cells are generated within the bone marrow, and the B-cells also maturate - become ready to act as an immune cell - within the bone marrow. T-cells are generated within the bone marrow but migrate to the thymus where they will eventually become immunocompetent cells. The thymus is also believed to be the site where the T-cells learn to recognize self, in a process known as negative selection of T-cells (Section 6.2.4).

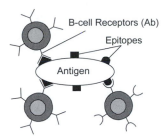

Figure 6.2: Immune cells, in particular B-cells, with their molecular receptors detached recognizing the epitopes on antigens. (Reproduced with permission from [de Castro and Timmis, 2002], © L. N. de Castro and J. Timmis.)

6.2.2. Pattern Recognition and Binding

The first step in promoting an *immune response*, that is, a response of the immune system against infection, is the recognition of the pathogen. Those portions of the pathogens that can promote an adaptive immune response are called *antigens*. Antigens present surface molecules, named *epitopes*, that allow them to be recognized by immune cells and molecules. Figure 6.2 depicts immune cells, in particular B-cells, and their receptors recognizing the epitopes of a given antigen. Note that B-cells are believed to be monospecific, i.e., present a single type of receptor on the surface. On the contrary, antigens can present various patterns of epitopes.

There are several types of immune cells and molecules. Most AIS focus on the adaptive immune system with emphasis either on B-cells or T-cells, which are the most important cells from the adaptive immune system. B-cells also have been the focus of most research on AIS because they are capable of altering their genetic composition during an immune response. The B-cell receptor is also known as *antibody* or *immunoglobulin*.

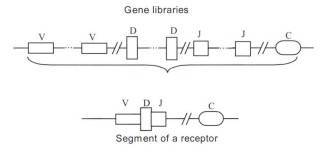

Figure 6.3: Gene rearrangement process that results in the production of part of the receptor molecule of a B-cell receptor. Segments from different libraries (V, D, J, and C) are joined together to form the antibody.

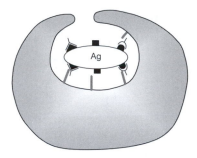

Figure 6.4: A phagocyte recognizes a pathogen covered with antibodies and thus destroys the whole complex.

The receptor molecules of both B-cells and T-cells are generated by the recombination of gene segments concatenated from several gene libraries, what allows the immune system to produce a large variety of receptors from a finite genome. The gene rearrangement process that results in the production of parts of a B-cell receptor molecule is illustrated in Figure 6.3. In this process, some genes are taken from the libraries (e.g., one gene from each of the libraries: V, D, J, and C) and concatenated so as to form the receptor molecule.

In order for the receptor on an immune cell to recognize an antigen, some portion of the antigen - the epitope - has to have a shape complementary to the shape of the receptor molecule of the immune cell (see Figure 6.2). However, these shapes do not have to be fully complementary for a recognition event to occur, a partial recognition (or match) between both molecules - receptor and epitope - is sufficient. The degree of recognition or interaction between these two molecules is termed *affinity*; the better the recognition, the higher the affinity, and vice-versa. Affinity is a very important concept in artificial immune system. Note that recognition is not only based on shape complementarity, it also depends upon a number of chemical cues, van der Waals interactions, etc.

It is important to have in mind that recognition by itself is not sufficient to eliminate an infectious agent; there must be a binding between the cell receptor and the antigen so as to signal other immune cells, mainly those of the innate immune system, to destroy the complex formed by the cell receptor bound with the antigen. Figure 6.4 illustrates the case of an antigen covered with antibodies (Y-shaped molecules attached to the antigen Ag). The complex formed by the antigen covered with antibodies signals other immune cells, in this case, big cell eaters known as *phagocytes*, to ingest and digest the complex, and thus destroy the antigen.

6.2.3. Adaptive Immune Response

Several theories were proposed as attempts to explain how the immune system copes with antigens. For instance, L. Pasteur suggested, in the late 1800s, that elements contained in the vaccine were capable of removing nutrients essential to the body and, thus, avoiding the growth and proliferation of pathogens. It were M. Burnet and D. Talmage who in the mid-1900s proposed and formalized the *clonal selection theory* of adaptive immunity (Burnet, 1959), broadly accepted as an explanation of how the adaptive immune system responds to pathogens. Together with the theory of *affinity maturation* of antibodies (Nossal, 1993; Storb, 1998), clonal selection forms the core of an adaptive immune response, and both have been used in the literature of artificial immune systems to design adaptive systems for problem solving.

According to the clonal selection theory, the immune system is composed of sets of discrete cells and molecules that remain at rest until a pathogenic agent invades the organism. After invasion, some subset of these immune cells, either B-cells or T-cells, is capable of recognizing the invading antigen and binding with it. This recognition process stimulates the immune cells capable of recognizing the antigen to start reproducing themselves. Reproduction in this case is asexual; it refers to cellular reproduction, a process of *mitotic* cell division, and is sometimes called *cloning* or *clonal expansion* (Figure 6.5). Thus, a subset of cells (*clone*) capable of recognizing a specific type of antigen is generated.

As with all reproductive events, the B-cell division (or reproduction) process may be subject to an error (*mutation*). A particular feature of the immune system is that mutation occurs in proportion to the cell affinity with the antigen that the

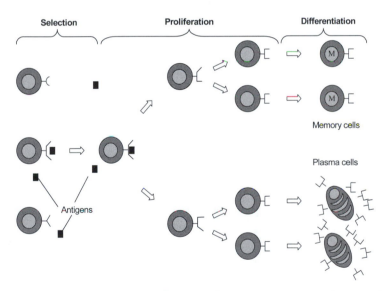

Figure 6.5: Clonal selection, expansion, and affinity maturation. From the repertoire of B-cells, the one(s) presenting higher affinities with the antigen are selected and stimulated to proliferate. During each proliferation stage, mutation may occur, thus allowing the cell receptor to become more adapted to the antigen presented. Those mutated immune cells with higher affinities with the antigen are selected to become memory and antibody secretor (plasma) cells.

parent cell bind. If the affinity between antigen (Ag) and antibody (Ab) is high, then the mutation rate is low; and if the Ag-Ab affinity is low, then high mutation rates are allowed. This is very interesting because it suggests that mutation in the immune system is regulated by the degree of recognition of an antigen by an antibody (Allen et al., 1987). In addition, there are authors (e.g., Kepler and Perelson, 1993) who suggest that a short burst of mutation is followed by periods of low mutation rates to allow for selection and clonal expansion (reproduction).

Another aspect of clonal selection is that it is local. As the immune system is distributed throughout the organism, only those cells located closer to the site of infection are recruited for combat. More cells from the neighborhood can be recruited, but not all immune cells are involved in an adaptive immune response. Figure 6.5 summarizes the clonal selection and affinity maturation processes. The B-cell(s) with higher affinity with the antigen are selected to proliferate. During mitosis, the cells can suffer somatic mutation, and a selection mechanism allows those mutated progenies with increased affinities to be retained in a pool of memory cells. Memory cells usually have very long life spans, on the order of years, even decades.

Adaptation via Clonal Selection

Clonal selection and affinity maturation suggest that the immune system becomes more adapted to the environmental stimulation provided by pathogens. Two degrees of plasticity (forms of adaptation) can be observed in these processes: structural and parametric plasticity. The immune cells successful in recognizing and binding with antigens are selected to proliferate, while those cells that do not perform any role whatsoever in an immune response die, and are eliminated from the repertoire of immune cells. Affinity maturation leads to a parametric adaptation in the molecular structure of antibodies, and thus contributes to the adaptability of the immune system, and the learning of molecular patterns present on pathogens.

The idea can be likened to that of a reinforcement learning strategy (Section 4.3.3), as illustrated in Figure 6.6. When an antigen Ag_1 invades the organism, a few specific antibodies capable of recognizing and binding with this antigen will be selected to proliferate. But some time is required until the immune cells start reproducing themselves so as to fight against Ag_1. After this *lag phase*, sufficient antibodies that recognize and bind with Ag_1 are generated so as to guarantee the elimination of the antigen. In a future or secondary exposition to this same antigen Ag_1, a faster and stronger response is promoted by the immune system. If a new antigen Ag_2 is presented to the immune system, then a pattern of response similar to that to Ag_1 occurs, in the sense that the primary response to Ag_2 is also slow and with a long lag phase. This is because the immune response is specific, meaning that antibodies successful in recognizing a given antigen are specific in recognizing that antigen or a similar one Ag_{11}. Therefore, the response of the antibodies initially primed with Ag_1 to a similar antigen Ag_{11} is similar to the secondary response to Ag_1.

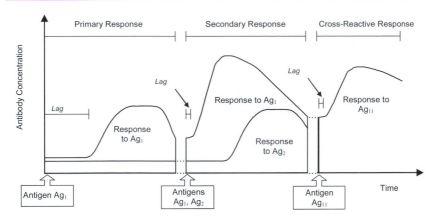

Figure 6.6: Primary, secondary and cross-reactive immune responses. After an antigen Ag_1 is seen once (primary response), subsequent encounters with the same antigen (secondary response), or a similar one Ag_1' (cross-reaction), will promote a faster and more effective response not only to Ag_1, but also to Ag_1'. Primary responses to antigens Ag_1 and Ag_2 are qualitatively equivalent. (Reproduced with permission from [de Castro and Timmis, 2002], © L. N. de Castro and J. Timmis.)

In the (artificial) immune system literature this is known as *cross-reactive response*, and can be likened to the generalization capability of neural networks. While discussing neurocomputing, we argued that generalization is important for the creation of models of the world. In the present context, cross-reactivity is also important for the creation of models of the antigenic universe. If you were not able of learning different types (e.g., Ag_1 and Ag_2) of molecular structures, then death from infection would be inevitable. Of course if the variation between these patterns is too great, cross-reactivity may not be enough for a successful secondary response and thus, a primary response has to be raised.

Thus, adaptation and learning in the immune system involve the variation in the population sizes and affinity of specific cells successful in recognizing and binding with pathogens. The primary adaptive responses may take some time to act, but the secondary response is fast and efficient. Generalization or cross-reactivity is also present in the adaptive immune response.

Clonal Selection and Darwinian Evolution

Chapter 3 presented the neo-Darwinian theory of evolution. Basically, it states that an evolutionary procedure can be defined as the one involving a population of individuals capable of reproducing, and subjected to genetic variation followed by natural selection. By looking at the clonal selection and affinity maturation principles of an adaptive immune response it is easy to note that clonal selection is a type of evolutionary process. It may come as no surprise, thus, the fact that clonal selection immune algorithms may represent a new type of evolutionary algorithm that borrows ideas from the immune system.

In (de Castro and Timmis, 2002), the authors suggested that natural selection can act on the immune system at two distinct levels. First, clonal selection and affinity maturation can be seen as a *microevolutionary process*, i.e., one that occurs within organisms in a time scale orders of magnitude smaller than that of the species (organism) evolution. Second, the immune system contributes to natural selection in a sense that individuals with malfunctioning immune systems have lesser probabilities of survival and thus reproduction. Finally, it was also suggested that the immune system is capable of generating and maintaining diverse sets of cells and molecules that cover the various niches of antigens presented to the immune system.

6.2.4. Self/Nonself Discrimination

One question that for long has intrigued scientists in various fields is that of how the immune system differentiates between the cells of the organism, known as *self*, and the foreign elements capable of causing disease, known as *nonself*. There are various theories that try to approach this question, and one of these involves the *negative selection of T-cells* within the thymus. Other less orthodox proposals are the idea that the immune system evolved to discriminate between infectious nonself and noninfectious self (Janeway, 1992), and the Danger theory that suggests the immune system is capable of recognizing stress or damage signals (Matzinger, 1994).

When an immune cell encounters an antigen, several outcomes might arise. For instance, it has been discussed that if the antigen is nonself, i.e., disease-causing, then the clonal expansion of those cells successful in recognizing and binding with the antigen will occur. But this is not the whole picture. In the case an antigen is recognized by an immune cell while it is patrolling the organism for nonself, a *second signal*, also called *co-stimulatory signal*, from other immune cells is required before an adaptive immune response can be launched.

What if the antigen is a self-antigen? There are a few possibilities in this case and only the negative selection of T-cells within the thymus will be considered here. In a simplified form, if a self antigen is recognized by an immature T-cell within the thymus, this cell is purged from the repertoire of T-cells, else it becomes an immunocompetent cell and is released to circulate throughout the body in the search for nonself antigens. This process, called *thymic negative selection of T-cells* or simply *negative selection*, is only possible because the thymus is protected by a blood-thymic barrier that filters out any molecule that does not belong to self. Thus, all molecules within the thymus are self molecules, and the immature T-cells learn to be *tolerant* (not respond to) to the self molecules while within the thymus.

6.2.5. The Immune Network Theory

The last immune theory to be reviewed in this chapter was formalized by N. Jerne in the mid 1970s. In contrast to the clonal selection theory that proposes an immune system composed of discrete cells and molecules at rest and stimulated by nonself antigens, the *immune network theory* suggests that all immune cells

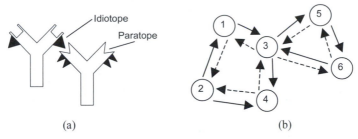

(a)   (b)

Figure 6.7: The immune network theory. (a) The recognition between two antibodies. The idiotope of an antibody is recognized by the complementary paratope of another antibody. (b) In the network depicted, cell 1 recognizes cell 3 and is thus stimulated, while cell 3 is suppressed; cell 2 is stimulated by the recognition of cells 1 and 4, which by themselves are suppressed by cell 2; and so forth. Solid arrow: stimulation. Dashed arrow: suppression.

and molecules recognize each other (Jerne, 1974). Thus, no foreign stimulation is necessary for a working (dynamic) immune system.

Jerne suggested that in and around the receptors of an immune cell there are molecular patterns, called *idiotopes*, that allow this cell to be recognized by other cells and molecules of the immune system. The idiotope of an immune cell can be recognized by a complementary paratope of another immune cell, as illustrated in Figure 6.7(a). This view suggests a dynamic immune system with behavior governed by the patterns of interactions of immune cells and molecules, and whose state is changed due to the perturbation caused by nonself antigens. As the immune cells are assumed to be monospecific, the recognition of an immune cell can be viewed as corresponding to the recognition of its receptor. When the receptor on one cell recognizes the receptor on another cell, the recognizing cell is stimulated, and the cell recognized is suppressed, as illustrated in Figure 6.7(b). The effects of stimulation and suppression vary from one model to the other, and will be discussed further.

Adaptation and Learning via Immune Network

To formally describe the immune network activity, F. Varela and collaborators (Varela et al., 1988) introduced three important network concepts: *structure*, *dynamics*, and *metadynamics*:

1. *Structure*: as with the case of neural networks, the immune network structure corresponds to the patterns of connectivity between the cells and molecules in the network, i.e., how the network components are linked together. However, immune networks do not usually have predefined architectures. Instead, the immune network structure tends to follow that of the antigenic environment and the molecular structure of the immune cells.

2. *Dynamics*: the immune network dynamics corresponds to how the network cells and molecules vary with time. This is one of the main features

that allow immune networks to adapt to the environment, what is a consequence of the interactions between the components of the network themselves and with the environment.

3. *Metadynamics*: one important feature of the immune system is the constant generation of immune cells and the death of unstimulated cells. New cells are constantly being generated and replacing cells that do not play any role in immune responses. The generation and death of immune cells results in the creation and elimination of network connections as well. This capability of having its structure varied with time is also referred to as the double plasticity of the immune system.

6.2.6. Danger Theory

With a conceptually different viewpoint, P. Matzinger (1994, 2002) introduced what came to be known as the *danger theory*. In essence, the danger model adds another layer of cells and signals to the self/nonself discrimination models. It proposes that antigen-presenting cells (APCs), such as those exposed to pathogens, toxins, mechanical damage, etc., are activated by the alarms caused by these phenomena.

The danger model tries to answer one of the fundamental questions in immunology: How is self-tolerance induced? It suggests that the immune system is more concerned with damage (preventing destruction) than with foreignness. It takes into account issues like what happens when bodies change (e.g., through puberty, pregnancy, aging, etc.); why are there T- and B-cells specific for self-antigens; why do we not mount immune responses to vaccines; why neonates are easily tolerizable; why silicone, well boiled bone fragments, or solitary haptens do not elicit immune responses; why do we fail to reject tumors; and so forth (Matzinger 1994, 2002).

A puzzling question is how to distinguish between dangerous and non-dangerous. At the same time there are several foreign things that are dangerous, such as bacterial toxins, viruses, worms, and others, there are also dangerous 'self', such as tumors, and nondangerous foreign, such as beneficial bacteria and nonlytic viruses. The new danger theory proposes that antigen presenting cells are activated by danger/alarm signals from injured cells, such as those exposed to pathogens, toxins, and mechanical damage. The mechanisms of immune activation would be a consequence of cell/tissue damage. Cell death is not always a result of parasitic attack. It is a normal event during embryonic development, formation and death of hematopoietic cells, maturation and expulsion of oocytes, etc. In such cases, death is controlled, usually apoptotic, and cells that die by these normal programmed processes, are usually scavenged before they desintegrate. By contrast, cells dying by stress or necrotically release their contents in the surroundings and these serve as (danger) signals, as illustrated in Figure 6.8. The key issue is that danger signals are not sent by normal (healthy) cells, only by injured tissues. The danger signals can be active or passive (Matzinger, 1994). Abrupt changes in the condition of a cell, like, for instance, temperature variation or infection, elaborate a series of heat chock proteins that

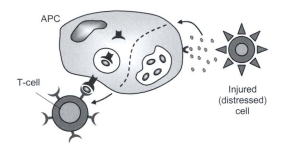

Figure 6.8: When a cell is injured (or stressed) it produces some damage (or stress) signals that are recognized by cells of the immune system.

aid their recovery, and serve as danger signals. Internal molecules, normally not secreted, may also serve as a danger signal; thus, any cell damage caused by a cut, bruise and infection, can be noted.

6.2.7. A Broader Picture

Although this chapter has reviewed a number of theories and principles that try to explain how the immune system behaves, the overall system is much more complex than this chapter could cover. To provide the reader with a reasonably broader picture of the basic recognition and activation mechanisms involved in an immune response, consider the illustration presented in Figure 6.9 (de Castro and Timmis, 2002).

Part (I) of Figure 6.9 shows how specialized *antigen presenting cells*, called APCs, such as phagocytes, circulate throughout the body ingesting and digesting antigens. These antigens are fragmented into *antigenic peptides* (Nossal, 1993). Part of these peptides bind with some molecules termed MHC and the whole complex is presented in the APC cell surface (II). The T-cells carry surface receptors that allow them to recognize different MHC/peptide complexes (III). Once activated by the MHC/peptide recognition, the T-cells become activated, divide, and secrete chemical signals that stimulate other components of the immune system to enter into action (IV).

In contrast to the T-cells, B-cells have receptors with the ability to recognize parts of the antigens free in solution (V). The surface receptors on these B-cells respond to a specific antigen. When a signal is received by these B-cell receptors, the B-cell is activated and will proliferate and differentiate into *plasma cells* that secrete antibody molecules in high volumes (VI). These released antibodies (which are soluble forms of the B-cell receptors) are used to neutralize the pathogen (VII), leading to their destruction. Some of these activated B- and T-cells will differentiate into *memory cells* that remain circulating through the organism for long periods of time, thus guaranteeing future protection against the same (or a similar) antigen that elicited the immune response.

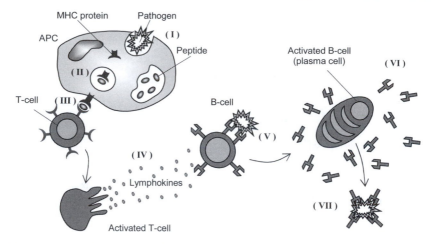

Figure 6.9: Broader picture of the basic immune recognition and activation mechanisms. (Reproduced with permission from [de Castro and Timmis, 2002], © L. N. de Castro and J. Timmis.)

6.3 ARTIFICIAL IMMUNE SYSTEMS

Differently from evolutionary algorithms that emerged from one main central idea and developed into several branches and variations, artificial immune systems were proposed taking into account various different features, processes, and models of the immune system. The number of applications and algorithms present in the literature of AIS is vast, but some core ideas have been broadly explored, namely, clonal selection and affinity maturation, negative selection, and immune networks. Many new proposals involving mainly concepts from innate immunity and the danger theory have appeared in the last ICARIS conferences (Timmis et al., 2003; Nicosia et al., 2004; Jacob et al., 2005), but still not with a common ground among them.

To design artificial immune systems, de Castro and Timmis (2002) have proposed a layered approach based on the *immune engineering* framework introduced by de Castro (2001, 2002, 2003). The term *immune engineering* refers to " … a meta-synthesis process that is going to define the tool for solving a given problem based upon the features of the problem itself, and then apply it to obtain the solution to the problem. Instead of trying to create accurate models of the immune system, the immune engineering must try to develop and implement pragmatic models inspired by the immune system. These must preserve some of the essential properties of the immune system which demonstrate to be implementable and efficient for the development of engineering tools." (de Castro, 2001; p. 44). Note that the distinction between this definition and that of artificial immune systems is sharp; immune engineering is basically concerned with the problem of designing AIS.

The immune engineering process that leads to a framework to design AIS is composed of the following basic elements (de Castro and Timmis, 2002):

- A *representation* for the components of the system.
- A set of *mechanisms to evaluate the interaction* of individuals with the environment and each other. The environment is usually simulated by a set of input stimuli or patterns, one or more fitness function(s), or other means.
- *Procedures of adaptation* that govern the dynamics and metadynamics of the system, i.e., how its behavior varies over time.

6.3.1. Representation

There are several types of immune cells and molecules composing the immune system. The first step toward designing an artificial immune system is to devise a scheme to create abstract models of these immune cells and molecules. Any immune response requires the recognition of an antigen by a cell receptor. Recognition in the immune system occurs mainly through shape complementarity between these two molecules: the cell receptor and the antigen. Note that some cell receptors (e.g., antibodies) can be released from the cell surface, and can thus be found free in solution.

To model this shape recognition process in the immune system, A. Perelson and G. Oster (1979) introduced the concept of *shape-space*. The shape-space approach assumes that all the properties of receptor molecules relevant to determining their interactions with each other and foreign antigens can be described by a *data structure*. Thus, the data structure, which usually depends upon the problem being studied, corresponds to the *generalized shape* of a molecule and is sufficient to quantify the affinity between molecules.

Although recognition within the immune system occurs through shape complementarity, most AIS quantify the degree of similarity instead of complementarity or dissimilarity between data structures. Thus, it is important to known if shape complementarity or similarity is going to be the target of recognition in the AIS developed. The most common type of data structure is an *attribute string*, which can be a real-valued vector, an integer string, a binary string, or a symbolic string. These result in diverse shape-spaces:

- *Real-valued shape-space*: the attribute strings are real-valued vectors.
- *Integer shape-space*: the strings are integers.
- *Hamming shape-space*: the strings are built out of a finite alphabet of length k.
- *Symbolic shape-space*: usually composed of different types of attribute strings where at least one of them is symbolic, such as 'age', 'height', etc.

Assume the case in which any receptor molecule of an immune cell is called an antibody and is represented by the set of coordinates $\mathbf{Ab} = \langle Ab_1, Ab_2, ..., Ab_L \rangle$; an antigen is given by $\mathbf{Ag} = \langle Ag_1, Ag_2, ... , Ag_L \rangle$, where boldface letters correspond to attribute strings. Without loss of generality, antigens and antibodies are assumed to be of same length L. The interaction of antibodies, or of an antibody and an antigen, can be evaluated via a *distance* or a *similarity measure*, also termed *affinity measure*, between their corresponding attribute strings, depending if shape complementarity or similarity is being sought. Thus, the affinity measure performs a mapping from the interaction between two attribute strings into a nonnegative real number that corresponds to their *affinity* or *degree of match*, $S^L \times S^L \to \Re^+$. In the case recognition is proportional to shape complementarity, the higher the distance (for instance, for Hamming shape-spaces) the better the recognition. By contrast, if recognition is proportional to similarity, the smaller the distance, the better the recognition.

6.3.2. Evaluating Interactions

Assuming the various types of attribute strings to represent the generalized shapes of molecules in the immune system, each one of these types will require a particular class of *affinity measure*. Real-valued shape-spaces require affinity measures that deal with real-valued vectors; Integer shape-spaces require affinity measures that deal with integer strings; and so forth. For real-valued shape-spaces, the most common affinity measures are the Euclidean and Manhattan distances, given by Equation (6.1) and Equation (6.2), respectively:

$$D = \sqrt{\sum_{i=1}^{L}(Ab_i - Ag_i)^2} \qquad (6.1)$$

$$D = \sum_{i=1}^{L}|Ab_i - Ag_i| \qquad (6.2)$$

For Hamming shape-spaces, the Hamming distance can be used to evaluate the affinity between two cells. In this case, the molecules are represented as sequences of symbols over a finite alphabet of length k. Equation (6.3) depicts the Hamming distance used to evaluate the affinity between two attribute strings of length L in a Hamming shape-space.

$$D = \sum_{i=1}^{L}\delta_i, \text{where} \quad \delta_i = \begin{cases} 1 & \text{if } Ab_i \neq Ag_i \\ 0 & \text{otherwise} \end{cases} \qquad (6.3)$$

If binary strings, termed *bitstrings*, $k \in \{0,1\}$, are used to represent the molecules, then one has a *binary Hamming shape-space*, or a *binary shape-space*. If ternary strings, i.e., $k = 3$, are used to represent the molecules, then one has a *ternary Hamming shape-space*, or *ternary shape-space*; and so on.

To illustrate the use of the Hamming distance to evaluate affinity, consider two arbitrary strings **Ag** = [1 0 0 0 1 1 1 0 1 0] and **Ab** = [1 0 1 0 1 0 1 0 1 0]. To perform a match between these strings means to compare them so as to evaluate their affinity, as illustrated in Figure 6.10. If affinity is being measured via the complementarity between the two strings, then their affinity is equal to their Hamming distance. Else, if similarity corresponds to affinity, then their affinity is given by $L - D(\mathbf{Ag},\mathbf{Ab})$, where $D(\mathbf{Ag},\mathbf{Ab})$ is the Hamming distance between **Ag** and **Ab**.

$$\textbf{Ag} = [1\,0\,0\,0\,1\,1\,1\,0\,1\,0]$$
$$\textbf{Ab} = [1\,0\,1\,0\,1\,0\,1\,0\,1\,0]$$

Match(**Ab**,**Ag**):   0 0 1 0 0 1 0 0 0 0

Complementarity: $D(\textbf{Ab},\textbf{Ag}) = \sum$ Match   (Affinity = 2)
Similarity: $L - D(\textbf{Ab},\textbf{Ag})$ (Affinity = 8)

Figure 6.10: Match between an antigen **Ag** and an antibody **Ab**, and their affinity assuming shape complementarity or similarity.

Two other affinity measures are broadly used in Hamming shape-spaces, namely, the so-called *r*-contiguous bits (*rcb*) and the *r*-chunks matching rules. In the *rcb* case, the number of contiguous matching bits (assuming the similarity between two strings is desired) determines the affinity between two strings. The match between two strings is performed and the affinity corresponds to the size of the longer sequence of matching (equal) bits. In *r*-chunks, only *r* contiguous positions are specified instead of specifying all *L* attributes of a string. An *r*-chunk detector can be thought of as a string of *r* bits together with its starting position within the string, known as its window (Esponda et al., 2004). Figure 6.11 illustrates the *r*-contiguous bit and the *r*-chunks affinity measures.

In Figure 6.11(a), if $r \le 4$, then the antigen is recognized by the antibody; otherwise no recognition takes place. The affinity is the number of contiguous bits, thus, 4. In Figure 6.11(b) an antigen is represented by a string of length $L = 5$ and three chunks of length $r = 3$ are presented. Each of these chunks could correspond, for instance, to portions of antibodies. Thus, for $r = 3$ any antibody containing one of the chunks illustrated would recognize the antigen.

In Symbolic shape-spaces, the attribute strings that represent the components of the artificial immune system are composed of at least one symbolic attribute. Figure 6.12 illustrates a Symbolic shape-space containing two antibodies that represent a database of trips.

One last concept has to be discussed before proceeding with the main immune algorithms, for the determination of the affinity between two attribute strings is not enough to indicate if the two molecules bind. For instance, one may want to qualify a recognition and binding between two molecules only in cases their affinity is greater than or equal to a given threshold ε, termed *affinity threshold*.

$$\textbf{Ag} = [1\,0\,0\,0\,1\,1\,1\,0\,1\,0]$$
$$\textbf{Ab} = [1\,0\,1\,0\,1\,0\,1\,0\,1\,0]$$

Match(**Ab**,**Ag**):   1 1 0 1 1 0 1 1 1 1

(a)

$$\textbf{Ag} = [1\,0\,1\,1\,0]$$
$$\textbf{d}[1] = [1\,0\,1]$$
$$\textbf{d}[2] = [0\,1\,1]$$
$$\textbf{d}[3] = [1\,1\,0]$$

(b)

Figure 6.11: (a) The *r*-contiguous bit matching rule. (b) The *r*-chunks matching rule.

	Description	Date	Flight	Country	From	To	Price (£)
Antibody (Ab$_1$):	Business	1996	212	Brazil	Campinas	Greece	546.78
Antibody (Ab$_2$):	Holiday	2000	312	U.K.	London	Paris	102.35
Antigen (Ag):	Holiday	2000	212	U.K.	London	Greece	546.78
Match Ag – Ab$_1$:	0	0	1	0	0	1	1
Match Ag – Ab$_2$:	1	1	0	1	1	0	0

Figure 6.12: Example of a Symbolic shape-space representation in which some attributes are symbolic and others are numeric. (Reproduced with permission from [de Castro and Timmis, 2002], © L. N. de Castro and J. Timmis.)

The affinity threshold is important because it allows artificial immune systems to incorporate the concept of cross-reactivity, and thus it is also called *cross-reactivity threshold*. To illustrate the role of the cross-reactivity threshold when the Hamming distance, $D = \sum$Match, is used to evaluate the affinity between the bitstrings proportionally to their degree of similarity $L - D$ (L is the bitstring length), consider the example of Figure 6.13. In this example, the affinity between **Ab** and **Ag**$_1$ is *aff*(**Ab**,**Ag**$_1$) = 8, and the affinity between **Ab** and **Ag**$_2$ is *aff*(**Ab**,**Ag**$_2$) = 6. If an affinity threshold $\varepsilon = 8$ is used, then the antibody **Ab** recognizes **Ag**$_1$, but does not recognize **Ag**$_2$.

$$\mathbf{Ab} = [1\ 0\ 1\ 0\ 1\ 0\ 1\ 0\ 1\ 0]$$
$$\mathbf{Ag}_1 = [1\ 0\ 0\ 0\ 1\ 1\ 1\ 0\ 1\ 0]$$
$$\mathbf{Ag}_2 = [0\ 1\ 1\ 1\ 1\ 0\ 1\ 1\ 1\ 0]$$

Match(**Ab**,**Ag**$_1$):   0 0 1 0 0 1 0 0 0 0
Match(**Ab**,**Ag**$_2$):   1 1 0 1 0 0 0 1 0 0

Affinity **Ab-Ag**$_1$: 8
Affinity **Ab-Ag**$_2$: 6

Figure 6.13: If the cross-reactivity threshold is $\varepsilon = 8$, then **Ag**$_1$ was recognized by **Ab**, while **Ag**$_2$ was not.

6.3.3. Immune Algorithms

The literature is rich with works using particular aspects and principles of the immune system to design new algorithms or improve existing techniques for problem solving. However, given a suitable representation for the immune cells and molecules, and how to evaluate their interactions, it is possible to identify some general-purpose immune algorithms. These algorithms can be separated into two classes: population-based and network-based. The first class involves all algorithms that do not take into account the immune network, and the network-based algorithms are all those inspired by the network theory of the immune system. Five main classes of algorithms will be reviewed here:

- *Bone marrow*: used to generate populations of immune cells and molecules to be used in AIS (and possibly in other approaches, such as genetic algorithms).

- *Negative selection*: used to define a set of detectors (e.g., attribute strings) to perform, mainly, anomaly detection.

- *Clonal selection*: used to generate repertoires of immune cells driven by antigens. It regulates the expansion, genetic variation, and selection of attribute strings.

- *Continuous immune network models*: used to simulate dynamic immune networks in continuous environments.

- *Discrete immune network models*: used to simulate dynamic immune networks in discrete environments.

6.4 BONE MARROW MODELS

The bone marrow is the site where all blood cells, including the immune cells, are generated. The simplest bone marrow model simply generates attribute strings of fixed length L using a (pseudo) random number generator. For real-valued shape-spaces, the interval in which a molecule m is going to be defined, e.g., $m \in [0,1]^L$, has to be chosen. In Hamming shape-spaces, the string that represents m is randomly generated from elements belonging to a predefined alphabet, e.g., $m \in \{0,1\}^L$ for binary strings (bitstrings). In the case of Integer shape-spaces, an algorithm to perform a random permutation of L elements can be used.

Some more complex and biologically plausible bone marrow models use gene libraries from which the immune cells and molecules are rearranged or evolved, as illustrated in Figure 6.14. In this picture, two libraries of genes are available and one gene from each library is chosen and concatenated with the others so as to form an antibody molecule. Assuming, for instance, that each gene has 6 bits, the final **Ab** molecule has a total length $L = 12$ bits.

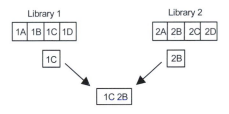

Figure 6.14: Generation of an antibody molecule by concatenating genes from different gene libraries. (Compare this figure with Figure 6.3.)

```
procedure [Ab] = gene_library(N)
   initialize L      //initialize all libraries
   for i from 1 to N do,           //N Abs will be generated
      for j from 1 to n do,        //for each library
            gene_index ← rand(L₁) //select a gene randomly
            Abᵢⱼ ← L(j,gene_index)//each gene of Abᵢ
      end for
   end for
end procedure
```

Algorithm 6.1: Procedure to generate antibodies from gene libraries.

One particular feature of bone marrow models based upon gene libraries is that not all genes present in the *genotype* (total collection of genes) are expressed in the *phenotype* (expressed antibody molecules). Thus, with a small finite genome it is possible to generate a very large number of different molecules. Besides, the use of specific genes in the libraries allows us to store previous and useful information in the gene segments, thus naturally introducing previous knowledge in the system. To illustrate this, take the idea of using an AIS to solve the traveling salesman problem. In this case, assume that you know some good partial routes in the TSP that you want to maintain or privilege in your system. These partial routes can be encoded in some gene segments to be used in the generation of initial candidate solutions to the AIS, or any other search strategy, including evolutionary algorithms.

Let L be the set of n libraries initialized either at random or using any predefined knowledge. The number of libraries is arbitrary and is usually defined by the problem at hand. As one gene is taken from each library and each gene has a length L_g, the number of libraries will be a function of the length desired for the molecules. For instance, if a molecule of length 24 is desired and each gene has a length $L_g = 8$, then three libraries are required. Algorithm 6.1 summarizes the procedure used to construct antibody molecules from gene libraries of length L_l assuming a number N of antibodies to be generated. Note that N antibodies are generated at random by concatenating one gene from each library; each gene has a length $L_g \geq 1$. The procedure returns a matrix of antibodies **Ab** of dimension $N \times (g.l)$, where g is the length of the genes and l is the number of libraries (assuming one gene is taken from each library).

6.4.1. Selected Applications from the Literature: A Brief Description

Although we suggested that more complex bone marrow models, such as the one that uses gene libraries to construct receptor molecules, can be used as means to naturally account for previous knowledge in artificial immune systems, not much has been done in this direction. Hart and Ross (1999) and Coello Coello et al. (2003) used gene libraries for scheduling problems, and Kim and Bentley (1999a) used concepts from gene libraries to generate detectors for network intrusion.

Most AIS use simple algorithms to generate the initial repertoires of molecules, and models based on gene libraries have basically been used to study the biological immune system and its theoretical models. So, this section briefly reviews two applications of gene libraries in contexts that are more related to biological modeling than to problem solving. The example of an application of gene libraries for solving scheduling problems is also briefly discussed. Some exercises in the end of the chapter explore the use of bone marrow models to generate candidate solutions for other problems.

Evolution of the Genetic Encoding of Antibodies

Hightower et al. (1995) and Perelson et al. (1996) used a genetic algorithm and a binary Hamming shape-space to study the effects of evolution on the genetic encoding for antibody molecules in the immune system.

The genetic material for one antibody molecule was stored in five separate gene libraries. Producing one antibody molecule begins with a random selection of a gene segment from each of the libraries. In their model, the bitstring representing the genotype of an individual was divided into four equal-size libraries of antibody segments (see Figure 6.14). Within each library there were eight elements, represented as bitstrings of length sixteen, so each individual genome had a total of 512 bits. The expressed antibodies had a length of 64 bits.

The experiments showed that the GA could optimize complex genetic information, being able to organize the structure of the antibody libraries. They also showed that the selection pressure operating on the phenotype as a whole could translate to selection pressure acting on individual genes, even though not all genes were expressed in the phenotype.

Antigenic Coverage and Evolution of Antibody Gene Libraries

Oprea and Forrest (1998) were intrigued by how the vertebrate immune system is capable of binding with any type of antigen, even if this antigen has never been encountered before. In order to better understand how this coverage is achieved, they used an evolutionary algorithm to explore the strategies that the antibody libraries may evolve in order to encode antigenic sets of various sizes. They derived a lower and an upper bound on the performance of the evolved antibody libraries as a function of their size and the length of the antigen string. They also provided some insights in the strategy of antibody libraries, and discussed the implications of their results for the biological evolution of antibody libraries.

Their work was based on that of Hightower et al. (1995) and also employed bitstrings to represent antibodies and antigens. They tested several different antibody repertoires with the same number of antibodies. The affinity between any two strings was proportional to the degree of similarity between two strings - length minus the Hamming distance - and each antigen is presented to all antibodies in the system. The fitness of an antibody is given by the average affinity over the whole set of antigens.

In their discussion, they suggested that (somatic) mutation allows the affinity maturation of antibodies that recognize antigens. Thus, there is an evolutionary trade-off between increasing the size of the antibody libraries and improving the efficiency of the somatic mutation process. From the perspective of AIS, this result is important for it suggests that affinity maturation can compensate for an inappropriate choice (or evolution) of antibody libraries.

Generating Antibodies for Job Shop Scheduling

Scheduling is a terminology used to refer to a class of planning problems in which a specific set of tasks has to be executed in a certain period of time subjected to limited resources and other types of constraints. In brief, scheduling is the problem of allocating resources through time (Rardin, 1998). *Job shop scheduling* (JSS) is a scheduling problem in which there is a production plant with different types of machines and a certain number of jobs to be completed, each one corresponding to a specific task. Solving a job shop scheduling problem corresponds to determining what tasks from the different jobs have to be executed in what order so as to minimize the production cost.

Coello Coello et al. (2003) used an artificial immune system based on gene libraries and clonal selection for solving the job shop scheduling problem. To use gene libraries to construct antibody molecules (candidate solutions) for solving the JSS problem, the authors employed a special representation. Each element of the library corresponds to the sequence of jobs processed by each of the machines. An antibody is thus an attribute string with the job sequence processed by each of the machines.

Their procedure works as follows. An antibody library (i.e., set of strings that encode different job sequences for each machine) is initially generated at random. An antigen, which is a candidate solution to the JSS problem (i.e., a sequence of jobs) is also generated randomly. Then, a single antibody is created by combining different segments from the library, in a way similar to that illustrated in Figure 6.14. This antibody is decoded such that one can read the job sequence for each machine and a local search algorithm is used to improve this potential solution. The solution encoded by the antibody is then compared with that of the antigen, and in case it is better it replaces the antigen. Note that the antigen is a sort of memory of the best solution found during the iterative process of adaptation. A clonal selection procedure (Section 6.6) is the applied and the best segments produced are used to update the antibody library.

6.5   NEGATIVE SELECTION ALGORITHMS

One of the main functions of the thymus is to promote the maturation of T-cells. Immature T-cells, generated in the bone marrow, migrate into the thymus where some of them differentiate into immunocompetent cells (cells capable of acting during an adaptive immune response), and others are purged from the repertoire due to a strong recognition of self. This process of eliminating cells whose receptors recognize self is known as *negative selection*. The thymic negative selection has to guarantee that the T-cell repertoire that leaves the thymus and goes to the periphery does not contain cells that recognize self cells and molecules. Thus, the immature T-cells that mature and leave the thymus become a sort of change or anomaly detectors.

The thymic negative selection has inspired the development of a negative selection algorithm with applications focusing, mainly, anomaly detection. There are two main versions of this algorithm: one for binary shape-spaces and another for real-valued shape spaces.

6.5.1. Binary Negative Selection Algorithm

Forrest et al. (1994) proposed a change detection algorithm inspired by the negative selection of T-cells within the thymus. This procedure was named *negative selection algorithm,* and its original application was in computational security, though many other applications were proposed. A single type of immune cell was modeled: T-cells represented as bitstrings of length L.

The algorithm is divided into two phases: a *censoring phase*, and a *monitoring phase*. In the censoring phase, given a known set of self patterns, named self-set **S**, the T-cell receptors will have to be tested for their capability of recognizing and binding with the self patterns. If the receptor on a T-cell (attribute string) recognizes a string from the self set, it is discarded; else it is selected as an immunocompetent cell and enters a set of *detectors*. The detectors are then used to monitor the system for anomalies, i.e., unusual patterns or behaviors.

The censoring phase of the negative selection algorithm is depicted in Figure 6.15(a) and can be summarized as follows:

1. *Initialization*: assuming a set **S** of self patterns (strings) to be protected, randomly generate attribute strings and place them in a set R_0.

2. *Affinity evaluation*: test if the strings that are being generated in R_0 match any of the strings in the self set **S** by determining their affinity.

3. *Generation of the detector set*: if the affinity of a string from R_0 with at least one string from **S** is greater than a given threshold, then this string recognizes a self pattern and has to be eliminated (negative selection); else the string is introduced into the detector set **R**.

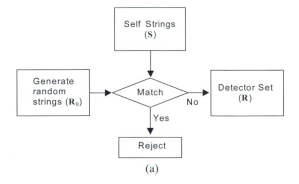

(a)

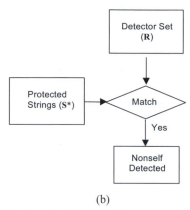

(b)

Figure 6.15: Negative selection algorithm. (a) Generation of valid detector set (censoring). (b) Monitoring of protected data.

After generating a set of detectors, the algorithm can be used to monitor changes in the self set or to track novelties. In a *monitoring* phase (Figure 6.15(b)), a set **S*** of protected strings is matched against the elements of the detector set. The set **S*** may be the set **S** with no change, a completely new set, or can be composed of elements of **S** perturbed with noise or added with new elements. If a recognition occurs between a protected string and a string from the detector set, then a nonself pattern (string) is detected. Figure 6.15(a) illustrates the censoring phase of the negative selection algorithm as introduced by Forrest et al. (1994).

```
procedure [R] = NSA_censor(S,ε,N,L)
    t ← 1
    while t <= N do, //N detectors will be generated
        r₀ ← rand(L) //randomly generate a string of length L
        flag ← 0
        for every s of S do, //for each element of S (self)
            aff ← affinity(r₀,s) //determine affinity
            if aff >= ε then
                flag ← 1; break
            end if
        end for
        if flag == 0,
            R ← add(R,r₀)    //add r₀ to the set of detectors R
            t ← t + 1
        end if
    end while
end procedure
```

Algorithm 6.2: Procedure to implement the censoring phase of the negative selection algorithm (NSA).

```
procedure [flag] = NSA_monitor(S*,R,ε)
    for every s of S* do,    //for all elements of S*
        for every r of R do,
            aff ← affinity(r,s)    //determine affinity
            if aff >= ε then
                flag ← nonself detected    //flag
            end if
        end for
    end for
end procedure
```

Algorithm 6.3: Procedure to implement the monitoring phase of the negative selection algorithm.

Let **S** be the self set, ε the affinity threshold (i.e., the minimum degree of match to qualify a recognition between two molecules), and N the total number of detectors of length L to be generated and placed in R_0. The censoring phase of the negative selection algorithm that returns a set **R** of detectors can be implemented using the pseudocode presented in Algorithm 6.2.

After a set of detectors has been generated, the system can be monitored for anomalies. Algorithm 6.3 describes a procedure to monitor anomalies. It receives as inputs the set of detectors **R**, a set of protected strings **S***, and the affinity threshold. Its output can be the patterns from **S*** that were detected as a nonself (anomaly), or simply a message (flag) indicating that nonself patterns were detected.

6.5.2. Real-Valued Negative Selection Algorithm

González and Dasgupta (2003) introduced a new negative selection algorithm based on a real-valued shape-space. The *real-valued negative selection algorithm* (RNS) is rooted in the same ideas as the original binary negative selection algorithm with the main difference that the self and nonself spaces are subsets of $\Re^L$, particularly $[0,1]^L$.

A detector (antibody) in the RNS algorithm is a hyper-sphere with radius ε in an L-dimensional real-valued shape-space $\Re^L$. The proposed matching between real-valued attribute strings (i.e., detectors, elements from the self set, and monitored data) is expressed as a membership function of the detectors **r** in relation to the elements $s \in S$, $\mu_r(s)$, and is directly proportional to their Euclidean distance $\|\cdot\|$ and inversely proportional to the radius of the detector (affinity threshold) ε, as follows

$$\mu_r(s) = \exp(-\|r-s\|^2/2\varepsilon^2) \tag{6.4}$$

Similarly to the binary version of the algorithm, the input is a set **S** of self patterns, but represented by real-valued vectors instead of binary strings. A number N of detectors is initially generated at random and updated so as to cover the nonself space by moving detectors away from self vectors and keeping detectors separated from each other in order to maximize the coverage of the nonself space. Note that the approach here is based on moving detectors about the space and varying their affinity threshold, instead of constantly generating random detectors and comparing them with the elements of the self set **S**.

In order to determine the match between a detector **r** and a self-point, the k-nearest neighbors from **S** in relation to **r** are determined; that is, a number k of points from **S** that present the smallest distance to **r** are selected. The median distance of these k neighbors is calculated and if this is less than the affinity threshold ε, then the detector is assumed to match self.

Each candidate detector that recognizes an element from the self set has an associated age that is increased at each iteration. When a candidate detector reaches a certain age T and has not been able to move away from the self set it is replaced by a new, randomly generated, candidate detector.

To ensure that the algorithm will converge to a stable state the authors proposed the use of a decaying adaptation rate $\eta(t)$ for the algorithm, where t is the time index

$$\eta(t) = \eta(0).\exp(-t/\tau) \tag{6.5}$$

where $\eta(0)$ is the initial value for $\eta(t)$ and τ is a parameter that controls its decay. The stopping criterion is a prespecified maximum number of iterations.

```
procedure [R] = RNS_censor(S,ε,N,L,T,η₀,k,max_it)
    t ← 1
    R ← rand(N,L)   //generate N detectors of length L
    while t <= max_it do,

        for every r of R do, //for each detector
                NearCells ← k-nearest neighbors of r in S
                NearestSelf ← median of ordered NearCells
                if distance(r,NearestSelf) < ε then
                    Δr = η(t)*(Σc∈NearCells(r-c))/|NearCells|
                    if age of r >  T
                        Replace r by a new random detector
                    else
                        Increase age of r
                        r ← r + Δr
                    end if
                else
                    age of r ← 0
                    Δr = η(t)*(Σr'∈Rμr(r')(r-r'))/Σr'∈Rμr(r')
                    r ← r + Δr
                end if
        end for
        t ← t + 1
    end while
end procedure
```

Algorithm 6.4: Procedure to implement the censoring phase of the real-valued negative selection algorithm (RNS).

Let **S** be the self set to be protected, ε the affinity threshold, N the number of detectors of length L, T the maturity age of the detectors, η_0 the initial value for the updating rate, k the number of neighbors for defining a self-match, and `max_it` the maximum number of iterations allowed. Algorithm 6.4 summarizes the real-valued negative selection algorithm as proposed by González and Dasgupta (2003). The equations used to move the detectors in the space are embodied in the procedure. After the detectors are generated, an anomaly can be detected by matching the monitored patterns, **S***, against the detectors. If an element from **S*** lies within the hypersphere of a detector; that is, if its distance to a detector is less than ε, then it is recognized as a nonself. Otherwise, it is a self.

6.5.3. Selected Applications from the Literature: A Brief Description

Negative selection algorithms are among the most well-known and applied immune algorithms. As they are usually used to generate a set of detectors to identify anomalies, their application areas have emphasized computer security against viruses and network intrusion detection. Besides, the application of artificial immune systems to computer security is very intuitive: "if we have an immune system capable of protecting us against viral infections, why don't we extract some of these ideas to develop an artificial immune system for computers?" Negative selection is the main algorithm used in this endeavor. This section reviews two applications of the negative selection algorithms to the problem of detecting anomaly. The first example describes the application of the binary negative selection algorithm to computer network traffic anomaly detection. The second example involves the use of RNS for cancer diagnosis.

Network Intrusion Detection

Hofmeyr and Forrest (2000) focused on the problem of protecting a local area broadcast network of computers (LAN) from network-based intrusions. The negative selection algorithm was the main immune principle employed in this application and will be the focus of this section. However, several other concepts were also incorporated such as co-stimulation, presentation of processed antigens by phagocytes, and the use of both B-cells and T-cells.

Broadcast LANs have the property that every computer in the network has access to each packet passing through the LAN. The aim of their artificial immune system for network security is to monitor the network traffic through the network connections. A connection is defined in terms of its *data-path triple*: the source internet protocol (IP) address, the destination IP address, and the service (port) by which the computers communicate (Figure 6.16). Each data-path triple corresponds to a network connection and is represented using a binary string of length $L = 49$ in a binary Hamming shape-space. Self is defined as the set of normally occurring connections observed over time on the LAN. By contrast, nonself consists of those connections not normally observed on the LAN, and are assumed not known *a priori*. In this case, the AIS is required to learn to discriminate between what is self (normally occurring connections) and what is nonself (intrusions).

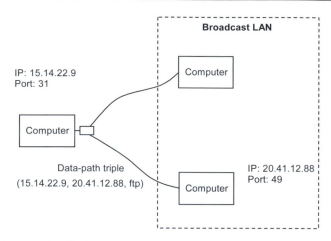

Figure 6.16: Example of a broadcast LAN and its connections. (Reproduced with permission from [de Castro and Timmis, 2002], © L. N. de Castro and J. Timmis.)

The detector set is generated based on the binary version of the negative selection algorithm as follows. Each binary string of length $L = 49$ representing a detector is created randomly and remains immature for a time period t, known as the tolerization period. During this period, the detector is exposed to the self strings, and if a recognition occurs, then the detector is eliminated. If it does not recognize any string during tolerization, it becomes a mature detector. Mature detectors need to exceed a cross-reactivity threshold (r-contiguous bits) in order to become activated; this increases its life span. Recognition is quantified by matching two strings using the r-contiguous bit rule.

These nonself detectors are constantly being matched against the bitstrings representing the data-path triples that correspond to the connections in the network. Each detector is allowed to accumulate matches through a *match count* (m), which itself decays over time. In order for a detector to become activated, it has to match at least δ strings within a given time period; this threshold δ was termed the *activation threshold*. Once a detector has been activated, its match count is reset to zero.

When a new packet (bitstring) enters the network, several different detectors are activated with distinct degrees of affinity (through the r-contiguous bit rule). The best matching detectors become memory detectors (memory cells). These memory detectors make copies of themselves that are then spread out to neighboring nodes in the network. Consequently, a representation of the memory detector is distributed throughout the network, accelerating future (secondary) responses to this specific nonself bitstring.

When a detector is activated, it sends an initial signal that an anomaly has been detected. As in the biological immune system, the detection of an anomaly is not enough to promote an immune response; a co-stimulatory signal is required from an accessory cell. The authors simulated the co-stimulatory (or second) signal by using a human operator, who is given a time period t_s to decide if the detected anomaly is really nonself. If the operator decides that this anomaly is indeed nonself, then a second signal is returned to the detector that identified it. This second signal increases the life span of this detector, maintaining it as a memory detector. The recognition of an anomaly in the absence of the second signal causes the death of a detector and its replacement by a new one. Figure 6.17 summarizes the life cycle of a detector.

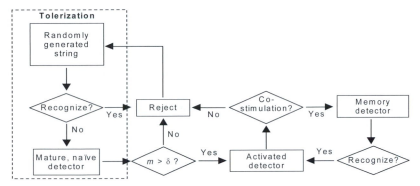

Figure 6.17: Life cycle of a detector. (Note that the negative selection algorithm is embodied in the tolerization period.) (Reproduced with permission from [de Castro and Timmis, 2002], © L. N. de Castro and J. Timmis.)

Breast Cancer Diagnosis

Among the many problems González and Dasgupta (2003) used to assess the performance of the real-valued negative selection algorithm is the breast cancer diagnosis dataset (Wolberg and Mangasarian, 1990), available at (Newman et al., 1998). This dataset was obtained by Dr. William H. Wolberg from the University of Wisconsin Hospitals and contains 699 samples. Each sample corresponds to different variations of 10 cell features (e.g., cell size, shape, etc.) and, based on these features, an assigned class label (benign or malignant). Thus, the data set is composed of 699 input patterns with 10 attributes each plus a class label. Among the 699 patterns, 16 had missing attributes and these were not used in the experiments. The data set was divided into two parts: one for training, with 271 benign samples; and one for testing, with 412 mixed benign and malignant samples.

In order to perform anomaly detection, the authors used a hybrid approach combining the real-valued negative selection algorithm with a supervised classification algorithm. In particular, they used a multi-layer feedforward network trained with the standard backpropagation algorithm (Section 4.4.4) as a classifier. The hybrid system is as follows. Using only the 271 benign samples, the RNS algorithm was employed to generate samples of anomalies (malignant samples), which altogether (sampled benign + generated malignant) served to train the MLP network. The MLP network was then responsible for classifying the new input data (**S***) into self (benign) or nonself (malignant).

When the goal is to assess the effectiveness of the anomaly detection process, two measures commonly used are the *detection rate* (*DR*) and the *false alarm rate* (*FAR*), which can be defined as follows:

$$DR = TP/(TP+FN) \qquad\qquad (6.6)$$

$$FAR = FP/(FP+TN) \qquad\qquad (6.7)$$

where *TP* is the number of *true positives* (anomaly identified as anomaly), *FP* is the number of *false positives* (normal identified as anomaly), *TN* is the number of *true negatives* (normal identified as normal), and *FN* is the number of *false negatives* (anomaly identified as normal).

400 samples of the malignant class were generated using RNS with $\varepsilon = 0.1$, $\eta_0 = 1$, $T = 5$ and $k = 1$. The RNS algorithm was allowed to run for 400 iterations, $\mathtt{max\_it} = 400$. A trade-off between the detection rate (*DR*) and the false alarm rate (*FAR*) was used to evaluate the performance of the algorithm.

6.6   CLONAL SELECTION AND AFFINITY MATURATION

The clonal selection principle is used to describe the basic features of an adaptive immune response to antigens. It establishes the idea that only those cells that recognize the antigens proliferate, thus being selected against those that do not. The selected cells are subject to an affinity maturation process, which improves their affinity to the selective antigens. Some features are important for the development of a clonal selection algorithm (Forrest et al., 1993; de Castro and Timmis, 2002):

- Antigens are typically encountered sequentially, and the immune system responds with a subset of its immune cells - those that come into contact with the antigen.

- An antigen selects several immune cells to proliferate (cloning). The proliferation rate of each immune cell is proportional to its affinity with the selective antigen: the higher the affinity, the higher the number of offspring generated; and vice-versa.

- The mutation suffered by each immune cell during reproduction is inversely proportional to the affinity of the cell receptor with the antigen: the higher the affinity, the smaller the mutation rate, and vice-versa.

Forrest and collaborators (Forrest et al., 1993) focused on the first feature listed - the locality of the antigenic presentation - and claimed that a genetic algorithm without crossover is a reasonable model of clonal selection, while the GA with crossover models genetic evolution. With an independent view, de Castro and Von Zuben (2002) focused on the affinity proportional reproduction and mutation of clonal selection and suggested that clonal selection is more than a GA without crossover, though it can be characterized as an evolutionary algorithm. Both algorithms can be applied to pattern recognition and will be reviewed separately. Other Clonal selection algorithms have been proposed in the artificial immune systems literature, such as CLIGA (Cutello and Nicosia, 2004) and the B-cell algorithm (Kelsey and Timmis, 2003), but these are not discussed here.

6.6.1. Forrest's Algorithm

Forrest et al. (1993) employed a binary Hamming shape-space to model clonal selection and affinity maturation. The goal was to study the pattern recognition and learning that takes place at the individual and species level of this artificial immune system. In their model, a population of antibodies was evolved in order to recognize a given population of antigens. A genetic algorithm was used to study the maintenance of diversity and generalization capability of their AIS, where generalization means the detection of common patterns shared among many antigens. The affinity function between an antibody and an antigen was evaluated by summing the number of complementary bits between their attribute strings (i.e., Hamming distance). This way, it was possible to study the ability of a GA in the detection of common patterns in the antigen population, and to discern and maintain a diverse population of antibodies.

As a result of their study about diversity in a model of the immune system, the authors proposed an immune algorithm that has been used in important applications of AIS, such as computer network security (Kim and Bentley, 1999b) and job-shop scheduling (Hart and Ross, 1999). The authors proposed the following procedure:

1. Build random populations of antigens and antibodies.
2. Match antibodies against antigens and determine their affinity.
3. Score the antibodies according to their affinity.
4. Evolve the antibody repertoire using a standard GA.

As can be observed from the procedure described above, the authors proposed to evolve a population of antibodies towards the antigens using a standard genetic algorithm, that is, a GA with crossover. However, as suggested by themselves, a GA without crossover is a reasonable model of clonal selection. Thus, they also performed some experiments with a GA without crossover, and claimed that similar performances were obtained when compared with a GA with crossover. However, to evolve a population of antibodies that maintains diversity under the GA, a fitness measure for the antibodies was proposed as follows:

1. Choose an antigen randomly.
2. Choose, without replacement, a random sample of μ antibodies.
3. Match each antibody against the selected antigen.
4. Add the match score of the highest affinity antibody to its fitness maintaining the fitness of all other antibodies fixed.
5. Repeat this process for many antigens (typically three times the number of antibodies).

The authors argued that this model allows the study of how the immune system learns which antibodies are useful for recognizing the given antigens. In addition, it was suggested that change of details such as the shape-space used and the sampling methods for the antibody molecules could create several interesting variations of this algorithm.

Let **S** be the set of *N* antigens to be recognized, `max_it` the maximum number of iterations for the GA, *pc* the crossover probability (if crossover is used), *pm* the mutation probability, and **P** the antibody repertoire. Algorithm 6.5 summarizes Forrest's clonal selection algorithm.

```
procedure [M] = CSA(S,N,max_it,pc,pm)
    initialize P
    gl ← 3*N                    //three times the number of Abs
    t ← 1
    while t <= max_it do, //for g iterations
      for j from 1 to gl do,
            sⱼ ← select_random(S) //select an Ag randomly
            Pμ ← select(P,μ)        //select μ Abs from P
            //best matching antibody w with match score m
            [w,m] ← affinity(Pμ,sⱼ)
            fitᵥ ← fitness(w,m)  //update the winner fitness
      end for
      P ← select(P,fit)
      P ← reproduce(P,fit,pc)
      P ← variate(P,fit,pm)
      j ← j + 1
    end while
end procedure
```

Algorithm 6.5: Procedure to implement Forrest's clonal selection algorithm. (Note that the last steps are those of an evolutionary algorithm.)

6.6.2. CLONALG

The clonal selection algorithm, named CLONALG, proposed by de Castro and Von Zuben (2002) takes into account the other two important features of clonal selection not accounted for by Forrest and collaborators; namely, affinity proportional selection and mutation. On the other side, CLONALG does not consider the locality of an immune response. CLONALG was initially proposed to perform pattern recognition and then adapted to solve multimodal function optimization tasks, as will be illustrated in the selected applications. Given a set of patterns to be recognized (**S**), the basic steps of the CLONALG algorithm are as follows (de Castro and Timmis, 2002):

1. *Initialization*: create an initial population of individuals (**P**).
2. *Antigenic presentation*: for each pattern from **S**, do:
   2.1. *Affinity evaluation*: present it to the population **P** and determine its affinity with each element of the population **P**.

2.2. *Clonal selection and expansion*: select n_1 highest affinity elements of **P** and generate clones of these individuals proportionally to their affinity with the antigen: the higher the affinity, the higher the number of copies, and vice-versa.

2.3. *Affinity maturation*: mutate all these copies with a rate inversely proportional to their affinity with the input pattern: the higher the affinity, the smaller the mutation rate, and vice-versa. Add these mutated individuals to the population **P** and reselect the best individual to be kept as the memory **m** of the antigen presented.

2.4. *Metadynamics*: replace a number n_2 of low affinity individuals by (randomly generated) new ones.

3. *Cycle*: repeat Step 2 until a certain stopping criterion is met.

```
procedure [M] = CLONALG_PR(S,max_it,n1,n2)
    initialize P                    //P ⊇ M
    t ← 1
    while t <= max_it do,  //for max_it iterations
        for every s of S do, //for all patterns (antigens)
            aff ← affinity(s,P)  //determine the affinity
            P1 ← select(P,n1,aff)//n1 highest affinity
            C ← clone(P1,aff)    //proportional to affinity
            C1 ← mutate(C,aff)   //inversely proportional
            aff ← affinity(s,C1) //affinity of clones with s
            //highest affinity is candidate to become memory
            m ← select(C1,aff)
            //m replaces the current memory cell for s if it
            //has a greater affinity
            M ← insert(M,m)
            //replace n2 low affinity cells
            P ← replace(P,n2)
        end for
        t ← t + 1
    end while
end procedure
```

Algorithm 6.6: The clonal selection algorithm, CLONALG, to pattern recognition. (There is one memory cell for each pattern to be recognized.)

Matrix $\mathbf{M} \subseteq \mathbf{P}$ with the difference that the elements of **M** are only replaced by new elements of higher affinity. This process, together with the affinity proportional mutation, promotes a greedy search of the affinity landscape.

Let **S** be the set of patterns (antigens) to be recognized, max_it the number of iterations to be performed, N the size of the population, n_1 the number of high affinity antibodies to be selected for cloning, and n_2 the number of low affinity antibodies to be replaced at each iteration ($n_1 + n_2 \leq N$). Algorithm 6.6 contains the pseudocode for implementing the clonal selection algorithm CLONALG to perform pattern recognition. The number of memory antibodies in this case is the same as the number of patterns to be recognized. The algorithm assumes one memory cell for each antigen, and a memory cell is only replaced by memory cells with higher affinity with a given antigen. It returns as output the set of memory cells **M**. The immune repertoire **P** is composed of the memory cells **M** plus a number of remaining cells **R**.

To apply the proposed CLONALG algorithm to optimization tasks, a few modifications have to be made in the algorithm as follows (de Castro and Von Zuben, 2002). **P** corresponds to a repertoire of antibodies (chromosomes in evolutionary algorithms), and $g(\cdot)$ is the objective function used to evaluate the quality (fitness) of each antibody.

- In Step 2, there is no explicit antigen population to be recognized, but an objective function $g(\cdot)$ to be optimized (maximized or minimized). This way, an antibody affinity corresponds to the evaluation of the objective function for the given antibody, i.e., it corresponds to the *fitness* of the antibody.

- As there is no specific antigen population to be recognized, the whole antibody population will compose the memory set and, hence, it is no longer necessary to maintain a separate memory set.

- Instead of selecting the single best antibody for each antigen, n antibodies are selected to compose the antibody set.

The CLONALG algorithm to solve optimization tasks can be described as follows:

1. *Initialization*: create an initial population of antibodies (**P**).
2. *Fitness evaluation*: determine the fitness of each element of **P**.
3. *Clonal selection and expansion*: select n_1 highest fitness elements of **P** and generate clones of these antibodies proportionally to their fitness: the higher the fitness, the higher the number of copies, and vice-versa.
4. *Affinity maturation*: mutate all these copies with a rate that is inversely proportional to their fitness: the higher the fitness, the smaller the mutation rate, and vice-versa. Add these mutated individuals to the population **P**.
5. *Metadynamics*: replace a number n_2 of low fitness individuals by (randomly generated) new ones.
6. *Cycle*: repeat Steps 2 to 5 until a certain stopping criterion is met.

If the optimization process aims at locating multiple optima within a single population of antibodies, then two parameters may assume default values. Assign $n_1 = N$, i.e., all antibodies from the population will be selected for reproduction; and the affinity proportionate cloning is not necessarily applicable, meaning that the number of clones generated for each of the N antibodies is assumed to be the same. This implies that each antibody will be viewed locally and have the same clone size (number of offspring) as the other ones, not privileging anyone for its affinity (fitness). The fitness will only be accounted for to determine the mutation rate of each antibody, which is still inversely proportional to the fitness.

```
procedure [P] = CLONALG_OPT(max_it,n1,n2)
    initialize P                        //P ⊇ P1
    t ← 1
    while t <= max_it do,               //for max_it iterations
        f ← eval(P)             //determine the fitness
        P1 ← select(P,n1,f)//select the n1 highest fitness
        C ← clone(P1,f)         //proportional to affinity
        C1 ← mutate(C,f)        //inversely proportional
        f1 ← eval(C1)           //fitness of the clones
        P1 ← select(C1,n1,f1)       //select the n1 best clones
        P ← replace(P,n2)    //replace n2 low fitness cells
        t ← t + 1
    end while
end procedure
```

Algorithm 6.7: The clonal selection algorithm, CLONALG, to optimization.

Following the same notation as before, Algorithm 6.7 summarizes the optimization version of the CLONALG algorithm. In this case, the subpopulation **P** could be viewed as the set of memory cells, but this is not explicitly accounted for in the algorithm.

6.6.3. Selected Applications from the Literature: A Brief Description

In order to illustrate the application of the CLONALG algorithm to perform learning and memory acquisition, it is going to be applied to a pattern recognition problem; the same problem used to illustrate the application of various artificial neural networks in Chapter 5. CLONALG is also going to be applied to a multimodal optimization task in order to evaluate its capability of locating and maintaining a stable population of multiple optima solutions.

Pattern Recognition

The goal of this example is to demonstrate that clonal selection together with affinity maturation can produce individuals with increasing affinities. In other words, this example shows that a blind search procedure is capable of extracting information from an unknown environment. Assume that the antigen population to be learned is represented by the benchmark set of eight binary characters we have been using in this book, as illustrated in Figure 6.18. Each character is represented by a bitstring of length $L = 120$ (the resolution of each picture is 12×10). The antibody repertoire is composed of $N = 10$ individuals, where $m = 8$ of them belong to the memory set **M**. The other running parameters were $max\_it = 500$, $n_1 = 5$, $n_2 = 0$ and $\beta = 10$, where β is a parameter that defines the number of clones to be generated for each antibody, as given by Equation (6.8):

$$Nc = (\beta.N)/i \qquad (6.8)$$

Figure 6.18: Set of antigens (input patterns) to be used to evaluate the CLONALG potential for pattern recognition.

where N is the total number of antibodies, and i is the index that determines the ranking of an antibody in relation to the others. Thus, the best matching (highest affinity) antibody from the repertoire is going to generate $\beta.N$ offspring, the second best matching antibody generates $(\beta.N)/2$, the third best antibody generates $(\beta.N)/3$, and so on. Equation (6.8) quantifies the affinity proportional reproduction of clonal selection.

Figure 6.19 illustrates the behavior of the pattern recognition version of CLONALG when applied to the problem depicted in Figure 6.18. Figure 6.19(a) depicts the initial memory set, and Figure 6.19(b) to (d) represents the maturation of the memory set (immune response) with time.

The affinity measure takes into account the length L of the strings minus the Hamming distance (D) between an antigen **Ag** and an antibody **Ab**, as illustrated in Figure 6.10. In this case, the smaller the distance, the lower the affinity, and vice-versa.

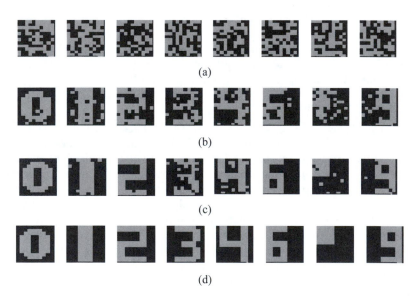

Figure 6.19: CLONALG applied to the pattern recognition task presented in Figure 6.18. (a) 0 iterations. (b) 100 iterations. (c) 200 iterations. (d) 500 iterations.

Multimodal Function Optimization

Consider the case of optimizing the function $g(x) = 2^{-2((x-0.1)/0.9)^2} \left(\sin(5\pi x)\right)^6$, studied in Chapter 3 and Chapter 4. Each antibody was represented using a binary string, with the same encoding scheme as the one adopted in Section 3.5.3. The running parameters adopted were `max_it` = 50, $n_1 = N = 50$, $n_2 = 0$, and $\beta = 0.1$, where $Nc = \beta.N$ defines the number of clones Nc to be generated for each antibody. For the parameters chosen, each antibody generates $Nc = \beta.N = 0.1 \times 50 = 5$ offspring. Figure 6.20 presents one simulation result obtained by CLONALG. Note that all the peaks of the function could be determined by the algorithm.

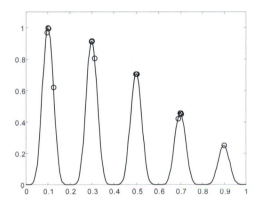

Figure 6.20: One simulation result of the optimization version of CLONALG when applied to function $g(x) = 2^{-2((x-0.1)/0.9)^2} \left(\sin(5\pi x)\right)^6$.

6.7 ARTIFICIAL IMMUNE NETWORKS

Following the network proposal of Jerne (1974), several network models were introduced by theoretical researchers interested in studying and modeling the immune system. The immune network is different from clonal selection in the sense that it assumes the immune system to be a dynamic system by its very nature. Under the network perspective, immune cells and molecules are capable of recognizing each other and thus presenting some pattern of dynamic behavior even if not in the presence of foreign stimulation.

With this dynamical view of the immune system, it is most natural that the pioneer models of the immune network theory were based upon Ordinary Differential Equations (ODEs) as means to quantify the dynamics of individual cells and molecules in the immune system. These *continuous* models were very important for the development of the field of AIS, and also served as basis for the proposal of *discrete* immune network algorithms. This section reviews both approaches separately. The focus will be on the specific continuous network model introduced by Farmer and collaborators (Farmer et al., 1986), and on a more generic discrete network model. The examples of application will, thus, focus on particular versions of both models.

The network approach is particularly interesting for the development of computational tools because it naturally provides an account of emergent properties such as learning, memory, self-tolerance, size and diversity of cell populations, and network interactions with the environment and the self components (de Castro and Timmis, 2002). What both models, continuous and discrete, have in common, is a general dynamics that can be summarized as follows:

$$RPV = NSt - NSu + INE - DUE \qquad (6.9)$$

where RPV is the rate of population variation, NSt is the network stimulation, NSu is the network suppression, INE is the influx of new elements, and DUE is the death of unstimulated elements. Note that the first two terms on the right hand side of Equation (6.9) are related to the immune network dynamics, and the last two terms correspond to the immune metadynamics.

The network activation embodies the stimulation of a cell or molecule by another cell or molecule of the immune system or a foreign antigen, while network suppression is due to self-recognition only.

6.7.1. Continuous Immune Networks

The paper by Farmer et al. (1986) is considered the pioneer work relating the immune system with other artificial intelligence approaches (de Castro and Timmis, 2002). They described a dynamical model for the immune system based upon the immune network theory. Their model consisted of a set of coupled differential equations to quantify the dynamics of the components of the immune network.

In the artificial immune system proposed by Farmer and collaborators, the cells and molecules are represented as variable-length binary strings in a Hamming shape-space and were allowed to interact in various different alignments. Let the *epitope* (*e*), also called *idiotope* (*i*), be the portion of the antibody molecule that can be recognized by another portion of the antibody attribute string termed *paratope* (*p*). Thus, an antibody is represented by a binary string with two parts, one corresponding to the paratope and another to the epitope, as illustrated in Figure 6.21. Epitopes and paratopes are mutually exclusive, in the sense that an epitope is not a paratope, and vice-versa.

The match between two bitstrings occurs in a complementary fashion in any possible alignment, accounting for the fact that two molecules may react in more than one way. Equation (6.10) specifies the *matrix of matching specificities* $\mathbf{M} = \{m_{ij}\}$, $\forall i,j$, containing the match value of each molecule (binary attribute string) with all the other molecules in the network; the first index of m_{ij} refers to the epitope and the second to the paratope. Matrix $\mathbf{M}$ is thus a function of the affinity (Hamming distance) among the network antibodies.

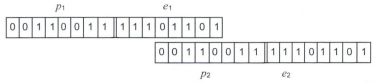

Figure 6.21: Bitstrings represent the epitope (or idiotope) and paratope of antibody molecules. (Reproduced with permission from [de Castro and Timmis, 2002], © L. N. de Castro and J. Timmis.)

$$m_{ij} = \sum_k G\!\left(D\!\left(\sum_n e_i(n+k), p_j(n) \right) - \varepsilon + 1 \right) \qquad (6.10)$$

where $p_j(n)$ is the n-th bit of the j-th paratope, $e_i(n)$ is the n-th bit of the i-th epitope, $D(\cdot,\cdot)$ corresponds to the Hamming distance between $e_i(\cdot)$ and $p_j(\cdot)$, given by Equation (6.3), and ε corresponds to the affinity or cross-reactivity threshold. A given alignment between a paratope and an epitope is specified by the parameter k. If a matching occurs in more than one alignment, its strength is summed to the strength of the other alignments, including the case of strings with different lengths.

Function $G(\cdot)$ measures the strength of a possible interaction between an epitope and a paratope: $G(x) = x$, if $x > 0$, and $G(x) = 0$, otherwise. For a given alignment k, G is zero if less than ε bits are complementary, while if ε or more bits are complementary, then $G = 1 + \delta$, where δ is the number of complementary bits in excess of ε.

The network dynamics takes into account *stimulation* and *suppression* events. Network stimulation is promoted by the recognition of an epitope by a paratope and results in antibody proliferation, while the antibody with the recognized paratope is suppressed and may be probabilistically eliminated from the repertoire. Assume N_1 antibody types with concentrations $\{c_1,...,c_{N_1}\}$, and N_2 antigens with concentrations $\{y_1,...,y_{N_2}\}$. The ODE that governs the dynamics of the antibody concentration, summarized in Equation (6.11), contains four parts: i) a first term corresponding to the stimulation of the paratope of an antibody type i by the epitope of an antibody type j; ii) a second term corresponding to the suppression of antibody type i when its epitope is recognized by the paratope of type j; iii) a third term that takes into account the concentration of antigens; and iv) a fourth term that models the natural death rate of antibodies.

$$\frac{dc_i}{dt} = k_1 \left[\sum_{j=1}^{N_1} m_{ji} c_i c_j - k_2 \sum_{j=1}^{N_1} m_{ij} c_i c_j + \sum_{j=1}^{N_2} m_{ji} c_i y_j \right] - k_3 c_i \qquad (6.11)$$

where parameter k_1 is a rate constant that depends on the number of collisions per unit time and the rate of antibody production stimulated by a collision; constant k_2 represents a possible inequality between network stimulation and suppression; and constant k_3 is the natural death rate.

Equation (6.11) describes how the concentration of antibodies varies with time. Antibodies that recognize antigens or other antibodies are stimulated and have their concentration increased, whereas antibodies that are recognized by other antibodies or do not recognize antigens are suppressed and have their concentration decreased. Very low concentration levels may lead to the death of a given cell type.

An important feature of this model is its metadynamics. That is, the repertoires of antibodies and antigens are dynamic, changing as new elements are added to or removed from the network. The continuous production of novel antibodies and their incorporation into the network provide the immune network with the ability to cope with unexpected (or unseen) antigens. To perform this, the authors used a minimum threshold on all concentrations, so that an element is eliminated when its concentration falls below the threshold. The new antibody types were generated by applying genetic operators to the bitstrings, such as crossover, inversion, and point mutations (Chapter 3).

6.7.2. Discrete Immune Networks

The main difference between the continuous immune network models and the discrete ones is that the former are based on the use of ordinary differential equations while the latter use either a difference equation or an iterative procedure to govern their dynamics. Most continuous models only account for changes in the concentration of antibodies in the network, such as the model described above. Discrete networks, however, commonly consider variations in the concentration of antibodies and in their structure (attributes) as well. This means that the elements of a discrete immune network can increase or decrease in number and can also change their attribute strings in order to improve their affinities with antigens. Finally, one last difference between them is that most continuous models were devised as means to simulate the immune network instead of as tools for problem solving. Most discrete immune networks, by contrast, are targeted at solving particular problems.

In the immune network theory introduced by Jerne (1974) the immune system is composed of a dynamic network whose behavior is perturbed by foreign antigens. Thus, there are basically two levels of interactions in the network: 1) the interaction with the foreign antigens, which correspond to the external environment; and 2) the interaction with other network elements, which represents the internal environment of the organism. As the foreign stimulation disturbs the network, it has to self-organize so as to cope with the perturbation. After the network self-organization, specific groups of cells and molecules correspond to specific antigenic groups.

Although there are several proposals of artificial immune networks in the literature, this section describes a single one. The description that follows is based on the discrete immune network proposed by de Castro and Von Zuben (2001), named aiNet (Artificial Immune Network). Other proposals that the students are strongly encouraged to look at are available in Timmis et al. (2000), Neal (2003), and in the comparative work performed by Galeano et al. (2005).

In aiNet, it is assumed a population **S** of antigens to be recognized and a randomly initialized set **P** of cells in the network. Each network element corresponds to an antibody molecule; there is no distinction between the antibody and the B-cell. A real-valued vector in an Euclidean shape-space is used to represent antibodies and antigens. Each antigenic pattern is presented to each network cell and their affinity is determined using the Euclidean distance (Equation (6.1)). A number of high affinity antibodies is selected and reproduced (*clonal expansion*) based on their affinity with the antigen: the higher the affinity, the higher the number of clones and vice-versa. The clones generated undergo somatic mutation inversely proportional to their antigenic affinity: the higher the affinity, the lower the mutation rate. A number of high affinity clones is selected to be maintained in the network, constituting what is defined as a clonal memory.

The affinity among the remaining antibodies is determined. Those antibodies whose affinity is less than a given threshold are eliminated from the network in a process known as *clonal suppression*. All antibodies whose affinity with the antigen is less than a given threshold are also eliminated from the network (*natural death rate*). Additionally, a number of new randomly generated antibodies are incorporated into the network (*metadynamics*). The remaining antibodies are incorporated into the network and their affinity with the existing antibodies is determined. All but one antibody whose affinity is less than a given threshold are eliminated.

The aiNet learning algorithm can be summarized as follows (de Castro and Timmis, 2002):

1. *Initialization*: create an initial random population **P** of network antibodies.
2. *Antigenic presentation*: for each antigenic pattern **s**, do:
   - 2.1 *Clonal selection and expansion*: for each network element, determine its affinity with the antigen presented. Select a number n_1 of high affinity elements and reproduce (clone) them proportionally to their affinity.
   - 2.2 *Affinity maturation*: mutate each clone inversely proportional to affinity. Re-select a number n_2 of highest affinity clones and place them into a clonal memory set.
   - 2.3 *Metadynamics*: eliminate all memory clones whose affinity with the antigen is less than the affinity (cross-reactivity) threshold ε.
   - 2.4 *Clonal interactions*: determine the network interactions (affinity) among all the elements of the clonal memory set.
   - 2.5 *Clonal suppression*: eliminate those memory clones whose affinity with each other is less than a prespecified suppression threshold σ_s.
   - 2.6 *Network construction*: incorporate the remaining clones of the clonal memory into the network.
3. *Network interactions*: determine the similarity between each pair of network antibodies.
4. *Network suppression*: eliminate all network antibodies whose affinity with each other is less than a prespecified suppression threshold σ_s.

5. *Diversity*: introduce a number n_3 of new randomly generated antibodies into the network.
6. *Cycle*: repeat Steps 2 to 5 until a prespecified number of iterations (max_it) is reached.

```
procedure [M] = aiNet(S,max_it,n₁,n₂,n₃,σₛ,ε)
    initialize P
    t ← 1
    while t <= max_it do,              //for max_it iterations
        vet_permut ← randperm(N)  //permutations of N
        for j from 1 to M do,          //S contains M antigens
            //select the index i of antigen sᵢ to be presented
            i ← vet_permut(j)           //present antigens randomly
            aff ← affinity(sⱼ,P)
            P₁ ← select(P,n₁)          //n₁ highest affinity
            C ← clone(P₁,aff)           //proportional to affinity
            C₁ ← mutate(C,aff)         //inversely proportional
            aff ← affinity(sⱼ,C₁)//affinity of clones with s
            Ms ← select(C₁,n₂)         //clonal memory for s
            for each m of Ms do,
                //death of unstimulated cells (metadynamics)
                if aff(Ms) > ε then,
                    remove m from Ms
                end if
                //clonal suppression
                clonal_affₘ ← affinity(m,Ms)
                if clonal_affₘ < σₛ then,
                    remove m from Ms
                end if
            end for
            //network construction
            M ← insert(P,Ms)       //introduce Ms in M
        end for
        //network suppression
        for each m of M do,
            affₘ ← affinity(M,m)   //network interactions
            if aff(M) < σₛ then,      //network suppression
                remove m from M
            end if
        end for
        R ← rand(n₃)                   //n₃ new random Abs
        P ← insert(M,R)                //diversity introduction
        t ← t + 1
    end while
end procedure
```

Algorithm 6.8: The aiNet adaptation procedure.

Let **S** be the set of M patterns (antigens) to be recognized, `max_it` the number of iterations to be performed, n_1 the number of high affinity antibodies to be selected for reproduction, n_2 the number of high affinity antibodies to become clonal memory, n_3 the number of new antibodies to be introduced into the population, σ_s the suppression threshold, and ε the cross-reactivity threshold. The suppression threshold is responsible for eliminating antibodies that are too similar to other antibodies in the network, and the cross-reactivity threshold eliminates all antibodies that do not have a minimal degree of affinity with the antigenic pattern presented. Algorithm 6.8 summarizes the adaptation procedure of the aiNet algorithm. It returns a matrix **M** of memory antibodies.

6.7.3. Selected Applications from the Literature: A Brief Description

This section illustrates the use of both types of immune networks: continuous and discrete. The potential of application of a continuous immune network model is illustrated by listing the main features of an AIS system to recommend videos, and the discrete network model, whose pseudocode was described in Algorithm 6.8, will be used to perform data compression and clustering.

A Recommender System

Based on information about the user's profile, a *recommender* or *recommendation system* is the one designed to predict items (e.g., books, movies, music, etc.) that this user may be interested in. Recommender systems based on *collaborative filtering* methods make predictions about the interests of (and recommendations to) a user by gathering information from many similar users, sometimes called neighbors.

Cayzer and Aickelin (2005) used a modified version of Farmer's continuous immune network model to the task of movie recommendation by collaborative filtering. The users were encoded as a set of 2-tuples, $user = \{\{id_1, score_1\}, \{id_2, score_2\}, \dots, \{id_n, score_n\}\}$, where id_i is the identifier of movie i, $score_i$ is the score given by the user to movie i, and n is the number of movies for recommendation. The main aspects to be decided in their proposal were: i) how to select the neighborhood of a given user; and ii) how to predict the rating of a movie by this user.

The authors chose the Pearson correlation coefficient to compare two users a and e, which is basically the dot product between the two normalized vectors:

$$p_{ae} = \frac{\sum_{i=1}^{N}(a_i - \bar{a})(e_i - \bar{e})}{\sqrt{\sum_{i=1}^{N}(a_i - \bar{a})^2}\sqrt{\sum_{i=1}^{N}(e_i - \bar{e})^2}} \tag{6.12}$$

where a_i is the vote of user a for movie i, e_i is the vote of user e to movie i, $\bar{a}$ is the average vote of user a over all movies including the overlapping votes, $\bar{e}$ is the average vote of user e over all movies including the overlapping votes, and N is the number of overlapping votes (i.e., movies for which both a and e have voted). In case there is no overlapping vote or the denominator of Equation (6.12) is zero, $p_{ae} = 0$. Their neighborhood selection scheme corresponded to choosing the best k absolute correlation scores, where k is the neighborhood size.

For the immune network predictor, a user for whom to make a prediction was encoded as the antigen **Ag** and a network of antibodies was initialized (the other users were the antibodies). At each step (iteration) an antibody's concentration was increased in a direct proportion to its matching score to the antigen (measured via the Pearson correlation coefficient), decreased in a direct proportional to its match in relation to other antibodies in the network, and decreased due to aging. To account only for antigenic stimulation, antibody suppression and aging, the authors simplified Equation (6.11), turning it into

$$\frac{dx_i}{dt} = k_1 p_i x_i y_j - \frac{k_k}{n} \sum_{j=1}^{n} p_{ji} x_i x_j - k_3 x_i \qquad (6.13)$$

where k_1, k_2 and k_3 are constants that rate antigenic stimulation, network suppression and natural death rate, respectively. Parameter p_i is the correlation between antibody i and the single antigen (user for whom to make the prediction), x_i is the concentration of antibody i, y is the concentration of the antigen, p_{ij} is the correlation between antibodies i and j, and n is the number of antibodies in the network. Note that the authors only use network suppression; no network stimulation is accounted for.

The rating r_i of a given movie i by the user a is determined by employing a weighted average of the neighborhood $k = N$ of the user, taken to be the whole network of antibodies. Let w_{ae} be the weight between users a and e, p_{ae} be the correlation score between a and e, and x_e be the concentration of the antibody corresponding to the user e, then

$$w_{ae} = p_{ae} x_e$$

$$r_i = \bar{a} + \frac{\sum_{e \in N} w_{ae}(e_i - \bar{e})}{\sum_{e \in N} w_{ae}}$$

The authors performed a number of experiments on a public dataset with their AIS, taking and not taking into account the immune network effects. As a comparative approach they used the simple Pearson (SP) predictor (Herlocker et al., 1999). The results showed that the AIS without network interactions performed similarly to the SP method in terms of prediction, but presented superior recommendation accuracy when network interactions were added.

Data Compression and Clustering

To illustrate the performance of the aiNet when applied to data compression and clustering, consider the two-spirals (SPIR) benchmark problem illustrated in Figure 6.22(a). This data set is composed of 190 samples in $\Re^2$ and aims at testing the aiNet capability to detect non-linearly separable clusters. Note that though the samples are labeled in the picture they are unlabeled for the aiNet. The following parameters were adopted in this simulation: $g = 40$, $n_1 = 4$, $n_2 = n_3 = 10$, $\alpha = 0.07$, and $\beta = 1.0$.

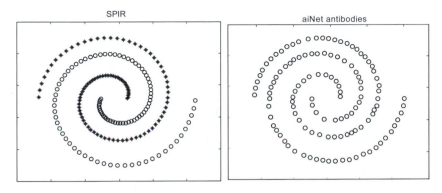

Figure 6.22: aiNet applied to the two spirals (SPIR) problem. (a) Input patterns. (b) Antibodies generated by the aiNet adaptation procedure presented in Algorithm 6.8.

Figure 6.22(b) depicts the final set of antibodies generated by the aiNet. In this case, the resultant memory matrix was composed of $m = 121$ antibodies, corresponding to a $CR = 36.32\%$ reduction in the size of the sample set. Note that the compression was superior in regions where the number of data (amount of redundancy) is larger, i.e., the centers of the spirals.

The next important question that remains is related to how to define the clusters of this network. The simplest answer to this question is to connect all cells in the network and remove those connections greater than a given threshold. Several authors still use this idea. Another simple, but more clever form of separating clusters of a trained aiNet, is to use the concept of the *minimal spanning tree* (MST). The MST of the resultant network antibodies is built and a criterion to eliminate some edges of the MST is used in order to define some clusters in the network (Appendix B.4.4).

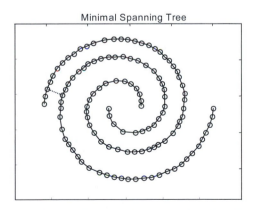

Figure 6.23: Minimal spanning tree for the antibodies produced by the aiNet adaptation procedure.

Building the MST and Identifying Clusters

For the two spirals problem under study, the resultant MST is illustrated in Figure 6.23. Assuming a factor $r = 2$, the MST is pruned in the dashed edge and the two clusters corresponding to the clusters observed in the original data set are identified.

6.8 FROM NATURAL TO ARTIFICIAL IMMUNE SYSTEMS

Table 6.1 summarizes a possible interpretation of (theoretical) immunology into the domain of artificial immune systems. For instance, cells and molecules are usually represented by attribute strings in a specific shape-space, affinity corresponds to the degree of match between strings or recognition between molecules, and so on, as summarized in Table 6.1.

Artificial immune systems borrow several ideas from the biological immune system. Most algorithms to date are inspired by some sort of immune principle, process or theoretical model. In some cases, such as with clonal selection algorithms, it is evident that the immune algorithms are either hybrids with evolutionary algorithms or can be viewed as such. However, the immune algorithms still have some particular features, such as distributivity, double plasticity, anomaly detection and diversity, that give them their identity and, more importantly, that perhaps would never had been introduced without insights from immunology.

Table 6.1: Interpretation from the immunological terminology into the computational domain of AIS.

Biology (Immunology)	Artificial Immune Systems
Cells and molecules	Data structure (attribute strings)
Affinity	Degree of match (interaction) between data structures
Fitness	Quality of a data structure for tasks that do not involve quantifying the degree of interaction between data structures
Bone-marrow models	Used to generate data structures
Affinity function	Quantify affinities
Somatic hypermutation	Used to introduce or maintain population diversity or genetic variation
Affinity maturation	Promotes learning (adaptation) through somatic hypermutation and selection
Clonal selection	Describes how the immune cells and molecules interact with antigens
Negative selection	Generates set(s) of nonself detectors for anomaly detection
Immune network	Performs the dynamics and metadynamics of the system structured in a network-like manner

6.9   SCOPE OF ARTIFICIAL IMMUNE SYSTEMS

One of the most remarkable features of the field of artificial immune systems is its broad applicability. There are AIS applied to domains such as:

- Machine-learning and pattern recognition.
- Anomaly detection and the security of information systems.
- Data analysis (knowledge discovery in databases, clustering, etc.).
- Agent-based systems.
- Scheduling.
- Autonomous navigation and control.
- Search and optimization.
- Artificial life.

There are a few survey papers, reports, and book chapters that can be used by the reader to find information on where and when AIS have been developed and applied. Chapter 4 and Chapter 6 of the book by L. N. de Castro and J. Timmis (2002) survey a vast amount of works on AIS. Y. Ishida (1996), D. Dasgupta and Attoh-Okine (1997, 1999), L. N. de Castro and F. Von Zuben (2000), Dasgupta et al. (2001), L. N. de Castro (2004), and U. Aickelin et al. (2004) all have published survey papers, reports or bibliographies of artificial immune systems and their applications.

6.10   SUMMARY

In the previous chapter we have seen that the cells (neurons) composing the nervous system are connected with one another via small junctions called synapses. The modulation of these synapses is believed to be the basic means by which an ensemble of nerve cells, called neural network, adapts and learns. The nervous system is known to be structured in a network-like manner. In the case of the immune system, however, three main perspectives can be found in the literature (de Castro, 2003). The first one assumes an immune system composed of discrete sets of cells and molecules targeted at performing self/nonself discrimination. The second one suggests an immune system composed of interconnected sets of cells and molecules, resulting in a network of affinities within the immune system. The last perspective is that the immune system works in concert with many other bodily systems, mainly the nervous and endocrine systems, in order to main *homeostasis* (i.e., the body's internal equilibrium state).

One important feature of a great number of artificial immune systems is the presence of adaptability in two levels: structural adaptability and parametric adaptability. In the first level, the architecture of the system is allowed to adapt to the environment, and in the second level, the parameters or attributes are also varied with time. This double-plasticity leads to high degrees of flexibility and robustness of AIS.

This chapter reviewed a framework to design artificial immune systems, including how to create abstract models of immune cells and molecules, quantifying their interactions, and using adaptation procedures. It became clear that there are much more to artificial immune systems than this single chapter could have reviewed. Other algorithms are available and some of these have been discussed in the literature cited throughout the chapter.

6.11   EXERCISES

6.11.1. Questions

1.   What is the difference between a vaccine and a serum?

2.   When the immune system mistakenly recognizes self as being nonself, it mounts an immune response against self, thus causing the so-called auto-immune diseases. Name three examples of auto-immune diseases and explain their main causes.

3.   If the immune system is capable of differentiating between self and nonself, why doesn't it reject the fetuses?

4.   It is known that there must be some compatibility between the mother's and the father's blood type. What is the relationship between blood type compatibility in parents and the immune system?

5.   The self/nonself discrimination issue occurs differently for B-cells and T-cells. Explain the positive selection of B-cells and T-cells, and the negative selection of B-cells and T-cells. How are these concepts related with the immune tolerance?

6.   Allergies are known to be related to a mal-functioning of the immune system. Explain what allergies are and why they are caused.

6.11.2. Computational Exercises

1.   Use a bone marrow algorithm to define genes for gene libraries to be used to generate the initial population of a genetic algorithm to solve the TSP problem presented in Chapter 3 and Chapter 5 (Figure 6.24). Assume the following structure for the gene libraries:

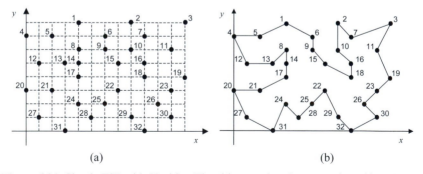

(a)   (b)

Figure 6.24: Simple TSP with 32 cities. The cities are placed on a regular grid on the *x-y* plane, in which each point (•) represents a city, the number next to each city is its index, and each square on the grid corresponds to one unit of distance (uod - e.g., Km).

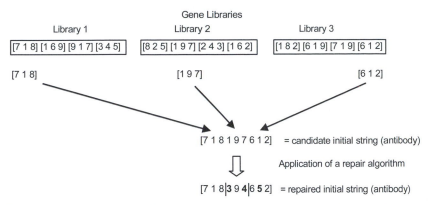

Figure 6.25: Generation of initial antibodies (chromosomes) and application of a repair algorithm.

Gene length $L_g = 4$, number of libraries $n = 8$, and library length (number of genes in each library) $L_l = 4$. As one gene from each library will be selected, the total chromosome length is $L = L_g \times n = 4 \times 8 = 32$, that corresponds to the number of cities in a tour.

Each gene will be defined as a sequence of four cities known to be part of an optimal route. For example, gene one from library A, g_{1A}, can be defined as $g_{1A} = [19\ 23\ 26\ 30]$, gene two from library A, g_{2A}, can be defined as $g_{2A} = [32\ 29\ 22\ 25]$, and so on.

It is important to note that in some (maybe most) cases, one or more elements of a gene will be repeated and the index of some cities will not appear in the final chromosome generated. Thus, a repair algorithm must be used in order to generate permutations of L integers while maintaining most of the genes intact. This is because the TSP problem has the constraint that no city can be visited more than once. To illustrate this problem and one possible way of solving it, consider the example presented in Figure 6.25.

In the example of Figure 6.25, the concatenation of genes from the libraries generated a chromosome (route) with some cities repeated. To fix this candidate string, a repair algorithm that seeks for repeating and missing numbers is used, and when a repetition occurs, this repeated number is exchanged by one number from the sequence $[1..L]$ that has not appeared yet.

Implement this bone marrow model to define an initial population of chromosomes to be used in an evolutionary algorithm to solve the TSP problem illustrated. Compare the performance of the algorithm with this type of initialization procedure and with the random initialization used in Project 1, Section 3.10.4.

2. The bone marrow model using gene libraries, described in Section 6.4, can also be used as a recombination scheme in evolutionary algorithms in place of the standard crossover operation, as follows.

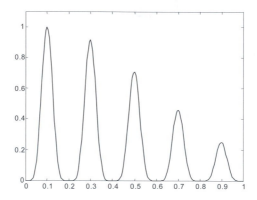

Figure 6.26: Graph of the function $g(x) = 2^{-2((x-0.1)/0.9)^2}(\sin(5\pi x))^6$ to be maximized.

At each generation, create libraries by taking sequences (attribute strings) of each individual chromosome from the population as genes to compose the libraries. The chromosomes from which the sequences of bits are going to be taken can be chosen using heuristics, such as their fitness value. If an individual from the GA is selected for recombination with a given probability pc, then this individual (or part of it) can be replaced by another one generated using the gene libraries.

Modify the standard genetic algorithm with binary representation to support the recombination scheme based on gene libraries, apply it to the function optimization problem described in Section 3.3.3 and discuss the results obtained.

Function to be optimized: $g(x) = 2^{-2((x-0.1)/0.9)^2}(\sin(5\pi x))^6$ (see Figure 6.26)

The problem corresponds to finding $\mathbf{x}$ from the range [0..1] which maximizes $g(\mathbf{x})$, i.e., to find $\mathbf{x}^*$ such that $g(\mathbf{x}^*) \geq g(\mathbf{x})$, $\forall\ \mathbf{x} \in [0..1]$.

3.   Given the data set illustrated in Figure 6.27, use the negative selection algorithm to generate a set of $N = 1,000$ detectors that recognize everything but the self patterns. Assume a cross-reactivity threshold $\varepsilon = 72$, corresponding to 60% of the length of each pattern. The affinity measure is given by *Affinity* $= L - D$, where L is the length of the strings ($L = 120$), and D is the Hamming distance between two strings (Equation (6.3)).

Figure 6.27: Set of input patterns (self set **S**).

Evaluate the set of detectors generated by randomly introducing noise in the samples, with various noise levels from 5% to 50%, and monitoring for the presence of unknown patterns. (Patterns with more than 40% of noise should be detected as nonself.)

4. Use the Animals data set, depicted in Table 6.2, as a self set to generate 1000 detectors using the binary negative selection algorithm.

   Use the nonself set depicted in Table 6.3 to calculate the detection rate (DR) and the false alarm rate (FAR), described in Equation (6.6) and Equation (6.7), respectively, of the 1000 detectors generated.

   Investigate the influence of different affinity measures (Hamming, r-contiguous and r-chunks) and affinity thresholds ε for the binary negative selection algorithm.

5. Apply the optimization version of CLONALG to the function optimization problem presented in Exercise 2. Assume both cases: a binary Hamming shape-space and a real-valued shape-space with Gaussian mutation (Section 3.3.3).

6. Implement the clonal selection algorithm CLONALG and apply it to the pattern recognition task of Section 6.6.2.

7. Apply Forrest's clonal selection algorithm (without crossover) to the problem above. Compare the performance of both CLONALG and Forrest's algorithm when applied to pattern recognition.

Table 6.2: Set of animals (names and their attributes) to be used to generate the detectors.

		Dove	Hen	Duck	Goose	Owl	Hawk	Eagle	Fox	Dog	Wolf	Cat	Tiger	Lion	Horse	Zebra	Cow
Is	Small	1	1	1	1	1	1	0	0	0	0	1	0	0	0	0	0
	Medium	0	0	0	0	0	0	1	1	1	1	0	0	0	0	0	0
	Big	0	0	0	0	0	0	0	0	0	0	0	1	1	1	1	1
Has	Two legs	1	1	1	1	1	1	1	0	0	0	0	0	0	0	0	0
	Four legs	0	0	0	0	0	0	0	1	1	1	1	1	1	1	1	1
	Hair	0	0	0	0	0	0	0	1	1	1	1	1	1	1	1	1
	Hooves	0	0	0	0	0	0	0	0	0	0	0	0	0	1	1	1
	Mane	0	0	0	0	0	0	0	0	0	1	0	0	1	1	1	0
	Feathers	1	1	1	1	1	1	1	0	0	0	0	0	0	0	0	0
Likes to	Hunt	0	0	0	0	1	1	1	1	0	1	1	1	1	0	0	0
	Run	0	0	0	0	0	0	0	0	1	1	0	1	1	1	1	0
	Fly	1	0	0	1	1	1	1	0	0	0	0	0	0	0	0	0
	Swim	0	0	1	1	0	0	0	0	0	0	0	0	0	0	0	0

Table 6.3: Set of nonself animals to evaluate detectors.

		Swan	Penguin	Pterodactyls	Bear	Sheep	Pig	Tyrannosaurus	Snail	Snake	Shark	Dolphin	Eagle	Rooster
Is	Small	0	0	0	0	0	0	0	1	1	0	0	0	1
	Medium	0	1	0	0	1	1	0	0	0	0	0	1	0
	Big	1	0	1	1	0	0	1	0	0	1	1	0	0
Has	Two legs	1	1	1	0	0	0	0	0	0	0	0	1	1
	Four legs	0	0	0	1	1	1	1	0	0	0	0	0	0
	Hair	0	0	1	1	0	1	1	0	0	0	0	0	0
	Hooves	0	0	0	0	1	1	0	0	0	0	0	0	0
	Mane	0	0	0	0	0	0	0	0	0	0	0	0	0
	Feathers	1	1	0	0	0	0	0	0	0	0	0	1	1
Likes to	Hunt	0	1	1	1	0	0	1	0	1	1	0	1	0
	Run	1	0	1	1	1	1	1	0	0	0	0	0	0
	Fly	0	0	1	0	0	0	0	0	0	0	0	1	0
	Swim	1	1	0	1	0	0	0	1	1	1	1	0	0

8. Apply the aiNet algorithm to the data set presented in Section 5.6.2 (Exercise 3). Make use of the MST edge inconsistency criterion in order to separate the network clusters and thus identify the clusters of the original data set. Use the same parameters as those used to solve the SPIR problem, including the factor $r = 2$. Compare the performance of both algorithms: ACA and aiNet.

9. Apply the aiNet to solve the chainlink problem illustrated in Figure 6.28. This data set is composed of 1000 data points in the $\Re^3$-space, arranged such that they form the shape of two intertwined 3-D rings, of whom one is extended along the *x-y* direction and the other one along the *x-z* direction. The two rings can be thought of as two links of a chain with each one consisting of 500 data points. The data is provided by a random number generator whose values are inside two toroids with radius R = 1.0 and r = 0.1.

The following parameters can be used to solve this problem with aiNet: $g = 40$, $n_1 = 4$, $n_2 = n_3 = 10$, $\alpha = 0.15$, and $\beta = 1.0$. Note that α is the only parameter different from those used in the SPIR problem.

6.11.3. Thought Exercises

1. How would you compare philosophically the clonal selection algorithms with the evolutionary algorithms? Hint: look at the basic steps and procedures of both theories.

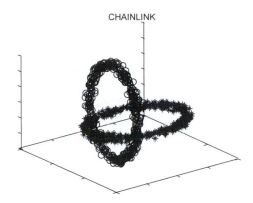

Figure 6.28: CHAINLINK problem with 1000 patterns.

2. How would you compare the generation of antibodies from gene libraries with a crossover procedure discussed in evolutionary algorithms?

3. Can you use a bone marrow model based upon gene libraries to initialize weight vectors for neural networks? If yes, how and what would be the benefits? If not, justify your answer.

4. What are the main differences between immune and neural networks? Tip: look at their design principles.

6.11.4. Projects and Challenges

1. Extend the clonal selection algorithm presented in Algorithm 6.6 to incorporate an affinity threshold ε. Write a pseudocode for the modified algorithm and test it with the data set illustrated in Figure 6.29. This data set is composed of 10 characters with resolution 20×20.

   The incorporation of a cross-reactivity threshold in the affinity measure of CLONALG leads to the possibility of having one single memory cell to represent groups of similar antigen patterns, and thus incorporates explicitly the notion of generalization in the AIS.

2. Implement the real-valued negative selection algorithm together with a multi-layer perceptron network to solve the breast cancer diagnosis experiment discussed in Section 6.5.3.

Figure 6.29: Data set to be used to test the generalization capability (cross-reactivity) of the extended CLONALG algorithm.

6.12   REFERENCES

[1]   Aickelin, U., Greensmith, J. and Twycross, J. (2004), "Immune System Approaches to Intrusion Detection - A Review", In G. Nicosia, V. Cutello, P.J. Bentley and J. Timmis, (Eds.), *Proceedings of the 3rd International Conference on Artificial Immune Systems*, Lecture Notes in Computer Science **3239**, Springer, pp. 316–329.

[2]   Allen, D. et al. (1987), "Timing, Genetic Requirements and Functional Consequences of Somatic Hypermutation during B-cell Development", *Imm. Rev.*, **96**, pp. 5–22.

[3]   Burnet, F. M. (1959), *The Clonal Selection Theory of Acquired Immunity*, Cambridge University Press.

[4]   Cayzer, S. and Aickelin, U. (2005), "A Recommender System based on Idiotypic Artificial Immune Networks", *Journal of Mathematical Modelling and Algorithms*, **4**(2), pp. 181–198.

[5]   Coello Coello, C. A., Rivera, D. C. and Cortés, N. C. (2003), "Use of an Artificial Immune System for Job Shop Scheduling", In J. Timmis, P. Bentley and E. Hart (Eds.), *Proc. of the 3rd Int. Conf. on Artificial Immune Systems*, Lecture Notes in Computer Science **2787**, Springer-Verlag, pp. 1–10.

[6]   Cutello, V. and Nicosia, G. (2004), "The Clonal Selection Principle for In Silico and In Vitro Computing", In L. N. de Castro and F. J. Von Zuben (Eds.), *Recent Developments in Biologically Inspired Computing*, Chapter VI, Idea Group Pub., pp. 104–146.

[7]   Dasgupta, D. (ed.) (1999), *Artificial Immune Systems and Their Applications*, Springer-Verlag.

[8]   Dasgupta, D. and Attoh-Okine, N., (1997), "Immunity-Based Systems: A Survey", *Proc. of the IEEE SMC*, **1**, pp. 369–374.

[9]   Dasgupta, D., Majumdar, N. and Niño, F. (2001), "Artificial Immune Systems: A Bibliography", CS Technical Report, CS-01-002 [On line]
        http://www.cs.memphis.edu/~dasgupta/AIS/AIS-bib.pdf

[10]   de Castro, L. N. (2001), "Immune Engineering: Design of Computational Tools Inspired in Artificial Immune Systems", (in Portuguese), *Ph.D. Thesis*, DCA – FEEC/UNICAMP, Campinas/SP, Brazil, 286 p., May.

[11]   de Castro, L. N. (2002), "Immune Engineering: A Personal Account", II Workshop on Computational Intelligence and Semiotics, CD ROM Proceedings. [On line] ftp://ftp.dca.fee.unicamp.br/pub/docs/vonzuben/lnunes/imeng.pdf.

[12]   de Castro , L. N. (2003) , "Immune Cognition, Micro-evolution, and a Personal Account on Immune Engineering", *S.E.E.D. Journal (Semiotics, Evolution, Energy, and Development)*, University of Toronto, **3**(3), pp.134–155. [On line]
        http://www.library.utoronto.ca/see/SEED/Vol3-3/Castro.pdf

[13]   de Castro, L. N. (2004), *Engineering Applications of Artificial Immune Systems*, Talk presented at ICARIS 2004. [On line]
        http://www.dca.fee.unicamp.br/~lnunes/AIS.html

[14]   de Castro, L. N. and Timmis, J. I. (2002), *Artificial Immune Systems: A New Computational Intelligence Approach*, Springer-Verlag: London.

[15]   de Castro, L. N. and Von Zuben, F. J. (2000), "Artificial Immune Systems: Part II – A Survey of Applications", *Technical Report – RT DCA 02/00*, p. 65. [Online] http://www.dca.fee.unicamp.br/~lnunes

[16]   de Castro, L. N. and Von Zuben, F. J. (2001), "aiNet: An Artificial Immune Network for Data Analysis", In Hussein A. Abbass, Ruhul A. Sarker, and Charles S. Newton (eds.), *Data Mining: A Heuristic Approach*, Idea Group Publishing, USA, Chapter XII, pp. 231–259.

[17]   de Castro, L. N. and Von Zuben, F. J. (2002), "Learning and Optimization Using the Clonal Selection Principle", *IEEE Transactions on Evolutionary Computation*, 6(3), pp. 239–251.

[18]   Esponda, F., Forrest, S. and Helman, P. (2004), "A Formal Framework for Positive and Negative Detection Schemes", *IEEE Trans. on Systems, Man and Cybernetics - Part B*, 34(1), pp. 357-373.

[19]   Farmer, J. D., Packard, N. H. and Perelson, A. S. (1986), "The Immune System, Adaptation, and Machine Learning", *Physica* 22D, pp. 187–204.

[20]   Forrest, S., A. Perelson, Allen, L. and Cherukuri, R. (1994), "Self-Nonself Discrimination in a Computer", *Proc. of the IEEE Symposium on Research in Security and Privacy*, pp. 202–212.

[21]   Forrest, S., Javornik, B., Smith, R. E. and Perelson, A. S. (1993), "Using Genetic Algorithms to Explore Pattern Recognition in the Immune System", *Evolutionary Computation*, 1(3), pp. 191–211.

[22]   Galeano, J. C., Veloza-Suan, A. and González, F. A. (2005), "A Comparative Analysis of Artificial Immune Network Models", *Proc. of the Genetic and Evolutionary Computation Conference*, pp. 361–368.

[23]   González, F. and Dasgupta, D. (2003), "Anomaly Detection Using Real-Valued Negative Selection", *Genetic Programming and Evolvable Machines*, 4(4), pp. 383–403.

[24]   Hart, E. and Ross, P. (1999), "An Immune System Approach to Scheduling in Changing Environments", *Proc. of the Genetic and Evolutionary Computation Conference*, pp. 1559–1566.

[25]   Herlocker, J. L., Konstan, J. A., Borchers, A. and Riedl, J. (1999), "An Algorithmic Framework for Performing Collaborative Filtering", *Proc. of the 22nd Annual International ACM SIGIR Conference on Research and Development in Information Retrieval*, pp. 230–237.

[26]   Hightower R. R., Forrest, S. A and Perelson, A. S. (1995), "The Evolution of Emergent Organization in Immune System Gene Libraries", *Proc. of the 6th Int. Conference on Genetic Algorithms*, L. J. Eshelman (ed.), Morgan Kaufmann, pp. 344–350.

[27]   Hofmeyr S. A. and Forrest, S. (2000), "Architecture for an Artificial Immune System", *Evolutionary Computation*, 7(1), pp. 45–68.

[28]   Ishida, Y. (1996), "The Immune System as a Self-Identification Process: A Survey and a Proposal", *Proc. of the ICMAS Int. Workshop on Immunity-Based Systems*, pp. 2–12.

[29]   Ishiguro, A., Kondo, T., Watanabe, Y. and Uchikawa, Y. (1997), "Emergent Construction of Artificial Immune Networks for Autonomous Mobile Robots", *Proc. of IEEE System, Man, and Cybernetics Conference*, pp. 1222–1228.

[30]   Jacob, C., Pilat, M. L., Timmis, J., and Bentley, P. J (2005), *Proceedings of the 4ᵗʰ International Conference on Artificial Immune Systems*, Lecture Notes in Computer Science **3627**, Springer.

[31]   Janeway Jr., C. A. (1992), "The Immune System Evolved to Discriminate Infectious Nonself from Noninfectious Self", *Imm. Today*, **13**(1), pp. 11–16.

[32]   Janeway, C. A., P. Travers, Walport, M. and Capra, J. D. (1999), "Immunobiology: The Immune System in Health and Disease", 4ᵗʰ Ed., Garland Publishing.

[33]   Jerne, N. K. (1974), "Towards a Network Theory of the Immune System", *Ann. Immunol.* (Inst. Pasteur) **125C**, pp. 373–389.

[34]   Kelsey, J. and Timmis, J. (2003), "Immune Inspired Somatic Contiguous Hypermutation for Function Optimisation", In Cantu-Paz et al. (Eds.), *Proc. of the Genetic and Evolutionary Computation Conference*, Lecture Notes in Computer Science **2723**, Springer, pp. 207–218.

[35]   Kepler, T. B. and Perelson, A. S. (1993), "Somatic Hypermutation in B Cells: An Optimal Control Treatment", *J. theor. Biol.*, **164**, pp. 37–64.

[36]   Kim, J. and Bentley, P. (1999a), "The Artificial Immune Model for Network Intrusion Detection", *Proc. of the 7th European Conference on Intelligent Techniques and Soft Computing*, Aachen, Germany.

[37]   Kim, J. and Bentley, P. (1999b), "Negative Selection and Niching by an Artificial Immune System for Network Intrusion Detection", *Proc. of the Genetic and Evolutionary Computation Conference*, pp. 149–158.

[38]   Leclerc, B. (1995), "Minimum Spanning Trees for Tree Metrics: Abridgements and Adjustments", *Journal of Classification*, **12**, 207–241.

[39]   Matzinger, P. (2002), "The Danger Model: A Renewed Sense of Self", *Science*, **296**, pp. 301–305.

[40]   Matzinger, P. (1994), "Tolerance, Danger and the Extended Family", *Annual Reviews of Immunology*, **12**, pp. 991–1045.

[41]   Neal, M. (2003), "Meta-Stable Memory in an Artificial Immune Network", In J. Timmis, P. Bentley, and E. Hart (Eds.), *Proc. of the 2ⁿᵈ International Conference on Artificial Immune Systems*, Lecture Notes in Computer Science **2787**, Springer, pp. 168–180.

[42]   Newman, D. J., Hettich, S., Blake, C. L. and Merz, C. J. (1998), *UCI Repository of Machine Learning Databases*, Irvine, CA: University of California, Department of Information and Computer Science, [On Line] http://www.ics.uci.edu/~mlearn/MLRepository.html.

[43]   Nicosia, G. Cutello, V., Bentley, P. J., and Timmis, J. (2004), *Proceedings of the 3ʳᵈ International Conference on Artificial Immune Systems*, Lecture Notes in Computer Science **3239**, Springer.

[44]   Nossal, G. J. V. (1993), "Life, Death and the Immune System", *Scientific American*, **269**(3), pp. 21–30.

[45]   Oprea, M. and Forrest, S. (1998), "Simulated Evolution of Antibody Gene Libraries Under Pathogen Selection", *Proc. of the IEEE System, Man, and Cybernetics*.

[46]   Perelson, A. S. and Oster, G. F. (1979), "Theoretical Studies of Clonal Selection: Minimal Antibody Repertoire Size and Reliability of Self-Nonself Discrimination", *J. Theor.Biol.*, **81**, pp. 645–670.

[47]   Perelson, A. S., Hightower, R. and Forrest, S. (1996), "Evolution and Somatic Learning in V-Region Genes", *Research in Immunology*, **147**, pp. 202–208.

[48] Prim, R. C. (1957), "Shortest Connection Networks and Some Generalizations", *Bell System Technology Journal*, pp. 1389–1401.

[49] Rardin, R. L., (1998), *Optimization in Operations Research*, Prentice Hall.

[50] Storb, U. (1998), "Progress in Understanding the Mechanisms and Consequences of Somatic Hypermutation", *Immun. Rev.*, **162**, pp. 5–11.

[51] Tarakanov, A. O., Skormin, V. A., and Sokolova, S. P. (2003), *Immunocomputing: Principles and Applications*, Springer.

[52] Timmis, J., Bentley, P. J., and Hart, E. (2003), *Proceedings of the 2nd International Conference on Artificial Immune Systems*, Lecture Notes in Computer Science **2787**, Springer.

[53] Timmis, J. and Bentley, P. J. (2002), *Proceedings of the 1st International Conference on Artificial Immune Systems*, UKC.

[54] Timmis, J., Neal, M. and Hunt, J. (2000), "An Artificial Immune System for Data Analysis", *BioSystems*, **55**(1), pp.143–150.

[55] Tizard, I. R. (1995), *Immunology: An Introduction*, 4th Ed, Saunders College Publishing.

[56] Varela, F. J., Coutinho, A. Dupire, E. and Vaz, N. (1988), "Cognitive Networks: Immune, Neural and Otherwise", *Theoretical Immunology*, Part II, A. S. Perelson (ed.), pp. 359–375.

[57] Wolberg, W. H. and Mangasarian, O. L. (1990), "Multisurface Method of Pattern Separation for Medical Diagnosis Applied to Breast Cytology", *Proceedings of the National Academy of Sciences U.S.A.*, **87**, pp 9193–9196.

[58] Zahn, C. T. (1971), "Graph-Theoretical Methods for Detecting and Describing Gestalt Clusters", *IEEE Transactions on Computers*, **C-20**(1), pp. 68–86.

PART II

THE SIMULATION AND EMULATION OF NATURAL PHENOMENA IN COMPUTERS

CHAPTER 7

FRACTAL GEOMETRY OF NATURE

"Why is geometry often described as 'cold' and 'dry'? One reason lies in its inability to describe the shape of a cloud, a mountain, a coastline, or a tree. Clouds are not spheres, mountains are not cones, coastlines are not circles, and bark is not smooth, nor does lightning travel in a straight line. ... The existence of these patterns challenges us to study those forms that Euclid leaves aside as being 'formless', to investigate the morphology of the 'amorphous'."
(Mandelbrot, The Fractal Geometry of Nature, W. H. Freeman and Company, 1983, p.1)

"In the past, mathematics has been concerned largely with sets and functions to which the methods of classical calculus can be applied. Sets or functions that are not sufficiently smooth or regular have tended to be ignored as 'pathological' and not worthy of study. Certainly, they were regarded as individual curiosities and only rarely were thought of as a class to which a general theory might be applicable. In recent years this attitude has changed. It has been realized that a great deal can be said, and is worth saying, about the mathematics of non-smooth objects. Moreover, irregular sets provide a much better representation of many natural phenomena than do the figures of classical geometry. Fractal geometry provides a general framework for the study of such irregular sets."
(K. Falconer, Fractal Geometry: Mathematical Foundations and Applications, John Wiley & Sons, 2003, p. xvii)

7.1 INTRODUCTION

Recent advances in computer graphics have made it possible to visualize mathematical models of natural structures and processes with unprecedented realism. The resulting images, animations, and interactive systems are useful as scientific, research and education tools in computer science, engineering, biosciences, and many other domains. Applications include computer-assisted landscape architecture, design of new varieties of plants, crop yield prediction, the study of developmental and growth processes, and the modeling and synthesis (and the corresponding analysis) of an innumerable amount of natural patterns and phenomena.

There are a number of modeling and synthesizing techniques for natural patterns (cf. Meinhardt, 1982; Murray, 1989), and this single chapter could not possibly cover a small sample of them. For instance, *reaction-diffusion models* (RDM) have been extensively studied in theoretical biology and computer science to provide plausible explanations of many observed phenomena, such as the generation of patterns covering models of animals and the development of pigmentation models generically; *diffusion-limited aggregation* (DLA) models have been used to capture diffusion of nutrients by simulating random movement of particles in a grid; *cellular automata* can be used to model a number of natural phenomena; and the list goes on.

One major breakthrough in the process of modeling and synthesizing natural patterns and structures was the recognition that nature is fractal and the development of *fractal geometry*. In a simplified form, fractal geometry is the geometry of nature with all its irregular, fragmented, and complex structures. This chapter starts by introducing the fractal geometry of nature, and the concept of fractal dimension. It follows with the presentation of a number of techniques for modeling natural phenomena. The focus is on the study of some well-known and easily implementable algorithms for the synthesis of natural phenomena, in particular, *cellular automata, L-systems, iterated function systems* (IFS), the *random midpoint displacement algorithm*, and *particle systems* (PS). Some high-level comments about the use of evolutionary algorithms for the generation of life-like patterns will be made in a purely descriptive and illustrative form.

An important aspect to have in mind while studying this chapter is that all techniques presented here are basically used to generate a sort of 'skeleton' or basic structure of some natural patterns, such as trees, grass, and other fractals. In order to generate an image that more closely resembles those patterns found in nature and some unknown patterns as well, rendering techniques for computer graphics are necessary. But these are out of the scope of this book.

7.2   THE FRACTAL GEOMETRY OF NATURE

"Why is geometry often described as 'cold' and 'dry'? One reason lies in its inability to describe the shape of a cloud, a mountain, a coastline, or a tree. Clouds are not spheres, mountains are not cones, coastlines are not circles, and bark is not smooth, nor does lightning travel in a straight line. ... The existence of these patterns challenges us to study those forms that Euclid leaves aside as being 'formless', to investigate the morphology of the 'amorphous'." (Mandelbrot, 1983; p. 1)

This is certainly the most quoted passage when talking about the fractal geometry of nature. The world we inhabit is neither smooth-edged nor uniform; many patterns in nature are irregular and fragmented. The geometry invented by Euclid describes ideal shapes, such as points, lines, spheres, circles, squares, and cubes. But these Euclidean shapes, or compositions of them, are more often found in man-made products than in nature. For instance, balls are spherical, boxes are cubic, and pencils are cylindrical. However, what shape is a snowflake, a coastline, a mountain, a cloud, a tree, and many other natural forms?

B. Mandelbrot (1983) coined the term *fractal* to identify a family of shapes that describe irregular and fragmented patterns in nature, thus differentiating them from the pure geometric forms from the Euclidean geometry. The word fractal comes from the Latin adjective *fractus*, whose corresponding verb *frangere* means "to break" or to create irregular fragments. *Fractal geometry* is the geometry of the irregular shapes found in nature, and, in general, fractals are characterized by infinite details, infinite length, *self-similarity, fractal dimensions*, and the absence of smoothness or derivative. Two of these properties will be emphasized here, namely, self-similarity and fractal dimension.

(a)   (b)

Figure 7.1: Examples of fractals commonly found in nature. (a) The Grand Canyon. (b) A dry tree.

Fractals are thus irregular all over; they have the same degree of irregularity on all scales. Fractals look the same when observed from far away or nearby; they are self-similar. As you approach a fractal, you find that small pieces of the whole, which seemed from a distance to be formless blobs, become well-defined objects whose shape is roughly that of the previously observed whole. Nature provides many examples of fractals: ferns, coastlines, mountains, cauliflowers and broccoli, and many other plants - each branch and twig is very much like the whole (Figure 7.1). The rules governing the growth ensure that small-scale features become translated into large-scale ones. It is important to note, however, that self-similar structures are not necessarily fractal. Take, for instance, some regular geometric figures (e.g., a line). It can be broken down into small copies of itself, but it is not a fractal.

7.2.1. Self-Similarity

Self-similarity is a core concept in fractal geometry; it is an underlying theme in all fractals, being more pronounced in some of them and in variations in others. But how precise should the concept of self-similarity be?

The word self-similar does not need a definition; it is intuitively self-explanatory. Let us start by giving an example of a natural structure that presents this property: a broccoli (see Figure 7.2). By removing some branches of the broccoli and comparing them with the whole broccoli, one can note that the branches are very much like the whole, but in a smaller scale. These branches can be further decomposed into smaller branches, which are again very similar to the whole broccoli. This self-similarity carries through for a number of dissections. In a mathematical idealization, the property of self-similarity of a fractal may be continued through infinitely many stages. This leads to new concepts, such as *fractal dimension*, which are useful for natural structures that do not have this infinite level of details (Peitgen et al., 1992).

(a)   (b)

Figure 7.2: Self-similarity of a broccoli demonstrated by dissection and successive enlargements. (a) Whole broccoli. (b) Smaller branch of a broccoli.

Despite this easy interpretation of self-similarity, a precise mathematical definition of it is hard to be obtained. While in natural structures the self-similarity holds only for a few orders of magnitude and with specific levels of details, mathematical fractals are usually thought of as objects with details in all scales. When looking at fractals whose small copies are variations of the whole, we have the so-called *statistical self-similarity* or self-similarity in the statistical sense. Furthermore, the reduced copies may be distorted in other ways, for instance squeezed. In such cases there is the concept of *self-affinity*.

The Sierpinski gasket, illustrated in Figure 7.3(a), has copies of itself nearly at every point of it. It is composed of small but exact copies of itself. In this case, one says that the Sierpinski gasket is *strictly self-similar*. The situation with the tree presented in Figure 7.3(b) is slightly different. The whole tree is made up of the stem and two reduced copies of the whole. Each branch of the tree constitutes a reduced copy of the whole tree; that is, the self-similarity property is restricted to the branches of the tree. The whole tree is not strictly self-similar, but *self-affine*. Now, having the idea in mind that the concept of self-similarity does not require strict self-similarity, the broccoli can be categorized as a self-similar, but not strictly self-similar, pattern.

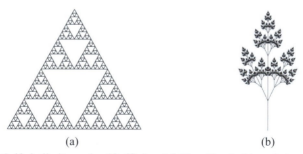

(a)   (b)

Figure 7.3: Self-similarity and self-affinity. (a) The Sierpinski gasket: a self-similar structure. (b) Fractal tree: a self-affine structure.

Figure 7.4: The first steps in constructing the Cantor set.

7.2.2. Some Pioneering Fractals

The first fractal was discovered in 1861 by K. Weierstrass. He discovered a no-where differentiable continuous function; that is, a curve consisting solely of corners. Other early-discovered fractals were introduced by G. Cantor, H. von Koch, and W. Sierpinski, among others. These fractals were considered 'mathematical monsters' because of some nonintuitive properties they possessed.

Take a line; remove its middle third, leaving two equal lines. Likewise, remove the middle third from each of the remaining lines. Repeat this process an infinite number of times, and you have (in the limit) the *Cantor set*. Figure 7.4 illustrates the first four steps in constructing the Cantor set.

The Cantor set was introduced by G. Cantor in 1883 and has several interesting features. First, it has no length or interior, it has zero measure; every part of it consists almost completely of holes. However, despite being formed by totally disconnected points, it is uncountable. It contains as many points as the whole curve it is carved from. Every point is a limit point, meaning that there are infinitely many other points from the set in any neighborhood of it, no matter how small. Conversely, the Cantor set contains all of its limit points. This fact that the Cantor set has as many points as the real line but has zero measure is perplexing.

Another fractal was devised by H. von Koch in 1904. He defined the curve as the limit of an infinite sequence of increasingly wrinkly curves. It was created as another example of a curve with no tangent at any point. The construction of the *Koch curve* is illustrated in Figure 7.5. Start with a line segment and iteratively apply the following transformation: 1) take each line segment of the Koch curve from the previous step and remove the middle third; and 2) replace the middle third with two new line segments, each with length equal to the removed part, forming an equilateral triangle with no base. Fitting together three suitably rotated copies of the Koch curve produces a picture called the *snowflacke curve*, the *Koch snowflacke* or the *Koch island*, illustrated in Figure 7.6.

In the limit, the Koch curve consists of no line segment. Instead, it has a corner coincident with every point in the curve - the curve consists entirely of corners. Thus, the Koch curve has no derivative (tangent) in any point. Another interesting fact about the Koch curve is that although it starts from a line segment with a fixed length l, its length is infinite. For instance, say $l = 1$um (um =

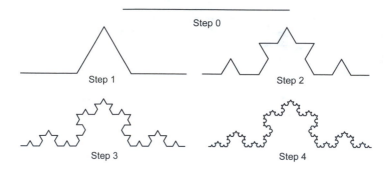

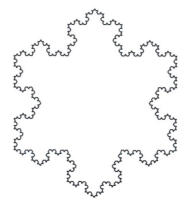

Figure 7.5: The first steps in constructing the Koch curve.

Figure 7.6: The Koch snowflake constructed from three Koch curves appropriately rotated.

unit of measure, e.g., centimeters - cm). At time step t, the curve consists of 4^t segments, each of which has a length $1/3^t$. Therefore, the total length of the curve is $(4/3)^t$, implying in an exponential increase in length with the number of iteration steps t. The case is even worse when one think that the Koch curve can be enclosed in a finite square. This means that a curve of infinite length can be placed within a finite area element.

The Polish mathematician V. Sierpinski introduced his fractal, known as the *Sierpinski gasket, Sierpinski triangle,* or *Sierpinski siege,* in 1916, though the underlying principles had been known for millennia. Generically, it can be constructed as follows. Start with an equilateral triangle (that can be filled or not), and divide it into four smaller equilateral triangles, as illustrated in Figure 7.7. Similarly to all the other fractals described in this section, in the limit we are going to have the Sierpinski gasket, a shape composed of several copies of itself.

Figure 7.7: The first steps in constructing the Sierpinski gasket.

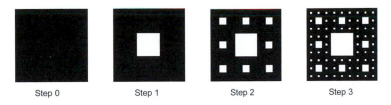

Figure 7.8: The first steps in constructing the Sierpinski carpet.

In addition to the Sierpinski gasket, V. Sierpinski added another pattern to the gallery of pioneering fractals, the *Sierpinski carpet*. To construct the Sierpinski carpet, start with a square in the plane, subdivide it into nine smaller congruent squares, and so on, as illustrated in Figure 7.8. Again, the Sierpinski carpet is the object obtained in the limit when an infinite number of steps is performed.

In the late 1800s G. Peano and D. Hilbert presented some curves lying in a plane but which demonstrated that our concepts of curve were quite limited. They presented curves capable of filling a plane; that is, given a patch of the plane, there is a curve that passes through every point in that patch. This is performed by constructing a curve that twists in such a complex way to visit every point in the entire plane. These curves came to be known as *space filling curves*. The Peano curve cannot only fit within an enclosed area, but it can also fill it.

Like most pioneering fractals, the Peano curve starts with an initial line segment. For each step in the construction process, all line segments of the previous step are replaced by a curve consisting of nine smaller line segments. Figure 7.9 depicts the first three steps in the construction process of the Peano curve. The Peano curve caused some discomfort among mathematicians due to some of its properties. It performs a one to one mapping from a line to a plane, although it is not continuous. However, if there is a mapping that is both continuous and one to one, the concept of dimension would have no topological meaning.

Living organisms are full of space- and volume-filling structures. An organ must be supplied with the necessary supporting substances such as water, blood, and oxygen. Additionally, space-filling structures promote a maximization of the space used for some functions. Take the case of the brain for instance. By twisting its structure it is possible to increase the volume of information stored and processed within the brain. The same holds true for the intestine, lungs, and

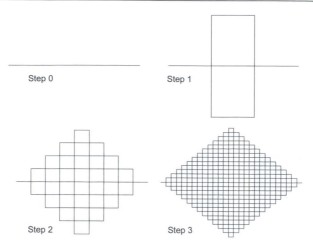

Figure 7.9: The first steps in constructing the Peano curve.

many other organs. In some cases, a number of substances have to be transported through vessel systems and must reach every point in the volume of a given organ; examples are the lungs, kidneys, etc. Fractals solve the problem of how to organize such complicated structures efficiently.

7.2.3. Dimension and Fractal Dimension

The discovery of space-filling curves, such as the Peano curve, had a major impact in the development of the concept of dimension. These structures questioned the intuitive perception of curves as one-dimensional objects because they filled the plane, and thus they are intuitively perceived as bi-dimensional objects. We learn early in the school days that points are zero-dimensional, lines and curves are one-dimensional, planes and surfaces are two-dimensional, and solids, such as cubes and spheres, are three-dimensional. Roughly speaking, a set is d-dimensional if d independent variables (coordinates) are needed to describe the neighborhood of any point. This notion of dimension is called the *topological dimension* of a set.

Topology is a branch of mathematics that deals with questions of form and shape from a qualitative perspective. It involves the study of properties of geometric figures or solids that are not changed by *homeomorphisms*, i.e., transformations, such as stretching or bending. Two objects are homeomorphic if there is a continuous one-to-one and onto mapping with a continuous inverse. Another basic notion of topology is *dimension*. From the point of view of topology, a straight line cannot be differentiated from the Koch curve, neither can the Koch island be differentiated from a circle. The transformations allowed to be applied to objects (e.g., stretching, twisting, etc.) are called homeomorphisms, and when applied they must not change the invariant properties of the objects. For example, a cube may be transformed into a cylinder, but not into a doughnut.

Despite this notion of topological dimension, mathematicians discovered in the late 1800s and early 1900s that a proper understanding of irregularity or fragmentation cannot be satisfied by defining dimension as a number of coordinates. For instance, what about the Koch curve described above? It has topological dimension one, but it is by no means a curve - the length between any two points on the curve is infinite. No small piece of it is a line, but neither is it like a piece of plane. In some sense, it is possible to say that it is too big to be a one-dimensional object, but too thin to be a two-dimensional object. Perhaps its dimension should be a number between one and two.

By the early 1900s, describing what dimension means and what are its properties was one of the major problems in mathematics. Since then, it has become even worse because mathematicians have come up with a number of different, but related, notions of dimension: topological dimension, Hausdorff dimension, fractal dimension, self-similarity dimension, box-counting dimension, capacity dimension, Euclidean dimension, and so forth. In some cases all of them are the same and in others not. It is difficult, even for a mathematician, to understand them all and their differences. This chapter reviews two special forms of *Mandelbrot's fractal dimension*, namely, the *self-similarity dimension* and the *box-counting dimension*.

The difficulty in assigning a dimension to an object such as the Koch curve, the Cantor set, and the Sierpinski gasket, is not only a problem of the fractals exemplified here. A similar phenomenon was reported in 1961 (posthumously) by the English meteorologist L. Richardson, when he attempted to measure the length of various coastlines, including the British coastline. He found that the apparent length of the coastline seemed to increase whenever the length of the measuring stick was reduced, because the greater the magnification, the greater the level of details found (small inlets, streams, peninsulas, etc.). Richardson concluded that the length of the coastline is not well defined, but he also found an empirical law that describes this increase in length with the level of detail accounted for. He noted that when the logarithm of the length of the measuring unit (stick) was plotted against the logarithm of the total length of a coastline, the points tended to lie on a straight line. The slope of the resulting line measures, in some ways, the amount of wrinkliness or meandering of the coastline. Mandelbrot (1983) found Richardson's works and realized that fractals could be classified in a similar form.

It has already been argued that fractals are self-similar objects, but how do we define the dimension of an object? If the dimension of the object is known, and this is the case for the Euclidean shapes, powers or exponents allow us to work out how many smaller copies of itself the object contains, of any particular size. In the case of regular shapes, such as lines, squares, and cubes, a d-dimensional shape is composed of $N \times (1/m)$ copies of itself, where each copy has a size m in relation to the original shape; m is called the reduction factor. The relationship is thus given by $N = (1/m)^d$. To illustrate, let us take the case of a square that can be divided into four ($N = (1/m)^d = 2^2 = 4$) one-half sized ($m = 1/2$) squares, or nine ($N = (1/m)^d = 3^2 = 9$) one-third sized ($m = 1/3$) squares (see Figure 7.10).

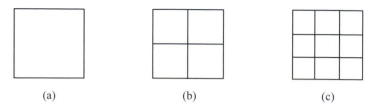

(a)   (b)   (c)

Figure 7.10: Relationship between the reduction factor *m* and the number of copies generated: $d = 2$. (a) A full square: $N = 1$, $1/m = 1$. (b) A square divided into four one-half sized squares: $N = 4$, $1/m = 2$. (c) A square divided into nine one-third sized squares: $N = 9$, $1/m = 3$.

This idea of the relation between the logarithm of the number of copies of an object and the logarithm of the size of its copies suggested a generalization of the concept of dimension, which allows fractional values. This dimension is known as the *self-similarity dimension*.

$$N = (1/m)^d \Rightarrow \log(N) = \log((1/m)^d)$$
$$\Rightarrow \log(N) = d.\log(1/m) \Rightarrow d = \frac{\log N}{\log 1/m} \tag{7.1}$$

Therefore, the dimension d_s of a self-similar shape is given by:

$$d_s = \frac{\log N}{\log 1/m} \tag{7.2}$$

where N is the number of copies of the object, and m is the reduction factor. For the examples of Figure 7.10, $d_s = \log 4/\log 2 = 2$, and $d = \log 9/\log 3 = 2$.

Mandelbrot noticed, in Richardson's works, that the slope of the resultant line plotted by the logarithm of the length of the measuring stick against the total length is but the *Hausdorff dimension* of the coastline. In 1919, F. Hausdorff extended the notion of similarity dimension, given by Equation (7.2), to cover all shapes, not just strictly self-similar shapes. This *fractal dimension* describes the fractal complexity of an object. For example, the coastline of Britain has a fractal dimension of approximately 1.26, about the same as the Koch curve, but slightly less than the outline of a typical cloud (about 1.35). On this scale, a dimension of one means totally smooth, while tending toward two implies increasing fractal complexity.

Although general in scope, the Hausdorff dimension is often difficult to calculate in practice. Another special form of the fractal dimension, named the *box-counting dimension*, can be determined by a method called the box-counting method (Peitgen et al., 1992; Lesmoir-Gordon et al., 2000; Falconer, 2003). Cover the shape whose dimension you want to measure with *boxes* and find how the number of boxes changes with the size of the boxes. Repeat this process with ever smaller squares. In the limit, for a fractal shape, the rate at which the proportion of filled squares decreases gives the box-counting dimension.

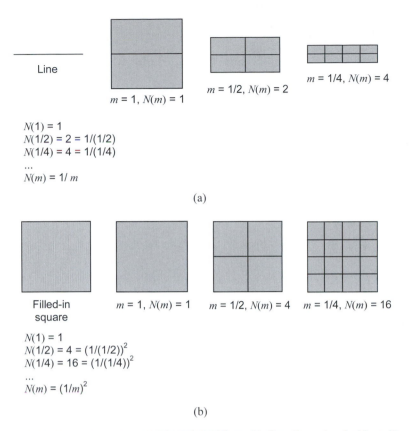

Figure 7.11: Determining how $N(m)$ varies with m. (a) One-dimensional object: line. (b) Two-dimensional object: square.

Assume $N(m)$ squares of side length m are necessary to cover the curve. Then, $N(m).m$ approximates the length of the curve, while $N(m).m^2$ approximates the area of the curve. The smaller the squares, the better the approximation, because more details can be detected. Finding the pattern of how $N(m)$ varies with m provides us some idea about the complexity of the shape. To illustrate the procedure, let us first take the known cases of some Euclidean shapes (Figure 7.11).

For more complicated objects, like the ones with fractal shapes, the relation between $N(m)$ and $1/m$ may be a power law: $N(m) = k(1/m)^d$. This leads to the definition of the box-counting dimension d. If the relation between $N(m)$ and $1/m$ is assumed to obey a power law, then the dimension d can be calculated by taking the logarithm of both sides of $N(m) = k(1/m)^d$:

$$\log(N(m)) = \log(k(1/m)^d) \Rightarrow$$
$$\log(N(m)) = d.\log(1/m) + \log(k) \tag{7.3}$$

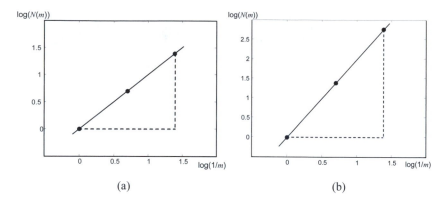

Figure 7.12: Plotting $\log(N(m)) \times \log(1/m)$ results in points lying approximately on a straight line with slope d_b, where d_b is the box-counting dimension. (a) Line segment. (b) Square.

Note that Equation (7.3) above corresponds to the equation of a straight line with slope d and that intercepts the y-axis at the point $\log(k)$. Therefore, plotting $\log(N(m)) \times \log(1/m)$ results in points lying approximately on a straight line with slope d_b, where d_b is the box-counting dimension; d in Equation (7.3). For the case of the line segment, for example, plotting $\log(N(m)) \times \log(1/m)$ will result in a straight line of slope $d_b = 1$, which corresponds to the topological dimension of the line.

This procedure can be applied to any type of curve. Figure 7.13 illustrates how the box-counting method can be applied to the Sierpinski gasket in order to estimate its fractal dimension. The approximate dimension (slope of the line plotted in Figure 7.13(b)) of the Sierpinski gasket is $d_b = 1.58$.

Another form of determining d_b is by isolating the variable d from Equation (7.3), as follows:

$$d_b = \frac{\log(N(m)) - \log(k)}{\log(1/m)} \tag{7.4}$$

Equation (7.4) is based on the expectation that the approximation becomes better for smaller m. For $m \to 0$, $\log(1/m) \to \infty$ and $\log(k)/\log(1/m) \to 0$ in the limit we have:

$$d_b = \lim_{m \to 0} \frac{\log(N(m))}{\log(1/m)} \tag{7.5}$$

If the limit exists, it corresponds to the *box-counting dimension*, d_b, of the object.

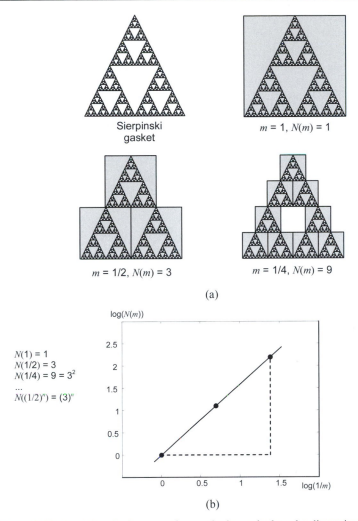

Sierpinski
gasket

$m = 1$, $N(m) = 1$

$m = 1/2$, $N(m) = 3$

$m = 1/4$, $N(m) = 9$

(a)

$\log(N(m))$

$N(1) = 1$
$N(1/2) = 3$
$N(1/4) = 9 = 3^2$
...
$N((1/2)^n) = (3)^n$

$\log(1/m)$

(b)

Figure 7.13: Applying the box-counting method to calculate the dimension of the Sierpinski gasket. (a) Varying m and determining $N(m)$. (b) Plotting $\log(N(m)) \times \log(1/m)$.

7.2.4. Scope of Fractal Geometry

We have said before that nature is lawful, and now we know that many of these laws or rules involve some roughness and a lot of irregularity; nature has a fractal geometry. Ferns, coastlines, mountains, snowflakes, cauliflowers, broccoli, plants, trees, clouds, fluids, etc., are all fractals. Shapes like the Cantor set and the Koch curve, far from being exceptional, are in fact ubiquitous in nature. Variations on the Cantor set occur in everything from frequencies of words and letters in language to noise on telephone lines, while Koch curves serve as models for natural coastlines, clouds, etc.

Moreover, organisms are fractals. Our lungs, our circulatory system, our brains, our kidneys, and many other body systems and organs are fractal. Fractal geometry allows bounded curves of infinite length and closed surfaces with an infinite area. This is how some of our organs, such as the lungs, kidneys, and brains manage to maximize their surface area. Research employing fractal rules has revealed many important facts about our bodies and has allowed us to understand how many organs are constructed.

But the scope and importance of fractals and fractal geometry go far beyond these. Forest fires have fractal boundaries; deposits built up in electro-plating processes and the spreading of some liquids in viscous fluids have fractal patterns; Brownian motion is fractal; complex protein surfaces fold up and wrinkle around toward three-dimensional space in a fractal dimension; antibodies bind to antigens through complementary fractal dimensions of their surfaces; fractals have been used to model the dynamics of the AIDS virus; cancer cells can be identified based on their fractal dimension; fractal geometry is being applied in the treatment of bone fractures; fractal fibers are optically efficient (and this has been used in jewelerry and fiber optic industries); statistical models using fractal geometry are in use in testing for stress loading on oil rigs and on aircraft in turbulence; fractal structures are the most effective shapes for mobile phone antennas; fractal geometry has been used effectively by the military for the detection of man-made features in natural environments; fractals have also been used in economics, astronomy, arts, image compression, in the study of growth and developmental processes, and so forth (Lesmoir-Gordon et al., 2000). A tiny fraction of these applications of fractals will be reviewed in this chapter.

7.3   CELLULAR AUTOMATA

Cellular automata (CA) are a class of spatially and temporally discrete, deterministic mathematical systems characterized by local interaction and an inherently parallel form of evolution in time (Ilachinski, 2001). For short, a cellular automaton can be defined as a dynamical system that is discrete in both space and time. Cellular automata are good examples of mathematical systems constructed from many identical components, each simple, but that together are capable of complex behavior. From their analysis it is possible to develop scientific models of particular systems and to abstract general principles applicable to a wide variety of complex systems.

The formalism for cellular automata was introduced by John von Neumann (with some suggestions from Stanislaw Ulam) in the late 1940s and early 1950s as a framework to study self-reproduction. Von Neumann was interested in the essence of reproduction and not in any particular implementation of the process. Thus, he purposely abstracted away all the details of how animals reproduce, and instead concentrated on the simplest mathematical framework that would allow 'information' to reproduce.

CA are prototypical models for complex systems and processes consisting of a large number of identical, simple, locally interacting components. The study of these systems has attracted a great interest over the years because of their ability to generate a rich spectrum of very complex patterns of behavior out of sets of relatively simple underlying rules. Furthermore, they appear to capture many essential features of complex self-organizing cooperative behavior observed in natural systems (Wolfram, 1984).

There are a wide variety of CA models, each one tailored to fit the requirements of a specific system. However, most CA models present five generic features (Ilachinski, 2001):

- *Discrete lattice of cells*: the system consists of a one-, two-, or three-dimensional *lattice* (or *grid*) of *cells* (or *sites*).

- *Homogeneity*: all cells are equivalent.

- *Discrete states*: each cell takes on one of a finite number of possible discrete states.

- *Local interactions*: each cell interacts only with cells that are in its local neighborhood.

- *Discrete dynamics*: at each discrete time step, each cell updates its current state according to a transition rule taking into account its current state and the states of cells in its neighborhood.

It is interesting to note that although these features might seem very simple, cellular automata are capable of generating very complex patterns. There are two general approaches to the study of patterns and behaviors in cellular automata (Langton, 1986):

- Start with specific behaviors in mind and derive *transition functions*, also called *transition rules*, that will support these behaviors.

- Specify transition functions and observe the resulting behavior.

The transition functions or rules define the automaton used by each cell; that is, they map the state of a local neighborhood to a new state for the cell at the 'center' of that neighborhood at the next iteration step.

7.3.1. A Simple One-Dimensional Example

To have an idea of what is a cellular automaton, consider the following example. The simplest cellular automaton is a *one-dimensional binary CA*; that is, a CA in which the grid corresponds to a row of cells and each cell can only assume a value of 0 or 1. The state s_i of a cell $\mathbf{i}$ ($\mathbf{i}$ is the vector indicating the position of a cell in the grid) is updated in discrete time steps according to an identical deterministic rule depending on a neighborhood of sites around it. In this simple example, consider the neighbors to be the cell on the right side and the one on the left side of the current cell, as illustrated in Figure 7.14. Thus, the state s_i of the center cell $\mathbf{i}$ at the next time step $t + 1$ is a function of the state of its neighbors at the current time step t: $s_i(t + 1) = f(s_{i-1}(t), s_i(t), s_{i+1}(t))$.

Figure 7.14: Neighborhood of the center cell **i** for a simple one-dimensional CA.

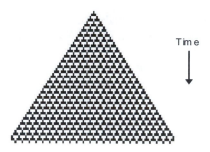

Figure 7.15: Time evolution of the cellular automaton defined by the rules of Table 7.1. Each row corresponds to one time step, and the algorithm starts from a single seed (initial condition).

Table 7.1: Transition rules for the cellular automaton exemplified.

$000 \rightarrow 0$	$100 \rightarrow 1$
$001 \rightarrow 1$	$101 \rightarrow 1$
$010 \rightarrow 1$	$110 \rightarrow 0$
$011 \rightarrow 0$	$111 \rightarrow 0$

For this example, the transition rules are provided in Table 7.1. In this table, if the center cell has a value 0 and both its neighbors (left and right) have values 0, then the state of the center cell at the next time step remains zero; else if the center cell has value 0, its left neighbor has value 0 and the right neighbor has value 1, then the state of the center cell at the next time step is 1; and so forth according to Table 7.1. Figure 7.15 shows the evolution in time of the one-dimensional binary cellular automaton following the transition rules of Table 7.1.

7.3.2. Cellular Automata as Dynamical Systems

For those readers not familiar with, a brief introduction to nonlinear dynamical systems is provided in Appendix B.4.3. Some discussion about this subject was also made in Chapter 4 while introducing Hopfield neural networks.

The behavior of a dynamical system in time can be represented as a trajectory through its state space. In general, when a system is started from some initial state, the point representing its state will travel around in some restricted region of the state space. There are three possibilities for the long-term behavior of the trajectory of the system in state space: 1) it will stop moving altogether; 2) it will fall into a closed cycle; or 3) it will not close on itself at all. In the first case, the system is said to have evolved into a *fixed* or *limit point*, in the second case to a *limit cycle*, and in the third case to something called a *strange attractor*. All three are kinds of *attractors* – regions of the state space to where the trajectories converge, given some initial condition.

The state space in a cellular automaton is analogous to the state space in the theory of nonlinear dynamical systems. At any one time there is a unique distribution of states over the cells of the automaton and this distribution is represented as a point in the state space. The time evolution of a cellular automaton can be studied by observing the trajectory that it follows in its state space. It turns out that attractors often abound in the state spaces of cellular automata as well. This feature of CA has motivated many physicists to its study. Cellular automata can exhibit behaviors characteristic of all three classes of attractors mentioned above and more. S. Wolfram identified four qualitative classes of cellular automata behavior (Wolfram, 1983):

- Class 1: evolution leads to a homogeneous state in which, for example, all sites have value '1'.

- Class 2: evolution leads to a set of stable or periodic structures that are separated and simple.

- Class 3: evolution leads to a chaotic pattern.

- Class 4: evolution leads to complex structures, sometimes long-lived.

From a dynamical systems' perspective, Wolfram found the following analogues for these classes of cellular automata behaviors:

- Class 1: cellular automata evolve to limit points.

- Class 2: cellular automata evolve to limit cycles.

- Class 3: cellular automata evolve to chaotic behavior of the type associated with strange attractors.

- Class 4: cellular automata effectively have very long transients. (No direct analogue for them has been found among continuous dynamical systems.)

Figure 7.16 illustrates the four classes of patterns generated by the evolution of a cellular automaton.

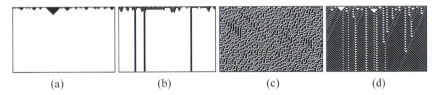

(a)   (b)   (c)   (d)

Figure 7.16: Classes of patterns generated by the evolution of cellular automata from various initial states. Successive rows correspond to successive time steps in the cellular automaton evolution. (a) Class 1. (b) Class 2. (c) Class 3. (d) Class 4.

7.3.3. Formal Definition

A d-dimensional cellular automaton consists of a finite or infinite d-dimensional grid of cells, each of which can take on a value from a finite, usually small, set of integers. The value of each cell at time step $t + 1$ is a function of the values of a small local neighborhood of cells at time t. The cells update their state simultaneously according to a given local rule (Mange and Tomassini, 1998).

Formally, a cellular automaton C can be defined as a quintuple

$$C = \langle S, s_0, G, d, f \rangle \tag{7.6}$$

where S is a finite set of states, $s_0 \in S$ are the initial states of the CA, G is the cellular neighborhood, $d \in Z^+$ is the dimension of C, and f is the local cellular interaction rule, also referred to as the *transition function* or *transition rule*.

Given the position of a cell $\mathbf{i}$, where $\mathbf{i}$ is an integer vector in a d-dimensional space ($\mathbf{i} \in Z^d$), in a regular d-dimensional uniform *lattice*, or *grid*, its neighborhood G is defined by

$$G_{\mathbf{i}} = \{\mathbf{i}, \mathbf{i} + \mathbf{r}_1, \mathbf{i} + \mathbf{r}_2, \dots , \mathbf{i} + \mathbf{r}_n\} \tag{7.7}$$

where n is a fixed parameter that determines the neighborhood size, and $\mathbf{r}_j$ is a fixed vector in the d-dimensional space. The local transition rule f

$$f : S^n \to S \tag{7.8}$$

maps the state $s_{\mathbf{i}} \in S$ of a given cell $\mathbf{i}$ into another state from the set S, as a function of the states of the cells in the neighborhood $G_{\mathbf{i}}$. In a uniform CA, f is identical for all cells, whereas in nonuniform CA, f may differ from one cell to another, i.e., f depends on $\mathbf{i}$, $f_{\mathbf{i}}$.

For a finite-size CA of size N, where N is the number of cells in the CA, a configuration of the grid at time t is defined as

$$C(t) = (s_0(t), s_1(t), \dots , s_{N-1}(t)) \tag{7.9}$$

where $s_{\mathbf{i}}(t)$ is the state of cell $\mathbf{i}$ at time t. The progression of the CA in time is then given by the iteration of the global mapping F

$$F : C(t) \to C(t+1), \qquad t = 0, 1, \dots \tag{7.10}$$

Through the simultaneous application in each cell of the local transition rule f, the global dynamics of the CA can be described as a directed graph, referred to as the CA's *state space*.

One- and bi-dimensional CA are the most usually explored types of cellular automaton. In the one-dimensional case, there are usually only two possible states for each cell, $S = \{0,1\}$. Thus, f is a function $f : \{0,1\}^n \to \{0,1\}$ and the neighborhood size n is usually taken to be $n = 2r + 1$ such that

$$s_{\mathbf{i}}(t+1) = f(s_{\mathbf{i}-r}(t), \dots , s_{\mathbf{i}}(t), \dots , s_{\mathbf{i}+r}(t)) \tag{7.11}$$

where $r \in Z^+$ is a parameter, known as the *radius*, representing the standard one-dimensional cellular neighborhood.

Assuming $r = 1$ and $S = \{0,1\}$, we have the so-called elementary CA, for which the neighborhood size is $n = 3$, such as the example of Section 7.3.1:

$$f: \{0,1\}^3 \rightarrow \{0,1\}, \qquad s_i(t+1) = f(s_{i-1}(t), s_i(t), s_{i+1}(t)) \qquad (7.12)$$

The domain of f is the set of all 2^3 3-tuples, which gives rise to $2^8 = 256$ distinct elementary sets of rules that can be used. For a two-state CA, in a grid with N cells, the configuration of the grid at time t is a binary sequence $C(t)$, $C(t) \in \{0,1\}^N$.

For finite size grids, spatially periodic boundary conditions are frequently assumed, resulting in a circular grid. It means that the edges of the arrays are wrapped around, yielding the topologies of a torus for bi-dimensional CA, and of a circle for one-dimensional CA.

7.3.4. Example of Application

This section illustrates one application of cellular automata. This example shows the rule table to generate a famous fractal pattern, known as the Sierpinski gasket or triangle. Further examples of the use of cellular automata will be provided in the next chapter, under the context of artificial life.

Fractal Patterns

Consider the simple CA example of Section 7.3.1, but now assume the transition rules defined in Table 7.2. Figure 7.17 shows the pattern, known as the Sierspinski gasket, generated after 500 time steps. The generation of this pattern required the application of the rules of Table 7.2 to a quarter of a million cell values.

Table 7.2: Transition rules for the cellular automaton illustrated in Figure 7.17.

$000 \rightarrow 0$	$100 \rightarrow 1$
$001 \rightarrow 1$	$101 \rightarrow 1$
$010 \rightarrow 1$	$110 \rightarrow 1$
$011 \rightarrow 1$	$111 \rightarrow 0$

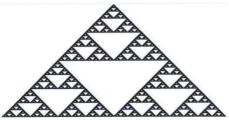

Figure 7.17: 500 time steps in the construction of the elementary CA following the rules of Table 7.2.

In this case, the $2r$ cells that are in the immediate neighborhood (left and right sides) of the center cell form a total neighborhood of $2r + 1$ cells. In general, for a CA with a neighborhood size of $2r + 1$ and with each cell having the ability to assume k possible states, a rule table that completely specifies the next-state function of a cell must have k^{2r+1} entries in it. Thus, for this example with $r = 1$ and $k = 2$, there are $2^3 = 8$ entries in the rule table. Since there are eight rows in the table, there are 256 different rule tables that could have been used.

Table 7.2 is still, perhaps, too lengthy, in the sense that the rule table can be compressed if some conventions or other types of representation for the rules are adopted. Another way of stating the rules in this table is the following:

- If all cells in a neighborhood are <u>on</u> (have value '1'), or all cells in a neighborhood are off (have value '0'), then the next state is <u>off</u>.

- Otherwise, the next state is <u>on</u>.

Taking the sum of the states of the cells in a particular neighborhood, it can be observed that this CA yields a '1' if, and only if, the sum is equal to 1 or 2, what corresponds to the sum modulo 2:

$$s_i(t + 1) = s_{i-1}(t) + s_{i+1}(t) \bmod 2$$

where mod 2 indicates that the 0 or 1 remainder after division by 2 is taken. This rule is also equivalent to the XOR Boolean function already presented for alphabets of length 2.

The pattern presented in Figure 7.17 is an intricate one but exhibits some striking regularities; one being *self-similarity*. As illustrated in Figure 7.18, when portions of the pattern are magnified, they are indistinguishable from the whole. Differences on small scales between the original and the magnified portion disappear when one considers the limiting pattern obtained after an infinite number of time steps. The pattern is therefore invariant under rescaling of lengths.

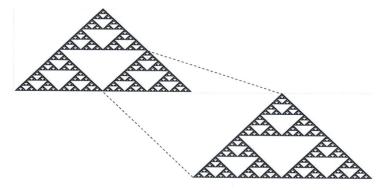

Figure 7.18: The pattern obtained in Figure 7.17 is self-similar. When a part of it is magnified, it is indistinguishable from the whole.

7.3.5. Scope of Cellular Automata

Cellular automata were proven great tools for simulating and studying massively parallel systems. CA are used in biology, chemistry, physics, in the design of parallel computers and image processing, as well as image generation. Cellular automata also became attractive to people doing research in artificial life, artificial intelligence, and theoretical biology. In summary:

- CA are *universal computers*.

- CA are used to study *artificial life*.

- CA can be used to perform *biological modeling* of insect swarms, wood fires, colonies of bacteria, clouds, development of organisms (e.g., cell and crystal growth), etc. (see further examples in the next chapter).

- CA are also important for *dynamical systems theory*, to study ordering, turbulence, chaos, fractals, and symmetry breaking.

- CA are important to study *population dynamics*, such as competition, co-operation, evolution, etc.

Cellular automata have proven to be extremely useful idealizations of the dynamical behavior of many real complex systems, including physical fluids, neural networks, plant growth, molecular dynamical systems, natural ecologies, immune system modeling, military command and control networks, forest fires, economy, among others. Because of their underlying simplicity, CA are also powerful conceptual tools with which to study general pattern formation. They have already provided critical insights into the self-organized patterns of chemical reaction-diffusion systems, crystal growth, seashells, and phase-transition-like phenomena in vehicular traffic flow, to name but a few examples.

7.4   L-SYSTEMS

Development at the multicellular level consists of the generation of structures by cell division, cell enlargement, cell differentiation, and cell death taking place at determined times and places in the entire life of the organism. It corresponds to the series of changes that animal and vegetable organisms undergo in their passage from the embryonic state to maturity, from a lower to a higher state of organization. A. Lindenmayer (1968) introduced a formalism to simulate the development of multicellular organisms, later referred to as Lindenmayer systems or simply *L-systems*. Due to their motivation, L-systems are also known as *developmental algorithms*. This formalism was closely related to abstract automata and formal languages, and quickly attracted the interest of many researchers. The development of the mathematical theory of L-systems was followed by its application to the modeling of plants, which is the main focus of this section. Those readers not familiar with formal languages, production systems, and grammars should refer to Appendix B.3.1 for a brief review.

7.4.1. DOL-Systems

The basic idea of an L-system is contained in the nature of formal languages. The geometrical forms to be presented are *words* in a formal language, a *parallel graph grammar*. The grammars in L-systems are similar to the formal grammar described in Appendix B.3.1, except that productions are applied simultaneously (in a parallel form), and there is no distinction between terminal and nonterminal symbols. All strings generated from an initial string, called *axiom*, are considered to be words in the language of the grammar. In formal grammars, productions are applied sequentially, one at a time, whereas in L-systems they are applied in parallel and simultaneously replace all letters in a given word.

The introduction of L-systems revived interest in the representation of images using string characters. Before proceeding with a more formal definition of an L-system, consider the following simple example that illustrates the main idea in intuitive terms (Figure 7.19). Consider an alphabet G composed of three letters a, b, and c, $G = \{a,b,c\}$, which may occur any number of times in a word. Each letter is associated with a production rule, as follows:

Rules:

1. $a \rightarrow c$

2. $b \rightarrow ac$

3. $c \rightarrow b$

The process of applying the rules, called *rewriting* or *derivation process*, starts from the *axiom*. Assume, in this example, that the axiom consists of a single letter c. The rewriting process basically functions by replacing parts of the simple initial object (the axiom) using the set of productions. In the first rewriting step, the axiom c is replaced by b by using production rule 3 ($c \rightarrow b$). In the second step, b is replaced by ac using production rule 2 ($b \rightarrow ac$). The word ac consists of two letters, both of which are simultaneously replaced in the next time step. Thus, a is replaced by c, and c is replaced by b, and the word becomes ac.

A *string* or *word OL-system* is defined as the ordered triplet $G = \langle V,\omega,P \rangle$, where V is the *alphabet* of the system, $\omega \in V^+$ is a nonempty word called the *axiom*, and $P \subset V \times V^*$ is a finite set of productions. Note that, formally, the main difference between this grammar and the formal grammar described in Appendix B.3.1 is that the present one makes no distinction between a terminal and a nonterminal symbol (Prusinkiewicz and Lindenmayer, 1990).

A production $(a,\chi) \in P$ is written as $a \rightarrow \chi$. The letter a and the word χ are called the *predecessor* and the *successor* of this production, respectively[1]. It is assumed that for any letter of the alphabet there is at least one word from the set V^* such that $a \rightarrow \chi$; that is, $\exists \chi \in V^* \mid a \rightarrow \chi, \forall a \in V$. The intuitive notion behind this formalism is very simple; basically, it assures that for any letter of the alphabet there is at least one production rule. If no production is specified by a

---

[1] In formal logic these are called *antecedent* and *consequent*, respectively.

Iteration	Word
0	*c*
1	*b*
2	*ac*
3	*cb*
4	*bac*
5	*accb*
6	*cbbac*
7	*bacaccb*
8	*accbcbbac*
9	*cbbacbacaccb*
10	*bacaccbaccbcbbac*

Figure 7.19: Example of a rewriting process in a DOL-system. Axiom: *c*, productions: $(a \to c)$, $(b \to ac)$, and $(c \to b)$.

given predecessor $a \in V$, then the identity production $a \to a$ is assumed to belong to the set of productions P. An OL-system is *deterministic*, noted *DOL-system*, if and only if for each $a \in V$ there is exactly one $\chi \in V^*$ such that $a \to \chi$.

Let $\mu = a_1 \ldots a_m$ be an arbitrary word over V. The word $v = \chi_1 \ldots \chi_m \in V^*$ is *directly derived* from (or *generated* by) μ, noted $\mu \Rightarrow v$, if and only if $a_i \to \chi_i$ for all $i = 1, \ldots, m$. A word v is generated by G in a derivation of *length* n if there exists a *developmental sequence* of words $\mu_0, \mu_1, \ldots, \mu_n$ such that $\mu_0 = \omega$, $\mu_n = v$ and $\mu_0 \Rightarrow \mu_1 \Rightarrow \ldots \Rightarrow \mu_n$.

Algorithm 7.1 presents a simple algorithm that can be used to generate a DOL-system. The algorithm receives as inputs the maximum number of iteration steps `max_it`, the axiom ω, and the set of production rules P. Function `rewrite` (simultaneously) replace all parts of the word according to the production rules. (Note that another way of implementing this system is by using recursion.)

```
procedure [word] = DOL_system(max_it,ω,P)
    word ← ω
    t ← 1
    while t < max_it do,
        word ← rewrite(word,P)
        t ← t + 1
    end while
end procedure
```

Algorithm 7.1: A simple DOL-system procedure.

7.4.2. Turtle Graphics

In the description presented above, L-systems were used to generate a sequence of words or strings. The *geometric interpretation* of these words can be used to generate schematic images of diverse natural patterns. A modified *turtle graphics* (Papert, 1980) language can be used to interpret the words generated by L-systems (Szilard and Quinton, 1979). The concept of *turtle graphics* was proposed by S. Papert as a simple computer language that children could use to draw graphical pictures. Under turtle graphics, plotting is performed by a smart little turtle that follows certain commands. Most of the commands involve simple forward movement or rotation.

The basic idea of turtle interpretation is as follows. The *state* of the *turtle* is defined as a triplet (x,y,α), where the Cartesian coordinates (x,y) represent the turtle's position, and the angle α, called the *heading*, is interpreted as the direction to which the turtle is facing. Given the *step size d* and the *angle increment* δ, the turtle can respond to commands represented by the following symbols:

F, G	Move forward a step of length d. The state of the turtle changes to (x', y', α), where $x' = x + d.cos\alpha$, and $y' = y + d.sin\alpha$. A line segment between points (x,y) and (x',y') is drawn.
f, g	Move forward a step of length d without drawing a line.
+	Turn right by angle δ. The next state of the turtle is $(x, y, \alpha + \delta)$. The positive orientation of angles is clockwise.
−	Turn left by angle δ. The next state of the turtle is $(x, y, \alpha - \delta)$. The negative orientation of angles is counterclockwise.

All other symbols are ignored by the turtle (the turtle preserves its current state). Given a string v, the initial state of the turtle (x_0,y_0,α_0) and fixed parameters d and δ, the *turtle interpretation* of v is the figure (set of lines) drawn by the turtle in response to string v. This method is particularly interesting to be applied to interpret words generated by L-systems. In the example shown in Figure 7.20, the turtle starts at coordinates $(x,y) = (1,1)$ and moves to coordinates $(2,4)$. Black (solid) turtles represent the initial and final position of the turtle, and dashed turtles correspond to its intermediary steps.

To illustrate the turtle interpretation of a word generated by an L-system, consider the case of the *quadratic Koch island* taken from Mandelbrot's book (Mandelbrot, 1982; p. 51). An L-system to implement this curve is as follows (Prusinkiewicz and Lindenmayer, 1990):

$\omega: F - F - F - F$

$p: F \rightarrow F - F + F + FF - F - F + F$

Figure 7.21 presents the turtle interpretation of this L-system for the first 3 rewriting processes. The only difference between this implementation in relation to the one presented in Algorithm 7.1 is the addition of a function `turtle` that receives as input the `word`, angle δ, and step length d, and interprets it as

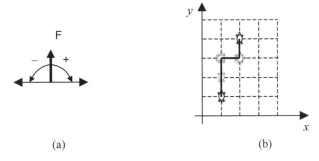

(a)   (b)

Figure 7.20: Turtle interpretation (angle increment $\delta = 90°$). (a) Interpretation of the symbols F, +, and −. (b) Illustrative figure drawn by the turtle for the word FF+F−F.

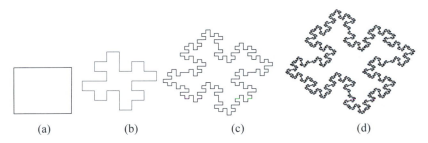

(a)   (b)   (c)   (d)

Figure 7.21: Turtle interpretation of the process of generating a quadratic Koch island. The angle increment $\delta = 90°$ and $d = 1$. (a) $t = 0$ rewriting (interpretation of the axiom). (b) $t = 1$ rewriting. (c) $t = 2$ rewritings. (d) $t = 3$ rewritings.

described above. This function can be added to Algorithm 7.1, right after the `while` loop, as described in Algorithm 7.2.

```
procedure [word] = DOL_turtle(max_it,ω,P,d,δ)
    word ← ω
    turtle(word,d,δ);          //interpretation of the axiom
    t ← 1
    while t < max_it do,
        word ← rewrite(word,P)
        turtle(word,d,δ);      //interpretation of rewritings
        t ← t + 1
    end while
end procedure
```

Algorithm 7.2: A simple DOL-system procedure with turtle interpretation at each iteration step, including the interpretation of the axiom, allowing the visualization of the developmental process.

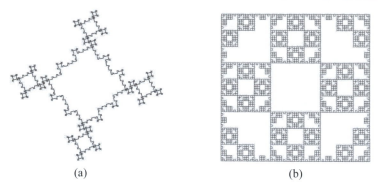

Figure 7.22: Different Koch curves generated from the axiom ω: *F+F+F+F*, *d* = 1, δ = 90°, and *t* = 4. (a) *p*: *F+F+F–FFF–F*. (b) *p*: *FF+F+F+F+FF*.

Note that the Koch curve generated above is a fractal curve. Many other fractals can be obtained by modifying L-systems. For instance, by simply using the axiom ω: *F+F+F+F* and different productions, it is possible to generate a number of distinct fractals, as illustrated in Figure 7.22.

7.4.3. Models of Plant Architecture

In his 1968 papers, Lindenmayer (1968) introduced a notation for representing graph-theoretic trees using strings with *brackets*. The *bracketed L-systems* extend an L-system alphabet by the set {[,]}. The motivation was to formally describe branching structures found in many plants, from algae to trees, using the general framework of L-systems. Subsequently, the geometric interpretation of bracketed L-systems was used with the purpose of presenting modeled structures in the form of computer-generated figures.

The two new symbols '[' and ']' are interpreted by the turtle as follows:

[Save the current state (x,y,α) of the turtle for later use onto a stack of saved states.

]   Remove the last saved state from the stack and use it to restore the turtle's last state. No line is drawn, although in general the position of the turtle changes.

A simple example is the L-system with alphabet $V = \{F,G,[,],+,-\}$, axiom ω = *F*, δ = 45°, and the following productions (Smith, 1984):

p_1: $F \rightarrow G[-F]G[+F]F$

p_2: $G \rightarrow GG$

p_3: $[\rightarrow [$

p_4: $] \rightarrow]$

The first three derivations (iterations) are:

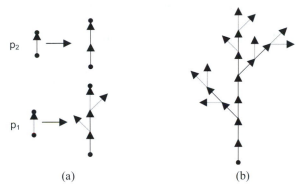

Figure 7.23: Graphic representation of a bracketed L-system. (a) Production rules. (b) Tree generated at time step $t = 2$.

Iteration	Word
0	F
1	$G[-F]G[+F]F$
2	$GG[-G[-F]G[+F]F]GG[+G[-F]G[+F]F] G[-F]G[+F]F$

An equivalent graphic representation of this L-system is depicted in Figure 7.23, where the letter F is represented by a dashed arrow and the letter G is represented by a solid arrow, but both correspond to move forward drawing a line.

There are a number of advanced commands that can be used in L-systems with various geometric interpretations. Some of these symbols control the turtle orientation in space, others allow the creation and incorporation of surfaces, and others change the drawing attributes, as listed below.

Symbols that control turtle orientation in space

| Reverse direction (i.e., turn by 180 degrees) without drawing a line.

& Pitch down by angle δ.

∧ Pitch up by angle δ.

\ Roll left by angle δ.

/ Roll right by angle δ.

Symbols for creating and incorporating surfaces

{ Start saving the subsequent positions of the turtle as the vertices of a polygon to be filled.

} Fill the saved polygon.

Symbols that change the drawing attributes

Increase the value of the current line width by the default width increment.

!	Decrease the value of the current line width by the default width decrement.
;	Increase the value of the index of the current color map by the default increment.
,	Decrease the value of the index of the current color map by the default decrement.

Figure 7.24 presents some examples of plant-like structures generated with bracketed OL-systems.

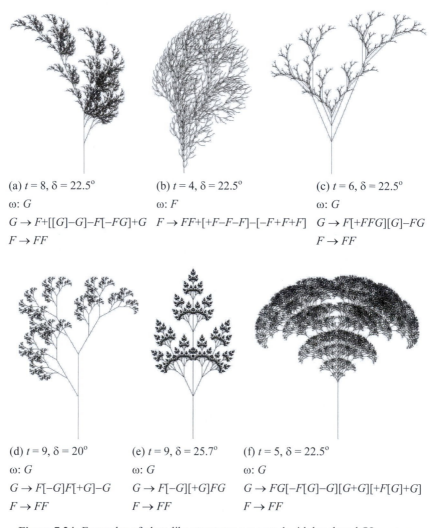

(a) $t = 8$, $\delta = 22.5°$

ω: *G*

$G \rightarrow F+[[G]−G]−F[−FG]+G$

$F \rightarrow FF$

(b) $t = 4$, $\delta = 22.5°$

ω: *F*

$F \rightarrow FF+[+F−F−F]−[−F+F+F]$

(c) $t = 6$, $\delta = 22.5°$

ω: *G*

$G \rightarrow F[+FFG][G]−FG$

$F \rightarrow FF$

(d) $t = 9$, $\delta = 20°$

ω: *G*

$G \rightarrow F[−G]F[+G]−G$

$F \rightarrow FF$

(e) $t = 9$, $\delta = 25.7°$

ω: *G*

$G \rightarrow F[−G][+G]FG$

$F \rightarrow FF$

(f) $t = 5$, $\delta = 22.5°$

ω: *G*

$G \rightarrow FG[−F[G]−G][G+G][+F[G]+G]$

$F \rightarrow FF$

Figure 7.24: Examples of plant-like structures generated with bracketed OL-systems.

7.4.4. Scope of L-systems

The realistic images generated by L-systems are useful in different applications, such as landscape design, ornamental design, and botanical illustration. But the usefulness of L-systems also goes beyond this. They provide models of the operation of other phenomena, including processes such as tropisms, abscission, signal propagation, or watering. Some of the disciplines that may benefit from these abilities are developmental biology, microbiology, biodiversity, agronomy, and ecology. There are a number of applications in which L systems play important roles as models of natural patterns (Room et al., 1996):

- Reconstruction of extinct plant species.

- Identification of plant responses to pest attacks.

- Structural models of trees integrated in more complex forest ecosystem simulations.

- Design of new varieties of plants.

- Synthesis of seashells and other natural patterns.

- Modeling of developmental processes.

- Classification of branching patterns in inflorescences.

- Crop yield prediction.

- Modeling of fractal patterns.

- Inflorescence description.

- Simulation of fungal growth.

- Computer aided learning for farm managers.

Also, they are present in other non-biological fields like formal language theory, tumor growth, and musical composition.

7.5   ITERATED FUNCTION SYSTEMS

The *iterated function systems* (IFS) were developed by J. Hutchinson, M. Barnsley, and S. Demko (Hutchinson, 1981; Barnsley and Demko, 1985; Barnsley, 1988) as a form of generating fractals through the use of a set of transformations, also called *contractive mappings*, of an image upon itself. Basically, it is constituted of the recursive application of a set of (simple) *affine transformations* (see Appendix B.4.6) to a set of initial points (image). After a number of iterations, the final or limit set is going to define a certain geometric configuration in the plane. This type of algorithm has been broadly used to generate images of complex structures found in nature.

7.5.1. Iterated Function Systems (IFS)

A contraction is a transformation f that reduces the distance between every pair of points in X. More formally, a transformation $f: X \rightarrow X$ on a metric space

($\mathbf{X}$,d) is called *contractive* or a *contraction mapping* if there is a constant $0 \le s < 1$ such that

$$d(f(x,y), f(x',y')) \le s.d((x,y), (x',y')), \quad \forall(x,y) \in \mathbf{X} \tag{7.13}$$

where s is called the *contractivity factor* for f.

Two important types of contractions have already been discussed: *similarity* and *affinity*. A similarity reduces all distances by the same number, $s < 1$; that is,

$$d(f(x,y), f(x',y')) = s.d((x,y), (x',y')), \quad \forall(x,y) \in \mathbf{X} \tag{7.14}$$

The transformation $f(x,y) = (s.x,s.y)$ is an example of a similarity with contraction factor s. An affinity reduces distances by different amounts in different directions. For example, $f(x,y) = (s.x,r.y)$, where $r < 1$ and $s < 1$, $r \ne s$.

A very interesting property of contractions is that independently of the starting point, the iterative application of a contraction mapping to it results always in the convergence to the same point, called the *attractor* of the contraction (see Appendix B.4.3 for attractors). This is the main result of the *Contraction Mapping Theorem* described as follows:

Let $f : \mathbf{X} \to \mathbf{X}$ be a contraction mapping on a complete metric space ($\mathbf{X}$,d). Then f possesses exactly one fixed point $x_f \in \mathbf{X}$ and moreover, for any point $x \in \mathbf{X}$, the sequence $\{f^{\circ n}(x) : n = 0, 1, 2, \ldots\}$ converges to x_f. That is,

$$\lim_{n \to \infty} f^{\circ n}(x) = x_f, \forall x \in \mathbf{X}. \tag{7.15}$$

An *iterated function system* (IFS) consists of a complete metric space ($\mathbf{X}$,d) together with a finite set of contraction mappings $w_n : \mathbf{X} \to \mathbf{X}$, with respective contractivity factors s_n, $n = 1, 2, \ldots, N$.

In most cases, the nomenclature IFS is used to simply mean a finite set of mappings acting on a metric space, with no particular conditions imposed on the mappings. It is then possible to render pictures of attractors of an IFS on the graphics display of a computer.

Let us restrict the attention to IFS of the form $\{\Re^2; w_n : n = 1, 2, \ldots, N\}$, where each mapping is an affine transformation. The algorithms to be presented are: 1) the Deterministic Algorithm, and 2) the Random Iteration Algorithm.

Deterministic Iterated Function System (DIFS)

Consider first the case of a *deterministic iterated function system* (DIFS). This algorithm is based on the idea of directly computing a sequence of sets $\{A_n = W^{\circ n}(A)\}$ starting from an initial set A_0. Let $\{\mathbf{X}; w_1, w_2, \ldots, w_N\}$ be an IFS. Choose a compact set $A_0 \subset \Re^2$ that will serve as an initial condition to the DIFS. Then compute successively $A_n = W^{\circ n}(A)$ according to

$$A_{n+1} = \cup_{j=1}^{N} w_j(A_n), \quad n = 1, 2, \ldots \tag{7.16}$$

Thus, a sequence $\{A_n : n = 0, 1, 2, \ldots\}$ will be constructed, and the sequence $\{A_n\}$ converges to the attractor of the IFS according to the contraction mapping theorem.

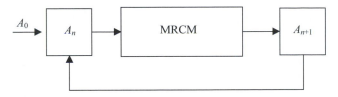

Figure 7.25: A schematic representation of the multiple reduction copy machine (MRCM).

A useful metaphor to understand how the deterministic iterated function system works is known as the *multiple reduction copy machine* (MRCM). In the literature, these terminologies, MRCM and DIFS, are sometimes used interchangeably. Basically, the copy machine takes an image as input and generates sequences of images by applying a set of contractions to the image obtained at the previous iteration, starting from the initial image. The crucial idea is that the MRCM runs in a feedback loop; its output is fed back as its new input again and again. Figure 7.25 illustrates a MRCM, where A_0 is the initial set of points (image) fed into the machine, and A_n is the set (image) at iteration n.

The MRCM (DIFS) is based on a collection of contractions (affine transformations) $w_1, w_2, \dots, w_N$, which are applied to the initial image A_0 producing small affine copies of this image $w_1(A_0), w_2(A_0), \dots, w_N(A_0)$. The machine brings all these copies together into a single new image:

$$A_1 = w_1(A_0) \cup w_2(A_0) \cup \dots \cup w_N(A_0). \tag{7.17}$$

The sequence of images generated by Equation (7.16), which is a generalization of Equation (7.17), converges to a unique image, A, which is the only image (of finite extent) invariant under the simultaneous application of the contractions $w_1, w_2, \dots, w_N$: $A = w_1(A) \cup w_2(A) \cup \dots \cup w_N(A)$. Because of this convergence property, A is called the *attractor* of the IFS $\{X; w_1, w_2, \dots, w_N\}$.

Figure 7.26 depicts the attractor of a DIFS initialized with a triangle. Despite the initialization with a triangle, the set of affine transformations used to generate this picture will always converge to the Sierpinski gasket. The code for generating this fractal is presented in Table 7.3(a) to be presented later.

Random Iterated Function System (RIFS)

Consider now the case of the *random iterated function systems* (RIFS). Let $\{X; w_1, w_2, \dots, w_N\}$ be an IFS, where a probability $p_i > 0$ has been assigned to each w_i, $i = 1, \dots, N$, $\sum_i p_i = 1$. Choose $x_0 \in X$ and then choose recursively and independently a new point x_1 obtained by applying only one of the transformations, chosen according to a given probability, to the current point x_0. Generically, given x_0:

$$x_n \in \{w_1(x_n - 1), w_2(x_n - 1), \dots, w_N(x_n - 1)\}, \forall n \tag{7.18}$$

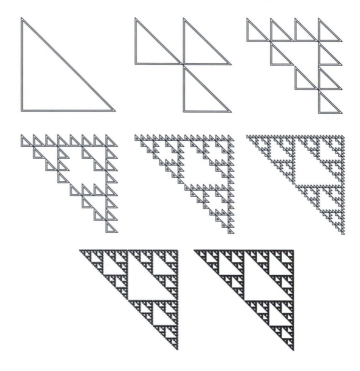

Figure 7.26: The result of running the deterministic iterated function system starting with a triangle. Each picture, from left to right and top to bottom, corresponds to one iteration (*n*) of the algorithm.

where the probability of an event $x_n = w_i(x_n - 1)$ is p_i. Thus, construct a sequence $\{x_n : n = 0, 1, 2, ...\} \subset \mathbf{X}$.

Let $\mathbf{x}_0$ be the vector indicating the initial point in $\Re^2$, $\mathbf{W}$ be the set of all affine transformations, `max_it` the number of iterations to be performed, and $\mathbf{p}$ the vector containing the probability of application of each affine transformation. Algorithm 7.3 provides the pseudocode of an algorithm to implement the random iterated function system. Note that the function select in this algorithm can be implemented using a *roulette wheel* procedure as described in Chapter 3.

In addition to the probabilistic nature of the RIFS, another important difference between the DIFS and the RIFS is that the latter updates a single point at each time step, while the former works on a sequence of points. Therefore, in the case of the DIFS there is an exponential increase in the number of images (points) plotted at each time step, while the RIFS generates a single new point at each time step. This means that a few number of time steps of the deterministic algorithm are usually equivalent to a very large number of steps obtained (points plotted) using the random algorithm.

```
procedure [] = RIFS(max_it,x0,W,p)
    x ← x0;
    t ← 1
    while t < max_it do,
        j ← select(pⱼ)  //select mapping j with probability pⱼ
        x ← wⱼ(x)        //apply mapping j to x
        draw(x)          //plot point x on the screen
        t ← t + 1
    end while
end procedure
```

Algorithm 7.3: A procedure to implement the random iteration function system (RIFS).

7.5.2. Creating Fractals with IFS

Table 7.3 presents various codes that can be used to generate the fractals illustrated in Figure 7.27. Each of these fractals was generated by plotting 10,000 points starting from the following initial condition $x = [0.01, 0.01]$.

7.5.3. Self-Similarity Revisited

In Section 7.2.1 the concepts of self-similarity and self-affinity were discussed and illustrated with the examples of a broccoli plant and other computer generated patterns. All the plant-like structures generated using L-systems, and depicted in Figure 7.24, also present either self-similarity or self-affinity. The fractals created with iterated function systems are also self-similar or self-affine in most cases.

Table 7.3: IFS codes for several fractals. (a) Sierpinski Gasket. (b) Square. (c) Barnsley fern. (d) Tree.

w	A	b	c	d	e	f	p
1	0.5	0	0	0	1	1	0.33
2	0.5	0	0	0	1	50	0.33
3	0.5	0	0	0	50	50	0.34

(a)

w	a	b	c	d	e	f	p
1	0.5	0	0	0.5	1	1	0.25
2	0.5	0	0	0.5	50	1	0.25
3	0.5	0	0	0.5	1	50	0.25
4	0.5	0	0	0.5	50	50	0.25

(b)

w	a	b	c	d	e	f	p
1	0	0	0	0.16	0	0	0.01
2	0.85	0.04	−0.04	0.85	0	1.6	0.85
3	0.2	−0.26	0.23	0.22	0	1.6	0.07
4	−0.15	0.28	0.26	0.24	0	0.44	0.07

(c)

w	a	b	C	d	e	f	p
1	0	0	0	0.5	0	0	0.05
2	0.42	−0.42	0.42	0.42	0	0.2	0.40
3	0.42	0.42	−0.42	0.42	0	0.2	0.40
4	0.1	0	0	0.1	0	0.2	0.15

(d)

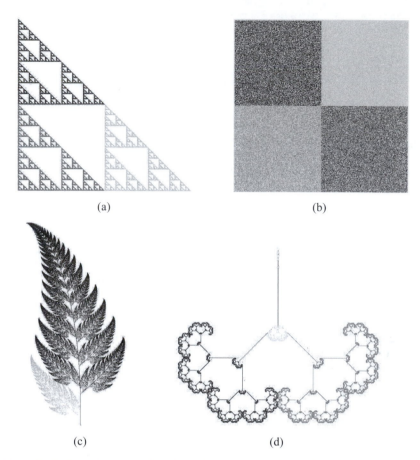

(a)   (b)

(c)   (d)

Figure 7.27: Fractals generated by the codes presented in Table 7.3. (Different gray levels correspond to different mappings.) (a) Sierpinski Gasket. (b) Square. (c) Barnsley fern. (d) Tree.

To emphasize the self-similarity and self-affinity concepts for the case of IFS, consider the pictures presented in Figure 7.28. In Figure 7.28(a), the Barnsley fern is plotted with a very high resolution. The pictures from left to right show a 10 times enlargement, from one picture to the other. The sites where the magnifying glasses were placed are depicted. Note that all enlarged pictures are copies of the whole fern. An infinite level of details can be observed. The same is performed with the picture of a tree depicted in Figure 7.28(b). Can you tell where were these enlargements taken from? The code to generate the tree presented in Figure 7.28(b) is provided in Table 7.4(d). Figure 7.29 shows some real ferns *in natura*.

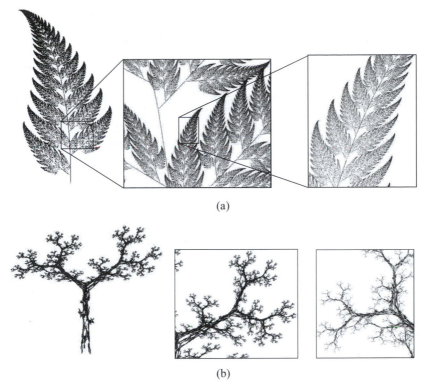

(a)

(b)

Figure 7.28: Self-similarity and self-affinity illustrated. (a) Barnsley fern. Each enlargement corresponds to a 10× increase over the selected box on the left picture. (b) Fractal tree. Can you figure out what parts of the pictures on the left were enlarged (10×) so as to generate the pictures on the right?

(a) (b)

Figure 7.29: Real ferns *in natura*.

7.5.4. Scope of IFS

The scope of iterated function systems is very similar to the scope of L-systems: landscape and ornamental design, botanical illustration and the synthesis of natural patterns. However, while L-systems are used for modeling growth and developmental processes, due to the contractive nature of IFS they are basically used for the generation of fractal shapes. The main advantage of iterated function systems when applied to the modeling and synthesis of natural patterns is their capability of representing patterns in a simple form, making them suitable for data compression. Actually, IFS have been broadly used for image compression (Barnsley and Hard, 1993; Fisher, 1995).

7.6   FRACTIONAL BROWNIAN MOTION

This section describes some methods for generating models of coastlines, mountains and other shapes. To model coastlines, it is necessary to have curves that look different when magnified but still possess the same characteristic impression. Thus, looking at the magnified version of the coastline should cause the impression that you are looking at a different part of the coastline drawn at the same scale, not at a magnification of the coastline. The text start by introducing the concept of Brownian motion and then follows with its generalization, fractional Brownian motion.

7.6.1. Random Fractals in Nature and Brownian Motion

To start the discussion, let us first describe the so-called *random walks*. A random walk is a path that can be generated by a random process as follows. Take the x and y coordinates of an object and add to each of these values a random step size. The new location corresponds to the previous one plus the random steps:

$$x(t+1) = x(t) + \Delta x$$
$$y(t+1) = y(t) + \Delta y$$

(7.19)

where Δx and Δy are random steps.

Repeat this process as many time steps as you wish. Figure 7.30 illustrates a random walk generated by adding independent Gaussian noise with zero mean and standard deviation one to the initial values of x and y. What is interesting about this picture is that you cannot say by inspection that Figure 7.30(a) is part - the first 100 steps - of Figure 7.30(b) or was independently generated, because random processes in nature are often self-similar on varying temporal and spatial scales; features akin to fractals. Such random walk is closely related to *Brownian motion*, which is found on the movement of particles in liquids and gases, and *white noise*, commonly used to describe other phenomena believed to be generated by random walk-like processes.

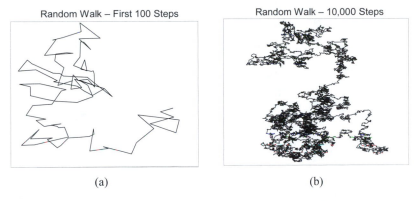

Figure 7.30: Random walk. (a) First 100 steps. (b) 10,000 steps.

Consider now an experiment which results in random fractal dendritic (tree-like) structures corresponding to an electrochemical aggregation process. A mathematical model of this aggregation process is based on Brownian motion and one important generalization of it, *fractional Brownian motion*, is the main topic of this section. With these tools in hands, fractal landscapes, coastlines, and many others nature-like phenomena can be simulated on a computer.

Electrochemical deposition has for long been one of the most familiar aggregation phenomena in chemistry and currently receives great attention from fractal geometry (Matsushita, 1989). Experimentally, controlled electrodeposition processes can be used to grow deposits exhibiting fractal structures. For instance, by placing a solution of zinc sulfate covered by a thin layer of n-butyl-acetate in a Petri-dish and applying a D.C. voltage to the experimental set up, it is possible to investigate the growth of fractal structures of electrodeposits (aggregates) and their morphological changes. Experiments on the aggregation of particles are relevant in polymer science, material science, immunology, and many other fields.

The mathematical modeling of the electrochemical deposition of zinc-metal is based on the fundamental concept of Brownian motion, which refers to the erratic movements of small particles of solid matter in liquid suspension (Peitgen et al., 1992). A simple technique for the simulation of this type of Brownian motion is known as *diffusion limited aggregation* (DLA) and was introduced by Witten and Sander (1981). To simulate a simple bi-dimensional DLA the following procedure can be adopted:

- Generate a square grid of cells, such as the grid in a cellular automaton.

- Fix a single cell, called the seed cell, at a given position of the grid, e.g. at its center. This cell is not allowed to move.

- Select a region of interest centered around the seed cell, e.g., a circular area of fixed radius, and introduce a new, moving particle at the boundary of the region and let it move about randomly. The neighborhood to which

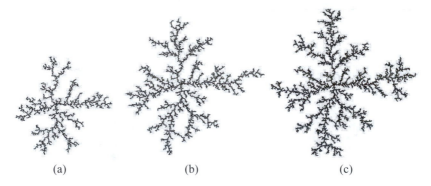

(a) (b) (c)

Figure 7.31: Three stages in the growth process of a simple 2-D diffusion-limited aggregation simulation. (a) 1,000 particles. (b) 2,500 particles. (c) 5,000 particles.

a particle moves can be of many types, including Moore's and von Neumann's types.

- If a moving particle leaves the area specified, then it is replaced by a new, randomly placed one; else, if a moving particle meets the seed cell or any cell belonging to the already fixed cells, then it attaches to the cluster becoming a fixed, nonmoving, particle.

This procedure is repeated until a cluster of connected particles is formed, resulting in a structure (Figure 7.31) that resembles the dendrites from diffusion-limited aggregation in electrochemical deposition. This process is called DLA because the growth of the cluster is governed by the diffusion of particles around the grid. Algorithm 7.4 contains the pseudocode for implementing a simple DLA algorithm. The grid can be generated as done with a cellular automaton and the stopping criterion can be either a fixed number of iteration steps or a given cluster size (cs).

In 1827, R. Brown observed that small particles suspended in a fluid behaved in continuous, erratic ways. The particles move about randomly because the fluid hits them in all directions. The same holds true for the movement of particles in gaseous media. In one-dimension, Brownian motion is characterized by a random process $X(t)$, which corresponds to a function X of a real variable t (time), whose values are random variables $X(t_1)$, $X(t_2)$, ... , where the increment $X(t_2) - X(t_1)$ has a Gaussian distribution and the mean square increments have a variance proportional to the time differences:

$$E[\,|X(t_2) - X(t_2)\,|^2\,] \propto |t_2 - t_1| \tag{7.20}$$

where $E[\cdot]$ denotes the mathematical expectation of a random variable.

The increments of X are statistically self-similar in the sense that

$$X(t_0 + t) - X(t_0) \text{ and } \frac{1}{\sqrt{r}}\big(X(t_0 + rt) - X(t_0)\big) \tag{7.21}$$

```
procedure [] = DLA(seed)
    C ← generate_C              //generate Grid
    Cr ← select_region(seed)    //region around seed
    p ← new_particle(Cr)        //new particle within Cr
    cs ← 1                      //cluster size
    while not_stopping_criterion do,
        p ← move_particle       //moving particle
        if p meets cluster,
            then attach p to cluster
                cs ← cs + 1
                p ← new_particle(Cr)
            else
                if p leaves Cr
                    p ← new_particle(Cr)
                end if
        end if
    end while
end procedure
```

Algorithm 7.4: A simple DLA algorithm.

have the same finite dimensional joint distribution functions for any t_0 and $r > 0$. By taking, for instance, $t_0 = 0$ and $X(t_0) = 0$, both random functions

$$X(t) \quad \text{and} \quad \frac{1}{\sqrt{r}}\left(X(rt)\right) \tag{7.22}$$

are statistically indistinguishable; the second one being just a properly rescaled version of the first (Peitgen and Saupe, 1988).

A popular method to generate Brownian motion is a *recursive subdivision algorithm*, introduced by Fournier et al. (1982), and currently known as the *random midpoint displacement* algorithm. The behavior of the algorithm is quite simple. If the process $X(t)$ is to be computed for time $t \in [0, 1]$, then start by setting $X(0) = 0$ and choosing $X(1)$ as a sample of a Gaussian value with mean 0 and variance σ^2. In the first step, the midpoint between $t = 0$ and $t = 1$ is determined as the average of $X(0)$ and $X(1)$ plus an offset D_1 of mean 0 and variance Δ_1^2:

$$X(\tfrac{1}{2}) - X(0) = \tfrac{1}{2}(X(1) - X(0)) + D_1 \tag{7.23}$$

As a sample of a Gaussian random variable has mean 0 and variance σ^2, it is expected that

$$\mathrm{var}(X(t_2) - X(t_1)) = |t_2 - t_1|\sigma^2 \tag{7.24}$$

Added to the fact that at each iteration the number of fragments doubles, the offset D_n, at time step n, must have a variance (see exercises)

$$\Delta_n^2 = \frac{1}{2^{n+1}}\sigma^2 \tag{7.25}$$

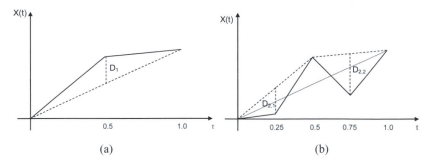

Figure 7.32: First steps of the random midpoint displacement algorithm. (a) In the first step the initial segment is divided in the middle. (b) In the second step each of the resulting segments is divided in the middle. The process continues until a given number of recursions have been performed.

Figure 7.32 summarizes the first two steps of the random midpoint displacement algorithm. In the first step, a displacement D_1 is added to the midpoint of the initial interval (dashed line). In the second step, each of the resulting segments from the previous step is divided in the middle and two new displacements $D_{2,1}$ and $D_{2,2}$ are determined and added to the values of the midpoints $X(\frac{1}{4})$ and $X(\frac{3}{4})$, respectively.

Algorithm 7.5 contains a pseudocode for the recursive subdivision (random midpoint displacement) algorithm. In this algorithm $\texttt{nrc}$ is the desired number

```
procedure [x] = randmidpoint1D(nrc,σ,seed)
    initialize_randn(seed)
    N ← 2^nrc
    x(0) ← 0
    x(N) ← σ*randn
    for i from 1 to nrc do
        Δ(i) ← σ*(0.5)^((i+1)/2)
    end for
    Recursion(x,0,N,1,nrc)          //call recursive procedure
end procedure

procedure Recursion(x,t0,t2,t,nrc)
    t1 ← (t0+t2)/2
    x(t1) ← 0.5*(x(t0)+x(t2))+Δ(t)*randn
    if t < nrc,
        Recursion(x,t0,t1,t+1,nrc)
        Recursion(x,t1,t2,t+1,nrc)
    end if
end procedure
```

Algorithm 7.5: Recursive subdivision algorithm (random midpoint displacement).

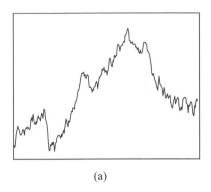 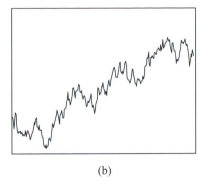

(a)   (b)

Figure 7.33: Two examples of Brownian motion generated with the random midpoint displacement algorithm described in Algorithm 7.5.

of recursions, σ is the initial standard deviation of the Gaussian distribution `randn`, Δ is the vector of variances for each point, and $\mathbf{x}$ is the vector containing the Brownian motion on one-dimension. Figure 7.33 depicts two simulations of the algorithm described. The parameter `seed` is the seed value for the random number generator. This is used to allow reproducing the same fractal.

7.6.2. Fractional Brownian Motion

Mandelbrot and van Ness (1968) introduced the term *fractional Brownian motion* (fBm) as a way to refer to a family of Gaussian random functions capable of providing useful models of various natural time series. Since then, a number of extensions and approximations have been developed to model a wide range of natural phenomena, from mountainous landscapes to clouds.

Fractional Brownian motion is a generalization of Brownian motion defined as a random process $X(t)$ with Gaussian increments and

$$\mathrm{var}(X(t_2) - X(t_2)) \propto |t_2 - t_1|^{2H} \tag{7.26}$$

where H is a parameter satisfying $0 < H < 1$. Note that $H = \frac{1}{2}$ results in ordinary Brownian motion.

In this more general case, the increments of X are statistically self-similar, with parameter H, in the sense that:

$$X(t_0 + t) - X(t_0) \quad \text{and} \quad \frac{X(t_0 + rt) - X(t_0)}{r^H} \tag{7.27}$$

have the same finite dimensional joint distribution functions for any t_0 and $r > 0$. By taking, for instance, $t_0 = 0$ and $X(t_0) = 0$, both random functions

$$X(t) \quad \text{and} \quad \frac{1}{r^H}\big(X(rt)\big)$$

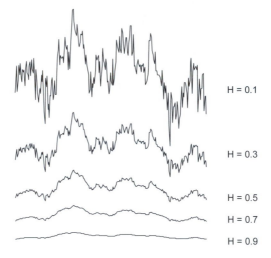

Figure 7.34: Fractional Brownian motion generated by the random midpoint displacement algorithm. The picture shows the influence of the parameter H on the motions generated.

are statistically indistinguishable; the second one being just a properly rescaled version of the first (Peitgen and Saupe, 1988).

In order to apply the random midpoint displacement algorithm to the more general case of fBm, Equation 18 has to be extended to

$$\text{var}(X(t_2) - X(t_1)) = |t_2 - t_1|^{2H} \sigma^2 \tag{7.28}$$

The midpoint displacements then become

$$\Delta_n^2 = \frac{\sigma^2}{(2^n)^{2H}}(1 - 2^{2H-2}) \tag{7.29}$$

Therefore, the only change to be made in Algorithm 7.5 in order to accommodate fBm corresponds to the update of the values of vector $\boldsymbol{\Delta}$.

The parameter H, called the Hurst exponent, describes the 'roughness' of the fraction at small scales, as depicted in Figure 7.34. This parameter is directly related with the fractal dimension, d, of the graph as follows

$$d = 2 - H \tag{7.30}$$

The extension of the random midpoint displacement algorithm to three or more dimensions is straightforward, and leads to algorithms capable of generating realistic landscape pictures. Assume that the fractal landscape is going to be generated starting from a square grid - other shapes, such as triangles, could be used instead. The idea is to apply the recursive subdivision algorithm until a certain granularity of the grid (picture) is obtained. The surface heights over the midpoints of all squares are computed by interpolating the heights of their four

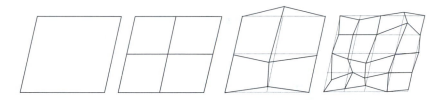

Figure 7.35: Main steps - subdivision and vertical perturbation - in the 3D version of the random midpoint displacement algorithm when a square grid is used. The picture shows a terrain surface with a grid of dimension 2 × 2 and then 4 × 4.

neighbor points plus an appropriate Gaussian random offset. The elevation of the remaining points also has to be determined by taking into account a Gaussian random offset. The algorithm can be summarized as follows (Figure 7.35):

- Determine the midpoints of the current grid.

- Vertically perturb each of the new vertices generated by an appropriate Gaussian distribution.

- Repeat this process for each new square generated, decreasing the perturbation at each iteration.

The scaling factor for the random numbers should be decreased by $1/2^H$ $(0 < H < 1)$ at each iteration. The parameter H again determines the fractal dimension, i.e., roughness, of the resulting fractal landscape: $d = 3 - H$.

The main control parameters for such an algorithm are: 1) a seed for the random number generator; 2) the roughness coefficient H; 3) the initial standard deviation; 4) the initial set of points; 5) a *sea level* - a 'flood' value below which the surface is assumed to be under water; 6) a color grid; and 7) the resolution of the grid (number of iterations). Figure 7.36 illustrates the influence of parameter H on the landscapes generated. Shown are the mesh and contour plots of the landscapes. The following parameters were used to generate these pictures: seed = −2, $\sigma_0 = 1$, and nrc = 6. All initial points were chosen to have Gaussian distribution.

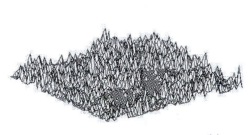 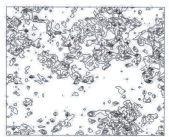

(a)

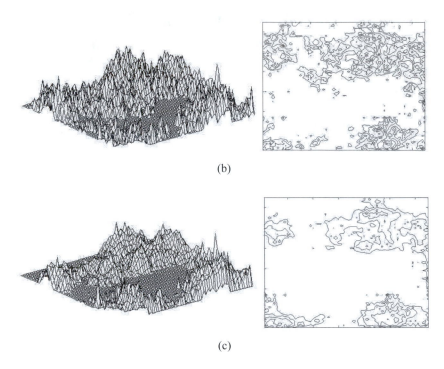

(b)

(c)

Figure 7.36: Fractal landscapes (mesh and contour plots) generated using the random midpoint displacement algorithm. Same initial parameters and varying values of H. The resolution of the grid is 65×65, what corresponds to a recursion of level 6. (a) $H = 0.1$. (b) $H = 0.5$. (c) $H = 0.9$.

7.6.3. Scope of fBm

Although this section focused on the application of Brownian motion to random walks, the simulation of diffusion limited aggregation and the creation of fractal landscapes, its applications go far beyond these. First and foremost, virtually every kind of dynamical process in nature contains Brownian motion, from the movement of butterflies to the movement of small particles in liquid or gaseous media. It also finds applications in fields such as economics (e.g., for the simulation of the behavior of stock prices and other commodities), ecology, material science, immunology, etc. Concerning computer graphics, Brownian motion can also be used to simulate the course of rivers, clouds, coastlines on maps, the folding and unfolding of some materials (e.g. paper), for the design of flight simulators, traffic generation, and computer animation in general. Figure 7.37 depicts one last picture for this section where the clouds, mountain, water and snowline were all generated using a variety of fBm.

Figure 7.37: Cool Afternoon II. The clouds, mountain, water and snowline were all generated using a variety of fBm. (Courtesy of © K. Musgrave.)

7.7 PARTICLE SYSTEMS

Some of the most beautiful physical phenomena are the flowing, dripping and pouring of liquids, the liquid mixing with other substances, gases in motion, explosions, clouds, fireworks, etc. The complex behavior of this very large number of phenomena can be modeled approximately or accurately using *particle systems* (PS). A particle system consists of a collection of particles (objects) with various properties and some behavioral rules they must obey. The precise definition of these properties and laws depends on what is intended to be modeled. Some particle systems may require complex rules and many properties, while others may be quite simple.

7.7.1. Principles of Particle Systems

A particle is a point in a space, usually the three-dimensional space, which has a number of attributes, such as position, velocity, color, lifetime, age, shape, size, and transparency. A particle system is a collection of particles that together represent an object. The particles are dynamically changing over time as a result of external forces or some stochastic process. Particle systems were introduced by W. Reeves as a method for modeling fractal patterns (what he called 'fuzzy' objects) such as clouds, smoke, water, and fire (Reeves, 1983). The standard PS consists of independent particles; that is, particles that do not interact with each other, only with the environment.

The representation of particle systems differs in three basic ways from the representation normally used in image synthesis: 1) a pattern is represented as clouds of primitive particles that define its volume, instead of being represented by a set of primitive surface elements; 2) particles are dynamic in the sense that they can change form, move over time, 'be born' and 'die'; 3) a pattern represented by particles is not deterministic, i.e., its shape and form are not completely prespecified.

Particle systems also have a number of interesting features from the modeling perspective: 1) a particle is a very simple primitive; 2) the model definition is procedural and is controlled by some random processes, thus allowing the creation of complex patterns with little human effort; 3) particle systems model dynamic (alive) patterns, such as burning fire and moving clouds; and 4) the level of detail can be easily controlled by regulating the number of particles in the system.

7.7.2. Basic Model of Particle Systems

Over a period of time, particles are generated and inserted into the system, move and change within the system, die and are removed from the system. The attributes of a particle are going to depend upon the pattern being modeled. In the original proposal of Reeves (1983), he applied a particle system to the wall of fire element from the Genesis Demo sequence of the film *Star Trek II: The Wrath of Khan* (Paramount, 1982). In this standard model, the particles had the following attributes:

- *Position*: every particle has a position in the space. If in the 2D space, two coordinates are used (x,y), and if in the 3D space, three coordinates are used (x,y,z).

- *Velocity* (speed and direction): velocity is a vector that tells each particle how fast and in what direction it is going to move. At each time step (frame), the velocity vector is added to the particle's position in order to move it.

- *Color*: the use of colors allows shading and other features that create life-like visual effects, such as the change in brightness and color in different parts of a cloud or a fire.

- *Lifetime or age*: the lifespan of a particle may be finite for several reasons, such as computing power restraints or the need for visual effects. For instance, one may want particles such as sparks from an explosion to burn out after a few seconds, in which case other attributes, such as color, may also change over time. The *age* indicates how old is a particle and is used as a parameter to remove particles from the system.

- *Shape*: the shape of a particle is a very important attribute used to display the final pattern. For instance, if a particle system is being used to model fireworks, then particles can be simply points.

- *Size*: the size of a particle has a similar importance to its shape. It is going to contribute in the definition of the final image.

- *Transparency*: same as above, but transparency also contributes to the superposition of colors and the visual effects of depth.

The procedure to compute each frame (time step) in a motion sequence involves the following sequence of steps: 1) new particles are generated; 2) each new particle is assigned its attributes; 3) too old particles are eliminated; 4) the remaining particles are transformed and moved according to their dynamic attributes; and 5) an image of the remaining particles is *rendered*.

Particle Generation

Particles are generated into a PS by means of controlled stochastic processes. Two forms of generating particles were originally proposed. In the first method, the mean number of particles and its variance are controlled at each time step. In the second method, the number of new particles depends on the screen size of the object. Let us consider only the first method:

$$N = m_p + r.\sigma_p \tag{7.31}$$

where N is the current number of particles at a given frame, m_p is the mean number of particles, r is a random number generated using a uniform distribution over the interval $[-1, +1]$, and σ_p is the variance of the number of particles.

The mean number of particles generated, m_p, can also be controlled:

$$m_p = m_{p0} + \Delta m_p.(f - f_0) \tag{7.32}$$

where m_{p0} is the initial mean number of particles, Δm_p is the rate of change in the mean number of particles, f is the current frame (time step) and f_0 is the initial frame during which the PS is working. (Note that for $\Delta m_p > 1$, the mean number of particles increases, and for $0 < \Delta m_p \le 1$, m_p decreases.)

Particle Attributes

Each new particle generated must have its attributes initialized. It means that the initial position, velocity, size, color and transparency have to be defined. In addition, the shape and lifetime of the particle have to be chosen.

A particle system has several parameters to control the initial position of the particles, such as their initial coordinates in the space, the angle(s) of rotation that give their orientation, and a *generation shape* (e.g., a circle or a sphere of radius R) that defines the region around the origin in which new particles are randomly placed and their direction of movement as well.

The initial speed s of a particle can be given by:

$$s = m_s + r.\sigma_s \tag{7.33}$$

where m_s is the mean speed of the particles, r is a random number generated using a uniform distribution over the interval $[-1, +1]$, and σ_s is the variance of the speed of the particles.

The initial color c can be given by the average color m_c and a maximum deviation from that color σ_c:

$$c = m_c + r.\sigma_c \tag{7.34}$$

The transparency and size are also determined by mean values and maximum variations, using expressions similar to the ones presented above. The shapes of the particles can be diverse, and vary according to the type of phenomenon being modeled. Their lifetime f_{max} is also a user-defined parameter.

Particle Extinction

All particles are given a lifetime in frames when generated. After the computation of each frame, the lifetime of a particle is decremented by one unit, or equivalently, its age is increased by one unit. When its lifetime reaches zero or its age reaches a maximal value, the particle dies and is removed from the particle system. Other killing mechanisms could be used, such as eliminating a particle when it hits the ground.

Particle Dynamics

Individual particles in a particle system move in the space and can change color, size, and transparency over time. A particle is moved by adding its velocity vector to its position vector:

$$\mathbf{x} = \mathbf{x} + \mathbf{v} \qquad (7.35)$$

where $\mathbf{x}$ is the position vector of the particle, and $\mathbf{v}$ its velocity vector ($\mathbf{v} = d\mathbf{x}/dt \approx \Delta\mathbf{x}$).

More complexity can be added by using an acceleration factor that alters the speed of the particles. This parameter allows the designer to simulate gravity and other external forces, thus causing particles to move in various forms:

$$\mathbf{v} = \mathbf{v} + \mathbf{a} \qquad (7.36)$$

where $\mathbf{v}$ is the velocity vector of the particle and $\mathbf{a}$ its acceleration ($\mathbf{a} = d\mathbf{v}/dt \approx \Delta\mathbf{v}$).

Particle Rendering

Once the particle positions have been determined, the last step is to render the scene. There are several methods that can be classified into two categories: rendering individual particles, or rendering the surface of the object the particles are representing. Rendering individual particles works well for modeling waterfalls or spray, and is usually faster but can only give a general idea about liquid or other continuous materials. A surface rendering will give a more accurate description of the object if it is made of some cohesive substance (i.e. slime vs. fireworks), but is generally much slower, especially with a system that is changing over time.

7.7.3. Pseudocode and Examples

Let `max_it` be the maximum number of frames to be run, **P** the data structure (e.g. a matrix) containing all the particles in the system and their attributes, d a global death threshold (particles older than the threshold are removed), and o all

```
procedure [] = PS(max_it,d,o)
    initialize P;      //generate particles; assign attributes
    t ← 1
    while t < max_it do,
        P ← destroy(P,d)     //destroy particles older than d
        P ← dynamics(P,o)    //change each remaining particle
        render(P)            //draw particles and plot
        initialize P'        //generate new particles
        P ← insert(P,P')     //insert the new particles into P
        t ← t + 1
    end while
end procedure
```

Algorithm 7.6: A procedure to implement the standard particle system (PS). (Note: several parameters are needed in functions initialize and render.)

other parameters necessary to run the PS algorithm. Algorithm 7.6 presents a procedure to implement the standard PS.

To illustrate how to design an elementary particle system, consider the following example of simulating a firework. The goal is very simple, after the particle is fired (launched), it is going to travel (fly up) for some time, and then it is going to explode. Several other particles will be generated to simulate the explosion. It is important not to forget that the particles (firework) are subjected to a simulated gravity force. In this simple example, the particles may have the following attributes (see Figure 7.38):

- *Position*: $\mathbf{p} = (x,y)$ on the plane.
- *Velocity*: (v,θ), where v is the speed of the particle and θ is the angle with which it (the firework) is launched.
- *Color*: the initial color of the particle (firework) may be different from its color after it explodes and during the flight.
- *Lifetime*: this firework has two global life spans, one before it explodes and another after explosion. The initial particle lives until it explodes, then the other particles, generated to simulate the explosion, have their own, usually much smaller, life spans.
- *Shape*: to simplify the explanation, all particles will have the shape of a small circle.
- *Size*: the particles after explosion will have half the size of the fired particle.
- *Transparency*: shadings and transparency can be used at will.

The initial position of the particle is anyone within the proposed area, and its initial height is assumed to be $y_0 = 0$. The initial velocity is $\mathbf{v}_0 = (v_0,\theta_0)$ and it is subjected to a vertical force simulating gravity. No horizontal force is assumed to act upon the particle. At each time step t (frame), the current position $\mathbf{p}$ of the particle is updated according to the following equation:

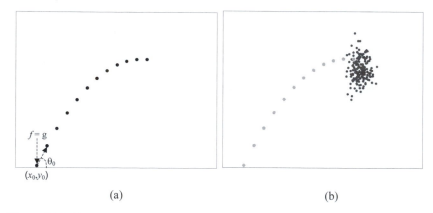

(a)   (b)

Figure 7.38: Simulating firework with particle systems. (a) Initial position and direction of the particle, and the particle's trajectory until the time frame it explodes. (c) Firework explosion.

$$\mathbf{v} = \mathbf{v} + \mathbf{a}$$
$$\mathbf{p} = \mathbf{p} + \mathbf{v} \qquad\qquad (7.37)$$

where $\mathbf{v}$ is the velocity of the particle, $\mathbf{a} = (0,-a)$ is the force acting upon it (only gravity a is assumed to act on the particle). Note that as the particle is suffering the action of a negative vertical force (gravity), its speed decreases with time (see Figure 7.39).

Figure 7.39 depicts a particle system simulating fireworks. During the explosion 200 particles are generated from each of the original particles. The trajectories of each particle are frozen so as to illustrate the whole process.

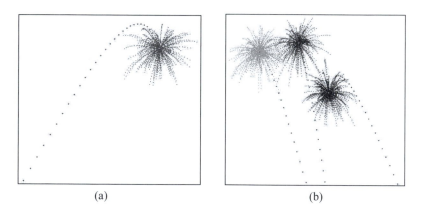

(a)   (b)

Figure 7.39: Particle system simulating fireworks. (a) A single firework. (b) Three different, but simultaneous, fireworks.

7.7.4. Scope of Particle Systems

Particle systems have been used to model waterfalls, ship wakes and breaking waves, fire, clouds, water, fountains, fireworks, explosions, schools of fish, gases, stars, rainbows, fluid dynamics, plants, trees, grass, forest scenes, flocks of birds, tornados, etc. The applications of PS also include screen savers for computers (e.g., http://www.mephzara.com/particle-systems/main_e.php), several motion pictures, such as *Twister* by Warner Bros Pictures, and *The Prince of Egypt* by Dreamworks Pictures, and computer games.

7.8 EVOLVING THE GEOMETRY OF NATURE

Chapter 3 introduced evolutionary algorithms as techniques for solving problems by performing a search in a search space. In the next chapter, Section 8.3.2, the *Blind Watchmaker* (Dawkins, 1986) will be briefly described as an application of a genetic algorithm to the creation of several artificial life forms that may resemble natural patterns. K. Sims (1991) was one of the pioneers in the use of evolutionary algorithms to create structures, textures, and motions for use in computer graphics and animation. This section makes a brief review of the application of an evolutionary algorithm to the design of L-systems.

One of the main essences of this type of application of evolutionary systems is that the user is the judge who chooses the short-term path of evolution. In this case, the user judgment must contribute to or replace the fitness function(s). Evolutionary algorithms modified this way are termed interactive evolutionary algorithms, or *collaborative evolutionary algorithms* (Bentley and Corne, 2002).

7.8.1. Evolving Plant-Like Structures

In the Blind Watchmaker, the simulated evolution of 2D branching structures made from sets of genetic parameters resulted in life-like patterns. The user selected the *biomorphs* that survive and reproduce to create the next generation. Since it is difficult and subjective to automatically measure the aesthetic visual appeal (beauty) of patterns, the fitness of each evolved individual is provided by a human observer based on visual perception. Some combinations of automatic selection and interactive selection can also be used. As a result, smaller population sizes may be employed and the evolutionary process may take longer to run because human interaction is required. But, note that this may not be viewed as purely a limitation of the approach; instead it is a form of guaranteeing the desired human appeal.

In all the procedural models described in this chapter, a set of simple productions, mappings, or movement rules described the new states of the system; that is, how they are going to develop with time. When using evolutionary algorithms to evolve patterns, however, a different design framework can be proposed containing five basic elements (Bentley and Corne, 2002) (see Figure 7.40):

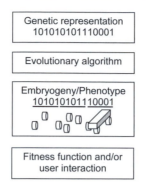

Figure 7.40: Main elements of the framework for the use of evolutionary systems to design natural patterns.

- *A genetic representation*: the genes in this case will represent, directly or indirectly, a variable number of components, and the search space may be of various sizes (although only the fixed size case will be discussed here).

- *An evolutionary algorithm*: whose operators are responsible for generating new patterns (genotypes) by combining features from existing patterns and introducing variation.

- *Embryogeny*: this is a special type of mapping process from genotype to phenotype in which the genotype is viewed as a set of growing or developmental instructions; that is, a recipe that defines how the phenotype will develop.

- *A phenotype representation*: once constructed by the embryogeny, the resulting solution is assessed by the phenotype representation. Typically this representation is specific to the application: if we are to evolve plants, then the representation might define their branching structures, etc.

- *A fitness function and/or user interaction*: the fitness function must provide an evaluation score for every solution. In the case of interactive selection, however, this process may be a little more complicated. This is mainly because evolution does need time to produce a pattern that may look appealing to us and, in most cases, the initial configurations are random and thus hard to choose from.

Note that the distinction between embryogeny and phenotype representation is sometimes blurred. Some researchers consider them to be the same thing.

In this new approach to generate natural patterns, the genetic operators will play a role in sampling the space of possible geometrical forms by randomly changing some of the attributes of each genotype. This allows the exploration of the 'pattern space' in incremental arbitrary directions without requiring any knowledge about the specific effects of each attribute of the data structures.

In (Bonfim and de Castro, 2005), the authors investigated the effect of applying an evolutionary algorithm to evolve derived bracketed L-systems. They studied the importance of evolving plants from the same species and showed that the crossing-over of plants from different species is equivalent to performing a graft between the parent plants. Therefore, if the parent plants are not appropriately selected, then mongrel offspring, termed 'FranksTrees', may result.

In their approach, a bracketed L-system plant was seen as a genetic programming (GP) tree, as described in Section 3.6.3, with each bracket representing the interconnection between tree nodes and the initial point of each bracket representing a potential cut point. However, the GP tree is not completely equivalent to an L-system tree (plant). For instance, in L-system plants there are not variables and operators as in computer programs represented as parse trees, only symbols that indicate the direction of movement of a turtle agent responsible for drawing the plant. In L-systems there is an inherent difficulty in distinguishing which symbols belong to terminals or nonterminal nodes. A method to appropriately recombine derived L-system words thus had to be designed. Figure 7.41 illustrates the evolutionary design of plants using GP. Note that when plants of the same species are crossed the result is a 'usual' offspring, but when plants from different species are crossed the result is a FranksTree.

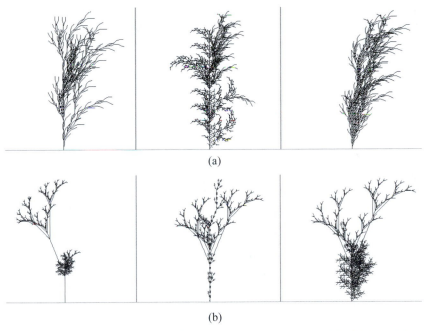

(a)

(b)

Figure 7.41: Aesthetic evolution of derived bracketed L-systems. (a) Evolution of three similar plants (of same 'species'). (b) Evolution of three different plants (of different 'species').

7.8.2. Scope of Evolutionary Geometry

It is interesting to note that this example of application of evolutionary algorithms uses evolution more as an explorer than an optimizer, which was the perspective presented in Chapter 3. Normally, guided by a human, the system is used to investigate many possible life-like forms, from trees and grass to whole forests. Nevertheless, the application domain of evolutionary geometry is not restricted to computer graphics. The subfield of evolutionary computing, now known as *creative evolutionary systems* (Bentley and Corne, 2002), has found applications in areas such as computer animation, engineering design, music composition, arts, artificial life, robotics, military applications, and, of course, the biosciences.

7.9   FROM NATURAL TO FRACTAL GEOMETRY

Fractals have helped reconnect pure mathematics research with both the natural sciences and computing. Over the last thirty years, fractal geometry, together with its concepts and applications, have become central tools in most natural sciences: physics, chemistry, biology, geology, meteorology, and materials science. Also, fractals are of interest to graphic designers and filmmakers for their ability to create new, but realistic shapes of natural patterns. Although fractal images appear complex, they can be constructed from simple rules and algorithms. The computer rendering of fractal shapes leaves no doubt of their relevance to the natural sciences and others. It also plays an important role in computer graphics.

This chapter described a number of models to generate natural patterns, such as plant-like structures and the fluid of particles. However, it could be noticed that all these algorithms generate some sort of 'skeleton' of the corresponding natural patterns, more complex procedures and some rendering techniques have to be used in order to generate patterns with a visual effect that more closely resembles nature. The fusion of fractal geometry with advanced computer graphics results in pictures of unprecedent realism. Figure 7.42 gives an idea of how much life-like the patterns generated by some of these systems and others can appear after some computer graphics' processing. Figure 7.42(a) illustrates the pattern of lightining generated with a random midpoint displacement algorithm (the background was inserted as a contextualization). Figure 7.42(b) displays a palm tree created using a parametric L-system (Prusinkiewicz and Lindenmayer, 1990). The use of fractal geometry to generate nature-like landscapes has been investigated for decades. K. Musgrave (www.kenmusgrave.com) has been an important contributor to this area. Another site that is worth looking at is the site by A. Brown (http://www.fractal-landscapes.co.uk/). Figure 7.42(c) describes a fractal landscape by K. Musgrave.

The images speak for themselves!

(a)

(b)

(c)

Figure 7.42: Patterns generated by some of the techniques discussed in this chapter. (a) 'Lightening' generated using a random midpoint displacement algorithm. (b) 'Palm tree' generated with a parametric L-system. (c) 'Lethe'; a landscape generated using fractional Brownian motion. [(a),(b) Courtesy of © N. M. Suaib from the Department of Computer Graphics & Multimedia, FSKSM, UTM; (c) Courtesy of © K. Musgrave from Pandromeda.com.]

7.10 SUMMARY

The present chapter reviewed some techniques to generate life-like patterns, focusing the case of plants. Basic notions of fractals, fractional Brownian motion, cellular automata, L-systems, iterated function systems, particle systems, and some comments about the evolutionary creation of plants were provided. Most techniques discussed here find diverse application areas, from computer graphics to the biosciences. The potential of these and other techniques for the generation of natural life-forms is illustrated in Figure 7.42. Other applications of some of the concepts described here is image compression (see Project 5, Section 7.11.4) and film-making in general. The next chapter will discuss the main philosophy of computationally inspired biology focusing behavioral patterns under the perspective of artificial life.

7.11 EXERCISES

7.11.1. Questions

1. Using Equation (7.2) determine the dimension of the Cantor set, the Koch curve, and the Sierpinski gasket.

2. Repeat the previous exercise using the box-counting method.

3. From the relation $N((1/2)^n) = 3^n$ for the Sierpinski gasket it is possible to compute the exact value of its box-counting dimension d_b. Determine d_b analytically using Equation (7.5).

4. The Peano curve starts with an initial line segment and is expanded iteratively with a simple rule. For each step in the construction, all line segments are replaced with a curve of nine smaller line segments. Each of the smaller line segments is one-third the length of the line segments from the previous step. There will also be eight 90-degree turns between each of the new smaller line segments. This curve has a very interesting property: it is a space-filling curve.

   Draw two iterations of the Peano curve and determine its dimension. Is it a fractal? Discuss.

5. Find an L-system to construct the Cantor set.

6. Given the definition of a metric space, show that the following are all metrics in the space $\mathbf{X} = \Re$:

   $d(x,y) = |x - y|$

   $d(x,y) = |x^3 - y^3|$

7. What is the contraction factor for the contraction (affinity) $f(x,y) = (x/4, y/3)$? Justify your answer.

8. Write equations for initializing all the parameters of a particle system (Section 7.7.2).

9. Contrast the particle systems (PS) described here with the particle swarm optimization algorithm (PSO) introduced in Chapter 4.

10. The boids algorithm developed by Reynolds (1987) is also considered a particle system. Discuss the main differences between the PS described here and the boids algorithm introduced by Reynolds (Section 8.3.1).

11. Evolutionary algorithms have been used in combination with L-systems, IFS, and particle systems. Perform a literature review and provide a list of references for each of these hybrid systems.

7.11.2. Computational Exercises

1. It is possible to generate a fractal with a simple game, named *the chaos game*. It is basically a dot drawing game where dots are plotted according to some simple rules. All you will need is a dice (or a random number generator).

    Plot (draw) three dots forming an equilateral triangle and a fourth dot in any random place within the triangle. This fourth dot is the starting point.

    Take a dice and roll it. If the top face is 1 or 2, draw another dot half way from the starting point to the first point. If the top face is 3 or 4, draw another dot half way from the starting point to the second dot. If the top face is 5 or 6, draw another dot half way from the starting point and the third dot. The new point becomes the new starting point, and the whole process iterates.

    Implement an algorithm to play the chaos game for no less than 5,000 plays, and print the fractal it generates. Is it familiar?

2. Implement the simple one-dimensional cellular automaton example of Section 7.3.1. Compare the evolution of your CA with the one presented in Figure 7.15.

3. Reproduce the fractal pattern presented in Section 7.3.4.

4. Suggest 10 different rule tables for the elementary CA of Section 7.3.1 and plot the evolution in time for 500 time steps of the CA.

5. Implement a DOL-system given the rules below and the axiom $\omega = c$. Compare the development of the system with the one presented in Section 7.4.1.

    Rules:
    $p_1: a \rightarrow c$
    $p_2: b \rightarrow ac$
    $p_3: c \rightarrow b$

6. One of the first applications of a DOL-system was to simulate the development of a fragment of a multicellular filament such as that found in the blue-green bacteria *Anabena catenula* and various algae. The symbols a and b can be used to represent the cytological states of the cells, i.e., their size and readiness to divide. Subscripts l and r will be used to indicate cell pola-

rity, specifying the positions in which daughter cells of type a and b will be produced. The development is described by the following L-system:

$\omega: a_r$

$p_1: a_r \rightarrow a_l b_r$

$p_2: a_l \rightarrow b_l a_r$

$p_3: b_r \rightarrow a_r$

$p_4: b_l \rightarrow a_l$

Implement a DOL-system to simulate the development of *Anabaena catenula*, and list the sequence of words obtained in the first 10 generations.

Knowing that under the microscope the filaments appear as a sequence of cylinders of various lengths, with e-type cells longer than b-type cells, draw a schematic image of filament development according to the result obtained with the DOL-system implemented. Hint: you can indicate the polarity of a cell with an arrow. For instance, an l polarity is indicated by a left arrow, and an r polarity is indicated by a right arrow.

7. Implement an L-system to derive the quadratic Koch island described in Section 7.4.1 and visualize the resultant words, up to derivation 5, using turtle interpretation. Compare your results with the pictures shown in Figure 7.21.

8. Implement an L-system to derive the Koch curves shown in Figure 7.22.

9. Propose an L-system to generate the Cantor set; that is, define the axiom ω, δ, and the productions to generate the Cantor set using an L-system. Implement your proposed L-system.

10. Implement a bracketed OL-system and reproduce all plant-like structures of Figure 7.24. Change some derivation rules and see what happens. Make your own portfolio with, at least, ten plants.

11. In the previous chapter we illustrated the capability of a cellular automaton to generate a fractal pattern with the Sierpinski-Gasket. A bracketed OL-system can also be used to draw the Sierpinski-Gasket as follows:

$\delta = 60°$

$\omega: F$

$p_1: F \rightarrow F - - F - - F - - GG$

$p_2: G \rightarrow GG$

Implement this OL-system and use the turtle interpretation to plot the resultant Sierpinski-Gasket.

12. Write a pseudocode for the deterministic iterated function system (DIFS) presented in Section 7.5.

13. Run the algorithm developed in the previous exercise for a DIFS whose attractor is the Sierpinski gasket. The code is presented in Table 7.3(a). As-

sume that the initial set is a square with side of arbitrary length (e.g., 100 pixels).

14. Repeat the previous exercise with an initial set corresponding to a triangle. What happens to the final picture? Is it the same as the one found with the square initialization? Justify.

15. Implement a RIFS to generate all the fractals whose codes are presented in Table 7.3.

16. Implement a RIFS to generate all the fractals whose codes are presented in Table 7.3, but now, plot the points generated by each mapping with a different color. Discuss the fractals generated.

17. Change the probabilities and some parameters of the codes provided in Table 7.3, generate the fractals, and discuss the influence of these parameters in the final geometry.

Table 7.4: Codes to generate fractals using RIFS (and DIFS).

w	a	b	c	d	e	f	p
1	0.387	0.430	0.430	−0.387	0.256	0.522	0.33
2	0.441	−0.091	−0.009	−0.322	0.422	0.506	0.33
3	−0.468	0.020	−0.113	0.015	0.400	0.400	0.34

(a)

w	a	b	c	d	e	f	p
1	0.5	0	0	0.75	0.25	0	0.52
2	0.25	−0.2	0.1	0.3	0.25	0.5	0.16
3	0.25	0.2	−0.1	0.3	0.5	0.4	0.16
4	0.2	0	0	0.3	0.4	0.55	0.16

(b)

w	a	b	c	d	e	f	p
1	0	0	0	0.5	0	0	0.15
2	0.02	−0.28	0.15	0.2	0	1.5	0.10
3	0.02	0.28	0.15	0.2	0	1.5	0.10
4	0.75	0	0	0.5	0	4.6	0.65

(c)

w	a	b	c	d	e	f	p
1	0.195	−0.488	0.344	0.443	0.443	0.245	0.2
2	0.462	0.414	−0.252	0.361	0.251	0.569	0.2
3	−0.058	−0.07	0.453	−0.111	0.598	0.097	0.2
4	−0.035	0.07	−0.469	−0.022	0.488	0.5069	0.2
5	−0.637	0	0	0.501	0.856	0.251	0.2

(d)

18. Table 7.4 presents four new codes to generate fractals using RIFS. Discuss what type of pattern is generated by each of these codes. Implement the algorithms and print the resultant fractals (for fractals (a) and (c) plot 20,000 points; for fractals (b) and (d) plot 100,000 points). Use different colors for each mapping so that you can visualize their importance.

19. Implement the random midpoint displacement algorithm to simulate Brownian motion and plot some pictures of Bm.

20. Repeat the previous exercise for fBm and study the influence of H on the motions generated.

21. Implement the random midpoint displacement algorithm in 3D and generate some fractal landscapes. Study the influence of H on the landscapes generated.

7.11.3. Thought Exercises

1. How can one find an affine transformation that approximately transforms one given set into another given set in $\Re^2$.

2. Demonstrate why the variance of the offset D_n, at time step n, to be used in the random midpoint displacement algorithm has to be

$$\Delta_n^2 = \frac{1}{2^{n+1}}\sigma^2$$

where σ is the standard deviation and n the iteration counter.

3. What is the scaling invariance in the graph of one-dimensional Brownian motion?

4. The *lbest* population in a PS algorithm (Chapter 5) suggests a relationship between particle swarms and cellular automata. Categorize particle swarms as a kind of CA.

7.11.4. Projects and Challenges

1. It has been discussed in the introduction that *reaction-diffusion models* (RDM) have been used to provide explanations of many observed phenomena, such as the generation of patterns covering models of animals, how ash forms, how water seeps through rocks, how cracks spread in a solid, and how the lightning discharges.

   Cellular automata can be considered as discrete-counterparts of RDM. Discuss.

2. The complexity apparent in the behavior of Class 4 cellular automata suggests that these systems may be capable of universal computation. A computer can be seen as a system in which definite rules transform an initial sequence of states into a final sequence of states. The initial sequence may be considered as a program and data stored in computer memory, and part of the final sequence may be considered as the result of the computation. CA may be considered computers: their initial configurations represent pro-

grams and initial data, and their configurations after a long time contain the results of computations.

A system is a universal computer if, given a suitable initial program, its time evolution can implement any finite algorithm (Chapter 9). A universal computer thus needs only be reprogrammed, not rebuilt to perform each possible calculation. Virtually every digital computer implements each of its operations in terms of the logical primitives AND, OR, and NOT. Strictly speaking, only one of AND or OR is needed, since each can be implemented by the other with the help of the NOT operation. If something can provide these basic logical primitives, then it is capable of universal computation.

Demonstrate that the Game of Life is capable of performing universal computation.

3. Discuss the main similarities and differences between cellular automata models and models created using differential equations.

4. Determine the L-systems necessary to generate all the pioneer fractals presented in Section 7.2.1. Provide the axiom ω and the productions p_i, $i = 1$, ... , n, necessary to construct each of these fractals and plot all the iteration steps from 0 (initial) to 10.

By simply scaling the triangle generated in the Koch curve and moving it to one side, for instance, the left side, it is possible to create a modified Koch curve. If you continue to iterate this new Koch curve, the results change dramatically. The attractor can look like a forest of bare trees, for example. Implement this change in order to achieve the end result suggested. Draw the resultant fractals and discuss.

5. All plants generated by the same deterministic L-system are identical. One form of preventing this effect, and reproduce individual-to-individual variation in natural plants, is by randomizing the L-system. A *stochastic OL-system* is an ordered quadruplet $G\pi = \langle V, \omega, P, \pi \rangle$. The new element in this grammar is the function $\pi : P \rightarrow (0,1]$, called the *probability distribution*, that maps the set of productions into the set of *production probabilities*. It is assumed that for any letter of the alphabet $a \in V$, the sum of probabilities of all productions with the predecessor a is equal to 1.

For the plant-like structures presented in Figure 7.24, mix up their productions so as to make a stochastic OL-system. For each original production there should be at least two new productions to be applied with varied probabilities.

Use the turtle interpretation to plot some plants generated with this stochastic OL-system and discuss their similarities.

6. Fractal image compression is a new technique to encode images compactly that builds on local self-similarities within images. Images blocks are seen as rescaled and intensity transformed approximate copies of blocks found elsewhere in the image. This results in a self-referential description of ima-

ge data. Survey the literature on fractal compression and discuss its most basic implementations.

7. Implement a particle system to simulate a fire or torch burning, in a form similar to the one described in Section 7.7.2.

   Change your implementation so that there are multiple generation points randomly created within a square boundary on the ground at random times during the sequence. The particles should be created in a cluster at each generation and should not be regenerated. What phenomenon does it look like?

7.12 REFERENCES

[1]   Barnsley, M. F. (1988), *Fractals Everywhere*, Academic Press.

[2]   Barnsley, M. F. and Demko, S. (1985), "Iterated Function Systems and the Global Construction of Fractals", *Proc. of the Royal Soc. of London*, **A339**, pp. 243–275.

[3]   Barnsley, M. F. and Hard, L. P. (1993), *Fractal Image Compression*, A. K. Peters, Wellesley, Mass.

[4]   Bentley, P. J. and Corne, D. W. (2002), *Creative Evolutionary Systems*, Morgan Kaufmann.

[5]   Bonfim, D. M. and de Castro, L. N. (2005), " 'FranksTree': A Genetic Programming Approach to Evolve Derived Bracketed L-Systems", In L. Wang, K. Chen and Y. S. Ong (Eds.), *Lecture Notes in Computer Science*, **3610**, pp. 1275–1278.

[6]   Dawkins, R. (1986), *The Blind Watchmaker*, Penguin Books.

[7]   Falconer, K. (2003), *Fractal Geometry: Mathematical Foundations and Applications*, John Wiley & Sons.

[8]   Ferrero, R. S. (1999), *A Skecth Book on L-systems*, EUITIO University of Oviedo, Spain [Online book] http://coco.ccu.uniovi.es/malva/sketchbook/.

[9]   Fisher, Y. (1995), *Fractal Image Compression: Theory and Applications*, Springer-Verlag.

[10]   Fournier, A., Fussell, D. and Carpenter, L. (1982), "Computer Rendering of Stochastic Models", *Communications of the ACM*, **25**(6), pp. 371-384.

[11]   Hutchinson, J. (1981), "Fractals and Self-Similarity", *Indiana Jorunal of Mathematics*, **30**, pp. 713–747.

[12]   Ilachinski, A. (2001), *Cellular Automata: A Discrete Universe*, World Scientific.

[13]   Langton, C. G. (1986), "Studying Artificial Life with Cellular Automata", *Physica 22D*, pp. 120–149.

[14]   Lesmoir-Gordon, N., Rood, W. and Edney, R. (2000), *Introducing Fractal Geometry*, ICON Books UK.

[15]   Lindenmayer, A. (1968), "Mathematical Models for Cellular Interaction in Development, Parts I and II", *Journal of Theoretical Biology*, **18**, pp. 280–315.

[16]   Mandelbrot, B. B. and van Ness, J. W. (1968), "Fractional Brownian Motions, Fractional Noises and Applications", *SIAM Review*, **10**(4), pp. 422–437.

[17]   Mandelbrot, B. (1983), *The Fractal Geometry of Nature*, W. H. Freemand and Company.

[18]   Mange, D. and Tomassini, M. (1998), *Bio-Inspired Computing Machines*, Presses Polytechniques et Universitaires Romandes.

[19]   Matsushita, M. (1989), "Experimental Observation of Aggregations", In D. Avnir (ed.), *The Fractal Approach to Heterogeneous Chemistry: Surfaces, Colloids, Polymers*, Wiley.

[20]   Meinhardt, H. (1982), *Models of Biological Pattern Formation*, Academic Press, London.

[21]   Murray, J. D. (1989), *Mathematical Biology*, Springer-Verlag, Berlin.

[22]   Papert, S. (1980), *Mindstorms: Children, Computers, and Powerful Ideas*, New York, Basic Books.

[23]   Paramount (1982), *Star Trek II: The Wrath of Khan* (film), June.

[24]   Peitgen, H.-O, Jürgens, H. and Saupe, D. (1992), *Chaos and Fractals: New Frontiers of Science*, Springer-Verlag.

[25]   Peitgen, H.-O and Saupe, D. (eds.) (1988), *The Science of Fractal Images*, Springer-Verlag.

[26]   Prusinkiewicz, P. and Lindenmayer, A. (1990), *The Algorithmic Beauty of Plants*, Springer-Verlag.

[27]   Reeves, W. T. (1983), "Particle Systems - A Technique for Modeling a Class of Fuzzy Objects", *ACM Transactions on Graphics*, **2**(2), pp. 91–108.

[28]   Room, P., Hanan, J. and Prusinkiewicz, P. (1996), "Virtual Plants: New Perspectives for Ecologists, Pathologists and Agricultural Scientists", *Trends in Plant Science*, **1**(1), pp. 33–38.

[29]   Sims, K. (1991), "Artificial Evolution for Computer Graphics", *Computer Graphics*, **25**(4), pp. 319–328.

[30]   Smith, A. R. (1984), "Plants, Fractals, and Formal Languages", *Computer Graphics*, **18**(3), pp. 1–10.

[31]   Szilard, A. L. and Quinton, R. E. (1979), "An Interpretation for DOL systems by Computer Graphics", *The Science Terrapin*, **4**, pp. 8–13.

[32]   Witten, T.A. and Sander, L. M. (1981), "Diffusion Limited Aggregation: A Kinetic Critical Phenomena", *Physical Review Letters*, **47**(19), pp. 1400–1403.

[33]   Wolfram, S. (1983), "Cellular Automata", *Los Alamos Science*, **9**, pp. 2–21.

[34]   Wolfram, S. (1984), "Cellular Automata as Models of Complexity", *Nature*, **311**, pp. 419–424.

CHAPTER 8

ARTIFICIAL LIFE

"The borders between the living and the nonliving, between the Nature-made and the human-made appear to be constantly blurring."
(M. Sipper, Machine Nature: The Coming Age of Bio-Inspired Computing, McGraw-Hill, 2002, p. 222).

"Molecular biologists seldom care to define life – they know it when they have it."
(C. Emmeche (1992), "Modeling Life: A Note on the Semiotics of Emergence and Computation in Artificial and Natural Living Systems", In A. Sebeok and J. Umiker-Sebeok (eds.), Biosemiotics: The Semiotic Web 1991, Mouton de Gruyter Publishers, pp. 77-99.)

8.1 INTRODUCTION

The essence of life has been sought for as long as human beings have been consciously thinking. We have had a great progress in understanding several aspects of the inanimate and animate worlds. Most experiments with living systems are geared toward understanding the living state based on the analysis or decomposition of natural organisms. *Artificial life* (ALife) performs one step toward the challenging achievement of synthesizing life-like phenomena and, ultimately, life. As put by C. Langton in the first of a series of workshops specifically devoted to the study of ALife:

"Artificial Life is the study of man-made systems that exhibit behaviors characteristic of natural living systems. It complements the traditional biological sciences concerned with the *analysis* of living organisms by attempting to *synthesize* life-like behaviors within computers and other artificial media. By extending the empirical foundation upon which biology is based *beyond* the carbon-chain life that has evolved on Earth, Artificial Life can contribute to theoretical biology by locating *life-as-we-know-it* within the larger picture of *life-as-it-could-be*." (Langton, 1988; p. 1, original emphasis)

However, artificial life does not have to be restricted to (life-like) behaviors, it can also encompass patterns and even living systems created by humans. Several other definitions for artificial life are available in the literature:

"Artificial Life (AL) is the enterprise of understanding biology by constructing biological phenomena out of artificial components, rather than breaking natural life forms down into their component parts. It is the synthetic rather than the reductionist approach." (Ray, 1994)

"Alife is a constructive endeavor: Some researchers aim at evolving patterns in a computer; some seek to elicit social behaviors in real-world robots; others wish to study life-related phenomena in a more controllable setting, while still others are interested in the synthesis of novel lifelike systems in chemical, electronic, mechanical, and other artificial media. ALife is an experimental discipline, fundamentally consisting of the observation of run-time behaviors, those complex interactions generated when populations of man-made, artificial creatures are immersed in real or simulated environments." (Ronald et al., 1999)

Some important ideas lie behind the study of artificial life:

- Life is viewed as a dynamic process with certain universal features that are not dependent upon the matter. Thus, life is an emergent property of the organization of matter, rather than a property of the matter itself.

- ALife employs a synthetic approach to the study and creation of life(-as-it-could-be).

- ALife involves life-like phenomena made by human beings rather than by nature, independently of the means used to synthesize life.

In summary, *ALife can be broadly defined as the synthetic or virtual approach to the study of life-like patterns (forms), behaviors, systems, and organisms.* As such, there are many systems and media that can be used to develop ALife products, such as digital computers, test tubes, mechanical devices, and so forth. These media can be used to perform experiments aimed at revealing principles, patterns, and organizations of known living systems (life-as-we-know-it) and unknown living systems (life-as-it-could-be).

Note that this definition of ALife could include organisms created or designed by humans, such as the Dolly sheep born on February 24, 1997, in Edinburgh, Scotland. Dolly was cloned from a cell of another sheep (Wilmut et al., 2001) and, prior to her, most scientists did not believe that mammals could be successfully cloned. Still, any organism which has been genetically engineered; that is, that has had its genetic material altered by humans (Nicholl, 2002), could be viewed as a form of ALife. Nevertheless, the focus of almost all ALife research is on systems synthesized under a computational platform; and this is the focus of this chapter as well.

ALife uses living and nonliving parts to study and generate patterns, behaviors, and organisms with the aim of increasing our understanding of nature and enhancing our insight into artificial models. These endeavors can be divided into two main areas or forms (Harnad, 1994; Adami and Brown, 2000):

- *Synthetic*: the creation of life using the classical building blocks of nature (carbon-based life) or any material radically different from those of natural living systems, but that are still capable of producing organisms with life-like patterns and behaviors.

- *Virtual*: the computer realization, emulation, or simulation of natural patterns, forms or behaviors. Virtual living systems can be systematically interpreted as if they were alive, but they are no real organisms; they live in a virtual environment.

The field of artificial life is sometimes divided into 'weak ALife' and 'strong ALife' (Sober, 1992; Rennard, 2004). In the so-called strong ALife, it is claimed that the outcomes of ALife research may really come to be living organisms; it represents the desire to build living things. Weak ALife by contrast, considers that models represent certain aspects of living phenomena; that is, weak ALife is concerned with building systems that act or behave as if they were alive.

8.1.1. A Discussion about the Structure of the Chapter

When the term *artificial intelligence* (AI) was coined, in the early to mid 1950s, researchers felt it was necessary to understand and define what was meant by 'intelligence', and what would the term artificial mean in the context of artificial intelligence. Although several attempts were made in this direction, there is still not much consensus in the literature as to what does intelligence mean. In the early days of AI, most research was concentrated on the human brain as a main source of intelligent processes. However, nowadays many intelligent systems are designed with the inspiration taken from much simpler organisms, such as insect societies (cf. Chapter 5). These techniques were studied in this book under the umbrella of computing inspired by nature.

Many artificial life systems are outcomes of the observation of natural phenomena as well. Thus, this chapter starts by discussing some similarities and differences between ALife and computing inspired by nature (Section 8.2.1). It then follows with a discourse concerning the difficulties in conceptualizing 'life' (Section 8.2.2). Similarly to AI, there are still many researchers trying to understand what is life in the context of ALife. Several references from the early days of ALife (late 1980s and early 1990s) address this issue.

Section 8.2.3 discusses the strict relationship between ALife and biology, and Section 8.2.4 provides some models and features of computer-based ALife. Most ALife systems are complex systems, and this is the theme of Section 8.2.5. Unlike most nature inspired techniques, it is very hard to provide a framework to design ALife systems. Thus, Section 8.3 visits several examples that reflect part of the philosophy of artificial life, while keeping in mind that the selection provided is neither complete nor even representative of the universe of approaches covered by the ALife community. The discussion is basically illustrative, no detail about each project is provided, because a complete description of a single of these ALife projects would certainly take most of the chapter. Thus, instead of making a broad and complete coverage of two or three projects, we felt it was better to include a much larger number of works as means of illustrating the richness of this broad field of investigation. In addition to the Artificial Life conference proceedings and journal (Appendix C), the recently released volume by Adamatzky and Komosinski (2005) brings a good selection of current research on ALife.

Throughout the text the reader will notice a strong relationship between computing inspired by nature and ALife. Several techniques (tools) of the former are used as part or are inspired by the latter. These aspects are discussed in due course, and the text is then concluded with the scope of ALife and a discussion of artificial life in the broader picture of life-as-we-know-it.

8.2   CONCEPTS AND FEATURES OF ARTIFICIAL LIFE SYSTEMS

8.2.1. Artificial Life and Computing Inspired by Nature

Together with the computational geometry of nature (Chapter 7), the field of Artificial Life constitutes the main component for the simulation and emulation of natural phenomena in computers. It uses several computer algorithms inspired by nature as methodological techniques for the simulation and emulation of life-like phenomena (see Section 8.3 for examples). However, ALife can be differentiated from techniques belonging to the computing inspired by nature branch, based on their different main rationales. While computing inspired by nature emphasizes problem-solving, artificial life focuses, in most cases, the understanding of life-as-we-know-it through the emulation and simulation of natural phenomena. Thus, one - computing inspired by nature - has a more technological (engineering problem solving techniques) motivation, while the other - ALife - is more concerned with the scientific development and biological understanding. Many applications of ALife can be found in the literature though, and some of these will be discussed later.

Therefore, the most important difference between ALife and computing inspired by nature techniques, such as evolutionary algorithms, artificial neural networks, swarm systems and artificial immune systems, is that ALife is not concerned with building systems to find solutions to problems. Instead, ALife focuses on the generation of life-like patterns, behaviors, systems, and organisms.

Despite that, there are also many similarities between these approaches. First, both involve complex systems exhibiting some sort of emergent phenomena and a strong relationship with nature. In the first case (computing inspired by nature), biology (nature) is the source of inspiration for the development of computational tools for problem solving. In the second case, by contrast, nature is usually used as a model for the creation or realization of known and new forms of life. Furthermore, there are some systems that can be viewed as belonging to both branches of natural computing, such as the use of ideas from ant retrieval of prey to engineer collective robotic systems (Chapter 5). More examples of this type will be given in Section 8.3.

8.2.2. Life and Artificial Organisms

One central question raised in artificial life regards how can we claim that a given system is a realization, emulation, or simulation of life (life-as-we-know-it) or even a new form of life (life-as-it-could-be). While simulations are assessed by how well they generate similar morphologies, patterns or behaviors of some specified aspects of a system, a realization is a literal, substantive and functional device that does not explicitly involve the specification of the global structure or behavior of the system (Pattee, 1988). The concept of emergent patterns and behavior is fundamental in ALife research.

To evaluate if a system is really an ALife system, not merely a simulation of life, some insights must be given as to what is meant by the word *life*, at least within the context of artificial life. What is life is a question that has been puzzling us for centuries. Definitions vary from philosophy to biology to psychology, and the literature on the subject is also vast (cf. Emmeche, 2000; Margulis et al., 2000; Schrödinger, 1992; Rennard, 2004). Consider first some dictionary's definitions of life:

life, n., pl. lives (Dictionary.com)

a. The property or quality that distinguishes living organisms from dead organisms and inanimate matter, manifested in functions such as metabolism, growth, reproduction, and response to stimuli or adaptation to the environment originating from within the organism.

b. The characteristic state or condition of a living organism.

Life\\ (l[imac]f), n.; pl. Lives (l[imac]vz). 1. The state of being which begins with generation, birth, or germination, and ends with death; also, the time during which this state continues; that state of an animal or plant in which all or any of its organs are capable of performing all or any of their functions; -- used of all animal and vegetable organisms.

M. Sipper quotes some alternative and poetic definitions of life (http://www.moshesipper.com):

"Life is one long process of getting tired." (Samuel Butler)

"Life's but a walking shadow, a poor player that struts and frets his hour upon the stage and then is heard no more: it is a tale told by an idiot, full of sound and fury signifying nothing." (Shakespeare, Macbeth V. v.)

Despite these attempts to define life, some ALife approaches involve assembling a list of properties of life, and then testing candidates on the basis of whether or not they exhibit any item of the list (Farmer and Belin, 1991; Ray, 1994). Another approach is to provide relationships between an artificial and a natural agent (Keeley, 1997). The question is how can we identify something as life-as-it-could-be (ALife) rather than something exceedingly interesting but not alive?

Keeley (1997) suggests three types of relationship between an artificial and a natural entity: 1) they can be *genetically* related; that is, they can share a common origin; 2) they can be *functionally* related in that they share properties when described at some level of abstraction; and 3) they can be *compositionally* related; that is, they can be made of similar parts arranged in similar ways.

Following another avenue, Farmer and Belin (1991) provided an eight-item list (argued to be imprecise and incomplete) of properties associated with life:

1. *Life is a pattern in space-time*, rather than a specific material or object.

2. *Life involves self-reproduction.* If not in the organism itself, at least some related organism must be capable of reproduction (mules cannot reproduce but none would argue they are not alive).

3. *Information storage of a self-representation.* For instance, in DNA molecules.

4. *Metabolism,* responsible for converting matter and energy from the environment into the pattern and activities of the organism.

5. *Functional interactions with the environment.* Living organisms can respond to or anticipate changes in the environment.

6. *Interdependence of parts.* The components of the living system depend on one another.

7. *Stability under perturbation* and insensitivity to small changes.

8. *The ability to evolve,* which is not a property of a single individual, but of a population of individuals.

Despite whether we consider a system as alive because it exhibits any or all of these properties of life, or because it relates to a natural organism (life-as-we-know-it), what is more important is the recognition that it is possible to create simple disembodied realizations or instances of (specific properties of) life in artificial media or systems. By separating the property of life to be studied from the many other complexities of natural living systems and focusing on the behaviors or patterns to be synthesized, it becomes easier to manipulate and observe the phenomena of interest. This capability is a powerful research tool employed by ALife researchers.

The products of artificial life, as differentiated from the scientific research field ALife, are usually called *artificial organisms.* They are automata or agents that exhibit life-like patterns, processes, and/or behaviors (Laing, 1988; Keeley, 1997). There are many possible media for artificial organisms. They might be made of carbon-based materials in a terrestrial or aqueous environment similarly to natural organisms; they might be robots, made of metal and silicon; they might be chemical, mechanical, or electronic devices; or they might be abstract forms, represented as patterns existing only within a given medium, for instance the computer (Farmer and Belin, 1991).

One important aspect to be learnt from ALife studies then, is the extent to which an emulated or simulated environment can provide these artificial organisms with the potential for realizing emergent patterns, phenomena, or processes.

8.2.3. Artificial Life and Biology

Biological research has largely concerned itself with the material basis of life and is *analytic* in most of its branches, though there are biosciences concerned with synthetic research, such as genetic engineering. In most cases, biology is reductionist, trying to break down complex systems and phenomena into their basic components, from organisms, to organs, to tissues, to cells, down to molecules. By contrast, ALife is concerned with the formal basis of life and is *synthetic*, attempting to construct patterns, behaviors and organisms from their elementary units (Langton, 1988; Sipper, 1995).

However, the many sub-fields of biology have greatly contributed to ALife studies, mainly microbiology, genetics, evolutionary theory, ecology, and developmental biology. Theoretical studies have also provided great insights into the essence of natural phenomena and how to emulate/simulate them in computers. To date, there are two main ways that artificial life has drawn on biology: crystallizing intuitions about life from the study of life itself, and by using (or developing) models that were originally devised to study a specific biological phenomenon. Note that this is the same line of thought that guides most research efforts on computing inspired by nature: the use of biological inspiration or theoretical models.

Biology has also influenced the problems studied in artificial life, since ALife's models provide answers to problems that are intractable by the traditional methods of theoretical biology. Mainstream biologists are increasingly participating in artificial life, and the methods and approaches pioneered in artificial life are increasingly more accepted in biology (Packard and Bedau, 2000).

Generally, ALife models choose a level of life-as-we-know-it to model or study. The lowest level is analogous to the chemical level; higher levels include modeling simple organisms such as bacteria, cells, complex organisms themselves, ecologies, and even planets. The most primitive phenomena explored by ALife research is self-organization and emergence. This may involve the interactions of cells with one another, organisms with one another, the influence of the environment in these interactions, and many other types of interactions.

As discussed previously, ALife products and artificial organisms involve human-made patterns, behaviors, processes and organisms that either present any property of natural life, or are genetically, functionally, or compositionally related with it. Thus, ALife complements the more traditional biological research by exploring new paths in the quest toward understanding life. When viewed as the study of all possible life, including life-as-it-could-be, ALife can be considered a generalization of biology (Moreno, 2000).

The use of the term 'artificial' in artificial life means that the systems in question are human-made instead of naturally evolved; that is, the basic components are not created and evolved by nature. However, the higher-level emergent phenomena are completely genuine, the difference lies mainly in the basic components and in their origin. The fact that true life-like phenomena are observed serves as a basis for ALife research - the underlying belief asserts that life is not necessarily carbon-based but can consist of other elements as well (Sipper, 1995).

8.2.4. Models and Features of Computer-Based ALife

If the goal of the ALife system being designed is to recreate some form of natural system, a good way to start the ALife project is by identifying the mechanisms, processes, properties, and principles underlying the natural system under study (just as done with all computing inspired by nature techniques discussed). The computer is an ideal synthesizer device to the study of life-as-we-know-it and life-as-it-could-be. After these basic elements were identified, the next step is to introduce them in the design of a computer simulation, realization or program in which only low level instructions (local rules) are made explicit, so that new patterns and behaviors may appear in an emergent higher (global) level.

When trying to develop (computational) artificial life, we must be careful not to explicitly or tacitly impose our rational theories or any other (cultural) constraints on how life attains its characteristic patterns, structures, and behaviors. Whether we use natural or artificial environments, we must allow only universal physical laws, local rules, theories of natural selection, immune cells interactions, etc., to restrict the resultant ALife system. This means that simulations that are dependent upon ad-hoc or special-purpose global rules and constraints for the mimicry of life cannot be seen as ALife. ALife systems will be a result of emergent properties of interacting elements, and imposing specific theories or constraints may not allow for emergent phenomena.

In the previous part of this text we observed that most computing techniques inspired by nature can be designed using a given engineering framework or design procedure. The creation of ALife systems, however, does not have any known generic model, platform, or technique. Instead, evolutionary algorithms, rule-based systems, standard software, specific programming languages (e.g., StarLogo), neurocomputing devices, cellular automata, etc., can all be used in isolation or as a hybrid to produce artificial life. Despite this, some essential features of most computer-based artificial life can be listed (Langton, 1988):

- They consist of ensembles or populations of simple programs, agents, or specifications.
- There is no single agent or program that directs all other agents or programs (there is no central control mechanism).
- Each program or agent details the way in which a simple entity reacts to local situations in its environment, including the interaction with other agents.
- There are no global rules in the system; that is, rules dictating global patterns or behaviors.
- Any phenomenon at a level higher than the individual level is an emergent phenomenon.

Of course artificial life also involves the study of single individuals. One such example, and probably the most pursuit of all, is the design of autonomous robots. A robot capable of autonomously moving in and around an environment, interacting with the environment and other agents, that may also be other robots or even humans, is an instance of ALife implemented in hardware.

8.2.5. ALife Systems as Complex (Adaptive) Systems

Most entities that exhibit life-like patterns and behaviors are complex systems made up of many elements interacting with one another and the environment. One way to understand the global behavior of a complex system is to model that behavior with a simple system of equations that describe how global variables interact. As discussed in Chapter 2, this is the approach usually adopted by theoretical biology. In ALife systems, however, the approach often taken is the construction of low-level models that themselves are complex (adaptive) systems, and then iterate the models and observe the emergent global phenomena. These lower-level models are usually agent-based models because the behavior of the whole system is only represented locally, and thus emerges out of the interaction of a collection of directly represented agents.

Therefore, ALife is often described as attempting to understand high-level patterns or behaviors from low-level rules, as most emergent phenomena. The characteristics of these patterns are not previously known by the designer. In this process, the emergence of new structures, clearly distinguishable in space/time terms, is fundamental. Some of these new patterns develop organized hierarchies, structures, and functionalities so that they can become subject of empirical study.

As the complex (adaptive) system changes over time, each element changes according to its state and those of some of its neighbors. These systems lack any central control or behavioral rule, but they may have boundary conditions. The elements of a complex adaptive system and the rules they follow are often simple when compared to the whole system and its global behavior. In rare cases this global behavior can actually be derived from the rules governing the entities' behaviors, but typically the behavior of a complex system cannot be discerned short of empirically observing the emergent behavior of its constituent parts. All these features are clearly distinguishable in most ALife systems.

8.3 EXAMPLES OF ARTIFICIAL LIFE PROJECTS

This section contains a number of examples of artificial life projects. These examples were chosen for various reasons. Some because they are classical works in the field (e.g., flocks, herds, and schools); others for bridging a gap with nature-inspired computing (e.g., biomorphs and creatures); others for bringing important discussions for computing (e.g., computer viruses); others for demonstrating that ALife is invading the great entertainment industry (e.g., the AIBO robot and creatures); others for being a more holistic approach to life-as-we-know-it (e.g., artificial fishes and framsticks); others for being of great interest to biologists (e.g., wasp nest building and the CAFUN simulations); and others for serving educational purposes (e.g., the StarLogo programming language). Most of these examples have several related and supporting web sites, including on line documentation, which can be found with a simple web search. Some directions to these sites are also provided within each specific section.

8.3.1. Flocks, Herds, and Schools

The motions of a flock of birds, a herd of land animals, and a school of fish are very beautiful social behaviors intriguing to contemplate. Producing an animated computer graphic portrayal of these social behaviors is not an easy task, mainly

if one tries to script the path of a large number of agents using the more traditional computer animation techniques; that is, by carefully describing the behavior of each bird, animal, or fish. In a classic ALife work, Reynolds (1987) demonstrated that the apparently intentional, centralized behavior of a flock of birds
can be described by a small set of rules governing only the behavior of individual agents, acting solely on the basis of their own local perception of the environment. The overall resultant behavior is an emergent property of a set of interacting elements following simple local rules.

In the approach proposed by Reynolds (1987), a flock is the result of the interactions between the behaviors of individual birds. For a bird to participate in a
flock, it must have behaviors that allow it to coordinate movements with those
of flock mates. These behaviors are not particularly unique; all birds have them
to some degree. Natural flocks seem to consist of two balanced opposing behaviors: a desire to stay close to the flock, and a desire to avoid collisions within the
flock. Reynolds also realized that flocks of birds have related synchronized
group behavior such as schools of fish or herds of land animals, and thus named
the generic virtual agents *boids*. This is a terminology largely used in ALife to
designate agents simulating the collective behavior of birds, land animals, and
fish.

To create a flocking behavior, a boid must be aware of itself and its nearest
neighbors, suggesting that it can flock with any number of flock mates. The basic flocking model consists of three simple steering behaviors (Reynolds, 1999),
which can be stated as rules describing how an individual boid maneuvers based
on the positions and velocities of its nearby flock mates, as illustrated in Figure
8.1:

- *Collision avoidance and separation*: attempt to avoid collisions with
  nearby flock mates.

- *Velocity matching and alignment*: attempt to match velocity with nearby
  flock mates, and to move towards the average heading of nearby flock
  mates.

- *Flock centering or cohesion*: attempt to stay close to the average position
  of nearby flock mates.

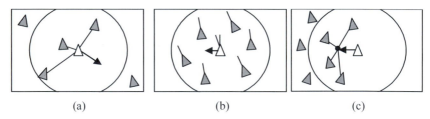

(a)   (b)   (c)

Figure 8.1: Illustration of the response generated by the three rules governing the behavior of each boid. A reference boid (empty triangle) has a limited vision range depicted by
the circle. (a) Collision avoidance and separation. (b) Velocity match and alignment.
(c) Flock centering or cohesion. (Courtesy of © C. Reynolds.)

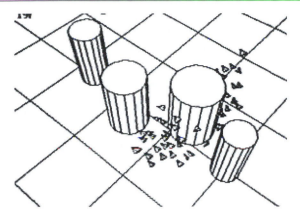

Figure 8.2: Boids flocking and avoiding obstacles (Courtesy of © C. Reynolds).

Collision avoidance and velocity matching are complementary. Together they ensure that the boids fly within the flock without crashing with one another. Collision avoidance contributes to boids steering away from imminent impact. The velocity matching separation between boids remains approximately invariant with respect to flight. Real flocks sometimes split apart to go around an obstacle. Flock centering induces each boid to remain close to the center of the nearby flock mates, thus allowing flocks to bifurcate for going around an obstacle and then getting close back together after the obstacle has been passed.

With a slightly more elaborate model, collision avoidance can be introduced. Very interesting movements of a flock are results of the interaction of boids with the environment. For instance, the conflicts caused by flock centering and collision avoidance increase the complexity of the behavior of the flock. Environmental obstacles are also important to model the scene in which the flock is going to be placed. If the flock is supposed to fly under a bridge or around a tree, the shape and dimension of the obstacles must be geometrically represented. Reynolds (1987) implemented two types of strategies for environmental collision avoidance: one based on the force field concept, and another one called steer-to-avoid, which is more robust and seems closer in spirit to the natural mechanism. Basically, the steer-to-avoid approach works by finding the silhouette edge of an obstacle at a point, and steering in case the object is at one body length beyond the silhouette edge. Figure 8.2 illustrates a boid flock avoiding collisions with cylindrical obstacles.

Discussion and Applications

The flocking model described gives boids a very close pattern of behavior to that observed in real flocks of birds. Boids released near one another start flocking together, cavorting, and jostling for position. They stay close to one another, but always maintain prudent distance from the neighbors, and the flock quickly becomes biased with its members heading in approximately the same direction at approximately the same speed. When they change direction, they do it in co-ordination. Solitary boids and smaller groups tend to cluster together in larger flocks and, in the presence of obstacles, larger flocks can (temporarily) split into smaller flocks.

The boid flocking model has many applications, from the visual simulation of bird flocks to fish schools and herds of land animals. Together with the Symbolics Graphics Division and Whitney/Demos Productions, the authors developed an animated short featuring movie of the boids called *Stanley and Stella in: Breaking the Ice*. Since 1987 there have been many other applications of the boids model in the realm of computer animation. *Batman Returns*, the 1992 Tim Burton's movie, presented bat swarms and penguin flocks created using modified versions of the original boids developed at Symbolics. Several other motion pictures used these or similar concepts in order to create life-like patterns of behavior for virtual organisms (animated creatures), for example:

- 1988: *Behave*, (short) Produced and directed by Rebecca Allen.
- 1989: *The Little Death*, (short) Director: Matt Elson, Producer: Symbolics, Inc.
- 1992: *Batman Returns*, (feature) Director: Tim Burton, Producer: Warner Brothers.
- 1993: *Cliffhanger*, (feature) Director: Renny Harlin, Producer: Carolco Pictures.
- 1994: *The Lion King*, (feature) Director: Allers/Minkoff, Producer: Disney.
- 1996: *From Dusk Till Dawn*, (feature) Director: Robert Rodriguez, Producer: Miramax.
- 1996: *The Hunchback of Notre Dame*, (feature) Director: Trousdale/Wise, Producer: Disney.

Reynolds maintains a website with a short movie about his model and also containing links to many applets with the most varied implementations of boids. Reynolds web site can be found at: http://www.red3d.com/cwr/.

It is important not to forget that, as with all other ALife techniques, one important application of the boids is in the biological sciences. Boids can aid the scientific investigation of real flocks, herds, and schools.

8.3.2. Biomorphs

With a program called *Blind Watchmaker*, R. Dawkins (1986) designed an *evolutionary algorithm* to generate figures from dots and lines. He created a world of two-dimensional artificial organisms on a computer. The artificial organisms generated may or may not bear some resemblance to a variety of objects known by humans. Dawkins gives some of them names of insects, animals, and objects, such as scorpion, bats, fox, lamp, spitfire, lunar lander, and tree frog (see Figure 8.4).

The artificial organisms created with the Blind Watchmaker were named *biomorphs*, after the painter Desmond Morris (1995) who had one of his paintings reproduced on the cover of the first Dawkins' book. The main goal of the Blind Watchmaker is to design an emergent biology of biomorphs based on the following neo-Darwinian processes of evolution: random mutation followed by selection (Dawkins, 1988).

In natural life, the form of each individual organism is defined during embryonic development. Evolution occurs because there are slight differences in embryonic development in successive generations, and these differences come about as results of genetic variations occurring during reproduction. Thus, the Blind Watchmaker also has something equivalent to genes that can mutate, and to embryonic development that performs the transformation from the genetic code to the phenotype of the biomorph.

Inspired by the recursive branching of tree-growing procedures, Dawkins used recursive branching as a metaphor for embryonic development. He developed a large program labeled *Evolution* containing two main modules: *reproduction* and *development*. The *reproduction* module is responsible for creating new biomorphs (offspring) from the existing ones (parents) through the application of a mutation operator that changed one gene (attribute) of the individual. As a result of this mutation, the offspring biomorphs had a form or phenotype, generated by the *development* module, different from its parent. We know, from Chapter 3, that the other step necessary for evolution to occur is *selection*. Selection in the Blind Watchmaker is performed by the user who chooses the individual biomorph to be the parent of the next generation.

Using a data structure to represent the chromosome of each individual in a way akin to genetic algorithms, he initially created individuals with nine genes. Each gene is represented as a number from a finite alphabet. Each gene exerts a minor quantitative influence on the drawing rule that is *Development*; the genotype of each biomorph drawn by *Development* is defined by the numerical values of the genes. By introducing more genes in the chromosome, more complex biomorphs can emerge. The procedure can be summarized as in Algorithm 8.1.

```
procedure [] = evolution()
    initialize
    t ← 1
    while not_stopping_criterion do,
        reproduction  //insert one mutation in each offspring
        development   //draw the biomorph
        selection     //the user chooses the new parent
        t ← t + 1
    end while
end procedure
```

Algorithm 8.1: The Blind Watchmaker algorithm.

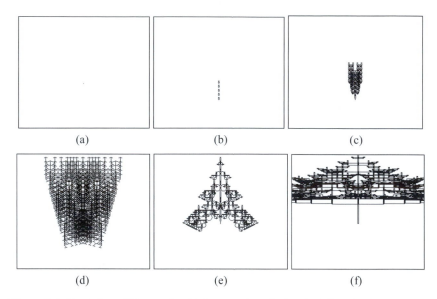

Figure 8.3: Evolution of biomorphs. (a) 0 generations (a tiny, invisible, dot in the middle of the picture). (b) 20 generations. (c) 40 generations. (d) 60 generations. (e) 80 generations. (f) 100 generations.

Consider the Blind Watchmaker with 15 genes as follows (Jones, 2001):

- Genes 1 to 8 control the overall shape of the biomorph.
- Gene 9 controls the depth of recursion.
- Genes 10 to 12 control the color of the biomorph.
- Gene 13 controls the number of segmentations.
- Gene 14 controls the size of the separation of the segments.
- Gene 15 controls the shape used to draw the biomorph (e.g., line, oval, rectangle, etc.).

To discuss the evolution of a biomorph and illustrate what types of form can be evolved, Figure 8.3 presents some biomorphs generated with the Applet written by M. Jones (2001). Note that biomorphs can have their shapes, colors, number of segmentations, etc., completely altered in a few number of generations.

Figure 8.4 presents a portfolio of biomorphs generated by some of my students from the graduate course on natural computing. Note the several biomorphs with forms familiar to us, such as the 'birds', the 'frog', the 'glasses', and the 'fat face'. The names attributed to the biomorphs were given by the students themselves.

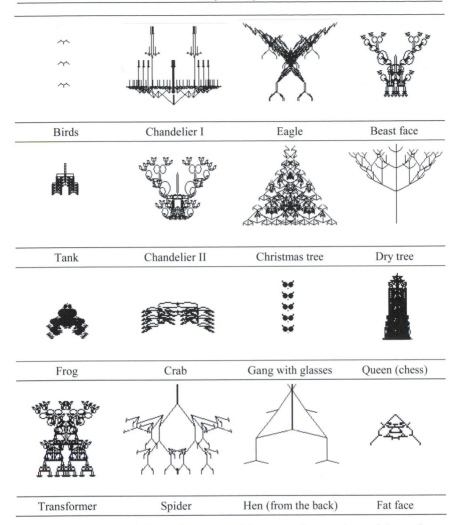

Birds	Chandelier I	Eagle	Beast face
Tank	Chandelier II	Christmas tree	Dry tree
Frog	Crab	Gang with glasses	Queen (chess)
Transformer	Spider	Hen (from the back)	Fat face

Figure 8.4: Portfolio of biomorphs generated by some of my students of the graduate course on natural computing. The names are the ones given by the students themselves.

Discussion and Applications

From an evolutionary perspective, important observations about the Blind Watchmaker can be made. First, as there is no exchange of genetic material between individuals during reproduction, biomorphs present an asexual type of reproduction. Second, selection here is not a natural process that leads to a better adaptation to the environment; instead, it is an artificial process where the user chooses the parents at will, according to how appealing, beautiful, etc., the biomorphs are to him/her.

Unlike the ALife model for reproducing the behavior of flocks, herds, and schools, biomorphs are most often not considered to be good models of life-as-we-know-it. However, they provide the very essence of life-as-it-could-be. For instance, by playing with the genotypes of biomorphs, Dawkins demonstrated that artificial organisms also follow some constraints of living organisms: the phenotype of an individual is restricted by its genotype (and the interactions with the environment). For example, he tried to write his name by successively selecting biomorphs created using a finite number of genes. He noticed that some letters could never be evolved with the chromosome length being tested.

The previous chapter provided an example of the power of evolutionary algorithms in reproducing natural patterns. Although the applications of biomorphs may not seem straightforward to many people, the forms that emerge from this simple evolutionary programs are a great example of the artificial life approach. As described by Dawkins himself

"When I wrote the program, I never thought that it would evolve anything more than a variety of tree-like shapes ... Nothing in my biologist's intuition, nothing in my 20 years' experience of programming computers, and nothing in my wildest dreams, prepared me for what actually emerged on the screen. I can't remember when it began to dawn on me that an evolved resemblance to something like an insect was possible. With a wild surmise, I began to breed, generation after generation, from whichever child looked most like an insect. My incredulity grew in parallel with the evolving resemblance." (Dawkins, 1986; p. 73)

In addition to M. Jones' site (Jones, 2001) that can be used to play with biomorphs, many other applets are available on the web. For instance, S. Maxwell presents an applet with a bouncing biomorph where you can control the number of mutations per second in a biomorph changing shape on line; available online at: http://home.pacbell.net/s-max/scott/biobounce.html. J.-P. Rennard has a biomorph viewer that allows you to change the genome of biomorphs directly; available online at: http://www.rennard.org/alife.

8.3.3. Computer Viruses

A biological *virus* is any of various simple submicroscopic parasites of plants, animals, and bacteria that often cause disease and that consist essentially of a core of RNA or DNA surrounded by a protein coat. The name *computer virus* was inspired by its analog biological virus, which originates from the Latin word *virus* meaning poison (Spafford, 1991). Viral infections are spread by the virus that injects its contents into the cells of another organism. The infected cells are converted into biological factories producing copies of the virus.

In a similar form, a computer virus is a fragment of computer code that embeds a copy of its code into one or more host programs when it is activated. When these infected programs are executed, the embedded virus is executed too, thus propagating the 'infection'. This normally happens invisibly to the user. Thus, a virus can be defined as a program (software) that can infect either host programs or the environment in which host programs exist. A possibly mutated copy of the virus gets attached to host programs (Cohen, 1987; Pieprzyk et al., 2003).

Viruses infect computer programs by changing their structure in a number of different forms. They can destroy data and even overwrite part of a file; they can steal CPU time; they can reduce the functionality of the infected program (file); and they can add new, not necessarily malicious, capabilities to the infected program.

Unlike a *worm*, a virus cannot infect other computers without assistance. Viruses are propagated by vectors such as humans trading programs, disks, CDs, and via the Internet. They can infect any form of writable storage medium, such as floppy disks, zip disks, optical media, and memory. To add themselves to other codes, viruses can become the original host program (and the original host program becomes an internal subroutine of the viral code), they can append their code to the host code, or they can replace some or all of the original code with the viral code.

The lifecycle of a virus involves basically three stages. First, the virus has to be executed so that it can spread. Then, the virus replicates itself infecting other programs. And finally, most viruses incorporate a manipulation task that consists of a variety of effects indicating their presence. Typical manipulations include altering screen displays, generating unusual effects, performing undesired system reboots, and reformatting the hard disk.

One problem encountered by viruses is the repeated infection of the host, thus leading to the depletion of memory and early detection. To prevent this unnecessary growth of infected files, many viruses implant a unique signature that signals infected files or sectors. While trying to infect a given host, the virus first looks for the signature. If it is not detected, then the signature is placed and the host infected; otherwise, the host is not reinfected. An important feature of virus signatures is that they allow for an easier virus scanning process by virus detection systems.

Taking into account the many properties of life described in Section 8.2.2, a computer virus satisfies most of them (Farmer and Belin, 1991; Spafford, 1991):

- A computer virus is represented as a pattern of computer code.

- A computer virus can copy itself to other computers.

- The code that defines the virus corresponds to its representation.

- A computer virus uses electrical energy of the host computer to survive.

- Viruses sense their hosts, altering interrupts, examining memory and disk architectures, and hide themselves to reproduce and support their lives.

- In most cases, the working code of a virus cannot be divided or have some of its parts erased without destroying the virus.

- Computer viruses can be executed on many different types of machines and operating systems, and many of them are capable of compromising or defeating antivirus software packages and copy protection.

- Computer viruses can evolve through the intermediary of human programmers; a study of the structure of computer viruses allows us to place
  them in a taxonomic tree defining lineages.

Discussion and Applications

Similarly to biomorphs, computer viruses are not life-as-we-know-it, they are
another good example of life-as-it-could-be. Computer viruses have some instability and require human intervention to evolve; they are not as fully alive as
their biological counterparts, which are also considered on the border of life.
However, as computers become more powerful and other techniques, such as
evolutionary computation, become more commonly used, it is possible that
computer viruses also become more powerful.

Entities responsible for the defense strategies of some countries have already
invested on the development of computer viruses to be used as war tactics
against any potential enemy. However, most of the applications of computer
viruses are for malefic reasons, though some of them are written as harmless
experiments. One important aspect to be learned from the study of computer
viruses is the realization that the development of ALife systems can sometimes
be dangerous to us. Computer viruses have caused great losses of money in repair, lost data, and maintenance of computer software over the world. But at the
same time computer viruses can cause us harm, ALife systems may be created
for the control of the more malicious types of artificial life. An example of this
is the development of artificial immune systems for computer security (Chapter
6).

8.3.4. Synthesizing Emotional Behavior

By observing how people behave according to their emotional state, D. Dörner
(1999) introduced a theory in which emotions are related to behavioral patterns.
For instance, when someone is angry, he/she becomes more activated (e.g., talk
faster, move the body faster), gets more extroverted (e.g., more likely to scream
and perform other actions), is less precise in his/her actions (e.g., may take the
first thing to appear and smash it), and gets more focused (e.g., focuses only on
a target and cannot see the surroundings). Under Dörner's theory, emotions can
be seen as behavior modulations. The theory identifies four different modulators
describing goal-directed behavior:

1) *Activation*: 'energy' spent while pursuing a goal.

2) *Externality*: amount of time used in performing external activities.

3) *Precision*: how much care is taken during the pursuit of a goal.

4) *Focus*: attention paid on the objective.

Similar to the characterization of anger, other types of emotional states can be
characterized according to Dörner's theory. For instance, an anxious person
tends to be highly activated, precise, focused, and introverted. By contrast, a sad
person is highly precise and focused, but is very introverted and inactive. Fear is

characterized very similarly to anger, i.e., high activation, externality, focus, and imprecision. Contentment is expressed by low activation, externality, focus, and a little precision. Excitement is characterized by activation and externality, low precision and focus.

K. Hille (2001) designed a virtual agent, called Alie, whose behavior could be varied in relation to the four different modulators: activation, externality, precision, and focus. Her goal was to verify if the virtual animation of modulation patterns (behaviors) could be identified as emotions. Alie is a virtual organism consisting of a circle with five smaller circles inside, who moves in a fixed virtual environment (Figure 8.5).

Six emotions (anger, excitement, contentment, anxiety, fear, and sadness) were embedded into Alie's behavior, and expressed by the pulsation of the circles within Alie. A high activation of Alie is represented by a fast pulsation of the inner circles; high externality was represented by a fast movement of Alie; a slow movement of the circles within Alie correspond to a high precision; and a wide spread of the circles indicate low focus.

In order to test the theory of emotions as behavior modulations, some experiments were performed with 70 undergraduate students of psychology (Hille, 2001). The students were asked to identify the emotional behavior of the Alie by knowing only its normal behavior and having general explanations about Alie and its embedded circles. No information about Dörner's theory of emotions, modulators, and behavioral patterns was provided. The students were given the list of six emotions and a video with Alie presenting each of these emotions.

Figure 8.5: Virtual environment and Alie for studying the relationship between behavior modulation and emotions. (Courtesy of © K. Hille.)

In the results obtained, five out of six animations were correctly matched with the emotions they represented. Nevertheless, students tended to confuse between anger and excitement, and anger and fear. Although this is not in accordance with the theory, the animations for anger and excitement are very similar. The same happens for the animations of anger and fear.

Discussion and Applications

Hille's experiment contributes to psychology by providing support to a behavioral theory of emotions. An applet with the simulation is available at http://get.to/alie. The important outcome of this simple experiment is the realization that emotions can be recognized from behavioral patterns. This means that emotions can be identified without reference to the culture, context, experiences, and psychological background of an individual. Elementary behavioral information may be sufficient for the identification of a set of basic emotions.

Such experiments illustrate the adaptiveness of emotions. Modulation patterns provide adjustment to acting and thinking. For instance, in situations that elicit fear, it is important to act quickly; happiness should be accompanied by a look around and broadened sight; anxiety should result in a better analysis of the source of anxiety; and sadness requires deep thinking and reconsideration (Hille, 2001). As evidenced by our everyday life and several ALife projects, emotion is one of the most complicated aspects of human beings. Its understanding and consequent modeling is a very difficult task.

8.3.5. AIBO Robot

AIBO is the name given to the Sony entertainment robot dog, developed to encourage the interaction between humans and robots. The name AIBO is coined from the words **A**rtificial **I**ntelligence Ro**bo**t, and means 'partner', 'pal', or 'companion' in Japanese. With good attention from its master, AIBO can develop into a mature and fun friend as time passes by. He goes through the developmental stages of an infant, child, teen, and adult. Daily communication and attention determines how he matures. The more interaction you have with him, the faster he grows up.

AIBO was designed with four legs, a head, a tail, and a total of twenty internal actuators that enable it to sit or lie down in quite a natural way. It can also use eye lights, tail and ear movement to express its emotions and entertain (Sony, 2000). AIBO responds to external stimuli by using several sensors: a touch sensor in its head, switches in its legs, a color CCD camera in its head, dual stereo microphones (one in each ear), dual heat detectors in its torso, an infra red range finder in its head, a spatial acceleration sensor in its torso, angular velocity sensors in its legs, and a miniature speaker in its mouth. Figure 8.6 illustrates one AIBO ERS-210 (second generation) and describes some of its specifications, including its height and the position of some sensors.

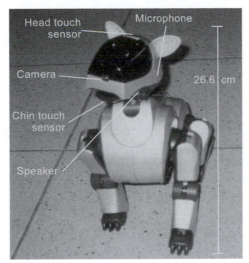

Figure 8.6: Some specifications of AIBO ERS-210.

Sony launched the first AIBO generation, the ERS-110, in June 1999. A limited edition of 5000 was available at the price U$ 2,500.00 each. A number of 3000 were produced to be sold in Japan and 2000 to be sold in America. The 3000 AIBOs for the Japanese market were sold in twenty minutes and the remaining 2,000 available in America were sold in four days; a phenomenon, to say the least! But the AIBO phenomenon is justifiable; it is considered to be the first autonomous robot with real learning capabilities based on artificial intelligence techniques and produced for market sales.

The first generation AIBO, the ERS-110, uses a 64-bit RISC processor and 16 MB of RAM. The software on the AIBO platform consists of a real-time operating system and an Application Programmers Interface (API) (Johansson, 2001):

- Aperios: object oriented (RTOS) real-time operating system developed by Sony.

- Open-R: an API for robots developed by Sony that can be seen as a middle layer between the Aperios operating system and the programmer. The Open-R (Open-Robot) was proposed by Sony as a standard architecture for robot entertainment systems and specifies standards for hardware, electronics, and software architectures.

- Programming language: C or C++.

Since AIBO is equipped with adaptive learning capabilities, each unit can develop its own personality, shaped by its owner. He is capable of simulating emotions, such as happiness and anger, and instinctively wants partnership. It has 18 degrees of freedom (DOF), which is the number of movable joints. Three DOF in each leg, three in the head, two in the tail, and one in the mouth. The ERS-210 has another DOF in each ear, having a total of 20 DOF (see Figure 8.6).

The AIBO ERS-210 has a special Memory Stick called 'AIBO Life' that endows it with instincts, emotions, and the ability to learn and mature that makes it autonomous. Using AIBO Life enables AIBO's voice recognition and imitation function. The AIBO by itself is fully equipped with the hardware it needs to act autonomously, including a 'brain', sensors and power source. To have a glimpse of AIBO's features, you can communicate with him through its sensors by touching him (he is capable of differentiating if he is being praised or scolded), and by talking to him (the stereo microphones allow him to recognize various sounds and words - while infant, he can only recognize a few words, but he learns more words as he matures); you can also give AIBO a name, and he will respond to it afterwards; he has the six emotions of happiness, sadness, anger, surprise, fear, and dislike, and expresses them in various ways, including sounds, melodies, body language and eye lights; AIBO has five instincts, including the need to interact with people, the need to search for something he likes (for instance the ball that comes with him), the need to move his body, the need to recharge, and the need to sleep or rest; finally, you can even ask AIBO how old (or mature) he is, and he will indicate his age using his eye and tail lights.

Discussion and Applications

AIBO is the most widely known and probably the most advanced robot pet, marketed by Sony® Corporation. For obvious commercial reasons not much is known about AIBO's architecture and design principles. It is certainly one of the most sophisticated entertainment robots that make use of near state-of-the-art artificial intelligence, ALife, and robotics technology in an attempt to generate complex and emergent behaviors much akin to life-as-we-know-it. Its most recent version is AIBO ERS 7M3, with the ability to record pictures, recognizing patterns, including voices and faces, obeying commands, playing MP3, dancing, and having wireless connectivity.

Even though AIBO is made of plastic, is powered by a battery, and has a 'nervous system' of integrated circuits, he is also cognizant, sensing, loving, and communicative (in his own means, of course). With eyes for seeing and stereo ears for hearing, AIBO is curious to know you and his environment. Although I could interact with one unit for only a few minutes at a Duty Free Shop in the Schipol airport, Amsterdam, I felt that I had been very close to some sort of ALife implemented in hardware (Figure 8.6 presents a picture I took from AIBO ERS-210). This feeling was even more strengthened after I performed a quick web search about the AIBO robot and found some of the many communities, movies, reports, and pictures available about it. See for instance the videos available at Aibosite.com (http://www.aibosite.com/sp/gen/video.html), and at AiboHack.com (http://www.aibohack.com/movies/index.html) and at AiboWorld-3.tv (http://www.aiboworld.tv/index.shtml); amusing and amazing! Figure 8.7 illustrates two other pictures of the AIBO ERS-210.

The AIBO's artificial life pattern of behavior resembles that of the Tamagochi pets, very popular a few years ago. Feed the pet, water it, play with it, or it will die. The AIBO robot will not die of course (mainly because it costs over a tho-

Figure 8.7: AIBO ERS 210 pictures.

usand dollars), but it may develop a surly attitude or be less attentive. Another good aspect about AIBO is that if you are willing to retrain your AIBO, you can replace his Memory Stick, and get a fresh AIBO with a new personality.

The use of ALife robotics and virtual organisms goes far beyond entertainment purposes, though the success of this application area is unquestionable. Companies like Matsushita Electric also came out with Tama, a robotic cat designed to be a conversation partner for elderly people. Unlike other robotic pets, like Tiger Electronic's Furby or Sony's AIBO, the catlike Tama will have more than just entertainment value, offering companionship and a variety of other services to the aged, said Matsushita (Drexler, 1999). Other robot pets may soon be released with entertainment, educational, assistential, and many other purposes.

8.3.6. Wasp Nest Building

The behavior of social insects such as termites and ants has already been discussed in previous chapters. Together with honeybees and social wasps, all these insects are capable of building nests. Social wasps have the ability to build nests with architectures ranging from extremely simple to highly complex, as illustrated in Figure 8.8. Most wasp nests are made of plant fibers chewed and cemented together with oral secretions. The resulting carton is then shaped by the wasps to build the various parts of the nest. Wasp nest architectures can be of many different types. A mature nest can have from a few cells up to a million cells packed in stacked combs, the latter being generally built in highly social species (Camazine et al., 2001).

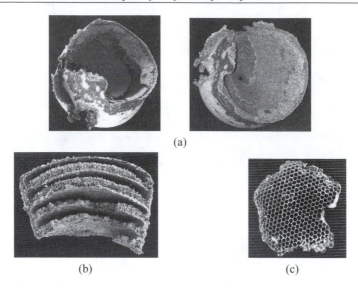

Figure 8.8: Part of a real wasp nest. (a) Top views. (b) Cross-section. (Note the layered architecture.) (c) Comb with hexagonal cells.

In *Vespidae*, the functional unit of a nest is the *cell*. Cells are generally hexagonal and sometimes circular, and provide a container for the development of only one offspring. There are three major architectural components that result in the sophisticated structures (see Figure 8.8): the clustering of isolated cells in organized *combs*, the detachment of the comb from the substrate by means of a *pedicel*, and the protection of the comb with an *envelope*. It can be observed, from Figure 8.8, that wasp nests have a modular architecture, which is a result of a cyclic building activity imposed by the structure of the nest itself, and does not result from any internal building cycle.

From a foraging and manufacturing point of view, it is less costly to produce adjacent cells that are in contact, sharing a common wall and organized into a comb, than to place isolated cells scattered on the substrate. In addition, it is believed that the pedicel plays an important role in protecting the nest against attack by walking predators, such as the army ants, because the contact surface between the nest and the substrate is reduced. The envelope is also important for protection and in some species it plays an important role in thermoregulation. In the vast majority of wasp species, paper is the main building material, hence their common name of paper wasps.

Establishing as the goal to design an algorithm for nest construction, what is important is to understand the underlying behavioral 'algorithm' used by the wasps. Individual building algorithms consist of a series of if-then and yes-no decisions. Such an algorithm includes several types of facts; the first one considers most often the attachment of the future nest to a substrate with a stalk-like pedicel of wood pulp. It seems that the placement of the first cells of the nest follows always the same rule within one given species. Then wasps initiate the nest by building two cells on either side of a flat extension of the pedicel. Subsequent cells are added to the outer circumference of the combs, each between two previously constructed cells. As more cells are added to the developing structure, they eventually form closely packed parallel rows of cells and the nest generally has radial or bilateral symmetry around these initial cells. One important point is that wasps tend to finish a row of cells before initiating a new row, and that rows are initiated by the construction of a centrally located cell first (the pedicel). These rules ensure that the nest will grow fairly evenly in all directions from the pedicel. There is therefore an *a priori* isotropy in space. Other rules ensure the enlargement of the nest by adding new combs with pedicels to the others (Bonabeau et al., 1994; Theraulaz and Bonabeau, 1995).

In Bonabeau et al. (1994) and Theraulaz and Bonabeau (1995), the authors presented a formal model of distributed building inspired by wasp colonies. They characterized a set of distributed stigmergic algorithms, called *lattice swarms*, that allow a swarm of simple agents to build coherent nest-like structures. The agents that constitute the swarm of builders move randomly on a 3-D lattice and can deposit elementary bricks (corresponding to cells) in portions of the lattice according to its local configuration. The agents do not communicate, they have no global representation of the architecture being built, they do not possess any plan or blueprint, and they can only perceive the local configuration of the lattice around them. Only a few of these configurations are stimulating; that is, trigger a building action.

The agents use a limited number of bricks of different types and shapes. They deposit these bricks according to a specified set of rules, embodied in a lookup table. The lookup tables specify what types of bricks must be deposited at a site from the current configuration of bricks in the local neighborhood of that site. They vary according to the structure being built and are usually too complex to be described here, but they allow for the coordinated construction of nest-like structures in a process similar to the one described above. Figure 8.9 illustrates some successive building steps of a nest with a cubic-lattice swarm. Note that first the pedicel is constructed and then the cells around the pedicel are inserted so as to form the circumference of the comb; each row is finished before another one is initiated.

Discussion and Applications

The lattice swarm model of wasp nest building was developed mainly by biologists with the hope to gain some understanding of the algorithmic processes involved in, and required by, the building behavior of social insects, through the development of minimal algorithms carried out by simple agents. At the same time, these algorithms provide some good insight into the engineering process of how to look at some structure and try and find some simple behavioral procedures that can be used to generate emergent patterns similar to the ones observed in nature.

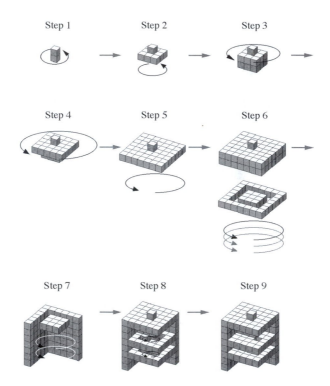

Figure 8.9: Successive building steps of the lattice swarm algorithm. (Reproduced with permission from [Bonabeau et al., 2000], © Elsevier.)

The lattice swarm algorithm is a good example that behavioral algorithms can produce coherent biological-like architectures. This study constitutes an important step towards a deeper understanding of the origins of natural shapes in social insects. It is another evidence that the coupled interactions of an insect society with its environment results in a complex dynamics whereby coherent functional global patterns emerge from the behaviors of simple agents interacting locally with one another or the environment. This model is one type of coordinated algorithm in which the shape to be built is decomposed into modular subshapes. Stimulating configurations do not overlap, so that at any time, all agents cooperate in the building of the current subshape. This model allowed researchers to conclude that, for nest wasps, algorithms with overlapping configurations yield structureless shapes, never found in nature.

8.3.7. Creatures

Creatures is a commercial home-entertainment software package that provides a simulated environment filled with a number of synthetic agents, the 'creatures', that a user can interact with in real-time. The internal architecture of the creatures is inspired by real biological organisms, embodying: 1) a variable-length *genetic code* containing the information necessary to define the behavioral and chemical features of the creature; 2) a *neural network* responsible for sensory-motor coordination, behavior selection, and learning; and 3) a *biochemistry* that models a simple energy metabolism along with a hormonal system that interacts with the neural network to model diffuse modulation of neuronal activity and development. The genetic code specifies not only the neural network and biochemistry of the creatures, but it also allows for the possibility of evolutionary adaptation through sexual reproduction (Grand, 1997; Grand et al., 1997).

The creatures inhabit a platform environment with multi-plane depth cueing so that objects can appear, relative to the user, to be in front of or behind other objects. Within the creatures' environment, there are a number of objects that the creature can interact with in a variety of ways. The system was written using object-oriented programming techniques in which the virtual objects have scripts attached, determining how they interact with one another, including the creatures and the static parts of the environment. The mouse pointer can be used as a means to interact with the creatures and the environment.

All creatures are bipedal, but minor morphological details are genetically specified. In addition, they can grow older and in size. They have simulated senses of sight, sound and touch. Creatures can learn simple verb-object language via keyboard input or interactions with the environment and other creatures. An illustration of a creature is presented in Figure 8.10. The software can be found at http://www.gamewaredevelopment.co.uk/ and several technical descriptions are available at http://fp.cyberlifersrch.plus.com/articles.htm.

A great part of the creatures' structure and function, including their neural networks and biochemistries, is determined by their genes. Primarily, the genome is provided to allow for the transmission of characteristics from parents to offspring creatures, but they also allow for evolutionary development with the possibility of introduction of new structures via genetic variation operators (crossover and mutation). The genome is a string of bytes divided into isolated

Figure 8.10: Illustration of two creatures. (Reproduced with permission from © [Gameware Development Ltd 2004].)

genes by means of punctuation marks. Genes of particular types have character-istic lengths and contain specific bytes (subject to mutation). The genome forms a single, haploid chromosome that, during mutation, can be crossed-over with the chromosome of another parent in a sexual reproduction process.

Each creature's brain is a heterogeneous artificial neural network, sub-divided into objects called *lobes*, as illustrated in Figure 8.11. Note that one lobe may correspond to more than one layer according to the terminology introduced in Chapter 4. The lobes define the electrical, chemical, and morphological charac-teristics of a group of neurons. The initial model contained approximately 1000 neurons, grouped in 9 lobes, and interconnected through roughly 5000 synaptic connections, though connections are allowed to be created and pruned (Grand et al., 1997; Grand, 2000). The overall model is behaviorist (based on observable and quantifiable aspects of behavior) and based on reinforcement by drive learn-ing. Changing some network weights in response to either a reward or a pun-ishment chemical provides delayed reinforcement learning. These chemicals are not generated directly by environmental stimuli, but during chemical reactions involved in drive level changes. Each creature maintains a set of chemical *drives*, such as 'the drive to avoid pain' and 'the drive to reduce hunger'.

Central to the function of the neural network is a simplified *biochemistry* that controls the flow of information. This mechanism is also used to simulate other endocrine functions, a basic metabolism, and a very simple immune system. The biochemistry is based on four classes of objects: chemicals, emitters, reactions, and receptors. The receptors and emitters provide the interface between chemis-try and physiology. The emitters secrete chemicals in response to certain activi-ties, and the receptors cause events in response to chemical levels. Receptors and emitters attached to synaptic strength parameters allow the setting up of chemical driven feedback loops for controlling synaptic strength.

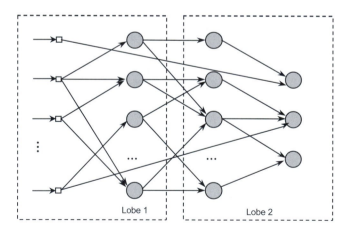

Figure 8.11: Sample network with two lobes.

Discussion and Applications

As with all emergent systems, the program that generates Creatures does not describe their behavior, but only their structure. The behavior of each creature is an outcome of its genetic information and of its interaction with other creatures and the environment (including the user). Inside the computer there is software to describe rules for Physics, Chemistry, Genetics and so on - a virtual universe. Composite structures are then built inside this universe modeling the constituent parts of the creatures: their senses, brains, digestive systems, reproductive mechanisms, and so forth; only how the creatures are is specified, not what they should do.

As a commercial-entertainment product, Creatures offers users an opportunity to engage with artificial life technologies. They belong to a class known as *software toys* as opposed to *computer games*. The use of the word toy, instead of game, aims at stressing a different style of user-machine interaction. A game is usually played in one (extended) session until an end condition or a given stage is reached, or a score is awarded. A toy by contrast, does not imply a score or an aim to achieve some end-condition or stage, it is based on a more creative, interactive, on-going and open-ended experience. Like the AIBO robot, Creatures is also the application itself of artificial life targeted at the entertainment market.

8.3.8. Artificial Fishes

Terzopoulos et al. (1994) implemented a virtual world inhabited by a variety of realistic fishes. As described by themselves:

"In the presence of underwater currents, the fishes employ their muscles and fins to swim gracefully around immobile obstacles and among moving aquatic plants and other fishes. They autonomously explore their dynamic world in search for food. Large, hungry predator fishes stalk smaller prey fishes in the deceptively peaceful habitat. Prey fishes swim around contentedly until the sight of predators compels them to take evasive action. When a dangerous predator appears in the distance, similar species of prey form schools to improve their chances of survival. As the predator nears a school, the fishes scatter in terror. A chase ensues in which the predator selects victims and consumes them until satiated. Some species of fishes seem untroubled by predators. They find comfortable niches and feed on floating plankton when they get hungry. Driven by healthy libidos, they perform elaborate courtship rituals to secure mates." (Terzopoulos et al., 1994)

The authors developed a virtual marine world inhabited by realistic *artificial fishes*. Their algorithms were used to emulate not only the appearance, movement, and behavior of individual fishes, but also to emulate the complex group behaviors observed in aquatic ecosystems. Each fish was modeled holistically; that is, an artificial fish is an autonomous agent situated in a simulated physical world. Each fish has: 1) a three-dimensional body with internal muscle actuators and functional fins, which deforms and moves in accordance with biomechanic and hydrodynamic principles; 2) sensors, including eyes that can see the environment; and 3) a brain with motor, perception, behavior, and learning centers.

The artificial fishes exhibit a repertoire of piscine behaviors that rely on their perceptual awareness of their dynamic habitat. Individual and emergent collective behaviors include caudal and pectoral locomotion, collision avoidance, foraging, preying, schooling, and mating. Furthermore, the artificial fishes can learn how to move through practice and sensory reinforcement. Their motor learning algorithms discover muscle controllers that produce efficient hydrodynamic locomotion. The learning algorithms also enable them to train themselves to accomplish higher level, perceptually guided motor tasks, such as maneuvering to reach a visible target.

To achieve such a goal, the artificial fishes are implemented in various stages. First, the primitive reflexive behaviors are implemented, such as obstacle avoidance, that directly couple perception to action. Then, the primitive behaviors are combined into motivational behaviors whose activation depends also on the fish's mental state, including hunger, libido, and fear. The artificial fishes are also provided with algorithms that enable them to learn automatically from first principles how to achieve hydrodynamic locomotion by controlling their internal muscle actuators. Figure 8.12 illustrates the three main modules or systems of the 'brain' of the artificial fish.

The motor system is responsible for driving the dynamic model of the fish. The perception system relies on a set of virtual sensors to provide sensory information about the dynamic environment, including eyes that can produce time-varying retinal images of the environment. The perception center also includes a perceptual attention mechanism that allows the fish to train its sensors, hence filtering out sensory information that is superfluous to its current behavioral needs. The learning system enables the artificial fish to learn how to locomote through practice and sensory reinforcement. This system is capable of discovering muscle controllers that produce efficient locomotion, and also allows the fishes to train themselves to accomplish higher-level sensorimotor tasks, such as maneuvering to reach visible targets.

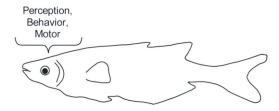

Figure 8.12: Main modules of the brain of the artificial fish.

Figure 8.13: Artificial fishes in their virtual world. (Courtesy of © D. Terzopoulos.)

Discussion and Applications

According to the authors, the long-term goal of their research is to devise a computational theory for the interplay of physics, locomotion, perception, behavior, and learning in higher animals. They were able to produce visually convincing results in the form of realistic computer graphics animation with little or no animator intervention. Some videos and references are available at Terzopoulos' website (http://mrl.nyu.edu/~dt/). Their videos are worth watching; they show a variety of artificial fishes foraging in water, a hook capturing a fish, a shark hunting smaller fish, and other interesting behaviors. Figure 8.13 shows a picture of some artificial fishes in their virtual environment.

The detailed motions of the artificial fishes emulate the complexity and unpredictability of movement of natural fishes, what enhances the visual beauty of the animations. The authors claim that the artificial fishes can also be seen as virtual robots, capable of offering a broader range of perceptual and animate abilities, lower cost, and higher reliability than can be expected from most physical robots currently available. It is another very good example of ALife playing life-as-we-know-it.

8.3.9. Turtles, Termites, and Traffic Jams

M. Resnick (1994) developed a new computational tool, named *StarLogo*, that people can use to construct, play with, and think about self-organized systems. To Resnick, the best way to understand self-organized or decentralized systems is by designing them; that is, by using a *constructionist* approach. An example of a constructionist tool is the programmable robotics system named LEGO/Logo. The combination of the LEGO bricks with a programming language named Logo, gave rise to the LEGO/Logo system. This allowed a simple mechanical robot connected to a computer by an 'umbilical cord' (wire) to be controlled. It gave students the opportunity to not only manipulate objects but to also construct personally meaningful objects.

Inspired by *cellular automata* (Chapter 7), Resnick extended the Logo into the StarLogo, a system designed to encourage constructivist thinking. Another aim was to design a system in which the actions of individual agents could arise from interactions of individual agents (from individual level to group level). This would result in the possibility of modeling many types of objects in the world, such as an ant in a colony, a car in a traffic jam, an antibody in the immune system, and a molecule in a gas.

StarLogo was thus designed as a massively parallel language with an environment capable of accommodating a great number of agents, named *turtles*, simultaneously. Many senses can be assigned to each turtle; for instance, a turtle can perceive and act on its surrounding environment. The environment in which the turtles are placed, termed *world*, is active, and is divided into square sections called *patches*. It means that patches can store information about the activities of the turtles and execute StarLogo instructions just as turtles do. Thus the environment is given equal status as being alive as the agents that inhabit it.

Several projects inspired by real-world or natural systems, such as ant colonies, traffic jams, and termite mound building, are presented in (Resnick, 1994), including the computer codes necessary to run them. These demo projects are also available with the software. In these cases, the natural systems serve only as inspiration, a departure point for thinking about self-organized systems. The idea is to study nature-like behavior, not the behavior of (real) natural systems. Some of these projects will be reviewed in this section. Details about programming and the computer codes necessary to implement them can be found in Resnick (1994) and in Colella et al. (2001). The illustrations to be presented are part of the StarLogo sample projects, available at http://education.mit.edu/starlogo/, and were all generated and pasted here with permission from © E. Klopfer. The StarLogo is freely available at the same website.

Predator-Prey Interactions

In nature, there are a number of enemies for an organism, including its natural predators, organisms that cause diseases, and parasites (parasitism). Predation, parasitism, and disease are overwhelmingly important aspects to life for all organisms. Many features of organisms are adaptations that have evolved to escape predators. In many cases the *predator-prey* interactions are unstable resulting in the extinction of one species. In other cases, one species is not extinct due to predation or parasitism by another, but they coevolve leading to the appearance of novel forms of individuals within the species.

Most of the predator-prey models stem either from the original model of Lotka and Volterra or some variation of it. This Lotka-Volterra model assumes that the system is closed involving coupled interactions between prey and predators. In a simple form, the model assumes that the number of predator depends on the size of the prey population, acting on the predators' birth rate. The predator population declines due to lack of prey, but increases due to an abundance of food. In contrast, the prey population declines because few escape predation, and it increases because many escape predation.

The project named *Rabbits*, on the Biology category, explores a simple ecosystem composed of rabbits and grass (Figure 8.14). The rabbits wander around the world randomly, and the grass grows randomly. Several parameters can be controlled, such as the growth rate of the grass and the number of rabbits. When a rabbit finds some grass, it eats the grass increasing its energy level. In case the

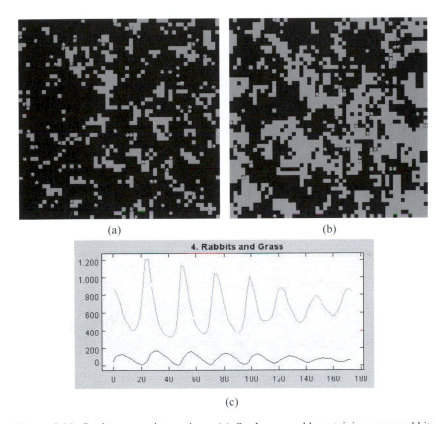

(a)   (b)

(c)

Figure 8.14: Predator-prey interactions. (a) StarLogo world containing many rabbits (little creatures) and little grass (rectangular patches). (b) Many rabbits died of starvation and thus the amount of grass increased in the environment, resulting in another growth of the population of rabbits. (c) Plot showing the dynamics of the rabbits' population (lower curve) and the amount of grass (upper curve).

rabbit eats enough grass (has an energy level above a given threshold), it reproduces asexually generating a clone of it. The offspring rabbit receives half the energy of its parent. If the energy level of the rabbit falls below a given threshold, then it dies. Figure 8.14(a) and (b) presents the StarLogo world for the rabbits-grass ecosystem in two different time steps and Figure 8.14(c) depicts the dynamics of the population.

By observing Figure 8.14, it is possible to note that initially there is plenty of food. As the rabbits eat grass, they get more energy and start reproducing. The large population of rabbits leads to a shortage of grass, and soon the rabbit population begins to decline. This allows the grass to grow more freely, providing an abundance of food for the remaining rabbits, and the process cycles. Note also that the amount of grass and the rabbit population oscillate out of phase. These dual oscillations are characteristic of predator-prey interactions.

Termites

Termites are among the master architects of the animal world. On the plains of Africa, termites construct giant mound-like nests rising more than ten feet tall, thousands of times taller than the termites themselves. Inside the mounds are intricate networks of tunnels and chambers. Certain species of termites even use architectural tricks to regulate the temperature inside their nests, in effect turning their nests into elaborate air-conditioning systems. Figure 8.15 presents two views of a real termite mound, pictured in Goiás, at the center of Brazil. This mound is around one meter tall and has several tunnels connecting the inner part with the outside environment.

Each termite colony has a queen. But, as in ant colonies, the termite queen does not 'tell' the termite workers what to do. On the termite construction site, there is no construction foreman, no one is in charge of a master plan. Rather, each termite carries out a relatively simple task. Termites are practically blind, so they must interact with one another and with the surrounding environment primarily through their senses of touch and smell. From local interactions among thousands of termites, impressive structures emerge (Figure 8.15).

      (a)   (b)

Figure 8.15: Views of a mound. (a) Top view. (b) Lateral view.

The project *Termites*, on the Biology category, is inspired by the behavior of termites gathering wood chips into piles. The termites follow a set of simple rules. Each termite starts wandering randomly. If it finds a wood chip, it picks it up and continues to wander randomly. When it finds another wood chip, it looks for a nearby empty space and puts this wood chip down. Note that this behavior is very similar to the termites' building behavior discussed in Chapter 2, and to the clustering of dead bodies and larval sorting in ant colonies reviewed in Chapter 5.

Figure 8.16 illustrates the behavior of the termites following the simple set of rules described above. Note that initially the wood chips are scattered around the environment and they are then grouped into a single cluster of wood chips.

With these simple rules, the wood chips eventually end up in a single pile. During the run it is possible to note that some piles emerge and then disappear. Usually, the number of piles decreases with time, mainly because there is no way a termite can start a brand new pile. This is another good example of a self-organized process; there is no global plan of how to make the pile or where to place it. Each termite follows a set of simple rules, but the colony as a whole accomplishes a rather sophisticated task.

Traffic Jams

Traffic flow is a rich domain for studying collective behavior. While discussing about the concept of emergence in Chapter 2, we sketched some ideas concerning traffic jams. Consider now a StarLogo project on *Traffic Flow*, available on the Social Systems category. The project demonstrates that no cause, such as a traffic light, a radar trap, or an accident, is necessary so as to form a traffic jam. Interactions among cars in a traffic flow can lead to surprising group phenomena.

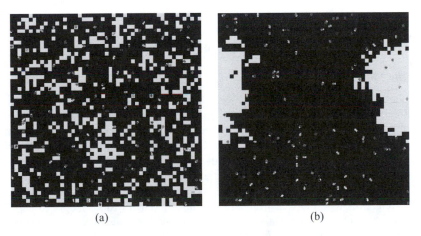

(a)   (b)

Figure 8.16: Termite clustering. Squares: wood chips. Small creatures: termites. (a) Initial configuration of the environment. (b) Environment after some iterations have passed.

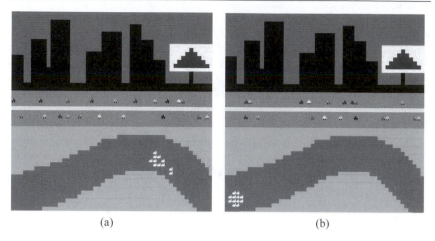

(a)   (b)

Figure 8.17: Traffic jam. Note the presence of some small jams in parts of the environment.

Each car follows a simple set of rules:

- If there is a car close ahead of you, slow down.
- If there is no car close ahead of you, speed up (unless you are already at speed limit).

Figure 8.17 depicts the sample project environment where some small jams are naturally formed following the simple rules above. The cars are initially placed in random positions and have random speeds. Note that traffic jams can start with a simple car slowing down in front of the others. When some cars are clustered together, they will move slowly, causing cars behind them to slow down resulting in a traffic jam. It is interesting to note also that although all cars are moving forward, the traffic jams tend to move backward. This behavior is common in wave phenomena: the behavior of the group is often very different from the behavior of its individual components.

Slime-Mold

Slime-mold cells exist as tiny amoebas when there is plenty of food. They move around, feed on bacteria in the environment, and reproduce simply by dividing into two. When food becomes scarce, the slime mold behavior changes drastically. The slime-mold cells stop reproducing and move toward one another, forming a cluster with tens of thousands of cells. At this point, the slime-mold cells start acting as a unified whole. Rather than behaving like lots of unicellular creatures, they act as a single multicellular creature. It changes shape and begins crawling, seeking a more favorable environment. When it finds a spot to its liking, it differentiates into a stalk supporting a round mass of spores. These spores ultimately detach and spread throughout the new environment, starting a new cycle as a collection of slime-mold amoebas.

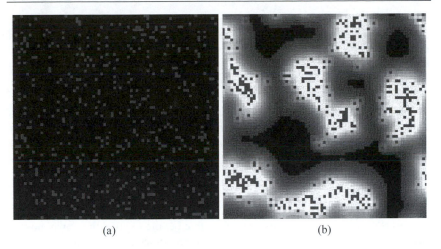

(a) (b)

Figure 8.18: Slime-mold aggregation. (a) Initial distribution of slime molds. (b) Clusters of slime molds after some time has elapsed.

The process through which slime-mold cells aggregate into a single multicellular creature has been a subject for scientific debate. Until 1980 or so, most biologists believed that specialized 'pacemaker' cells coordinated the aggregation. But scientists now view slime-mold aggregation as a decentralized process. According to the current theories, slime-mold cells are homogeneous: none is distinguished by any special feature or behavior. The clustering of slime-mold cells arises not from the commands of a leader, but through local interactions among thousands of identical cells. In fact, the process of slime-mold aggregation is now viewed as one of the classic examples of self-organizing behavior.

How do slime-mold cells aggregate? The mechanism involves a chemical called 'cyclic AMP' or cAMP, which can be viewed as a sort of pheromone. When the slime mold cells move into their aggregation phase, they produce and emit cAMP in the environment. They are also attracted to the very same chemical. As the cells move, they follow the gradient of cAMP. That is, they test around themselves, and they move in the direction where the concentration of cAMP is highest. Each cell can sense cAMP only in its immediate vicinity.

The project *Slime Mold*, in the Biology category, is inspired by the aggregation behavior of slime-mold cells (Figure 8.18). It shows how agents can aggregate into clusters in a self-organized form. In this example, each agent drops a chemical pheromone (cAMP) that can be smelt by other agents. The agents can smell ahead, trying to follow the gradient of the chemical. Meanwhile, the patches diffuse and evaporate the chemical.

With a small number of slime molds no clustering emerges, because as the slime-molds wander around dropping pheromone, it evaporates and diffuses fairly quickly, before an aggregation emerges. With a larger number of slime molds, following exactly the same rules, the result is qualitatively different.

When a few slimes happen to wander near one another, they create a 'puddle' of chemical. The creatures sniff the chemical and try to stay nearby. They then deposit more pheromone in the puddle, so the puddle expands and attracts more slime molds. Together with the Termites project, this project is a very good example of stigmergic behavior. Individual agents act according to the environmental configuration, which is modified by themselves.

Figure 8.18 presents the slime mold aggregation in two different stages. Initially (Figure 8.18(a)) the slime molds are spread all over the environment. After some time it is possible to observe clusters of molds being created. The slime molds are in the center of the clusters and a gradient of pheromone is observed surrounding the molds.

Discussion and Applications

The StarLogo language was designed to enable people to build their own models of complex dynamic systems. It supports the processes of designing, analyzing, understanding, and describing natural systems, and does not require advanced mathematical or programming skills. Tools such as StarLogo are very useful for the study of self-organized systems, in particular artificial life systems. Resnick (1994) views StarLogo projects as *explorations of microworlds*, not simulations of reality. Microworlds are simplified worlds specially designed to highlight particular concepts and ways of thinking. They are always manipulable, encouraging us to explore, experiment, invent, and revise.

Examples of many emergent phenomena can be demonstrated and studied with simple StarLogo programs. The whole package is available for download and has been used by many High Schools and Universities as a computational tool to study self-organization and ALife: http://education.mit.edu/starlogo/. A current description of StarLogo, including some explanations about the StarLogo tool (e.g., virtual machine, Interface, process scheduler, and processes) is provided in Begel and Klopfer (2005).

8.3.10. Cellular Automata Simulations

The previous chapter introduced cellular automata as a tool to create fractal patterns. In addition, CA can be used to generate a variety of artificial life phenomena as well, and some instances will be provided in this section. The first example explains how to implement J. Conway's Game of Life, the second one describes the Langton's loops, and the last examples introduce CAFUN - a cellular automaton simulator – and some of its demos.

The Game of Life

In the late 1960s, John Conway was motivated to extend von Neumann's work on cellular automata. Von Neumann had succeeded in describing a self-reproducing CA, but the machine itself was enormous and took too many pages to describe. Conway wanted to refine the description of a CA to the simplest one that could still support universal computation (see Appendix B.3.2 for universal computers). He proposed a solitaire game called 'Life', currently known as 'The Game of Life'. This name was chosen because of its analogies with the rise, fall, and alternations of a society of living organisms.

Gardner was one of the main responsible for the spread of Conway's Life game (cf. Gardner, 1970). Life was originally played on a large checkerboard and with a plentiful supply of flat counters of two colors to be placed in the board. Nowadays, Life is most often 'played' using a bi-dimensional cellular automaton implemented in computers. The main differences between the bi-dimensional CA and the one-dimensional CA, studied in the previous chapter, are the definition of the neighborhood of a cell and the evolution of the CA. Now, each time step corresponds to one frame of the CA on the screen. Figure 8.19 illustrates two possible types of neighborhood for a bi-dimensional grid. In both cases, the radius is $r = 1$ because only the next 'layer' is considered.

In Life, Conway wanted to produce transition rules that were both simple to write down and difficult to predict the overall behavior of the CA. Although concepts such as emergence, complexity, and self-organization had not yet been proposed at that time, he had these concepts very much in mind. Conway concentrated on meeting the following criteria:

1. There should not be any initial pattern for which there is a simple proof that the population can grow without limit.

2. There should be initial patterns that apparently do grow without limit. Not all initial states should immediately yield trivial final states.

3. There should be simple initial patterns that grow and change for a considerable period of time before coming to an end in three possible ways: fading away completely, settling into a stable configuration that remains unchanged thereafter, or entering an oscillating phase in which they repeat an endless cycle of two or more periods.

In brief, the rules should be such as to make the behavior of the population unpredictable. After a great deal of experimentation with many sets of rules, Conway finally settled on to what is arguably the single-most interesting elementary two-dimensional rule set on a regular lattice.

Life uses the Moore neighborhood: the eight cells around the center cell are its neighbors. Calling cells with state value $s = $ '1' 'alive', and cells with state $s = $ '0' 'dead', the following rules define Life:

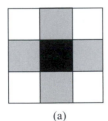
(a)

(b)

Figure 8.19: Examples of possible neighborhoods of the center cell in a bi-dimensional grid. (a) Von Neumann neighborhood. (b) Moore neighborhood.

- *Birth*: a previously dead cell comes alive if three neighbors are alive.

- *Death*: isolated living cells with no more than one live neighbor die; those with more than three neighbors die of overcrowding.

- *Survival*: living cells with two or three live neighbors survive.

There are many things special about Life, and one of these is that it can serve as a *general-purpose computer*. Another important feature of life is its capability of giving rise to complex ordered patterns out of an initially disordered state, sometimes referred to as *primordial soup*. To have an idea of what can happen within Life given a certain initial condition, Figure 8.20 shows a large grid initialized with the three letters (LND) using two cells as the width of each letter. The initial configuration is depicted in Figure 8.20(a), and Figures 8.20(b) to (d) show the evolution of the CA for 20, 40, and 300 generations, respectively. At generation 300 most of Life was frozen, but some oscillatory patterns remained.

Since its popularization by Gardner in the early 1970s, the temporal fates of literally thousands of various initial configurations have been analyzed and catalogued. You can find the population of cells constantly undergoing unusual, sometimes beautiful and unexpected, changes. In some cases the whole society dies out, and in others some stable figures are obtained - Conway calls them *still lifes* – or patterns that cannot change or oscillate forever. Patterns with no initial symmetry tend to become symmetrical, and once it happens it cannot be lost.

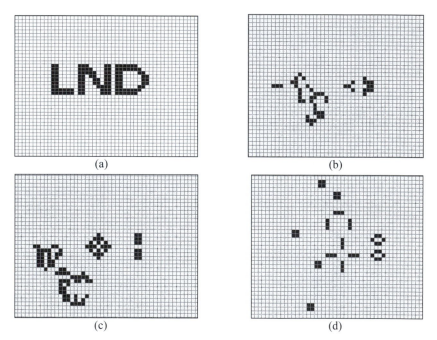

        (a)   (b)

        (c)   (d)

Figure 8.20: Snapshots of Life with the letters LND as starting condition. (a) Initial configuration. (b) 20 generations. (c) 40 generations. (d) 300 generations.

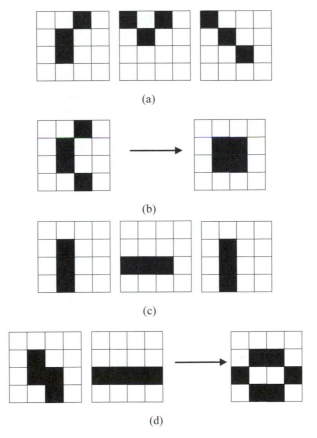

(a)

(b)

(c)

(d)

Figure 8.21: Fates of some initial states. (a) All these initial configurations die out after two steps. (b) This configuration becomes a *block* (two by two square). (c) Three live cells in a row or a column becomes an oscillating pattern of period 2, called a *blinker*. (d) Both configurations lead to a still pattern called *beehive*.

Figure 8.21 illustrates what happens to some simple seeds and introduces some of the most commonly obtained 'life forms' in Life. Some of these life forms can be observed in the resulting pictures presented in Figure 8.20(d).

Langton's Loops

Von Neumann constructed a self-reproducing machine M consisting of three parts: a universal constructor A; a tape copier B; and a control C. This machine is given a tape containing a description of itself: $\Phi(M) = \Phi(A + B + C)$. When M is started, it reads the description $\Phi(M)$, builds a copy of itself M', makes a copy of the description $\Phi'(M)$, and attaches the copy of the description $\Phi'(M)$ to the copy of itself M'. Thus, the machine has reproduced itself completely, and its copy M' can go on to reproduce itself completely, and so forth.

The important property that results from this strategy is that the information contained in the description on the tape must be used *twice*, in two fundamentally different ways. First, the information must be interpreted, or *translated*, as instructions for building a machine. Second, the information must be copied, or *replicated*, without interpretation, in order to provide the offspring with a copy of the description so that it is also able to reproduce itself.

This dual use of information is also found in natural self-reproduction. The information contained in the DNA is *transcribed* into messenger RNA (mRNA) and then *translated* into polypeptide chains at the ribosome, involving *interpreting* the information as instructions for constructing a polypeptide chain. The information of the DNA is also replicated to form two copies of the original information. This involves merely copying the information without interpretation (Langton, 1986).

Langton (1984) proposed a compact structure that also makes dual use of information contained in a description to reproduce itself. To this end, he used a bi-dimensional cellular automaton with the von Neumann's neighborhood (eight cell neighborhood - Figure 8.19(a)), and eight states per cell. The structure consists of a looped pathway with a construction arm projecting out of it. The loops consist of a sheath within which circulates the information necessary to construct a new loop. The description consists of a sequence of *virtual state machines* (VSM) - a structure, set of automata, that spans more than one cell of the grid - that cycle around the loop.

When a VSM encounters the junction between the loop body and the construction arm, it is *replicated*, with one copy propagating back around the loop again, and the other copy propagating down the construction arm, where it is *translated* as an instruction when it reaches the end of the arm. The first six of the eight VSMs that constitute the description extend the arm by one cell each. The final two VSMs cause a left hand corner to be constructed at the end of the arm. Thus, with each cycle of the VSM sequence around the loop, another side of the offspring loop is constructed. Figure 8.22 illustrates some stages in the development of a self-reproducing loop starting from a single initial loop.

CAFUN

CAFUN is an acronym for 'Cellular Automata Fun', a software that picks up the idea of cellular automata to perform its simulations. Although it adopts the concept of a CA, it introduces some new ideas as well. For instance, a slightly different nomenclature: the expression lattice or grid was substituted by 'universe' because it represents the whole virtual world in which the simulation takes place, and the notion of neighborhood was changed to 'cell context' because it accounts for everything that has an outer impact on a cell. CAFUN follows an object oriented approach to describe the laws of a cellular automaton. Instead of referring to a table of rules CAFUN summarizes them in a set of cell types which have individual rules assigned to them. Each cell in a universe of CAFUN

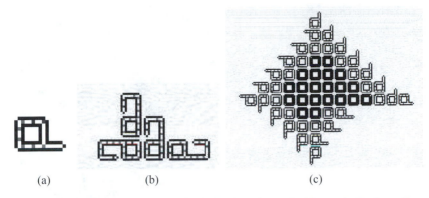

(a)   (b)   (c)

Figure 8.22: Three stages in the development of a colony of self-reproducing loops from a single initial loop.

is considered to be an instance of such a cell type. Unlike the states of standard CA, which play a role in the application of rules, CAFUN's cell types decide on some features of their cells.

CAFUN enables you to create interesting simulations of complex systems, such as living organisms, social groups, companies, physical processes, and chemical structures. Similarly to StarLogo, CAFUN has the advantage of being freely available for download at www.cafun.de. Figure 8.23 illustrates some simulations performed with CAFUN. These were generated and pasted here with permission from © A. Homeyer.

Figure 8.23(a) depicts the result of a simulation of a chemical reaction called Beloussov-Zhabotinsky (BZ) reaction. This reaction shows how a mixture of organic and inorganic chemicals, put in solution in a Petri dish, generates concentric rings that travel slowly from the center outward. It is a good example of an excitable medium generating patterns in space and time when stimulated. Another interesting fact about this reaction is that the emerging rings disappear after encountering one another, instead of interfering with one another. Similar patterns of activity can arise in systems that differ greatly from one another, such as chemical reactions, aggregating slime mould amoeba, heart cells, neurons, or ants in a colony. All of them show similar dynamic activities: rhythms, waves that propagate in concentric circles, spirals that annihilate when they collide, and chaotic behavior (Goodwin, 1994).

Figure 8.23(b) presents three stages of a simulation of bush fire using CA-FUN. The environment is composed of bushes and lakes. There are basically three levels of gray in the left picture (original environment before the fire). The lighter gray areas surrounded by a white ring correspond to some lakes in the field. A medium gray level indicates regions filled with bushes, and dark gray areas have few bushes (or green grass) to be burnt. Note from the picture that the fire is going to be started at different parts of the environment, and its spread can be accompanied in the sequence of images. The fire spreads quickly within the

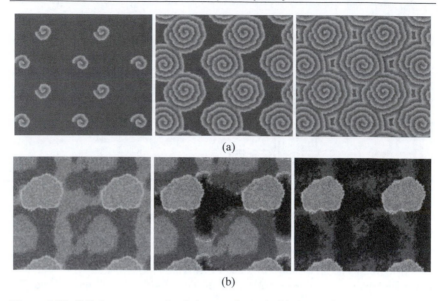

(a)

(b)

Figure 8.23: Cellular automata simulations performed with CAFUN 1.0. (a) Three stages of a simulation of the Beloussov-Zhabotinsky (BZ) reaction. Note that the reaction starts in several different places and the spirals disappear after encountering other spirals. (b) Three stages of a forest fire. Some fire start close to the small lakes (closed regions) and spread all over the green field.

areas full of bushes and burns down all the forest in a few seconds. Observe also that the lakes and regions with few bushes serve as a sort of barrier to the fire; an approach that is actually used by firemen while combating forest fires.

8.3.11. Framsticks

The original objective of the Framsticks Project was to study the evolution of artificial creatures in a computer simulated three-dimensional virtual world presenting Earth-like conditions. An artificial creature in the Framsticks Project is built with a body and a brain. The body is composed of material points, called parts, connected by elastic joints. The brains are made of artificial neurons with receptors, effectors, and neural connections (Komosinski and Ulatowski, 1998, 1999, 2000).

The system was designed so that it did not introduce restrictions concerning the complexity and size of the creatures, and the complexity and architecture of their brains (neural networks). The whole creature was to be evolved by an evolutionary algorithm. Several types of interaction between the physical objects in the world were considered: static and dynamic friction, damping, action and reaction forces, energy losses after deformations, gravitation, and uplift pressure - buoyancy (in water environments).

Although the evolutionary process can be guided by some predefined criteria, such as prey catching, it is possible to study spontaneous open-ended evolution by assuming the fitness as the life span of the artificial organism. The creatures are capable of interacting with one another (locating, pushing, hurting, killing, eating, etc.) and the environment (walking, swimming, ingesting, etc.).

Architecture of the Framsticks and Its Environment

Body

The basic element of a Framstick is a stick made of two flexibly joined parts, as illustrated in Figure 8.24. The authors used finite elements to perform the step-by-step simulation of the Framsticks. The parts and joints have some fundamental properties, like position, orientation, weight and friction. Other properties may also be added, such as the ability to assimilate energy and the durability of joints under collisions. The endpoints between sticks are connected by unrestricted articulations that can bend in two directions and twist as well. According to the authors, one reason why the Framsticks are made of sticks is because it allows for fast simulation, thus letting evolution to occur in a reasonable time.

Brains

The brains of the Framsticks are made of neurons and their connections. A neuron may be a signal processing unit and can also act upon the Framstick's body as a receptor or effector unit. There are some predefined types of neurons, and one of these is the standard neurcomputing neuron studied in Chapter 4. The connections between neurons generate a neural network that can have any topology and complexity, and this will be evolved during the Framstick's lifetime.

www.frams.alife.pl

Figure 8.24: Creatures are made of sticks (limbs). Muscles (red) are controlled by a neural network, which makes them bend and rotate. (Courtesy of © M. Komosinski.)

Receptors and Effectors

The receptors and effectors make the interface between the body and the brain of the Framstick. The receptors receive stimuli from the environment and transmit them to the organism's brain, and the effectors, on the other side, allow the Framstick to act upon the environment by sending stimuli from the brain to the body.

The Framsticks have three basic receptors: some for orientation in space (equilibrium sense, gyroscope), others for the detection of physical contact (touch), and some for the detection of energy (smell). The effectors play the role of muscles, which can rotate and bend, and are placed on stick joints. Changes of muscle control signals result in movements; thus how to move is a capability that has to be learned with time.

Environment

The virtual world for the Framsticks can be built out of a number of basic primitives, such as planes and curves. It can also be filled with water, so that swimming creatures can emerge. The world can be infinite (with no boundaries), it can be finite (bounded by fences or walls), and can be wrapped, in which case the crossing of a boundary means teleportation to another world. Figure 8.25 illustrates three screen shots of one of the several movies generated by the authors and available at the website (http://www.frams.alife.pl).

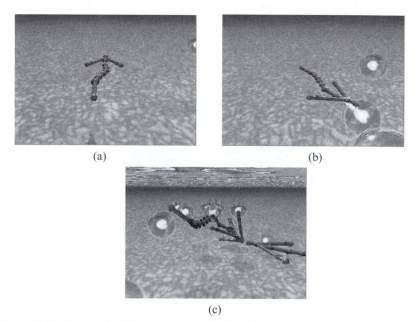

(a)   (b)

(c)

Figure 8.25: Framstick video screenshots. (a) The virtual creature wanders around the environment in search for food or other framsticks. (b) When the creature smells food, it heads toward it for feeding itself. (c) Another creature (on the left) enters the environment and takes all the attention of the framstick. (Courtesy of © M. Komosinski.)

The pictures presented in Figure 8.25 emulate underwater creatures seeking food (bubbles) and one another. Food sources and other creatures are detected by the framsticks via a smell receptor. When another framstick is perceived on the environment, it attracts the attention of the existing framstick, which moves towards it ignoring the food.

Evolving the Framsticks

The Framsticks have various possible genotype symbolic encodings combining the body shape and brain design. Thus, both morphology (body shape) and control (brains) evolve simultaneously. The project provides a basic direct encoding scheme, a recurrent direct encoding scheme and a developmental encoding scheme, each one allowing the creation, design or development of Framsticks. New individuals are created by selecting the genotypes of existing individuals using a standard roulette wheel selection method. The authors also used some niching techniques to increase the diversity of creatures in the virtual world.

In order to estimate the fitness of each creature in the virtual world, the framsticks have to be simulated. In Framsticks, it is possible to choose the number n of creatures in the environment that can be inserted out of the possible number N of genotypes ($n \ll N$). While evaluating the fitness of a Framstick, several features are measured, such as age, average velocity, center of gravity for movements, etc. After selecting the parent creatures, they are reproduced subjected to crossover and mutation operators.

Discussion and Applications

The Framsticks now became a general tool for modeling, simulating and optimizing virtual organisms. The system is quite versatile and can be simplified if needed. It can be employed for research experiments and in education, for the study of virtual worlds, the emergence of complex behaviors, the study of (spontaneous open-ended) evolution, and the experimentation of many other natural phenomena. The Framsticks model is complex enough to allow the emergence of sophisticated, life-like dependencies and phenomena, and simple enough to be simulated on existing home computer systems.

One interesting aspect studied by the developers of the Framsticks was the phenotypic representation of various different genetic encodings for the organisms. Framsticks with significantly different body shapes can emerge and with different physical capabilities as well. The authors claim that the most important part of this research is the study and evaluation of capabilities of various evolutionary processes, including those concerning undirected evolution. For instance, when sticky creatures are evolved under water environments, these learn how to swim in a warm or snake-like fashion, which suggests that evolution can indeed be emulated within computers. The whole project, including publications, software, documentation, pictures and videos, is available at the web site http://www.frams.alife.pl/.

8.4 SCOPE OF ARTIFICIAL LIFE

This chapter presented examples of ALife applications in various domains, from entertainment (cf. Maes, 1995) to scientific studies. Artificial life is of interest to biologists mainly because its models can shed light on biological phenomena. It is relevant to engineers because it offers methods to generate and control complex behaviors that are difficult to generate or regulate using traditional techniques. It is important to people in general because it allows the creation of novel technologies that can be used as entertainment, scientific tools, and to assist humans in the accomplishment of difficult or unpleasant tasks. Artificial life has many other faces involving various aspects of cognitive science, economics, art, physics, chemistry, and even ethics. In summary, ALife has the following main goals:

- Increase the understanding of nature.

- Enhance our insight into artificial models and possible new forms of life.

- Develop new technologies: software evolution, sophisticated robots, ecological monitoring tools, educational systems, computer graphics, etc.

Its major ambition is to build living systems out of nonliving parts; that is, to accomplish strong ALife (Rennard, 2004). ALife, in most cases, emphasizes the understanding of nature, and applications are often left in second plan. However, we have seen through many examples that its applications are almost straightforward from its products. For instance, by developing a strategy to simulate bird flocking, it became possible to use this same strategy in motion pictures.

8.5 FROM ARTIFICIAL LIFE TO LIFE-AS-WE-KNOW-IT

Understanding the relationship between ALife and natural systems promises to provide novel solutions to complex real-world problems, such as disease prevention, stock-market prediction, data mining, and population control. An important point though, is that life is not necessarily an all-or-nothing affair. This chapter presented several examples of artificial life, including biomorphs and computer viruses, because they provide good samples of how ALife forms (life-as-it-could-be) can appear to be fundamentally different from the more familiar and contemporary natural life forms (life-as-we-know-it).

Most historical definitions of life, such as the ones reviewed here, rely upon listing certain common features among things that people had already agreed to be alive. This includes the ability to reproduce, a basis in the chemistry of carbon compounds, the presence of functional interactions with the environment and interdependence of parts, and so forth. If something shares these features, it is considered alive; if it does not, then it is not alive. This approach has enough problems as it stands: is a mule not alive because it cannot reproduce? If inhabitants from the Moon landed on Earth and turned out to be undoubtedly alive, but quite clearly different from us in major respects, then we would have to revise our definitions substantially. So it is with many ALife products (e.g., the biomorphs): just because they are not carbon-based and do not even have (in most cases) any similarity with known life forms, it does not mean they do not represent life-as-it-could-be.

Most artificial life products do and will challenge our definitions of life in important and practical ways. One long overdue revision is to stop treating life as an all-or-nothing affair, and start considering that things may be more or less alive than one another, or equally alive but in different ways. Also, the outcomes of ALife may represent new patterns, behaviors, and forms of life previously unimaginable to us. The creatures, the AIBO robot dog, the biomorphs, the artificial ants and termites, etc., are not as alive as you and I, but they are certainly more alive than a stone. How they compare to known forms of life, such as real dogs, ants, termites, people, bacteria, and to rats, however, is an interesting question, and one which we hope people will enjoy debating.

One important aspect of natural computing, and more specifically of computational ALife systems, is that it is usually possible to extract some minimal (behavioral) rules that can allow for the generation of patterns, behaviors, and artificial organisms. Although these rules are not always obtained in a straightforward manner, they can in most cases be obtained by studying the biological and/or theoretical models available in the literature. In return, ALife has been serving, among other things, as good and sometimes even accurate models of natural phenomena. This has provided new insights into how nature works, and opened several new avenues to study life-as-we-know-it and life-as-it-could-be.

8.6   SUMMARY

If you wanted to create the impression of life on a computer, the most obvious and commonly used approach would be to define a set of rules to describe a specific behavior: 'if this happens, do that', etc. However, the definitive characteristic of natural living things is that they clearly do not work in this way. Life certainly has structure and logic, but it is not driven by blind and unflexible rules. Trying to formulate a rule for how to react in every conceivable situation obtainable in even a mildly complex environment is totally out of the question, for humans or for computers.

Artificial life provides a new approach to what is life, how we see it, and how it can be synthesized. It is a naturally interdisciplinary subject of direct interest to all fields of investigation, from chemistry to physics to biology to engineering. The small sample of ALife systems presented in this chapter may have been sufficient to give you the feeling of what is ALife, where and when it can be used. There is, however, much more to be seen and discussed. I am particularly keen on the *Tierra* system (Ray, 1992), *Avida* (Adami, 1997), the *Floys* (Dolan, 1998), and the *Evolved Artificial Creatures* (Sims, 1994). Other good websites where to find general information about ALife are:

- The International Society for Artificial Life: http://www.alife.org

- Zooland (The Artificial Life Resource): http://zooland.alife.org/

- The Temple of ALife: http://alife.fusebox.com/
- ALife FAQ: http://www.faqs.org/faqs/ai-faq/alife

It is usually difficult to provide results and specific algorithms for ALife, because most of them are essentially projects that involve a number of techniques, computing paradigms, platforms, and observations from nature and its corresponding models. The overall objective is to create synthetic life-like systems whose pattern or behavior is sufficiently complex to be classified as life-as-it-could-be or maybe life-as-we-know-it. It is sometimes difficult to establish how much of the systems developed is genuinely emergent and how much is a result of ad-hoc procedures conferred by the designer, though some tests for emergence have already been suggested in the literature (Ronald et al., 1999).

8.7   EXERCISES

8.7.1. Questions

1. To cope with the potential for robots to harm people, Asimov, in 1940, in conjunction with science fiction author and editor John W. Campbell, formulated the three Laws of Robotics (Asimov et al., 1983):

    - A Robot may not harm a human being, or allow a human being to be harmed.

    - A Robot must obey the orders given by the human beings, except where such orders would conflict with the First Law.

    - A Robot must protect his own existence, as long as such protection does not conflict the First or Second Law.

    This chapter has pointed out that artificial life can provide us with novel forms of viewing and studying life, but it can also do us some harm when artificial organisms, such as computer viruses, are created. Thus, it is important to try and provide a Code of Ethics or some laws for ALife and artificial organisms. Name some important components of such Code of Ethics.

2. It has been argued that viruses are in the borderline between living and nonliving beings. Explain why.

    Name another example of an element that is in the borderline between living and nonliving.

3. Choose an applet about boids and discuss how faithful to life the behavior studied is. Ground your discussion on related works from the literature of behavioral ecology, theoretical biology, animal behavior and the like. Carefully detail the boids studied and the biological foundation for the discussion, and list the consulted bibliography, including papers, websites, journals, magazines, monographies, and reports.

4. Why do animals tend to aggregate in flocks, herds, and schools?

5. Compare Algorithm 8.1 with the pseudocode for a standard evolutionary algorithm presented in Chapter 3.

6. It was discussed, in Section 8.3.8, that the behavior of termites while gathering wood chips is similar to the dead bodies clustering behavior of some ant species. Discuss the similarities and differences between these two types of behavior.

7. Ant Colony Optimization (ACO) algorithms are inspired by the collective trail-laying, trail following foraging behavior of ants (Chapter 4). Individual ants deposit *pheromone* as they move from a food source to their nest and foragers follow such pheromone trails. An ACO algorithm involves the following main steps:

   - A probabilistic transition rule that defines the next step of an "artificial ant". This is a function of the pheromone trail.

   - After completing a tour, an ant lays some pheromone trail on the tour it performed. This is inversely proportional to the length of the tour performed.

   - The pheromone trail evaporates with time.

   Consider the following three rules that could be implemented to create an ALife simulation of ant foraging:

   - If an ant does not know where there is food, then forage.

   - If an ant sees another ant, ask if it knows where there is food. If it does, go for it. Else, run away from that ant.

   - If an ant finds food, it must take it back to the nest, and return to the place where it found food. If an ant does not find any food where it was supposed to be, then keep foraging.

   Explain the behaviors that would emerge from this rule set.

   Discuss what would probably happen if there is a single food source on the grass, and if there are more than one food source.

   Do you think the ants would privilege food sources that are nearer to the nest? If yes, why? If not, should they?

8. Visit the StarLogo website (http://education.mit.edu/starlogo/) and discuss the main similarities and differences between StarLogo and cellular automata.

9. The pattern generated by some cellular automata is strongly reminiscent of the pattern of pigmentation found on the shells of certain mollusks. It is quite possible that the growth of these pigmentation patterns follow cellular automata rules. Find and list some studies in the literature relating the patterns of seashells with cellular automata.

10. Search the literature for the Game of Life, and illustrate ten life forms typically found in it.

11. What is the GAIA hypothesis?

8.7.2. Computational Exercises

1. Using any of the biomorph applets available online (see Section 8.3.2), appropriately select or watch the biomorphs evolved in order to find some forms that closely resemble known patterns, such as the ones illustrated in Figure 8.4. Put some of these forms together and name them, making your own portfolio of Biomorphs.

2. Using StarLogo (http://education.mit.edu/starlogo/), implement the ant-clustering algorithm (ACA) described in Chapter 4. Apply it to the same benchmark problem and compare the results with those obtained using the bi-dimensional grid developed previously.

3. Choose one of the sample projects of StarLogo and solve its exploration tasks (http://education.mit.edu/starlogo/projects.html). Write a brief report with the results obtained including any theoretical background knowledge that may eventually be necessary to perform the explorations.

4. Implement a bi-dimensional CA following the rules of 'The Game of Life'.

8.7.3. Thought Exercise

1. When John Conway proposed 'The Game of Life', he conjectured that no pattern could grow without limit; that is, no configuration with a finite number of live cells can grow beyond a finite upper limit to the number of live cells in the cellular automaton. He offered the prize of US$50 to the first person who could prove or disprove this conjecture. (At that time, US$50 was good money!) How would you disprove it?

8.7.4. Projects and Challenges

1. Implement a 2-D simulation of a bird flock (boid) based on the description provided in Section 8.3.1 and on further reading that can be obtained at: http://www.red3d.com/cwr/. Other Java applets of boids can be found at: http://staff.aist.go.jp/utsugi-a/Lab/Links.html.

2. Based on the study of how slime mold cells cluster together and the background knowledge on swarm intelligence (Chapter 4), develop a slime mold clustering algorithm. Discuss the aggregation behavior of the slime molds.

3. Chapter 5 discussed with reasonable detail the behavior of ants when foraging for food. Develop an ALife project for ant foraging, including bibliographical documentation and computer implementations. Verify if your ants do privilege food sources that are closer to the nest, and allow some food sources to be richer than other food sources. Discuss the behavior of the ants.

4. Conway calls the speed a chess king moves in any direction the 'speed of light', because it is the highest speed at which any kind of movement can occur on the board. No pattern in Life can replicate itself rapidly enough to move at such speed. Conway has proved that the maximum speed dia-

gonally is a fourth the speed of light. A *glider* replicates itself in the same orientation after four moves and travels one cell diagonally after replicating itself. Thus, it glides across the field at a fourth the speed of light. Demonstrate that the movement of a finite figure horizontally or vertically into empty space in Life cannot exceed half the speed of light.

5. Implement a CA system to simulate a forest fire. Tip: use the ideas presented in CAFUN.

8.8 REFERENCES

[1]   Adamatzky, A. and Komosinski, M. (2005), *Artificial Life Models in Software*, Springer.

[2]   Adami, C. (1997), *Introduction to Artificial Life*, Telos Pr.

[3]   Adami, C. and Brown, T. (2000), "What is Artificial Life", [On line] http://www.alife7.alife.org/whatis.shtml

[4]   Asimov, I., Warrick, P. S. and Greenberg, M. H. (eds.) (1983), *Machines that Think*, Holt. Rinehart and Wilson.

[5]   Begel, A. and Klopfer, E. (2005), "StarLogo: A Programmable Complex Systems Modeling Environment for Students and Reachers", In A. Adamatzky and M. Komosinski, *Artificial Life Models in Software*, Springer, Chapter 8, pp. 187–210.

[6]   Bonabeau, E., Guérin, S., Snyers, D., Kuntz, P. and Theraulaz, G. (2000), "Three-Dimensional Architectures Grown by Simple 'Stigmergic' Agents", *BioSystems*, **56**(2000), pp. 13–32.

[7]   Bonabeau, E., Theraulaz, G., Arpin, E. and Sardet, E. (1994), "The Building Behavior of Lattice Swarms", in R. A. Brooks and P. Maes (eds.), *Artificial Life IV*, pp. 307–312.

[8]   Camazine, S., Deneubourg, J.-L., Franks, N. R., Sneyd, J., Theraulaz, G. and Bonabeau, E. (2001), *Self-Organization in Biological Systems*, Princeton University Press.

[9]   Cohen, F. (1987), "Computer Viruses: Theory and Experiments", *Computers & Security*, **6**, pp. 22–35.

[10]   Colella, V., Klopfer, E. and Resnick, M. (2001), *Adventures in Modelling: Exploring Complex Dynamic Systems with StarLogo*, Teachers College Press.

[11]   Dawkins, R. (1986), *The Blind Watchmaker*, Penguin Books.

[12]   Dawkins, R. (1988), "The Evolution of Evolvability", in C. Langton (ed.), *Artificial Life*, Addison-Wesley, pp. 201–220.

[13]   Dolan, A. (1998), Artificial Life website [On line] http://arieldolan.com/.

[14]   Dörner, D. (1999), *Bauplan für eine seele* (Blueprint for a soul), Reinbek: Rowohlt.

[15]   Drexler, M. (1999), "Matsushita Releases Electronic Pet", *IDG News Service*, [On line] http://www.idg.net/idgns/1999/03/24/MatsushitaReleasesElectronicPet.shtml.

[16]   Emmeche, C. (2000), "Closure, Function, Emergence, Semiosis and Life: The Same Idea? Reflections on the Concrete and the Abstract in Theoretical Biology", In J. L. R. Chandler and G. Van de Vijver (Eds.), *Closure: Emergent Organiza-*

tions and Their Dynamics. Annals of the New York Academy of Sciences, New York: The New York Academy of Sciences, **901**, pp. 187–197.

[17] Farmer, J. D. and Belin, A. d'A. (1991), "Artificial Life: The Coming Evolution", in C. Langton, C. Taylor, J. D. Farmer and S. Rasmussen (eds.), *Artificial Life II*, pp. 815–840.

[18] Gardner, M. (1970), "Mathematical Games: The Fantastic Combinations of John Conway's New Solitaire Game 'Life'", *Scientific American*, **233**, pp. 120–123.

[19] Goodwin, B. (1994), *How the Leopard Changed its Spots*, Princeton Science Library.

[20] Grand, S. (1997), "Creatures: An Exercise in Creation", *IEEE Intelligent Systems Magazine*, **4**.

[21] Grand, S. (2000), *Creatures: Life and How to Make It*, Harvard University Press.

[22] Grand, S., Cliff, D. and Malhotra, A. (1997), "Creatures: Artificial Life Autonomous Software Agents for Home Entertainment", *Proc. of the 1st Int. Conf. on Autonomous Agents*, ACM Press, pp. 22–29.

[23] Harnad, S. (1994), "Artificial Life: Synthetic vs. Virtual", in C. Langton (ed.), *Artificial Life III*, Addison-Wesley, pp. 539–552.

[24] Hille, K. (2001), "Synthesizing Emotional Behavior in a Simple Animated Character", *Artificial Life*, **7**(3), pp. 303–314.

[25] Johansson, J. (2001), *An Electric Field Approach: A Strategy for Sony Four-Legged Robot Soccer*, Master Thesis, Department of Software Engineering and Computer Science, Bleckinge Institute of Technology, Ronneby, Sweden. [On line]: http://www.student.bth.se/~pt96jjh/EFA/AnElectricFieldApproach.pdf.

[26] Jones, M. (2001), *Blind Watchmaker Applet*, Department of Physics, Syracuse University, NY/USA. [On line] http://suhep.phy.syr.edu/courses/mirror/biomorph.

[27] Keeley, B. L. (1997), "Evaluating Artificial Life and Artificial Organisms", in C. Langton and K. Shimohara, *Artificial Life V*, Addison-Wesley, pp. 264–271.

[28] Komosinski, M. and Ulatowski, S. (1998), "Framsticks – Artificial Life", *Proc. of the 4th European Conf. on Machine Learning*, pp. 7–9.

[29] Komosinski, M. and Ulatowski, S. (1999), "Framsticks: Towards a Simulation of a Nature-Like World, Creatures and Evolution", In D. Floreano and F. Mondada, *Lecture Notes in Artificial Intelligence 1674* (*Proc. of the 5th European Conf. on Artificial Life*), pp. 261–265.

[30] Komosinski, M. (2000), "The World of Framsticks: Simulation, Evolution, Interaction", In *Proceedings of the 2nd Int. Conf. on Virtual Worlds (VW2000)*, Springer-Verlag (LNAI **1834**), pp. 214–224. [On line] http://www.frams.alife.pl/.

[31] Laing, R. (1988), "Artificial Organisms: History, Problems, Directions", in C. Langton (ed.), *Artificial Life*, Addison-Wesley, pp. 49–61.

[32] Langton, C. G. (1988), "Artificial Life", in C. Langton (ed.), *Artificial Life*, Addison-Wesley, pp. 1–47.

[33] Langton, C. G. (1986), "Studying Artificial Life with Cellular Automata", *Physica 22D*, pp. 120–149.

[34] Langton, C. G. (1984), "Self-Reproduction in Cellular Automata", *Physica 10D*, pp. 135–144.

[35] Maes, P. (1995), "Artificial Life Meets Entertainment: Lifelike Autonomous Agents", *Communications of the ACM*, **38**(11), pp. 108–114.

[36]   Margulis, L., Sagan, D. and Eldredge, N. (2000), *What is Life?*, University of California Press.

[37]   Moreno, A. (2000), "Artificial Life as a Bridge Between Science and Philosophy", In M. A. Bedau, J. S. McCaskill, N. H. Packard, and S. Rasmussen (eds.), *Artificial Life VII*, pp. 507–512.

[38]   Morris, D. (1995), *The Secret Surrealist: The Paintings of Desmond Morris*, Phaidon Press.

[39]   Nicholl, D. S. T. (2002), *An Introduction to Genetic Engineering*, Cambridge University Press.

[40]   Packard, N. H. and Bedau, M. A. (2000), "Artificial Life", *Encyclopedia of Cognitive Science*, Macmillan.

[41]   Pattee, H. H. (1988), "Simulations, Realizations, and Theories of Life", in C. G. Langton, *Artificial Life*, Addison-Wesley, pp. 63–77.

[42]   Pieprzyk, J., Hardjono, T. and Seberry, J. (2003), *Fundamentals of Computer Security*, Springer-Verlag.

[43]   Ray, T. S. (1992), "An Approach to the Synthesis of Life", in C. G. Langton, C. Taylor, J. D. Farmer and S. Rasmussen (eds.), *Artificial Life II*, Addison Wesley, pp. 371–408.

[44]   Ray, T. S. (1994), "An Evolutionary Approach to Synthetic Biology", *Artificial Life*, **1**(1/2), pp. 179–209.

[45]   Rennard, J.-P. (2004), "Perspectives for Strong Artificial Life", In L. N. de Castro and F. J. Von Zuben, *Recent Developments in Biologically Inspired Computing*, Chapter 12, Idea Group Incorporation.

[46]   Resnick, M. (1994), *Turtles, Termites, and Traffic Jams: Explorations in Massively Parallel Microworlds*, MIT Press.

[47]   Reynolds, C. W. (1987), "Flocks, Herds, and Schools: A Distributed Behavioral Model", *Computer Graphics*, **21**(4), pp. 25–34.

[48]   Reynolds, C. W. (1999), "Steering Behaviors for Autonomous Characters", *Proc. of the Game Developers Conference*, [On line] http://www.red3d.com/cwr/papers/1999/gdc99steer.pdf

[49]   Ronald, E. M. A., Sipper, M. and Capcarrère, M. S. (1999), "Design, Observation, Surprise! A Test of Emergence", *Artificial Life*, **5**(3), pp. 225–239.

[50]   Schrödinger, E. (1992), *What is Life?: With Mind and Matter and Autobiographical Sketches*, Cambridge University Press.

[51]   Sims, K. (1994), "Evolving Virtual Creatures", Computer Graphics (Siggraph '94 Proceedings), pp.15–22. [On line] http://web.genarts.com/karl/.

[52]   Sipper, M. (1995), "An Introduction to Artificial Life", *Explorations in Artificial Life – Special Issue of AI Expert*, pp. 4–8.

[53]   Sober, E. (1992), "Learning from Functionalism – Prospects for Strong Artificial Life", in C. G. Langton, C. Taylor, J. D. Farmer and S. Rasmussen (eds.), *Artificial Life II*, Addison Wesley, pp. 749–766.

[54]   SONY® Corporation (2000), *Entertainment Robot AIBO: Operating Instructions ERS 210*, [On line] http://www.sony.com.

[55]   Spafford, E. H. (1991), "Computer Viruses – A Form of Artificial Life?", in C. G. Langton, C. Taylor, J. D. Farmer (eds.), *Artificial Life II*, Addison-Wesley, pp. 727–746.

[56] Terzopoulos, D., Tu, X. and Grzeszczuk, R. (1994), "Artificial Fishes: Autonomous Locomotion, Perception, Behavior, and Learning in a Simulated Physical World", *Artificial Life*, **1**(4), pp. 327–351.

[57] Theraulaz, G. and Bonabeau, E. (1995), "Modeling the Collective Building of Complex Architectures in Social Insects with Lattice Swarms", *Journal of Theoretical Biology*, **177**, pp. 381–400.

[58] Wilmut, I., Campbell, K. and Tudge, C. (2001), *The Second Creation: The Age of Biological Control by the Scientists who Cloned Dolly*, Headline.

PART III

COMPUTING WITH NEW NATURAL MATERIALS

CHAPTER 9

DNA COMPUTING

"The filament of DNA is information, a message written in a code of chemicals, one chemical for each letter. It is almost too good to be true, but the code turns out to be written in a way that we can understand...[it] is a linear language, written in a straight line ... As you grow up and accumulate experiences, the influence of your genes <u>increases</u> ... you gradually express your own innate intelligence and leave behind the influences stamped on you by others ... This proves two vital things: that genetic influences are not frozen at conception and that environmental influences are not inexorably cumulative."
(M. Ridley, Genome: The Autobiography of a Species in 23 Chapters, Fourth Estate, 1999, p. 13; 84–85)

"Molecular biology had come a long way in its first twenty years after the discovery of the double helix. We understood the basic machinery of life, and we even had a grasp on how genes are regulated. But all we had been doing so far was observing; we were molecular naturalists for whom the rain forest was the cell - all we could do was describe what was there. The time had come to become proactive. Enough observation: we were beckoned by the prospect of intervention, of manipulating living things. The advent of recombinant DNA technologies, and with them the ability to tailor DNA molecules, would make all this possible."
(J. Watson, DNA: The Secret of Life, Arrow Books, 2003, p. 81–82)

9.1   INTRODUCTION

One of the reasons for the great progress in science over the last century was the conception of life as a kind of information processing. The processes that transform matter and energy in living systems do so under the direction of a set of symbolically encoded instructions. The 'machine' language that describes the objects and processes of living systems contains four letters {A,C,T,G}, and the 'text' that describes a person has a great number of characters. These form the basis of the DNA molecules that contain the genetic information of all living beings. DNA is also the main component of *DNA computing*.

DNA computing is one particular component of a broader field called *molecular computing* that can be broadly defined as the use of (bio)molecules and biomolecular operations to solve problems and to perform computation. It constitutes a unique combination between computer science and molecular biology. DNA computing was introduced by L. Adleman in 1994 when he solved an NP-complete problem (Appendix B.3.3) using DNA molecules and biomolecular techniques for manipulating DNA. The basic idea is that it is possible to apply operations to a set of (bio)molecules, resulting in interesting and practical performances. To date, most molecular computing techniques are based on a brute force strategy in which the operations are simultaneously (in parallel) applied to all molecules being used.

DNA computing can be viewed as a novel approach for solving complex problems, or as a completely new computing paradigm that may eventually complement or supplement the current silicon-based computers. In order for such goals to be met, there are several questions concerning DNA computing that have to be answered. Two important ones are related to the potential of this new paradigm and to the feasibility of physically implementing it (Zingel, 2000):

1. Can any algorithm be simulated by means of DNA computing? In other words, is the DNA computing computationally complete? Is there a universal DNA system in the same sense as there is a universal Turing Machine: given a computable function, can it simulate the actions of that function for any argument?

2. Is it possible to design a programmable molecular computer? In other words, how feasible is it to construct a molecular computer? What difficulties should be overcome for the construction of a real (physical) molecular computer?

Several models of DNA computing have been proposed to answer these and other questions, and some of these models will be reviewed here. It is possible to divide the DNA computing models in two major classes (Zingel, 2000). A first class, commonly referred to as *filtering models* (Amos, 1997), which includes models based on operations that are successfully implemented in the laboratory (Section 9.3). A second class composed of the so-called *formal models*, such as the *splicing systems* and the *sticker systems*, whose properties are much easier to study, but only the first steps have been taken toward their practical implementation (Section 9.4).

In brief, DNA computing is based on the use of DNA molecules as data structure, and the application of DNA manipulation techniques to compute with DNA. This chapter focuses on the pioneer DNA computing models of the filtering type. As these models were originally proposed as novel problem solving techniques based on DNA, some basic concepts from computer science are reviewed in Appendix B.3, in particular 'Universal Turing Machines' and 'Complexity Theory'. These will serve as basis to provide answers to the questions listed above concerning the potentiality of DNA computing and its implementation. These concepts are also very important for a proper understanding of quantum computing, to be introduced in the next chapter.

Based upon standard biological operations to manipulate DNA (Section 9.2), it has been possible to introduce what can be called *DNA programming languages*, with specific operators and data structures. Two of these languages are discussed in Section 9.3, namely, the *test tube programming language*, and the *DNA Pascal*. A brief description of some formal models is provided in Section 9.4, and Section 9.5 presents one possible answer to the universal computing capability of DNA. Further references about the DNA universality are also provided.

9.1.1. Motivation

The main advantages of DNA computing are its high speed, energy efficiency, and economical information storage. There are, of course, a possibility for error and difficulties in implementing a real DNA computer. When compared with the currently known silicon-based computers, DNA computing offers some unique features (Ruben and Landweber, 2000):

- It uses DNA as data structures. Thus, data is stored using strings built out of a quaternary alphabet, {A,C,G,T}, instead of a binary alphabet {0,1}. The structure of DNA and how it is manipulated is completely different from the data structure and operations used in today's computers.

- DNA molecules, and thus computers, can work in a massively parallel fashion.

- Computation can be performed at a molecular level, potentially a size limit that may never be reached by the semiconductor industry.

- DNA computers can potentially work in an extraordinary way of high energy efficiency and economical information storage.

- DNA computers are highly effective in solving NP-complete problems; that is, DNA computers can be used to solve problems that cannot be (practically) solved using standard computers.

An overall, striking observation about DNA computing is that, at least theoretically, there seem to be many diverse ways of constructing DNA-based universal computers. In (Păun et al., 1998), the authors claim that *Watson-Crick complementarity* (see Section 9.2) guarantees universal computation in any model of DNA computers having sufficient capabilities for handling inputs and outputs. What is important to be stressed here is that there may be different forms of building DNA computers, and this chapter will discuss some of the pioneering models and briefly review some of the formal models.

9.2 BASIC CONCEPTS FROM MOLECULAR BIOLOGY

DNA computing is based on two aspects of molecular biology: the use of DNA molecules as a type of data structure to perform computation, and the use of DNA manipulation techniques to compute with DNA molecules. This section introduces some basic concepts about the structure of DNA and how to manipulate them so as to perform computation.

9.2.1. The DNA Molecule

It has been discussed in Chapter 3 that the genetic material is contained in the cell nucleus and is complexed with proteins and organized into linear structures called *chromosomes*. Chromosomes are composed of genes, which, by themselves, are segments of a helix molecule called *deoxyribonucleic acid*, or DNA for short (see Figure 9.1).

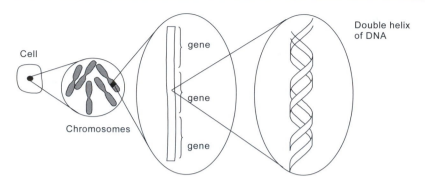

Figure 9.1: The DNA molecules are contained within the chromosomes of eukaryotic cells.

All the genetic information in cellular organisms is stored in DNA, which consists of *polymer chains*, commonly referred to as *DNA strands*, of four simple nucleic acid units, called *deoxyribonucleotides* or simply *nucleotides*. There are four nucleotides found in DNA. Each nucleotide consists of three parts: one *base molecule*, a *sugar* (deoxyribose in DNA), and a *phosphate group*. The four bases are *adenine* (A), *guanine* (G), *cytosine* (C), and *thymine* (T). As the nucleotides differ only by their bases, they are often called bases.

Numbers from 1′ to 5′ are used to denote the five carbon atoms of the sugar part of the nucleotide. The phosphate group is attached to the 5′ carbon, and the base is attached to the 1′ carbon. To the 3′ carbon there is attached a hydroxyl (OH) group. Each strand has, according to chemical convention, a 5′ and a 3′ end, thus any single strand has a natural orientation. This orientation, and the notation used here, is due to the fact that one end of the single strand has a free (i.e., unattached to another nucleotide) 5′ phosphate group, and the other has a free 3′ hydroxyl group.

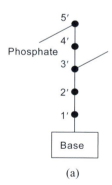

(a)

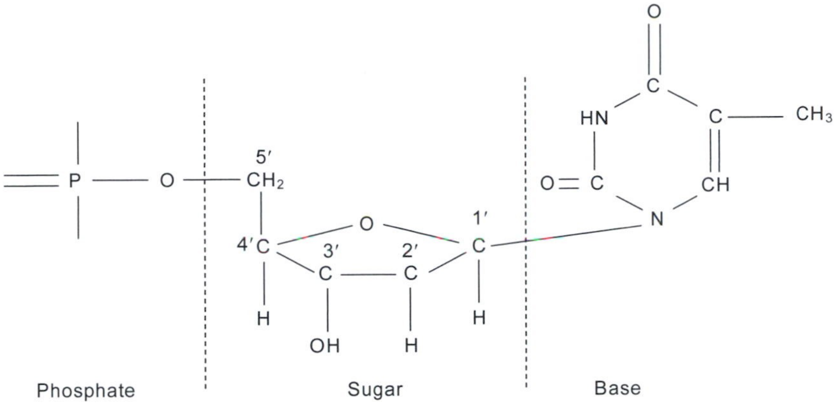

Figure 9.2: The structure of a nucleotide. (a) Schematic representation. (b) Chemical structure of a nucleotide with a thymine (T) base. (Reproduced with permission from [Păun et al., 1998], © Springer-Verlag.)

Figure 9.2(a) brings a schematic representation of a nucleotide and Figure 9.2(b) depicts the chemical structure of a nucleotide with thymine base. In Figure 9.2(a), the carbons of the sugar base are enumerated from 1′ to 5′. Note that the phosphate group is attached to the 5′ carbon, the hydroxyl is attached to the 3′ carbon, and the base is attached to the 1′ carbon, as discussed above.

Nucleotides can link together in two different ways (Păun et al., 1998), as illustrated in Figure 9.3:

- The 5′ phosphate group of one nucleotide is joined with the 3′ hydroxyl group of the other forming a *covalent bond*.

- The base of one nucleotide interacts with the base of the other to form a *hydrogen bond*, which is a bond weaker than the covalent bond.

Another important feature of the nucleotide bonding is that any two nucleotides can link together to form a sequence in the same way as several symbols form a string. As for symbols on a string, there is no restriction on nucleotides in the sequence. However, the bonds between the bases can only occur by the pairwise attraction of the following bases: A binds with T, and G binds with C. This is called the *Watson-Crick complementarity*, after J. D. Watson and F. H. C. Crick who discovered the double helix structure of DNA.

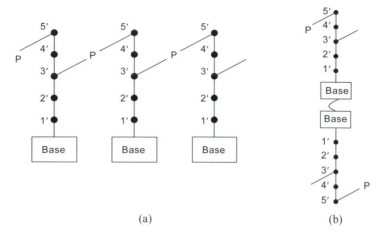

(a)   (b)

Figure 9.3: The two types of bonding between nucleotides: (a) covalent or phosphodi-ester bond; and (b) hydrogen bond, which has to obey the Watson-Crick complementarity (A-T and C-G).

Since DNA consists of two complementary strands bond together, these units are often called *base pairs*. The length of a DNA sequence is often measured in thousands of bases, abbreviated kb. Nucleotides are generally abbreviated by their first letter, and appended into sequences, written, e.g., GTACAGTT. The nucleotides are linked to each other in the polymer by phosphodiester bonds. It can be observed from Figure 9.3, that the bond is directional; a strand of DNA has a head (the 5′ end) and a tail (the 3′ end). One well known fact about DNA is that it forms a double helix; that is, two *helical* (spiral-shaped) strands of the polypeptide, running in opposite directions, held together by hydrogen bonds (Figure 9.4).

<!-- Figure 9.4(a) -->
5′ ───── 3′

C ≣ G
T -- A
G ≣ C
A --- T

3′ ───── 5′

(a)

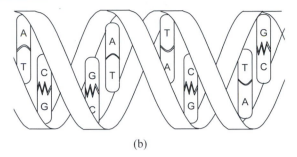

(b)

Figure 9.4: Other representations of the DNA molecule. (a) A schematic representation of a DNA molecule depicting the sugar-phosphate backbone, the complementary bonding between bases forming the base pairs, and the directional bonding. (b) The double helix of DNA. A is complementary to T, and C is complementary to G.

The single stranded sequences of nucleotides have a directionality given by the carbons used by the covalent bonds. The hydrogen bonds, based on complementarity, can bring together single stranded sequences of nucleotides only if they are of opposite directionality. Although the sequence in one strand of DNA is completely unrestricted, because of the bonding restrictions the sequence in the complementary strand is completely determined. It is this feature that makes it possible to produce high fidelity copies of the information stored in the DNA. Despite the many schematic representations of DNA molecules, such as the ones depicted in Figure 9.3 and Figure 9.4, in the DNA computing context, most DNA molecules are simply represented as linear strings of characters, such as the one depicted in Figure 9.5.

$$5' - T\,C\,G\,A\,T\,T\,G\,A\,A\,C\,C - 3'$$
$$3' - A\,G\,C\,T\,A\,A\,C\,T\,T\,G\,G - 5'$$

Figure 9.5: Schematic representation of a DNA molecule in the context of DNA computing.

Note the directionality and complementarity in this representation. Note also that the two strands are drawn one over the other, with the upper strand oriented from left to right in $5'-3'$ direction, and the lower strand oriented in the opposite direction, $3'-5'$.

The DNA molecules not necessarily have to have the same number of bases in the upper and lower strands. In some cases, there may be 'free' bases on the left or right end of the molecules. In this case we say that the molecule has *sticky ends*, as illustrated in Figure 9.6. These ends are called sticky because any complementary sequence to the free end can bind with it.

$$5' - T\,C\,G\,A\,T\,T\,G\,A - 3'$$
$$3' - A\,A\,C\,T\,T\,G\,G - 5'$$

Figure 9.6: DNA molecule with sticky ends.

9.2.2. Manipulating DNA

All DNA computing techniques apply a specific set of biological operations to a set of strands. These operations are all commonly used by molecular biologists, and the most important of them, in the context of DNA computing, will be reviewed here. Each operation will be presented by first providing its name (when appropriate) and, within brackets, its role; the explanation follows next. As will be seen, many operations with DNA can be mediated by *enzymes*, which are proteins that catalyze some chemical reactions where DNA is involved.

Denaturation (separates DNA strands): by heating double-stranded DNA, it becomes possible to separate the two strands into single strands. It is possible to separate the two strands without breaking the single strands because the hydrogen bonds between complementary nucleotides are much weaker than the covalent bonds between adjacent nucleotides in the two strands. Denaturation is also called *melting*.

Annealing (fuses DNA strands): annealing is the reverse of melting, whereby a solution of single strands is cooled down so as to allow for complementary strands to bind together. When the single strands in the solution are not complementary in their entirety, the result of annealing is DNA molecules with sticky ends. Annealing is also called *renaturation*. Figure 9.7 illustrates the processes of melting and annealing.

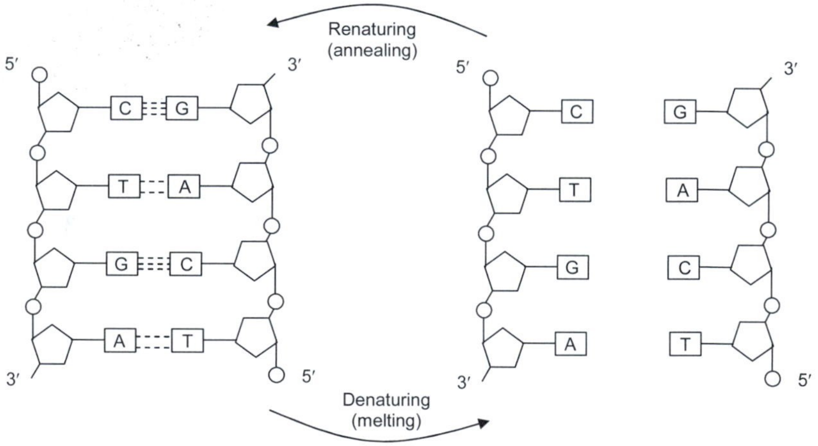

(a)

5′ – T C G A T T G A A – 3′   (single strand)

3′ – A A C T T C – 5′   (single strand)

↓ (Annealing)

5′ – T C G A T T G A A – 3′
3′ – A A C T T C – 5′   } (double strand)

(b)

Figure 9.7: Denaturing and annealing DNA. (a) The annealing process allows the bonding of two single strands into a double strand, and the melting allows the separation of a double strand into two single strands. (b) The annealing of two strands that are not fully complementary may result in a double strand with sticky ends.

Polymerase extension (fills in incomplete strands): a class of enzymes called *polymerases* is able to add nucleotides to an incomplete DNA molecule, such as the one shown in Figure 9.7(b). Thus, the molecule can be completed to a double strand without sticky ends. The polymerases are able to add nucleotides in the 5′–3′ direction until pairing each nucleotide with its Watson-Crick complement. This process requires an existing single strand that acts as a *template* prescribing the chain of nucleotides to be added (by Watson-Crick complementarity), and an existing sequence, called *primer*, which is bonded to a part of the template with the 3′ end available for extension in the 5′–3′ direction (Figure Figure 9.8(a)).

Nuclease degradation (shortens DNA molecules): the DNA *nucleases* enzymes are capable of removing nucleotides from the strands (Figure 9.8(b)). There are two classes of nucleases, *exonucleases* and *endonucleases*. Exonucleases shorten DNA by removing nucleotides from the ends of the DNA molecule.

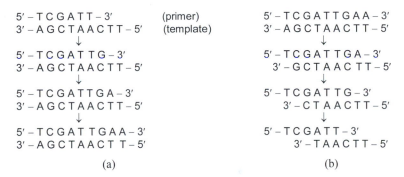

Figure 9.8: Polymerase extension and nuclease degradation of DNA molecules. (a) Lengthening DNA molecules via polymerase extension. (b) Shortening DNA molecules using nuclease degradation.

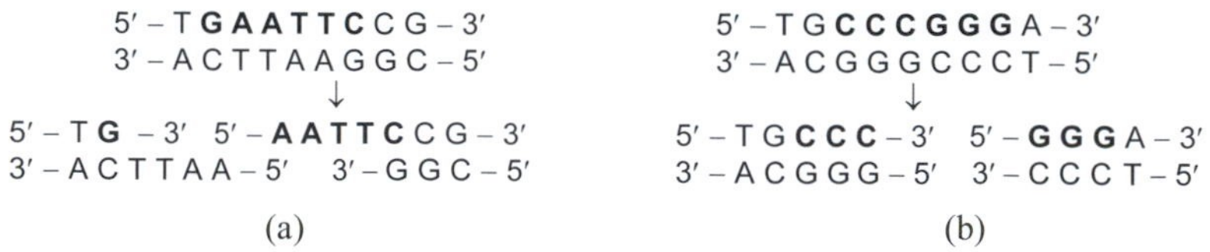

```
        5′ – T G A A T T C C G – 3′              5′ – T G C C C G G G A – 3′
        3′ – A C T T A A G G C – 5′              3′ – A C G G G C C C T – 5′
                      ↓                                        ↓
   5′ – T G – 3′  5′ – A A T T C C G – 3′     5′ – T G C C C – 3′   5′ – G G G A – 3′
   3′ – A C T T A A – 5′  3′ – G G C – 5′     3′ – A C G G G – 5′   3′ – C C C T – 5′

             (a)                                        (b)
```

Figure 9.9: Endonucleases in action. The bold nucleotides are the sites where the cut is going to be made; these are recognized by specific enzymes. (a) The staggered cut promoted by an enzyme called *Eco*RI. (b) The blunt cut promoted by the enzyme *Sma*I.

Some exonucleases remove nucleotides from the 5′ end, others remove from the 3′ end, and others remove nucleotides from single strands. Figure 9.8 illustrates the polymerase extension and nuclease degradation of DNA molecules.

Endonucleases (cut DNA molecules): endonucleases are able to cut DNA molecules by destroying the covalent bonds between adjacent nucleotides. They can be specific as to what, where, and how they cut. For instance, endonucleases, called *restriction enzymes*, can cut DNA molecules at specific sites. The cut can be *blunt*, i.e., straight through both strands, or *staggered*, i.e., leaving sticky ends (Figure 9.9). There are also endonucleases that cut only single strands, or within single strand pieces of a mixed DNA molecule containing single stranded and double stranded pieces.

Ligation (links DNA molecules): a class of enzymes can be used to link together, or ligate, fragments of DNA molecules. Ligation is often performed after an annealing operation is used to concatenate DNA strands. The difference between annealing and ligation is that the former allows for the bonding of complementary bases, while the latter performs the phosphodiester bonding between two consecutive nucleotides. The hydrogen bond keeps complementary sticky ends together, but a gap (called *nick*) remains in each of the strands. A ligase is,

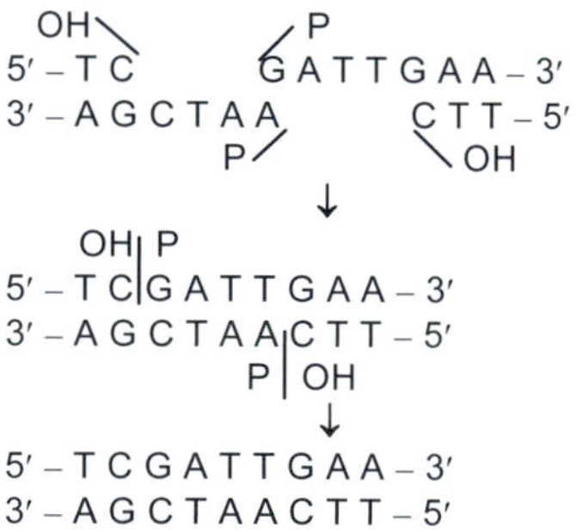

```
          OH\        /P
      5′ – T C    G A T T G A A – 3′
      3′ – A G C T A A      C T T – 5′
                   P/        \OH
                        ↓
         OH| P
      5′ – T C|G A T T G A A – 3′
      3′ – A G C T A A|C T T – 5′
              P | OH
                   ↓
      5′ – T C G A T T G A A – 3′
      3′ – A G C T A A C T T – 5′
```

Figure 9.10: Complementary base pairing followed by ligation. Although the single strands were annealed, a nick (gap) remains in each of the strands.

thus, used to establish this bond (see Figure 9.10). Although it is possible to use some ligases to concatenate free floating double stranded DNA molecules with blunt ends, it is much easier to allow single strands with sticky ends to bind together.

Modifying nucleotides (inserts or deletes short subsequences): it is possible to add, substitute, or delete certain subsequences from DNA molecules. The enzymes used in these operations are called modifying enzymes.

Amplification (multiplies DNA molecules): given some DNA molecules, a technique called *polymerase chain reaction* (PCR) can be used to make multiple copies of a subset of the strands present. PCR requires a start and an end subsequence, called *primers*, that are used to identify the sequence, called *template*, to be replicated. The PCR was devised in 1985 by Kary Mullis and revolutionized molecular biology. It is very sensitive, simple and efficient, consisting basically of three steps: denaturation, annealing, and polymerase extension.

Step 0: To start with, prepare a solution containing the molecule m to be copied and the primers.

Step 1: (Denaturation) Then, heat the solution so that the hydrogen bonds between the two strands are destroyed, and the molecule m denatures into two strands.

Step 2: (Annealing) Now, cool down the solution so that the primers will anneal to their complementary subsequences.

Step 3: (Polymerase extension) Finally, apply the polymerase extension technique to fill in the sticky ends so as to form double strands. The whole process is depicted in Figure 9.11.

Gel electrophoresis (measures the length of DNA molecules and separates them by length): the length of a single stranded molecule is the number of nucleotides composing the molecule. For example, if a molecule has 12 nucleotides, it is a 12 mer; that is, a polymer consisting of 12 monomers. The length of a double stranded molecule is the number of base pairs. For example, a double stranded molecule with 12 bp in each strand has length 12 base pairs. *Gel electrophoresis* is a technique that can be used with two main purposes: 1) to measure the length of a DNA molecule, and 2) to sort (separate) DNA strands by length. Gel electrophoresis (separation by length) is one of the ways often used to read out the results of DNA computing, with the advantage of being quite precise.

Electrophoresis is the movement of charged molecules in an electric field (Amos, 1997). As DNA molecules are negatively charged, when placed in an electric field they will move (migrate) towards the positive electrode. In *gel electrophoresis*, the migration rate of a molecule is mainly affected by its size. Smaller molecules migrate faster through the gel, sorting them according to size. Therefore, it is possible to do both, check the presence of molecules of a given length, as well as separate the molecules according to their length.

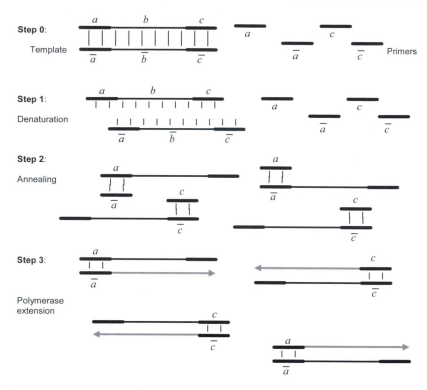

Figure 9.11: DNA Amplification via Polymerase Chain Reaction (PCR). Step 0: The solution is prepared with the double stranded DNA molecule (template) and the primers to be used in the polymerase extension. Step 1: Denaturation is used to separate the double strand into single strands. Step 2: The primers anneal to the single strands. Step 3: A polymerase enzyme is used to complete the single strands so as to form the double strands that are copies of the templates. This process is repeated resulting in an exponential growth in the number of templates (2^n copies are generated in n steps).

In the gel electrophoresis process, a gel is prepared and will function as a support for the separation of DNA fragments. Holes are created in the gel and will serve as reservoirs to hold the solution containing DNA. An electrical charge is then applied to the set up and since smaller molecules travel faster through the gel, larger molecules will lag behind. As the separation process continues, the distance between larger and smaller molecules becomes more apparent. After the gel electrophoresis has been run it is necessary to visualize the results. To do so, the DNA is stained with a dye and then the gel is visualized under ultraviolet light. The gel is then photographed, as illustrated in Figure 9.12. Note that the photograph contains several lanes corresponding to different DNA samples. The lane on the right is known as the *marker lane* and works as a standard for the *molecular weight marker*. DNA fragments of same length cluster together producing visible horizontal bands.

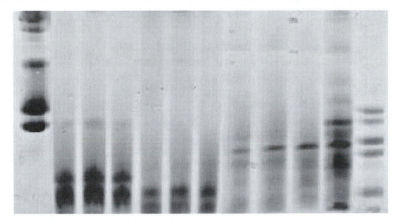

Figure 9.12: Photograph of gel electrophoresis of a protein molecule. A protein is any of a group of complex organic macromolecules that contain carbon, hydrogen, oxygen, nitrogen, and usually sulfur, and are composed of one or more chains of amino-acids. (Courtesy of © Elizabete L. da Costa.)

Filtering (separates or extracts specific molecules): there are a few techniques that can be used to separate some particular strands from a solution. One of the simplest methods is as follows. If a single stranded molecule of type o is to be separated from those of other types in a given solution S, one can attach its complement ($\bar{o}$) to a filter and pour the solution S through the filter. Then, o molecules will bind to $\bar{o}$ molecules while the others will just flow through the filter. The annealing between o and $\bar{o}$ results in a collection of double-stranded molecules fixed to the filter, and a solution S^* results from S by removing the o molecules. Finally, the filter is transferred to a container where the double stranded DNA is denatured, resulting only in the molecules to be separated.

Synthesis (creates DNA molecules): it has already been discussed how to synthesize double strands from single strands via polymerase extension. However, it still remains the question as to how can we synthesize single stranded molecules. It is possible to chemically synthesize single stranded molecules, called *oligonucleotides* (or simply *oligos*), using a particular machine. This synthesizer is supplied with the four nucleotide bases in solution and adds nucleotide by nucleotide following a prescribed sequence entered by the user.

Sequencing (reads out the sequence of a DNA molecule): determining the exact sequence of nucleotides comprising a given DNA molecule is essential for the interpretation of the results obtained in most DNA computing approaches. The most popular *sequencing* technique is based on the extension of a primed single stranded template. For our present purposes, all we have to know is that there are techniques able to read a DNA molecule nucleotide by nucleotide. Further details are left as an exercise for the reader.

9.3 FILTERING MODELS

In all filtering models, a computation consists of a sequence of operations on finite multi-sets of strings that usually starts and ends with a single multi-set. By initializing a multi-set and applying specific operations to it, new or modified multi-sets are generated. The computation then proceeds by *filtering out* strings that cannot be a solution (Amos, 1997). This section describes the two pioneering works on DNA computing, which are basically filtering models, and follows with a more formal description of multi-sets and operations that can act upon them. The latter will serve as a general framework to devise and study filtering DNA computing models.

9.3.1. Adleman's Experiment

Speculations about the possibility of using DNA molecules to perform computation date back to the early 1970s and maybe even earlier (cf. Bennett, 1973; Conrad, 1963). However, none of these insights was followed by practical attempts at real world implementations. The first successful experiment involving the use of DNA molecules and DNA manipulation techniques for computing was reported by L. M. Adleman in 1994 (Adleman, 1994). In that paper, Adleman solved a small instance of the Hamiltonian path problem (HPP) in a directed graph using purely biochemical means.

In general, the HPP consists of deciding whether or not an arbitrarily given graph has a Hamiltonian path. HPP can be solved by an exhaustive search and various algorithms have been proposed to solve it. Although these algorithms are successful for some special classes of graphs, they all have an exponential worst-case complexity for general directed graphs. Therefore, in the general case, all known algorithms essentially amount to exhaustive search. The Hamiltonian path problem has been shown to be an NP-complete problem; that is, intractable in the traditional computing paradigm (see Appendix B.3.3). With Adleman's DNA computing solution to the HPP, the number of the laboratory steps was linear in terms of the size of the graph (number of vertices), although the problem itself is known to be NP-complete.

The Hamiltonian path problem can be explained as follows. A directed graph G with designated vertices v_{in} and v_{out}, is said to have a Hamiltonian path if and only if there is a sequence of compatible directed edges $e_1, e_2, \ldots e_z$ (i.e., a path) that begins at v_{in}, ends at v_{out}, and passes through each vertex exactly once. Figure 9.13 illustrates the instance of the Hamiltonian path problem used by Adleman (1994) in his pioneering implementation of DNA computing. In this graph, $v_{in} = 0$ and $v_{out} = 6$, and a Hamiltonian path is given by the following sequence of edges: $0 \rightarrow 1 \rightarrow 2 \rightarrow 3 \rightarrow 4 \rightarrow 5 \rightarrow 6$. This can be easily verified by inspection.

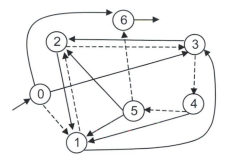

Figure 9.13: Instance of the Hamiltonian path problem solved by Adleman's pioneering experiment. The dashed arrows correspond to the Hamiltonian path of this instance.

To solve this problem Adleman used the following deterministic algorithm:

Step 1: generate random paths through the graph.

Step 2: keep only those paths that begin with v_{in} and end with v_{out}.

Step 3: if the graph has n vertices, then keep only those paths that enter exactly n vertices.

Step 4: keep only those paths that enter all the vertices of the graph at least once.

Step 5: if any path remains, say YES; else, say NO.

Adleman translated this algorithm step by step into molecular biology. Before generating the random paths through the graph, it is necessary to decide how these are going to be encoded using DNA molecules. Adleman chose to encode each vertex of the graph by a single stranded sequence of nucleotides of length 20 (a 20mer). The codes were constructed at random, and the length 20 was chosen so as to ensure different codes. A large number of these oligonucleotides were generated by PCR and placed in a test tube. The edges are encoded as follows: if there is an edge from vertex i to vertex j, and the codes of these vertices are $v_i = a_i b_i$ and $v_j = a_j b_j$, where a_i, b_i, a_j, b_j are sequences of length 10, then the edge $i \rightarrow j$ is encoded by the Watson-Crick complement of the sequence $b_i a_j$, as illustrated in Figure 9.14.

In order to link the edges to form paths, oligonucleotides $\bar{O}_i$ complementary to those representing the edges (O_i) had to be synthesized. An enzymatic ligation reaction was then carried out, linking the 20mer strands encoding the edges so as to form random paths through the graph, corresponding to Step 1 of the algorithm.

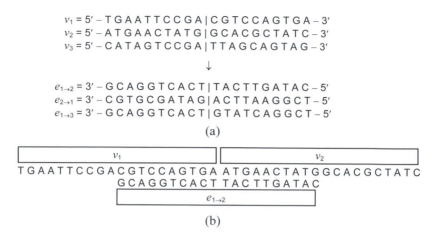

$v_1 = 5' - T\ G\ A\ A\ T\ T\ C\ C\ G\ A\ |\ C\ G\ T\ C\ C\ A\ G\ T\ G\ A - 3'$
$v_2 = 5' - A\ T\ G\ A\ A\ C\ T\ A\ T\ G\ |\ G\ C\ A\ C\ G\ C\ T\ A\ T\ C - 3'$
$v_3 = 5' - C\ A\ T\ A\ G\ T\ C\ C\ G\ A\ |\ T\ T\ A\ G\ C\ A\ G\ T\ A\ G - 3'$

$\downarrow$

$e_{1\to2} = 3' - G\ C\ A\ G\ G\ T\ C\ A\ C\ T\ |\ T\ A\ C\ T\ T\ G\ A\ T\ A\ C - 5'$
$e_{2\to1} = 3' - C\ G\ T\ G\ C\ G\ A\ T\ A\ G\ |\ A\ C\ T\ T\ A\ A\ G\ G\ C\ T - 5'$
$e_{1\to3} = 3' - G\ C\ A\ G\ G\ T\ C\ A\ C\ T\ |\ G\ T\ A\ T\ C\ A\ G\ G\ C\ T - 5'$

(a)

v_1	v_2

T G A A T T C C G A C G T C C A G T G A A T G A A C T A T G G C A C G C T A T C
          G C A G G T C A C T T A C T T G A T A C

$e_{1\to2}$

(b)

Figure 9.14: Encoding of nodes and edges in Adleman's experiment. a) Each node is encoded using a random 20mer, and the edges are encoded by taking the Watson-Crick complement of half parts of the encoding used for the edges. Note that directionality is taken into account by this encoding scheme ($e_{1\to2} \neq e_{2\to1}$). b) The complement of the last ten nucleotides of the encoding of the first node followed by the complement of the first ten nucleotides of the second node determine the encoding of the edge.

To implement Step 2, the product of Step 1 was amplified by polymerase chain reaction (PCR) using as primers $\bar{O}_0$ and $\bar{O}_6$. Therefore, only those molecules encoding paths that begin with vertex 0 and end with vertex 6 were amplified. A filtering operation can then be used to separate the strands that start with vertex 0 and end with vertex 6.

Then, the DNA molecules were separated according to their length by gel electrophoresis. The band on the gel, which by comparison with a molecular weight marker was identified as consisting of strands with 140bp (7 vertices) was separated from the gel and the DNA was extracted. Repeated cycles of PCR and electrophoresis were used to purify the product further. At the end of this step, there is a set of molecules that start with 0, end with 6, and pass through 7 vertices. Note however, that this does not ensure such a path is Hamiltonian, for example, the path 0, 3, 2, 3, 4, 5, 6 satisfies all these conditions but is not Hamiltonian.

To implement Step 4 of the algorithm, single stranded DNA was probed with complementary oligonucleotides attached to magnetic beads. Thus, with one step for every vertex, the molecules containing the required sequence could be literally pulled out of the solution. This process can also be explained as follows: for each vertex i melt the result of Step 3, add the complement of the code $\bar{O}_i$ of vertex i ($i = 1, 2, \dots, 5$), and let it anneal; then remove all molecules that do not anneal. Finally, to obtain the YES/NO answer of Step 5, one still has to amplify the product of Step 4 by PCR and analyze it by gel electrophoresis.

Discussion

The practical details of Adleman's experiment are not relevant to the present discussion. They depend on the laboratory techniques available currently, and can also be performed using other strategies. Actually, the algorithm itself can be changed. In the particular case of Adleman's work, the experiment took seven days of lab work to be completed. However, this technique has a remarkable and distinctive advantage: the number of oligonucleotides needed will increase linearly in relation to the number of vertices involved. Therefore, an NP-complete problem that requires more computational time as it gets bigger, can be solved in linear time due to the massive parallelism of DNA computing.

One of the main weaknesses of Adleman's procedure, from a practical perspective, is the number of single strands necessary to encode the vertices and edges of generic HPP problems, which is of the order $n!$, where n is the number of vertices in the graph (Calude and Păun, 2001). This imposes drastic limitations on the size of the problems that can be solved using Adleman's procedure. The question of increasing the size of the graphs that can be solved by DNA computing has been approached by some authors. One general strategy is to diminish the set of candidate solutions to be generated by a basic step where the parallelism of DNA is used. This has been accomplished by using, for instance, evolutionary algorithms (Barreiro et al., 1998).

Since the Hamiltonian path problem is NP-complete and the DNA computing technique proposed by Adleman can be used to solve this problem, such technique can be used to solve any problem from NP. However, this does not mean that any instance of an NP-problem can be solved in a feasible way. Adleman solved the HPP using brute force: he designed a system that tried out all possible paths in the graph. As the speed of any computer, biological or not, is determined by two factors, parallelism and the number of steps performed per unit time, the strength of Adleman's approach was its massive parallelism (Lipton, 1995).

At the time Adleman carried out his experiment, a typical desktop computer could execute 10^6 operations per second and the fastest supercomputer available could execute approximately 10^{12} operations per second. By contrast, Adleman's DNA computer, if each ligation counts as an operation, did over 10^{14} operations per second. Scaling up the ligation step could push the number over 10^{20} operations per second. In addition, it used an extremely small quantity of energy, only 2×10^{19} operations (ligations) per joule, whereas the second law of thermodynamics dictates a theoretical maximum of 34×10^{19} operations per joule. Modern supercomputers only operate at 10^9 operations per joule. Finally, one bit of information can be stored in a cubic nanometer of DNA, which is about 10^{12} times more efficient than existing storage media. So, molecular computers have the potential to be far faster and more efficient than any electronics developed so far. In summary, DNA computing was approximately 1,200,000 times faster than the fastest supercomputers; it could permit information storage 10^{12} times more effectively; and take 10^{10} times less energy than the existing computers (Adleman, 1994).

9.3.2. Lipton's Solution to the SAT Problem

Lipton (1995) showed how to use DNA procedures to solve the *satisfiability problem for propositional formulas* or *propositions* (see Appendix B.4.2), known as the SAT problem. In contrast to Adleman's experiment, Lipton did not solve the SAT problem in the lab, but he showed how this could be accomplished theoretically.

SAT is a search problem known to be NP-complete that can be described as follows. Given a finite set of logical variables $E = \{e_1, e_2,..., e_n\}$, define a *literal* as a variable, e_i, or its complement $\bar{e}_i$. If e_i is TRUE, then $\bar{e}_i$ is FALSE, and vice-versa. Let a *clause* C_j be a set of literals $\{e_1^j, e_2^j,..., e_n^j\}$. An instance I of the SAT problem consists of a set of clauses, more specifically, a Boolean formula of the form $C_1 \wedge C_2 \wedge ... \wedge C_m$, where each clause is a proposition that can be built from propositional variables e_i, $i = 1, ... , n$ and the *connectives* AND ($\wedge$), OR ($\vee$), and NOT ($\neg$). (Note that here we use a bar over a character to represent negation.) The SAT problem then corresponds to assigning a Boolean value to each variable $e_i \in E$, $i = 1, ... , n$, such that the whole formula is TRUE.

A key aspect of the SAT problem explored by Lipton was the possibility of representing it as a graph search problem. His solution had basically two phases: 1) the generation of all paths in the graph, and 2) the search for (or filtering of) a truth assignment satisfying the formula. Note that he used the same principles as Adleman: first generate all possible solutions and then filter out those that satisfy specific criteria.

Assume a propositional formula containing n variables. The generic graph illustrated in Figure 9.15 can be used to represent the truth assignment of this formula, where the variables e_i^j, $i = 1, ... , n$ can assume either the value '0' (FALSE), e_i^0, or '1' (TRUE), e_i^1. It is easy to note that from one vertex to another there are only two possible paths - either through e_i^0 or through e_i^1, thus there are 2^n possible paths from v_0 (v_{in}) to v_n (v_{out}). According to this picture, a generic path can be represented as a sequence $e_1^{i_1} v_0 e_1^{i_1} v_1 e_2^{i_2} ... v_{n-1} e_n^{i_n} v_n$, where the variable e_j can assume the truth value $i_j, j = 1, ... , n$.

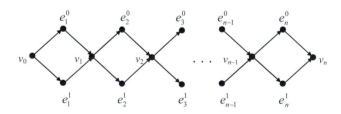

Figure 9.15: Graph associated with a truth assignment.

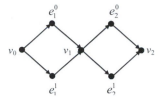

Figure 9.16: Graph associated with a propositional formula with two variables e_1 and e_2.

In the graph of Figure 9.15, all paths that start at v_0 and end at v_n correspond to a binary n-bit string. For instance, the path $v_0 e_1^1 v_1 e_2^0 v_2 e_3^0 \ldots v_{n-1} e_n^1 v_n$ encodes the binary string $100\ldots1$; there are only two choices for a path to go from one vertex to another, either using the edge with value '1' or using the edge with value '0'.

Lipton proposed to encode such graphs into a test tube using ideas from Adleman's experiment. Each vertex of the graph was encoded by assigning to each one of them a random nucleotide from the alphabet {A,C,G,T} of a given length L. However, the way Lipton proposed his solution was essentially different from Adleman's experiment, in the sense that Lipton proposed to work with 'test tube operations', which will be formally introduced in the next section. To illustrate how this can be accomplished, Lipton described a simple example: verify the satisfiability of the following propositional formula:

$$F = (e_1 \vee e_2) \wedge (\bar{e}_1 \vee \bar{e}_2) \qquad (9.1)$$

Figure 9.16 depicts the graph associated with the truth assignment of a propositional formula with two input variables.

Lipton constructed a series of test tubes, where the first one, t_0, is the tube containing all two bit (binary) strings. He proposed, among others, an extract test tube operation $E(t,i,a)$ that *extracts* all sequences in test tube t whose i-th bit is equal to a, $a \in \{0,1\}$. (Remember that the binary strings are encoded using a word composed of nucleotides, but it is simpler to explain the procedure by looking at binary strings instead of a quaternary alphabet with the letters A, C, T, G.) Then, he operated as follows:

Step 1: Let t_1 be the test tube corresponding to $E(t_1,1,1)$. Let the remainder be t_1', and let t_2 be $E(t_1',2,1)$. Then pour t_1 and t_2 together to form t_3.

Step 2: Let t_4 be the test tube corresponding to $E(t_3,1,0)$. Let the remainder be t_4', and let t_5 be $E(t_4',2,0)$. Then pour t_4 and t_5 together to form t_6.

Step 3: Check to see if there is any DNA in the last test tube. If any path remains, say YES; else, say NO.

Figure 9.17 presents the sequences of steps followed by the three-step algorithm described above. Note that tube t_3 is composed of all strings that satisfy the first clause $(e_1 \vee e_2)$, and the tube t_6 is composed of all strings from t_3 that satisfy the second clause $(\bar{e}_1 \vee \bar{e}_2)$. Tube t_6 contains the solutions to this SAT problem.

Test tube	Strings present
t_0	00, 01, 10, 11
t_1	10, 11
t_1'	00, 01
t_2	01
t_3	01, 10, 11
t_4	01
t_4'	10, 11
t_5	10
t_6	01, 10

Figure 9.17: Sequence of steps used by Lipton to solve the SAT problem for the formula: $(e_1 \vee e_2) \wedge (\bar{e}_1 \vee \bar{e}_2)$.

Lipton extended his DNA-based solution to the general case of the SAT problem as follows. Let C_1, C_2, ... , C_m be the clauses of a given propositional formula in the form: $C_1 \wedge C_2 \wedge C_m$. Construct a series of test tubes t_0, t_1, ... , t_m so that t_k is the set of n-bit strings e such that $C_1(e) = C_2(e) = ... = C_k(e) = 1$, where $C_i(e)$ corresponds to the value of the clause C_i on the setting of the variables to e. For t_0 use the set t_{all} containing all possible n-bit strings.

Assume all clauses C_i, $i = 1$, ... , m of the disjunctive form $o_1 \vee ... \vee o_l$, where o_i is a literal and $\bar{o}_i$ its complement. For each literal o_i operate as follows: if o_i is equal to e_j, then form $E(t_k, j, 1)$; if it is equal to $\bar{e}_j$, then form $E(t_k, j, 0)$. Each extraction is performed and the remainder is used for the next one. Put all these tubes together to form t_{k+1}, then do one detect operation on t_m to decide whether or not the clauses are satisfied.

Discussion

In the general case, no method essentially better than *exhaustive search* is known to solve the SAT problem. One has to search through all possible 2^n truth-value assignments given a formula with n variables. This makes the problem computationally intractable; it is already computationally infeasible, say, in the case of 200 variables (Păun et al., 1998). The SAT problem is known to be NP-complete. It is intuitively among NP-complete problems in the sense that it constitutes a suitable reference point for NP-complete problems. The reduction of a given problem to the satisfiability problem is sometimes natural.

Like Adleman's solution to the Hamiltonian path problem, Lipton's solution to the SAT problem made exhaustive search computationally feasible by the massive parallelism of DNA molecules and manipulation techniques. One of the main results of Lipton's solution was the verification that using his procedure any SAT problem on n variables and m clauses can be solved with, at most, order m extract steps and one detect step. Order m means that the number of steps is linear in m.

9.3.3. Test Tube Programming Language

The practical details of Adleman's and Lipton's DNA computing procedures depend on the current laboratory technologies. This implies that they could have been performed by other techniques. Even the algorithms for solving the problems could be changed, improved, or completely replaced by others. What is important in this case is the proof that such a computation is possible. Purely biochemical means were used to solve NP-complete problems, actually intractable problems, in linear time as the number of laboratory operations (Calude and Păun, 2001). These operations, in an abstract formulation, constitute another main output of Adleman's experiment and of the thought about it, leading to a sort of programming language based on test tubes and the manipulation of DNA molecules.

Such a *test tube programming language* was proposed by (Lipton, 1995), developed in (Adleman, 1995) and later discussed in many other works. This section provides a brief description of the test tube programming language as formalized by Adleman (1995). What is important for the reader to pay attention to, while reading the description of the few 'test tube operations', is that these can be implemented using some of the DNA manipulation techniques described in Section 9.2. Therefore, the test tube programming language will provide a guidance to what DNA manipulation techniques to use and when.

This section introduces the test tube programming language together with one of its extensions. It then follows with a number of examples of how to write 'test tube programs' for solving problems, including the Hamiltonian path problem and the satisfiability problem already solved by DNA computing. A brief description of another DNA computing programming language, named DNA-Pascal, will also be provided. DNA-Pascal can be viewed as a sort of hybrid between the Pascal programming language and the test tube programming language.

The Unrestricted DNA Model

A test tube is defined as a set of DNA molecules; that is, a multi-set of finite strings over the alphabet {A,C,G,T}. Given a tube, one can perform four basic operations: separate (also called extract), merge, detect, and amplify:

1. Separate (extract): given a tube t and a word w (string of symbols w over {A,C,G,T}), produce two tubes $+(t,w)$ and $-(t,w)$, where $+(t,w)$ consists of all DNA strands in t that contain w as a consecutive subsequence, and $-(t,w)$ consists of all DNA molecules in t that do not contain the consecutive subsequence w.

2. Merge: given a set of tubes $t_1, t_2, \ldots, t_m$ produce a tube with the contents of all tubes: $\cup(t_1, t_2, \ldots, t_m) = t_1 \cup t_2 \cup \ldots \cup t_m$.

3. Detect: given a tube t, say YES if t contains at least one DNA molecule, and say NO if it contains none.

4.   Amplify: given a tube t, produce two copies t_1 and t_2 of it: $t = t_1 = t_2$.

These four operations (separate, merge, detect, and amplify) can then be used to write programs that receive a tube as input and return as output either YES, NO or a set of test tubes.

In addition to the four operations described above, Adleman's experiment makes use of Watson-Crick complementarity and the following modifications of the operation separate (Calude and Păun, 2001):

1.   Length-separate: given a tube t and an integer n, produce the tube $(t, \leq n)$ consisting of all strands in t with length less than or equal to n.

2.   Position-separate: given a tube t and a word w, produce the tube $B(t,w)$ consisting of all strands in t that begin with the word w; or produce the tube $E(t,w)$ consisting of all strands in t that end with the word w.

Examples of Application

To exemplify, consider the simple program presented in Algorithm 9.1. This program receives as input a test tube t, and the bases A, T and G. It returns as output YES if tube t contains a DNA molecule composed entirely of the base C, and NO otherwise.

```
procedure [out] = extract(t,A,T,G)
     t ← -(t,T)
     t ← -(t,G)
     t ← -(t,A)
     out ← detect(t)
end procedure
```

Algorithm 9.1: An illustrative test-tube program that checks for the presence of at least one DNA molecule composed only of the base C.

Given the two modifications of the operation separate, it is now possible to write a simple test tube program to solve the Hamiltonian path problem following Adleman's experiment. This program receives as input the tube t, and the initial v_{in} and the end v_{out} vertices, as described in Algorithm 9.2. This algorithm receives as input a tube t with DNA strands representing random paths through the graph (Step 1); it then proceeds by selecting those molecules that begin with v_{in} and end at v_{out} (Step 2); next it selects all paths entering 7 vertices (Step 3); next it selects all paths that enter each vertex once (Step 4); finally, it detects if any path remains (Step 5).

In Lipton's solution to the SAT problem an operation extract $E(t,i,a)$ that *extracts* all sequences in test tube t whose i-th bit is equal to a, $a \in \{0,1\}$, was defined. This operation can also be represented using the notation proposed for the unrestricted DNA model:

```
procedure [out] = HPP(t,vin,vout)
    t ← B(t,vin)              //Step 2
    t ← E(t,vout)
    t ← (t,≤ 140)            //Step 3
    for i=1 to 5 do           //Step 4
        t ← +(t,si)
    end for
    out ← detect(t)          //Step 5
end procedure
```

Algorithm 9.2: An illustrative test-tube program that checks for the presence of at least one DNA molecule composed only of the base C. *si* corresponds to the DNA strand that represents node *i*.

$$E(t,i,a) = +(t,e_i^a) \qquad (4.2)$$

It is also possible to define the complementary operation:

$$E^-(t,i,a) = -(t,e_i^a) \qquad (4.3)$$

where $E^-(t,i,a)$ extracts all sequences in the test tube t whose i-th bit is complementary to a, $a \in \{0,1\}$.

Algorithm 9.3 contains a test tube program for solving the SAT problem for the formula $F = (e_1 \vee e_2) \wedge (\bar{e}_1 \vee \bar{e}_2)$. This program receives as input the test tube t with all possible paths over the graph depicted in Figure 9.16. The solution to the problem is obtained by an exhaustive search approach. The tube t_3 contains the assignments that satisfy the first clause of F, and these are further filtered to result in tube t_6 containing assignments that satisfy the second clause of F as well. Algorithm 9.4 depicts a general-purpose algorithm to solve the SAT problem.

```
procedure [out] = SAT(t)
    t1  ← +(t,e₁¹)
    t1' ← -(t,e₁¹)
    t2  ← +(t1',e₂¹)
    t3  ← merge(t1,t2)
    t4  ← +(t3,e₁⁰)
    t4' ← -(t3,e₁⁰)
    t5  ← +(t4',e₂⁰)
    t6  ← merge(t4,t5)
    out ← detect(t6)
end procedure
```

Algorithm 9.3: An illustrative test-tube program to solve the SAT problem for the formula $F = (e_1 \vee e_2) \wedge (\bar{e}_1 \vee \bar{e}_2)$.

```
procedure [out] = SAT(t)
    for a = 1 to m do                    //for each clause
        for b = 1 to n do                //for each literal
            if e_b^a = x_j
                then   tb ← +(t,e_b^a=1)
                else   tb ← +(t,e_b^a=0)
            end if
        end for
        t ← merge(t1,t2,…,tb)
    end for
    out ← detect(t)
end procedure
```

Algorithm 9.4: A test-tube program to solve the general case of the SAT problem.

An Extension of the Unrestricted DNA Model

Amos (1997) extended the test tube programming language proposed by Adleman (1995) and Lipton (1995) as an attempt to provide a formal framework for the description of DNA algorithms for solving any problem in the complexity class NP. In this model, all computation starts with the construction of the initial set of strings and follows with the application of a number of basic legal operations on sets within the model. This choice was based on what the authors believed could be implemented by precise and complete chemical reactions within the DNA implementation.

This model, termed *parallel filtering model* (PFM), is differentiated from the ones previously described by its implementation of the removal of strings (molecules). While all other models propose separation (extract) steps, where strings are conserved and may be used later in the computation, the parallel model strings that are removed are discarded, and play no further role in the computation. The four basic operations of the PFM, assumed to take constant time when executed in parallel, are:

1. Remove(t,w): given a tube t and a word w (string of symbols w over $\{A,C,G,T\}$), remove from the tube t any molecule containing as a subsequence at least one occurrence of each element of w.

2. Union($\{t_1, \dots, t_k\}, t$): given a set of tubes $t_1, \dots, t_k$, create a tube t by merging all tubes $t_1, \dots, t_k$.

3. Copy($t,\{t_1, \dots, t_k\}$): produce k copies of t, denoted by $t_1, \dots, t_k$.

4. Select(t): if t is non-empty, then select an element of t at random; otherwise, return *empty*.

The DNA Pascal

In order to suggest a high level model of molecular computation, D. Rooß and K. Wagner (1996) introduced a molecular programming language named *DNA-Pascal*. DNA-Pascal is similar to the test tube programming language and, thus, constitutes an abstract language to write programs using procedures that correspond to DNA manipulation techniques. However, DNA-Pascal uses standard Pascal instructions, particular data types, and some procedures based on those of the test tube programming language. One of their main goals was to demonstrate the computational power of DNA (Section 9.5).

In DNA-Pascal, test tubes filled with DNA are mapped into set variables containing words over the alphabet $\{0,1\}$. Let t, t_1, t_2 be such set variables (test tubes). DNA manipulation techniques are transformed into operations on these set variables. Some of the main operations introduced are:

- Initialization: fills a set variable t with $\{0,1\}^n$, $t := \text{In}(n)$.

- Empty word: $t := \{\varepsilon\}$.

- Union: the union of two set variables t_1 and t_2 contains all the words in t_1 and t_2, $t := t_1 \cup t_2$.

- Extraction: filters out all words of a set variable t_1 that have a special pattern. The authors proposed two types of an extraction procedure: 1) a bit extraction, and 2) a sub word extraction.

  o Bit extraction: looks for a special bit b at a particular position k, $t := \text{Bx}(t_1,b,k)$.

  o Sub word extraction: extraction looks for a special sub word w anywhere in a word, $t := \text{Sx}(t_1,w)$.

- Concatenation: the concatenation of two set variables t_1 and t_2 is $t := t_1.t_2$.

- Right cut: $t := t_1/$, where $t_1/ = \{z/ \mid z \in t_1\}$ and $za/ = z \ \forall a \in \{0,1\}$ and $\varepsilon/ = \varepsilon$.

- Left cut: $t := /t_1$, where $/t_1 = \{/z \mid z \in t_1\}$ and $/za = z \ \forall a \in \{0,1\}$ and $/\varepsilon = \varepsilon$.

- Right append: $t := t_1.a$, where $t_1.a = \{z.a \mid z \in t_1\}$.

- Left append: $t := a.t_1$, where $a.t_1 = \{a.z \mid z \in t_1\}$.

To define new conditions that can be verified on DNA-Pascal variables, three conditional tests were proposed:

- Subset test: $t_1 \subseteq t_2$.

- Detect test: $t = 0$.

- Membership test: $x \in t$.

9.4 FORMAL MODELS: A BRIEF DESCRIPTION

Most DNA computing techniques are based on initially generating a large (if not the whole) set of candidate solutions and then removing those that do not satisfy a given criterion. This computing paradigm is not usual in traditional computer science, though it can be seen in some natural computing approaches such as evolutionary algorithms and immunocomputing. In the case of DNA computing, the complete set of possible solutions is usually generated and is iteratively filtered so as to detect a solution (if it exists!) to the problem. This strategy has been called *computing by carving* by Calude and Păun (2001), and was formulated in terms of formal language theory. Appendix B provided some basic concepts from formal languages and grammars that may be useful for the brief descriptions of the formal DNA computing approaches to be provided here.

Virtually every researcher on DNA computing has its own way of using DNA for computation. This is a symptom indicating that the field is still exploring the possibilities and has hardly begun to settle on a preferred technique for making DNA do useful computations. However, a number of formal models, some introduced before DNA computing emerged, have been proposed. This section briefly reviews some of these models, in particular, *sticker systems*, *splicing systems*, *insertion/deletion systems*, and the *PAM model*. More comprehensive surveys and descriptions can be found in Păun et al. (1998), Calude and Păun (2001), Gramß et al. (2001), Amos (2003), and in the cited literature.

9.4.1. Sticker Systems

Roweis et al. (1996) introduced a DNA computing model called the sticker model. It makes use of DNA strands as the physical substrate in which information is represented and stored, and of separation by hybridization as a central mechanism. The stickers' model has a random access memory that requires no strand extension, uses no enzymes, and (at least in theory) has reusable materials. Two different kinds of single DNA strands are employed: *memory strands*, and *sticker strands* or *stickers*. Memory strands are long strands used to encode, or represent, binary strings. A DNA sequence of length m of a memory strand is used to encode one bit of a binary string. The stickers are single DNA strands of length m that are the Watson-Crick complement of regions of the memory strands. The bonding of a sticker with a portion of the memory strand means that the corresponding bit of the memory strand is set to '1'; otherwise it is set to '0'. A memory strand that is partially double-stranded; that is, which has some stickers bound to it, is known as a *complex*. Manipulating data encoded in such a way is quite efficient, and the following operations can be used to manipulate complexes: *union*, *extraction*, *set* of a bit, *reset* of a bit, and *clear*. A computation with this model corresponds to the application of these operations to an initial test tube until an 'output' tube is obtained.

9.4.2. Splicing Systems

In a simple form, *splicing* two strings corresponds to cutting them at specific points and to concatenate the obtained fragments crosswise. This operation is similar to the crossover operation used in genetic algorithms (Chapter 3). In the case of DNA molecules, restriction enzymes are used to cut the DNA strands in such a way that matching sticky ends are produced, and the obtained fragments are pasted together (ligated) such that new molecules are formed. T. Head (1987) introduced a model, based on formal language theory, of this DNA recombination operation, called *splicing*. The computing models based on it are referred to either as *H systems* or as *splicing systems*. Splicing systems were later suggested to represent DNA computations. Informally, a splicing system consists of a set of words (axioms) and a set of *splicing rules* that determine how and where the strands are going to be cut and pasted together. The *splicing language*, that is the language generated by the splicing system, is formed by all words, which can be derived from axioms and/or words obtained in the previous steps by using splicing rules. As the system uses only cutting and pasting, there is only a restricted number of each DNA strand. So it is proper to use multisets in the system. A *multiset M* on Σ^* can be viewed as the set Σ^* with a function M : $\Sigma^* \rightarrow N \cup \{\infty\}$ (N is the set of natural numbers), where $M(w)$ represents the number of occurrences of the word $w \in \Sigma^*$ in the multiset M. The set $supp(M) = \{w \in \Sigma^* : M(w) \neq 0\} \subseteq \Sigma^*$ is the *support* of M. A splicing system can thus be defined as a *quadruple* $\sigma = (\Sigma, T, A, R)$, *where Σ is an alphabet, $T \subseteq \Sigma$ is the terminal alphabet, A is a multiset on Σ^* (axioms), and R is the set of splicing rules.*

9.4.3. Insertion/Deletion Systems

The operations of *insertion* and *deletion* are fundamental in formal language theory. Kari and Thierrin (1996) introduced the *insertion/deletion* (or *insdel*) systems as a model of DNA computing. Given a pair of words (x,y), called a context, the (x,y)-contextual insertion of a word v into a word u is performed as follows. For each occurrence of xy as a sub word in u, the word v is inserted into u between x and y. On the contrary, the (x,y)-contextual deletion of a word v from a word u is performed as follows. For each occurrence of v in u between x and y, the word v is removed from the word u. Note that these insertion and deletion operations can be performed using the standard DNA manipulation techniques described previously. Let u and v be words over an alphabet Σ and (x,y) $\in \Sigma^* \times \Sigma^*$ be a pair of words called a context. The (x,y)-*contextual insertion* of v into u is defined as $u \leftarrow (x,y) v = \{u_1xvyu_2 \mid u = u_1xyu_2, u_1, u_2 \in \Sigma^*\}$. Thus, the insertion of a word only happens if a given context is present. If the word u does not contain xy as a sub word, the result of the (x, y)-contextual insertion is the empty set. In a similar manner the contextual deletion is defined. The (x,y)-*contextual deletion* can be defined analogously. An *insertion* (*deletion*) *rule* is a triple (x,z,y) where (x,y) is the context and z is the word to be inserted (deleted). An insdel system can thus be defined as *a quintuple insdel = (Σ, T, A, I, D), where Σ is an alphabet, $T \subseteq \Sigma$ is the terminal alphabet of insdel, $A \subseteq \Sigma^*$ is the set of axioms, and I and D are the sets of insertion and deletion rules, respectively.*

9.4.4. The PAM Model

Reif (1995) introduced one of the first proposals to define operations on DNA strands to promote DNA computation. The *parallel associative memory model* (PAM model) basically describes a parallel associative matching (PAM-match) operator. It also takes into account standard DNA computing operators, such as *union*, *extraction*, and *deletion*. The main component of the PAM-match operator is a restricted form of splicing; such that, given two strings $s = s_1s_2$ and $t = t_1t_2$, the restricted splicing operation $Rs(s,t)$ returns the string s_1t_2 only provided that $s_2 = t_1$. This proposal was the first one to introduce how to compute with DNA with a nonbrute force approach. The PAM model is a formalism that expresses the molecular parallelism. Reif (1995) demonstrated that his model could simulate a nondeterministic Turing machine, and could be used to solve a number of complex problems.

9.5 UNIVERSAL DNA COMPUTERS

The capability of solving NP-complete problems using DNA strands and genetic engineering operations demonstrate the computational power of molecules. One question that still arises regards the computational completeness of DNA. As a universal Turing machine can, in principle, compute any computable function, the design of a Turing machine using DNA molecules is a step in the direction of proving the universal computation capability of DNA. It specifies a molecular computing system capable of maintaining a state and a memory, and of executing an indefinite number of state transitions.

The verification of universal computation by a DNA computer has been performed in various forms by a number of researchers using different universal computing abstractions, such as Turing machines, cellular automata, Boolean circuits, and Chomsky grammars (Beaver, 1995; Rooß and Wagner, 1996; Rothemund, 1996; Winfree, 1996; Kari and Thierrin, 1996; Păun et al., 1998; Csuhaj-Varjú et al., 1997; Ogihara and Ray, 1998). This section provides a description of one of the pioneering proposals by D. Beaver (1995), who simulated a Turing machine using a DNA computer. This may suffice to provide the reader with some flavor of the universal computing capability of DNA computing, but alternative proofs can be found in the literature cited.

D. Beaver (1995) designed a DNA-based Turing machine consisting of a single DNA molecule, where the chemical mechanisms for state transitions allow parallel, synchronized, and heterogeneous computation. At each step of the computation proposed, a DNA molecule is used to encode a configuration of the Turing machine: the content of the tape, the current state, and the head position. Each state transition requires $O(1)$ laboratory steps to be carried out (Pisanti, 1998). As a single strand of DNA is used to encode a configuration of a Turing machine, to describe a molecular universal computer Beaver first showed how to implement a molecular *context sensitive text substitution*. The idea is that of substituting the portion of the DNA strand that is interested by the transition.

The biological operation used to substitute a portion of a DNA strand can be described as follows (Beaver, 1995). Let A, E, U, and O be sequences of DNA nucleotides, and L and R be the sequences to the left and right, respectively, of a strand M that will suffer substitution. To implement the substitution

$AUE \rightarrow AOE$, first anneal single-stranded $\bar{A}\bar{O}\bar{E}$ and $\overline{R}$ to M. Generate the remaining of the double-strand using polymerase extension ($\bar{A}\bar{O}\bar{E}$ and $\overline{R}$ are used as primers) and then ligase to seal the nicks. The result is a double-stranded DNA molecule $\overline{M'}$ not fully paired: the sequence U is not complementary to $\bar{O}$. The next step is to denature the double strand so as to form single-stranded molecules. Mix them with L, then anneal and (polymerase) extend so as to obtain the complement of $\overline{M'}$. M remains single-stranded because L does not anneal with it. Finally, use some enzymes to destroy the single-stranded M, leaving only the double-stranded copy of $\overline{M'}$. The resultant molecule contains M with U substituted by O. The whole process is depicted in Figure 9.18.

Assuming that the substitution operation described above is available, it is possible to simulate a Turing machine using DNA molecules and manipulation techniques. The idea is to follow the sequence of configurations, always keeping the current configuration in a test tube. Due to the local action of a Turing machine all that is necessary to perform a state transition is to be capable of replacing a constant length substring of DNA at a specified position. Therefore, a step of computation is simulated by substituting the significant area of a configuration according to the transition function.

Beaver (1995) encoded the states and symbols of a universal Turing machine using the DNA alphabet of nucleotides {A,C,T,G}. Let S be the length of the tape of the Turing machine, and let the encoding e be:

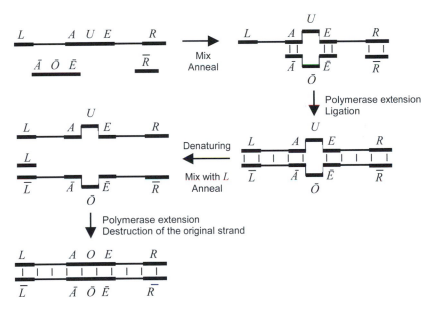

Figure 9.18: Molecular substitution of a DNA sequence (text) used for the state transition of a molecular universal Turing machine.

$$e : (Q \cup \Sigma) \times \{1..S\} \rightarrow \{A,C,G,T\} \qquad (9.4)$$

This allows the encoding of the configurations $C = (x_1 \ldots x_{k-1}qx_kx_{k+1} \ldots x_m)$ by $C^e = e(x_1,1) \ldots e(x_{k-1}, k-1)e(q,k)e(x_k,k) \ldots e(x_m,m)$, where $e(x_k,k)$ indicates that the symbol $x_k \in \Sigma$ is in the k-th position of the tape, and $e(q,k)$ indicates the current state of the machine. Note that, in this encoding scheme, the contents of the Turing machine, together with the current state q of the machine, are encoded one by one and concatenated to form the DNA strand that encodes the configuration of the machine.

To perform a transition, the molecule is isolated in a tube according to state q, position of tape head k, current symbol read x_k, and the symbols x_{k-1} and x_{k+1} to the left and right of the tape head, respectively. The values of q and x_k determine the state transition to be performed; either $\delta(q,x_k) = (q',x_k',L)$ or $\delta(q,x_k) = (q',x_k',R)$ if the transition is a left L or right R move, respectively. For example, if the transition is a left move, $\delta(q,x_k) = (q',x_k',L)$, then the substring $e(x_{k-1},k-1)e(q,k)e(x_k,k)$ is replaced by the substring $e(q',k-1)e(x_{k-1},k-1)e(x_k',k)$, as illustrated in Figure 9.19.

The substitution that performs this state transition is $LAUER \rightarrow LAOER$, where:

- $A = e(x_{k-2},k-2)$
- $U = e(x_{k-1},k-1)e(q,k)e(x_k,k)$
- $E = e(x_{k+1},k+1)$
- $O = e(q',k-1)e(x_{k-1},k-1)e(x_k',k)$

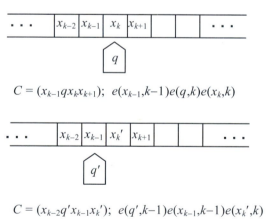

$$C = (x_{k-1}qx_kx_{k+1}); \; e(x_{k-1},k-1)e(q,k)e(x_k,k)$$

$$C = (x_{k-2}q'x_{k-1}x_k'); \; e(q',k-1)e(x_{k-1},k-1)e(x_k',k)$$

Figure 9.19: A state transition in a Turing machine and its equivalent DNA-based Turing machine. The current state of the machine x_k is changed by a new state x_k', and the head is moved to the left of the tape.

L and *R* are the left and right side of the encoding of the rest of the configuration, respectively.

When a halting state is reached in this way, it is necessary to test whether it starts with $e(h,1)e(1,1)$ and accept; otherwise, reject. This can be realized by using the following test tube operation: `detect(extract(`$t,e(h,1)e(1,1)$`))`, where *t* is an arbitrary tube.

Beaver (1995) also suggests extending his method to simulate nondeterministic Turing machines. In this case, the DNA strands that encode the configurations can be amplified at each step so as to reach a sufficient number to perform all possible state transitions. The configurations to which a transition is applicable are selected before applying the transition.

9.6   SCOPE OF DNA COMPUTING

DNA computing was originally proposed as a problem solving technique. This chapter has reviewed the works of Adleman and Lipton that solved the Hamiltonian path problem and the satisfiability problem, respectively. The massive parallelism and miniaturization of DNA suggest a number of problems that can be solved by DNA computing, in particular NP-complete problems. Since any instance of any NP-complete problem can be expressed in terms of an instance of any NP-complete problem, the DNA computing solutions to the NP-complete problems described in this chapter implicitly provide sufficient computational power to solve any problem in NP.

The basic strategy used by DNA computing is almost always the same: assemble a large number of potential solutions to the problem (each solution is encoded using a DNA strand), filter out (using DNA manipulation techniques) the non-solutions, and then check for the existence of a solution. Based on this idea, DNA computing has been used to solve search problems like *graph coloring, shortest common superstring, integer factorization, protein conformation, maximum clique*, and many others (see Section 9.9 for some exercises). Works can also be found in the literature proposing the realization of associative memories using DNA (Baum, 1995); the solution of cryptographic problems (Boneh et al., 1996); the development of DNA-based algorithms for addition and matrix multiplication (Guarnieri et al., 1996; Oliver, 1998); parallel machines (Reif, 1995); and the DNA implementation of, or hybridization with, computational intelligence approaches (Deaton et al., 1997; Mulawka et al., 1998; Maley, 1998; Mills et al., 2001).

In addition to all these "computational" applications of DNA computing, it can also be applied in the biological domain. For instance, it can be used for the processing of natural (biological) DNA. In this case, it may provide (alternative) techniques for the sequencing and fingerprinting of DNA; decoding the genetic material of humans and other living things; creating and searching *wet data bases* of DNA; and the detection of mutations. Possible outcomes could be the eradication of certain diseases, a fast and reliable identification of criminals, and the implantable biochips (those that can detect toxins in a body and release chemicals to destroy them).

9.7   FROM CLASSICAL TO DNA COMPUTING

DNA provides a completely novel computing paradigm. The data structures used in DNA computers are DNA strands instead of bitstrings. While in classical computers information is processed by using logic gates to perform logic operations, in DNA computing information (DNA) is processed using techniques for the genetic manipulation of DNA, such as DNA extension, cutting and joining together. Besides, the storage of information at the molecular level results in a high parallelization of information processing, higher storage capacity and lower power consumption. Table 9.1 presents a high-level comparison between classical and DNA computing.

Table 9.1: Comparison between classical and DNA computing.

Classical Computing	DNA Computing
Information is represented as a bitstring, i.e., string of either 0 or 1	Information is represented as a quaternary string, i.e., a string of the four nucleotides A, C, T, G
Sequential computation	Parallel computation
Computation performed by moving bits through logic gates	Computation performed by manipulating DNA molecules
Information can be copied and read without being disturbed	Information can be copied and read without being disturbed

9.8   SUMMARY AND DISCUSSION

Adleman's DNA based solution to an NP-complete problem was immediately followed by several generalizations and extensions to other NP-complete problems. It is interesting to note however, that several authors have proposed experiments aiming at solving hard problems from a theoretical perspective, but not all these experiments were effectively performed in the laboratory. Actually, most DNA computing approaches are of a theoretical basis only. G. Rozenberg coined a term for describing this activity, *menmology*, from **men**tal **mol**ecular bio**logy** (Calude and Păun, 2001). This chapter reviewed some pioneering DNA experiments and solutions to hard computational problems, including Adleman's experimental solution to the Hamiltonian path problem, and introduced the fundamental concepts of the test tube programming language. A brief discussion about the more formal models was also provided.

DNA computing may or may not be a wave of the future, paving the way for technological advances in chemistry, computer science and biology. The question that remains no longer concerns the computational power of DNA. Instead, it is about how to build a molecular computer. To think about building a real DNA computer, attention has to be focused on two major problems: *errors*, and the *realization and automation of DNA manipulation techniques*.

First, *errors* are incredibly frequent in biological reactions, and can take place during the manipulation of DNA. The extraction operation for instance, is very much prone to error. Extraction may result in *false positives* (strands that survive selection, but should not have been selected) and *false negatives* (strands that do not survive selection, but should have been selected). The annealing operation is also subject to errors. For instance, a strand *e* may anneal with another strand that is similar to *ē*, but which is not the right one. Like these, other DNA manipulation techniques are subject to errors.

Second, test tube programs are easy to write in terms of operations, but the feasibility of their implementation is a matter of DNA manipulation techniques. For instance, a natural form of realizing the operation merge is to pour the contents of one tube into another. The operation separate however, requires much more sophisticated approaches, and is error prone. The same holds true with the operation detect.

Much work has been done to find practical ways to describe DNA operations and to take advantage of the existing knowledge in mathematics and molecular biology to design DNA computers. Most of the work done is theoretical, but the great diversity of models gives a hope that DNA computers can be constructed. At its present stage of development molecular computing has several challenges (Ruben and Landweber, 2000; Maley, 1998):

1. The material used, whether DNA, RNA or proteins, are not reusable. While a silicon computer can operate indefinitely with electricity as its only input, a molecular computer would require periodic refueling and cleaning.

2. The molecular components used are specialized. The components are normally designed for a specific experiment.

3. When DNA is used to compute, one has to be always dealing with errors. Computational problems, in most cases, require very low error rates when compared with those presented by DNA computing. To rival conventional computers, all the DNA manipulation procedures would need to lower their error rates to possibly unattainable levels.

4. If an algorithm cannot be transformed to exploit the inherent parallelism of DNA, it will not run quickly on a DNA computer.

5. Providing input and reading the output of a DNA computation is not always an easy task.

6. The time required to perform an experiment (solve a problem and obtain a result) should be much reduced. Although the time for computation is short, the DNA manipulation techniques take quite long and have to be

automated. For instance, Adleman's experiment took seven days of lab work to be completed.

The main features of DNA computing serve to counterbalance the many difficulties in building a DNA computer. DNA computing presents very high speed when the DNA manipulation techniques are automated, it is massively parallel, it is highly energy efficient, and it provides economical and reduced information storage. The massive parallelism of DNA computing allows hard problems to be searched through using brute force; DNA molecules provide much shorter and cheaper information storage than silicon.

9.9 EXERCISES

9.9.1. Questions

1. Name four problems that cannot be solved by a Turing machine.

2. Name four NP-complete and four NP-hard problems.

3. The filtering technique was described as a means to separate specific molecules from a solution. There are other known and more efficient techniques for this purpose. Describe two of them.

4. Explain how you can perform the filtering operation to separate double stranded molecules.

5. The two most basic DNA sequencing techniques are known as a) Maxam-Gilbert and b) Sanger, after their proponents. Explain how each of these techniques work and contrast them.

6. Describe how to perform the extract operation, $E(t,i,b)$, proposed by Lipton, where t is a test tube, i is a given position on the strand, and b is a basis. This operation *extracts* all sequences in test tube t whose i-th symbol is equal to b, $b \in \{A,C,G,T\}$.

7. Lipton's solution to the generalized SAT problem assumes that the problem is represented in the restricted form: $C_1 \wedge C_2 \wedge \ldots \wedge C_m$, known as the conjunctive normal form. Explain how to place any formula in this form.

9.9.2. Computational Exercises

1. Write a test tube program to verify if a given tube t contains DNA strands with the following sequence of bases: ACTGG.

2. Write a test tube program that extracts from a given test tube t all strands containing at least one of the bases A and G, preserving at the same time the multiplicity of these strands.

3. Another NP-complete problem is the *3-vertex colourability problem*. In order to obtain a proper coloring of a graph $G = (V,E)$, three colors are assigned to the vertices in such a way that no adjacent vertices are similarly

colored. Write a test tube program to solve this problem assuming an initial tube *t*. Describe its functioning.

4. Write a test tube program to generate a set *t* of all permutations of *n* integers. Describe its functioning.

5. Write a test tube program to solve the *subgraph isomorphism problem* (SIP): given two graphs G_1 and G_2, determine whether G_2 is a subgraph of G_1 or not. Discuss its functioning.

6. A clique K_i is the complete graph on *i* vertices. Given a graph $G = (V,E)$, determine the largest *i* such that K_i is a subgraph of *G*. This problem is known as the *maximum clique problem* (MCP). Write a test tube program to solve it, and discuss its functioning.

7. The *maximum independent set* of a graph is the largest *i* such that there is a set of *i* vertices in which no pair is adjacent. Write a test tube program to solve this problem, and discuss its functioning.

9.9.3. Thought Exercises

1. Section 9.3.3 introduced the test tube programming language (TTPL) and one extension of it due to Amos (1997). Explain what types of DNA manipulation techniques, described in Section 9.2.2, can be used to perform each of the basic operations in the TTPL and its extension.

2. Describe how would you perform the insertion/deletion operations of an insdel system (Section 9.4) using the DNA manipulation techniques described.

3. Prove the NP-completeness of the satisfiability problem.

9.9.4. Projects and Challenges

1. There are a number of works in the literature hybridizing DNA computing with other natural computing paradigms, such as genetic algorithms and artificial immune systems.

    Review the literature in the search for these hybrid works. Write a paragraph briefly discussing each work and provide a list of references.

2. The traveling salesman problem studied in several chapters of this book, amounts to finding the shortest Hamiltonian path in an undirected graph, where the edges are provided with lengths.

    Reduce the Hamiltonian path problem to the traveling salesman problem, write a DNA computing algorithm to solve the TSP problem and describe how each of the steps can be performed using DNA strands and manipulation techniques.

    Discuss the computational complexity of your DNA solution to the TSP.

3. Given the description of a Turing machine provided in Appendix B.3.2, and the DNA-Pascal programming language described in Section 9.3.3, write a DNA-Pascal algorithm to simulate a Turing machine.

9.10 REFERENCES

[1] Adleman, L. M. (1994), "Molecular Computation of Solutions to Combinatorial Problems", *Science*, **226**, November, pp. 1021–1024.

[2] Adleman, L. M. (1995), "On Constructing a Molecular Computer", In R. J. Lipton and E. B. Baum (eds.), *DNA Based Computers*, Proc. of a DIMACS Workshop, pp. 1–22.

[3] Amos, M. (1997), *DNA Computation*, Ph.D. Thesis, Department of Computer Science, The University of Warwick.

[4] Amos, M. (2003), *Theoretical and Experimental DNA Computation*, Springer-Verlag.

[5] Barreiro, J. M., Rodrigo, J. and Rodriguez-Paton, A. (1998), "Evolutionary Biomolecular Computing", *Romanian Journal of Information Science and Technology*, **1**(4), pp. 289–294.

[6] Baum, E. B. (1995), "Building an Associative Memory Vastly Larger than the Brain", *Science*, **268**, pp. 583–585.

[7] Beaver, D. (1995), "Molecular Computing", Technical Report TR 95-001, Pennsylvania State University, Pennsylvania, USA, January.

[8] Beigel, R. and Fu, B. (1998), "Solving Intractable Problems with DNA Computing", *Proc. of the 13th Annual IEEE Conference on Computational Complexity*, pp. 154–168.

[9] Bennett, C. H. (1973), "Logical Reversibility of Computation", *IBM Journal of Research and Development*, **17**, pp. 525–532.

[10] Boneh, D., Dunworth, C. and Lipton, R. J. (1996), "Breaking the DES Using a Molecular Computer", In R. J. Lipton and E. B. Baum (eds.), *DNA Based Computers – Proc. of the DIMACS Workshop 1995*, pp. 37–66.

[11] Calude, C. S. and Păun, G. (2001), *Computing with Cells and Atoms*, Taylor & Francis.

[12] Conrad, M. (1972), "Information Processing in Molecular Systems", *Currents in Modern Biology*, **5**, pp. 1–14.

[13] Csuhaj-Varjú, E. Freund, R., Kari, L. and Păun, G. (1997), "DNA Computation Based on Splicing: Universality Results", In G. Rozenberg and A. Salomaa (eds.), *Lecture Notes in Computer Science*, **1261**, SpringerVerlag, pp. 353–370.

[14] Deaton, R, Garzon, M, Rose, J. A., Murphy, R. C., Stevens Jr., S. E. and Franceschetti, D. R. (1997), "A DNA Based Artificial Immune System for Self-Nonself Discrimination", *Proc. of the IEEE SMC'97*, pp. 862–866.

[15] Freund, G. P R., Kari, L. and Păun, G. (1999), "DNA Computing based on Splicing: The Existence of Universal Computers", *Theory of Computing Systems*, **32**, pp. 69–112.

[16] Gramß, T., Bornholdt, S., Groß, M., Mitchell, M. and Pellizzari, T. (2001), *Non-Standard Computation: Molecular Computation – Cellular Automata – Evolutionary Algorithms - Quantum Computers*, Wiley-VCH.

[17] Guarnieri, F., Fliss, M. and Bancroft, C, (1996), "Making DNA Add", *Science*, **273**(5272), pp. 220–223.

[18] Head, T. (1987), "Formal Language Theory and DNA: An Analysis of the Generative Capacity of Specific Recombinant Behaviors", *Bulletin of Mathematical Biology*, **49**, pp. 737–759.

[19] Kari, L. and Thierrin, G. (1996), "Contextual Insertion/Deletion and Computability", *Information and Computation*, **131**(1), pp. 47–61.

[20] Lipton, R. J. (1995), "Using DNA to Solve NP-Complete Problems", *Science*, **268**, pp. 542–545.

[21] Maley, C. C. (1998), "DNA Computation: Theory, Practice, and Prospects", *Evolutionary Computation*, **6**(3), pp. 201–229.

[22] Mills Jr., A. P., Turberfield, M., Turberfield, A. J., Yurke, B. and Platzman P. M. (2001), "Experimental Aspects of DNA Neural Network Computation", *Soft Computing*, **5**(1), pp. 10–18.

[23] Mulawka, J. J., Borsuk, P. and Weglenski, P. (1998), "Implementation of the Inference Engine Based on Molecular Computing Technique", *Proc. of the IEEE Int. Conf. on Evolutionary Computation (ICEC'98)*, pp. 493–498.

[24] Ogihara, M. and Ray, A. (1998), "The Minimum DNA Computation Model and Its Computational Power", In C. S. Calude, J. Casti, and M. J. Dinneen (eds.), *Unconventional Models of Computation*, Springer, pp. 308–322.

[25] Oliver, J. S. (1998), "Computation with DNA: Matrix Multiplication", In L. Landweber and E. Baum, *DNA Based Computers II*, Proc. of DIMACS, **44**, pp. 113–122.

[26] Păun, G., Rozenberg, G. and Saloma, A. (1998), *DNA Computing: New Computing Paradigms*, Springer-Verlag.

[27] Pisanti, N. (1998), "DNA Computing: A Survey", *Bulletin of the European Association for Theoretical Computer Science*, **64**, pp. 188–216.

[28] Pisanti, N. (2000), "DNA Computing: A New Computational Paradigm Using Molecules", In R. Lupacchini and G. Tamburrini (eds.), *Grounding Effective Processes in Empirical Laws: Reflections on the Notions of Algorithms*, pp. 101–116.

[29] Reif, J. (1995), "Parallel Molecular Computation", *Proc. of the 7th ACM Symposium on Parallel Algorithms and Architecture PSAA*, pp. 213–223.

[30] Rooß, D. and Wagner, K. W. (1996), "On the Power of DNA-Computing", *Information and Computation*, **131**(2), pp. 95–109.

[31] Rothemund, P. (1996), "A DNA Restriction Enzyme Implementation of Turing Machines", In E. B. Baum and R. J. Lipton (eds.), *DNA Based Computers*, American Mathematical Society, pp. 75–119.

[32] Roweis, S., Winfree, E., Burgoyne, R., Chelyapov, N. Goodman, M., Rothemund, P. and Adleman, L. (1996), "A Sticker Based Model for DNA Computation", In E. Baum, D. Boneh, P. Kaplan, R. Lipton, J. Reif and N. Seeman (eds.), *DNA Based Computers*, Proc. of the 2nd Annual Meeting, pp.1–27.

[33]   Ruben, A. J. and Landweber, L. F. (2000), "The Past, Present and Future of Molecular Computing", *Nature Reviews, Molecular Cell Biology*, **1**, pp. 69–72.

[34]   Winfree, E. (1996), "On The Computational Power of DNA Annealing and Ligation", In E. B. Baum and R. J. Lipton (eds.), *DNA Based Computers*, American Mathematical Society, pp. 199–221.

[35]   Zingel, T. (2000), "Formal Models of DNA Computing: A Survey", *Proc. of the Est. Ac. of Sci. Phys. Math.*, **49**(2), pp. 90–99.

CHAPTER 10

Quantum Computing

"An explanation is usually required to make sense in terms of things you already know about and quantum physics doesn't do that. It seems to make nonsense but it works."
(R. Gilmore, Alice in Quantumland, Copernicus, 1995, p. 43)

"Might I say immediately, so that you know where I really intend to go, that we always have had (secret, secret, close the doors!) we always have had a great deal of difficulty in understanding the world view that quantum mechanics represents."
(R. Feynman, Simulating Physics with Computers, Int. J. of Theor. Physics, 21(6/7), p. 471)

"Those who are not shocked when they first come across quantum mechanics cannot possibly have understood it."
(Attributed to N. Bohr)

10.1 INTRODUCTION

When we want to calculate the position, speed, energy and mass of a moving object, we apply Newton's laws of physics. To make a similar study at atomic and molecular scales a new type of physics, called *quantum physics*, has to be employed. Quantum physics is thus the theory from physics that explains the behavior of objects with atomic scales. *Quantum mechanics* is a terminology used to refer to the mathematical framework that describes quantum physics (Hirvensalo, 2001; Nielsen and Chuang, 2000). It incorporates new principles into *classical physics*, such as *energy quantization*, *duality*, *uncertainty*, and *probability*.

The application of quantum mechanics has a dramatic impact in computer science and engineering. This is because the technology used nowadays to design and explain the functioning of a computer is based on *classical mechanics* (*physics*). A 'classical' bit can only assume the values '0' or '1' in a mutually exclusive form. As the miniaturization of electronic components will naturally lead to atomic dimensions, quantum mechanics will become one of the few theories capable of providing further advances in computing technology. In addition to biomolecular computers, *quantum computers* are one of the major promises of a novel computing paradigm.

Quantum computing is about computing with *quantum systems*, leading to the so-called quantum computers, not about computational quantum physics. A quantum computer, however, can be used to model a number of quantum phenomena much more efficiently than by using classical computers. Similarly to DNA computers, quantum computers can also be used to solve otherwise intractable problems in other areas as well, such as breaking unbreakable codes.

487

The basis for quantum computing is quantum physics, which is concerned with small-scale material phenomena on the scale of electrons in atoms and electromagnetic radiation like light, *x*-rays, and gamma rays. The logic upon which quantum computing stands is named quantum logic, which differs from classical logic in many respects. Thus, an important fact about quantum mechanics that we have to prepare the reader to is that the behavior of quantum objects is unlike anything most of us have ever seen. One first and major feature of a quantum computer is the use of *quantum bits* instead of classical bits. A quantum bit can assume both values '0' and '1' simultaneously, thus allowing for massive parallel computation.

Part of the mathematical background necessary for an introductory understanding of quantum computing is reviewed in Appendix B. In particular, the reader should go through linear algebra with emphasis on vector spaces, matrices and complex numbers, statistics, and the theory of computation. This chapter starts with a brief description of some physical experiments to give an idea of the types of phenomena that occur in the quantum world. It then follows with an introduction to principles from quantum mechanics and, in the following section, presents the quantum analogues to the basic building blocks of standard computers: quantum bits, quantum gates, and quantum circuits. The main quantum algorithms are presented and the chapter is concluded with a brief description of the some physical realizations of quantum systems.

10.1.1.   Motivation

Similarly to DNA computing, some of the main advantages of quantum computing are a massive increase in computing power, due mainly to quantum parallelism, and the economical information storage, as we are now working at atomic scales. When compared with the currently known silicon-based computers, quantum computing offers some unique features:

- It uses quantum bits, *qubits*, as data structures. A quantum bit can assume any logical value 0, 1 or a superposition of them, thus promoting an exponential increase in the information storage capacity.

- Quantum computers can work in a massively parallel fashion.

- Computation can be performed at an atomic level, a size that will potentially limit the current technology of the semiconductor industry.

- Quantum technology can support an entirely new type of computation, with qualitatively new algorithms based on quantum mechanics.

- Quantum computers are highly effective in solving NP-complete problems; that is, they can be used to solve problems that cannot be (practically) solved using classical computers.

An overall, striking observation about quantum computing is that, at least theoretically, there seem to be many diverse ways of constructing quantum computers, as will be briefly discussed in Section 10.7.

10.2 BASIC CONCEPTS FROM QUANTUM THEORY

Quantum computing is based on the use of quantum systems to perform computation, and quantum systems are subjected to the laws of quantum physics. Thus, this section introduces quantum mechanics by placing it in the context of classical physics, and illustrates quantum phenomena with one example.

10.2.1. From Classical to Quantum Mechanics

Until the early 20th century, physicists were able to explain natural phenomena using two wide-ranging theories: Newton's Mechanics and Maxwell's Electromagnetism. The former studies the motion of bodies, such as the orbit of planets and the trajectory of projectiles, and the latter is used to explain electromagnetic phenomena, such as light and radio waves. Together they constitute what is now known as *classical physics*. Its general principles were first enunciated in 1687 by I. Newton in his *Philosophiae Naturalis Principia Mathematica*, simply referred to as the *Principia*. Classical mechanics is not only the first branch of Physics to be discovered, but also has many important applications in other fields of investigation, from Astronomy (e.g., movement of planets), Chemistry (e.g., dynamics of molecular collisions), Geology (e.g., propagation of waves), to Engineering (e.g., equilibrium and stability of structures).

The power of classical mechanics seemed so great during the 18th and 19th centuries that one of Newton's successors, P. S. Laplace, made the assertion that given complete knowledge of the dispositions of all particles at some instant of time and unlimited calculating power, it would be possible to retrodict the past and predict the future. However, classical physics was not sufficient to explain all types of phenomena. A vast number of experimental observations seemed to present contradictory results. One classical example is the observation that the same radiation that produces *interference patterns*, and therefore must consist of *waves*, also results in the *photoelectric effect*, and thus must consist of moving *particles*. This particular example and others will be discussed in the sequence.

The attempts to describe atomic events using classical mechanics led to contradictions, for atoms and molecules at very small distances do not behave following classical mechanics. Although by the early 1920s physicists were used to these difficulties, they were able, to some extent, to predict what would be the outcome of specific experiments at the atomic scale. A consistent new theoretical formulation in physics was thus under development in order to explain these phenomena, and this theory is known as *quantum theory*. In addition, Einstein's Relativity theory was proposed as another new branch of physics to deal with objects moving at very high speeds, such as the light speed (3×10^8m/s), or objects with astronomical masses, such as galaxies. To have an idea of the magnitude of the light speed, one can perform approximately nine laps around the Earth in only one second at light speed.

By the end of the 19th century, the German physicist Max Planck made an extraordinary discovery. He suggested that radiation was emitted or absorbed from time to time in packets of energy of a definite size, called *quanta*. He proposed that the energy content, E, of one of these quanta would be proportional to the frequency of radiation, v, with a constant of proportionality h, now known as Planck's constant:

$$E = h.v \qquad\qquad h = 6.626075 \times 10^{-34} \text{ J.s}$$

A few years later, in 1905, A. Einstein confirmed Planck's theory by using it to explain the *photoelectric effect*, another 'anomaly' in physics. The electrons contained in a metal can move about within the metal but do not have enough energy to escape from the metal. The photoelectric effect happens when a beam of light ejects electrons from within a metal. This occurs because the radiation transfers energy to electrons trapped inside the metal and, if the gain is sufficient, an electron can then escape from the forces that constrain it. In classical thinking, the coefficient of emission of electrons would depend on the intensity of the beam of light bombarding the metal, because light was viewed as an electromagnetic wave. However, the experimental findings indicated that the emission of electrons was dependent upon the frequency, not intensity, of light. Einstein insight was that this puzzle could be explained by assuming that light consists of little particles or quanta, called *photons*, with an energy proportional to Planck's constant times the frequency. Under this perspective, an electron would be ejected after the collision with a photon and then given up all of its energy. The intensity of the beam of light influences the number of photons emitted and thus the number of electrons ejected, but only the frequency would influence the ejection of electrons.

Several other experimental findings and theories contributed to the development of quantum theory during the 20th century. Among these, N. Bohr postulated a model of the hydrogen (H) atom with electrons of quantized angular momentum radiating energy between circular orbits; Compton studied the spectrum of monochromatic x-rays going through a thin target; L. de Broglie proposed that electrons have wavelike properties that relate momentum and wavelength; W. Pauli presented the exclusion principle (no two electrons in the same atom may have identical quantum numbers); W. Heisenberg introduced matrix mechanics and the uncertainty principle; E. Schrödinger developed the wave mechanics; and P. A. M. Dirac proposed the relativistic matrix mechanics, an even more elaborate formulation of quantum mechanics. Many Nobel prizes were awarded for these and other physicists involved in the development of quantum theory.

10.2.2. Wave-Particle Duality

Before presenting the formalism necessary to a basic understanding of quantum computing, let us first discuss in greater detail another unusual quantum mechanical behavior typical of objects at atomic scales. This will serve to illustrate many of the problems and paradoxes of quantum physics. In particular, let us consider three different versions of a famous experiment that will serve to illustrate the differences between the macro world, governed by classical mechanics, and the world at atomic scales, governed by quantum mechanics: the *double-slit experiment.* The experiment will be discussed using bullets, water waves and electrons, as described in (Hey and Walters, 1987).

Double-Slit with Bullets

In the first scenario, there is a wobbly machine gun that fires bullets with same speed in random directions. An armor plate with two parallel slits is placed in front of the gun and a detector plate is placed behind the double-slit plate to detect where the bullets arrive. After sometime running the experiment (shooting), it is possible to observe a certain pattern of hits in the detector. By closing one slit and keeping the other one open, it is also possible to observe the pattern that can be obtained with each of the slits open. If both slits are open simultaneously, it is possible to observe that the resultant pattern is the sum of the patterns with each of the slits open separately (Figure 10.1). Let P_{12} be the probability of arrival of bullets with both slits open, P_1 the same probability with only slit 1 open, and P_2 the probability with only slit 2 open, then:

$$P_{12} = P_1 + P_2$$

In this case we say that there is no interference between each term.

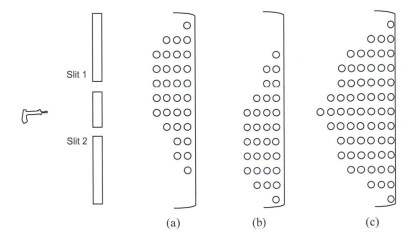

Figure 10.1: The double-slit experiment with bullets. (a) Distribution of bullets when only slit 1 is open. (b) Distribution of bullets when only slit 2 is open. (c) Distribution of bullets with both slits open.

Double-Slit with Water Waves

In this second scenario, assume there is a lake with a source of waves, a barrier with two gaps (slits) on it, and a detector after the barrier that can measure the amount of energy of the wave at that position. Assuming a single gap open for some time it is possible to observe one specific wave pattern at the detector. When both gaps are open simultaneously, however, it is possible to observe

some *interference* between the waves (Figure 10.2). This is a known phenomenon to us all; for instance, when we throw stones into a lake we observe that the
waves produced by them interfere with each other. The energy of water waves at
any given position is related to how big they are at that point: it depends on the
square of the maximum height of the wave. Let I be the intensity of the wave
(i.e., amount of energy arriving per second) and H be the maximum height of the
wave, then:

$$I = H^2$$

By labeling H_1 the height of the wave coming from slit 1 with slit 2 closed,
and H_2 the height of the wave coming from slit 2 with slit 1 closed, then we have
the following relation:

$$I_{12} = H_{12}^2 = (H_1 + H_2)^2$$

where I_{12} is the intensity of the wave with both slits open, which is different
from $I_1 + I_2$, and $H_{12} = H_1 + H_2$ is the height of the wave when both slits are
open.

Unlike the experiment with bullets, the same pattern of wave cannot be obtained by summing up the patterns obtained with a single slit open and the other
closed at a time, there is clearly an interference effect between the waves.

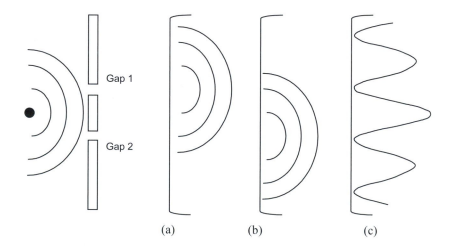

Figure 10.2: The double-slit experiment with water waves. (a) Pattern of waves when
only slit 1 is open. (b) Pattern of waves when only slit 2 is open. (c) Interference pattern
when both slits are open.

Double-Slit with Electrons

Consider lastly, the case of an electron gun (an apparatus that shoots electrons)
mounted in front of a thin metal plate (barrier) with two very narrow slits in it,
and a detector screen coated with a chemical substance that flashes every time

an electron arrives at it. An important fact about electrons is that they are parti-cles; they have a well-defined mass, electric charge, spin, etc. (Remember that a wave spreads out distributing its energy.)

In this set up, it is possible to count the number of flashes at given positions of the detector during some period of time, a pattern of behavior similar to that of bullets. For each of the slits open independently of the other, the pattern ob-tained is exactly the same as the one obtained for bullets. When both slits are open simultaneously, however, one obtains an interference pattern alike that of water waves (Figure 10.3). There are places on the detection screen where no electrons hit, and other places where more electrons arrive than the expected number from adding the contributions from each slit acting alone. This result is the central mystery of quantum mechanics: quantum objects, such as electrons, have attributes of both wave and particle motion but do not behave like neither.

Based on this duality in behavior, in the case of electrons one has to measure their probability of arrival and also take into account something like the height of an electron wave. Since the square of this height must give the corresponding probability, it is called a *quantum probability amplitude*. Thus, the equation for the probability of arrival of electrons will have the same form as the one for wa-ter waves, but the intensity will be replaced by probability P and the height will be substituted by the amplitude a:

$$P_{12} = (a_1 + a_2)^2$$

and, in this case, $P_{12} \neq P_1 + P_2$.

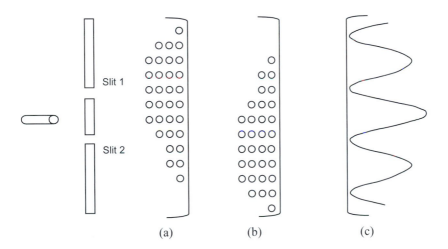

(a)   (b)   (c)

Figure 10.3: The double-slit experiment with electrons. (a) Distribution of flashes at given positions of the screen when only slit 1 is open. (b) Distribution of flashes at given positions of the screen when only slit 2 is open. (c) Pattern obtained when both slits are open.

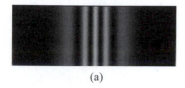 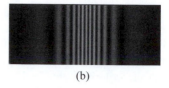

(a)   (b)

Figure 10.4: Interference patterns generated with a laser beam in a double-slit experiment. (a) Narrow separation between the slits. (b) Wider separation between the slits.

The wave-particle duality, thus, corresponds to the fact that electrons (and light) show wave-like interference in their arrival despite the fact that they arrive in lumps. Figure 10.4 depicts two experimental observations obtained with a laser beam in the double-slit experiment for a narrow and a wider separation between the slits.

10.2.3.   The Uncertainty Principle

The deterministic view of classical physics, that given complete knowledge of the dispositions of all particles at some instant of time and unlimited calculating power it would be possible to predict the future, has been proved wrong by quantum physics. At the quantum level there is a fundamental limit to the accuracy that can be achieved while observing (measuring) something. To illustrate this, let us return to the double-slit experiment with electrons.

In this experiment it is only possible to predict the relative chance (probability) of an electron landing at a particular position on the screen, because whenever a measurement is made the system is disturbed. This happens because to observe an electron after it passes through one of the slits it is necessary to shine some light on it and observe the reflected light. As electrons are delicate quantum objects, when the light interacts with the electrons it gives them a jolt thus disturbing their motion significantly. This disturbance is enough to remove the interference pattern between electrons in the experiment, allowing us to determine accurately through which slit the electron is passing. In this case, the resultant pattern is exactly the one observed using bullets, meaning that when we shine some light on the electron it behaves like a particle.

Broadly speaking, the *uncertainty principle* means that the more precisely the position of an object is determined, the less precisely its momentum (mass times velocity) is known in a given instant of time, and vice-versa. Therefore, the laws of quantum mechanics imply a fundamental limitation to the accuracy of experimental measurements. Of course that in the macroscopic world it is possible to measure position and momentum simultaneously with great precision, but this does not hold for quantum objects. For objects of microscopic dimensions the influence promoted by the measuring devices are not negligible, they significantly disturb the objects.

10.2.4. Some Remarks

Based on the double slit experiment and on the discussions presented above, several important conclusions regarding quantum physics can be made. First, light may present wave-like or particle-like behavior, depending on the observation. Second, observation affects the state of a quantum system. Third, the result of a future measurement of a quantum system cannot be predicted with certainty, only probabilistically. Fourth, there is a limit to the measuring accuracy as well. Fifth, the evolution of a system refers to how it behaves over time, being different from its measurement (Pittenger, 2001). A mathematical framework to describe quantum systems must take into account these five aspects. First, it must represent physical states as mathematical entities. Second, it must model interference phenomena. Third, it must account for nondeterministic behaviors. Fourth, it must model the effect of measurement on the system. Fifth, it must treat the dynamics (evolution) of the system differently from measurement (Pittenger, 2001).

10.3 PRINCIPLES FROM QUANTUM MECHANICS

Quantum mechanics is usually formulated in one of two forms: 1) using wave mechanics, introduced by E. Schrödinger, based on partial differential equations to describe waves; or 2) using matrix mechanics, introduced by W. Heisenberg, which involves the use of matrices that are associated with the properties of matter. Although the two formulations differ substantially from a mathematical perspective, Schrödinger has shown that they are equivalent.

Another commonly used formulation of quantum mechanics is the transformation theory proposed by P. Dirac, which unifies matrix and wave mechanics. In this formulation, to be presented here, the instantaneous state of a quantum system is described by a quantum state that encodes the probabilities associated with all *observables*, i.e., measurable properties. The mathematical framework necessary to understand quantum mechanics and quantum computing is based on linear algebra and statistics. The reader not much familiar with these concepts is invited to carefully go through Appendix B, which has sections dedicated to review the mathematical concepts used in this chapter. The discussion to be presented here is based on many references, in particular: Nielsen and Chuang (2000), Hirvensalo (2001), Pittenger (2001), Meglicki (2002), Rieffel and Polak (2000), and Perry (2004).

10.3.1. Dirac Notation

In quantum mechanics, the standard notation used to represent a vector in a vector space is slightly different from the one used so far. A vector is represented by the following object:

$$|\mathbf{x}\rangle \tag{10.1}$$

where $\mathbf{x}$ is a label[1] for the vector. The notation $|\cdot\rangle$, usually referred to as *ket*, indicates that $\mathbf{x}$ is a (column) vector. Every ket $|\mathbf{x}\rangle$ has a dual *bra* $\langle\mathbf{x}|$, which is the conjugate transpose of $|\mathbf{x}\rangle$: $\langle\mathbf{x}| = |\mathbf{x}\rangle^{\dagger}$.

The state of a quantum system is described by a unit-length vector in a Hilbert space, i.e., an n-dimensional complex vector space H^n corresponding to the *state space* of the system. The state space of a quantum system, known as a *quantum state space*, can be described in terms of vector and matrices or using a standard notation called *bra-ket (bracket) notation* or *Dirac notation* due to its proponent, P. Dirac (1982). The terminology 'bracket notation' is so called because the *inner product* of two states is represented by a *bracket*, $\langle\mathbf{y}|\mathbf{x}\rangle$, which consists of a left part $\langle\mathbf{y}|$, the *bra*, and a right part $|\mathbf{x}\rangle$, the *ket*.

We have seen above that the combination of $\langle\mathbf{y}|$ and $|\mathbf{x}\rangle$, $\langle\mathbf{y}||\mathbf{x}\rangle$ also represented as $\langle\mathbf{y}|\mathbf{x}\rangle$, corresponds to the inner product between the two vectors, and results in a complex number. The notation $|\mathbf{x}\rangle\langle\mathbf{y}|$ represents the *outer product* of vectors $|\mathbf{x}\rangle$ and $\langle\mathbf{y}|$, and results in a matrix.

To illustrate the use of the bracket notation, consider the orthonormal basis formed by the vectors $|0\rangle$ and $|1\rangle$, $\{|0\rangle,|1\rangle\}$. In standard linear algebra notation this basis can be expressed as $\{(1,0)^T,(0,1)^T\}$ or, alternatively, $\{(0,1)^T,(1,0)^T\}$. In the former case, the inner product between $\langle 0|$ and $|1\rangle$ is 0, $\langle 0|1\rangle = 0$; and the inner product between $\langle 0|$ and $|0\rangle$ is 1, $\langle 0|0\rangle = 1$.

The outer product of vectors $|1\rangle$ and $\langle 0|$ is

$$\mathbf{A} = |1\rangle\langle 0| = \begin{bmatrix} 0 & 0 \\ 1 & 0 \end{bmatrix}$$

Note that the result of the outer product is a matrix $\mathbf{A}$ that corresponds to a transformation on quantum states responsible for dictating what happens to the basis vectors. For instance, the outer product $|1\rangle\langle 0|$ maps $|0\rangle$ into $|1\rangle$ and $|1\rangle$ into $(0,0)^T$.

10.3.2. Quantum Superposition

Any state of a quantum system can be written as a linear combination of a number of *basis states* or basis vectors:

$$c_1|\mathbf{x}_1\rangle + c_2|\mathbf{x}_2\rangle + \ldots + c_n|\mathbf{x}_n\rangle \tag{10.2}$$

where c_i, $i = 1, \ldots, n$ are complex numbers called *amplitudes*, and $\sum_i |c_i|^2 = 1$, $\forall i$. Each value $|c_i|^2$ corresponds to the probability that the system is found at state $\mathbf{x}_i$.

Note that $\mathbf{x}_i$, $i = 1, \ldots, n$, denote *observables* from a physical object. For instance, $\mathbf{x}_1$ may represent the speed of a particle and $\mathbf{x}_2$ may represent its momentum. Equation (10.2) is known as a *superposition of basis states*.

---

[1] Symbols like φ and ψ are commonly found labels for vectors in the quantum mechanics literature, but we chose to use $\mathbf{x}$, $\mathbf{y}$, etc., to maintain the standard notation of the book.

10.3.3.   Tensor Products

The *tensor product* of $|x\rangle$ and $|y\rangle$, represented by $|x\rangle \otimes |y\rangle$ and abbreviated as $|x\rangle|y\rangle$, or even more compactly as $|xy\rangle$, is one form of putting together vector spaces to form larger vector spaces; it allows the combination of quantum states. Thus, the state space of a composite system is the tensor product of the state spaces of the component physical systems. For example, given the basis states $\{|0\rangle, |1\rangle\} = \{(1,0)^T, (0,1)^T\}$, we have the following tensor products:

$$|0\rangle \otimes |0\rangle = |00\rangle = (1\ 0\ 0\ 0)^T$$

$$|0\rangle \otimes |1\rangle = |01\rangle = (0\ 1\ 0\ 0)^T$$

$$|1\rangle \otimes |0\rangle = |10\rangle = (0\ 0\ 1\ 0)^T$$

$$|1\rangle \otimes |1\rangle = |11\rangle = (0\ 0\ 0\ 1)^T$$

For an arbitrary scalar c of the underlying field, and some vectors x, x_1, x_2, y, y_1, and y_2, from the appropriate spaces, the tensor product satisfies three basic properties:

a) $c(|x\rangle \otimes |y\rangle) = (c|x\rangle) \otimes |y\rangle = |x\rangle \otimes (c|y\rangle)$.

b) $(|x_1\rangle + |x_2\rangle) \otimes |y\rangle = |x_1\rangle \otimes |y\rangle + |x_2\rangle \otimes |y\rangle$.

c) $|x\rangle \otimes (|y_1\rangle + |y_2\rangle) = |x\rangle \otimes |y_1\rangle + |x\rangle \otimes |y_2\rangle$.

More on tensor products will be presented while talking about multiple qubit systems.

10.3.4.   Entanglement

Two classical bits can be put together (combined) in a composite system as 00, 01, 10 and 11. The value of any combination can be written as the product of the individual bits. There are some quantum composite systems that cannot be written as a tensor product of states of its component systems, a property called *entanglement*.

The Hilbert space of a system composed of two systems A and B is $H_A \otimes H_B$; that is, the tensor product of the respective spaces. Assuming that the first system is in an arbitrary state $|x\rangle_A$ and the second system in another arbitrary state $|y\rangle_B$, if the state of the composite system cannot be written as the tensor product $|x\rangle_A \otimes |y\rangle_B$, then the states are said to be entangled; otherwise, they are called *separable*, *decomposable*, or *product states*. The general state of a composite system $H_A \otimes H_B$ has the form

$$\Sigma_{x,y}\ c_{xy}\ |x\rangle_A\ |y\rangle_B \tag{10.3}$$

Examples of famous entangled two-qubit states are the Bell states:

$$\beta_{00} = |\ x_0\rangle = \frac{1}{\sqrt{2}}\left(|\ 00\rangle_{AB} + |\ 11\rangle_{AB}\right)$$

$$\tag{10.4}$$

$$\beta_{01} = |\ x_1\rangle = \frac{1}{\sqrt{2}}\left(|\ 01\rangle_{AB} + |\ 10\rangle_{AB}\right)$$

$$\beta_{10} = |\mathbf{x}_2\rangle = \frac{1}{\sqrt{2}} \left(|00\rangle_{AB} - |11\rangle_{AB} \right)$$

$$\beta_{11} = |\mathbf{x}_3\rangle = \frac{1}{\sqrt{2}} \left(|01\rangle_{AB} - |10\rangle_{AB} \right)$$

An example of a decomposable state is the one formed by a linear combination of all basis states in H^4: $\frac{1}{2}(|00\rangle + |01\rangle + |10\rangle + |11\rangle)$. This can be verified by finding the coefficients $\mathbf{a}_1$, $\mathbf{a}_2$, $\mathbf{b}_1$, $\mathbf{b}_2$, such that:

$$\frac{1}{2}(|00\rangle + |01\rangle + |10\rangle + |11\rangle) = (\mathbf{a}_1|0\rangle + \mathbf{a}_2|1\rangle)(\mathbf{b}_1|0\rangle + \mathbf{b}_2|1\rangle) =$$

$$= \mathbf{a}_1\mathbf{b}_1|00\rangle + \mathbf{a}_1\mathbf{b}_2|01\rangle + \mathbf{a}_2\mathbf{b}_1|10\rangle + \mathbf{a}_2\mathbf{b}_2|11\rangle.$$

In this case, $\mathbf{a}_1\mathbf{b}_1 = \mathbf{a}_1\mathbf{b}_2 = \mathbf{a}_2\mathbf{b}_1 = \mathbf{a}_2\mathbf{b}_2 = \frac{1}{2}$, what corresponds to $\mathbf{a}_1 = \mathbf{a}_2 = \mathbf{b}_1 = \mathbf{b}_2 = 1/\sqrt{2}$.

10.3.5. Evolution (Dynamics)

The *evolution* of a quantum system corresponds to its dynamics, i.e., how it changes over time. If the quantum system is not interacting with any other system (i.e., is *closed*), its evolution can be described by a *unitary transformation* or *operator* represented as a matrix. In other words, the state $|\mathbf{x}_2\rangle$ of the system at time t_2 is related to the state $|\mathbf{x}_1\rangle$ at time t_1 by a unitary transformation $\mathbf{A}$ that is a function of t_1 and t_2

$$|\mathbf{x}_2\rangle = \mathbf{A}|\mathbf{x}_1\rangle \qquad (10.5)$$

The effect of a unitary transformation $\mathbf{A}$ on a state $\mathbf{x}$ is described by the corresponding rotation of the vector $|\mathbf{x}\rangle$ in the appropriate Hilbert space. Thus, the unitary transformation stands for the quantum mechanical operation as well as for the unitary rotation. Note that the order of applying the transformations matters: $\mathbf{A}.\mathbf{B} \neq \mathbf{B}.\mathbf{A}$, where $\mathbf{A}$ and $\mathbf{B}$ are unitary transformations. Furthermore, a unitary time evolution means that the time evolution of a quantum system is *invertible* (*reversible*); that is, $|\mathbf{x}_1\rangle$ can be perfectly recovered from $|\mathbf{x}_2\rangle$.

Examples of unitary transformations are the Pauli matrices presented in Appendix B.1.5, and repeated here for convenience

$$\sigma_0 = \mathbf{I} = \begin{bmatrix} 1 & 0 \\ 0 & 1 \end{bmatrix} \qquad \sigma_1 = \mathbf{X} = \begin{bmatrix} 0 & 1 \\ 1 & 0 \end{bmatrix}$$

$$\sigma_2 = \mathbf{Y} = \begin{bmatrix} 0 & i \\ -i & 0 \end{bmatrix} \qquad \sigma_3 = \mathbf{Z} = \begin{bmatrix} 1 & 0 \\ 0 & -1 \end{bmatrix} \qquad (10.6)$$

The application of operator $\mathbf{X}$, for instance, to the basis vectors $\{|0\rangle, |1\rangle\}$ maps $|0\rangle$ into $|1\rangle$, and $|1\rangle$ into $|0\rangle$, respectively; thus acting as a sort of quantum NOT gate. Thus, a unitary operator can be viewed as a *quantum gate*. Other examples of quantum gates will be provided in Section 10.4.3.

In the continuous time case, the state evolution of a quantum system is described by the *Schrödinger equation*:

$$i\hbar \, d|\mathbf{x}\rangle/dt = \mathbf{H}|\mathbf{x}\rangle \qquad (10.7)$$

where $\hbar$ is Planck's constant, and $\mathbf{H}$ is the *Hamiltonian* of the closed system. Note that the Hamiltonian generates the time evolution of the quantum system. As this description is out of the scope of this chapter, the details are left as further reading.

10.3.6. Measurement

Sometimes it may be of interest to *measure* or to *observe* the system in order to obtain its characterization, e.g., its state. Quantum measurements are described by a set of measurement operators that act on the state space of the system being measured. Any device measuring a quantum system has an associated orthonormal basis with respect to which the measurement takes place. The result is a projection of the state of the system prior to measurement onto the subspace of the state space compatible with the measured values; that is, the measurement projects (*collapses*) the quantum state onto one of the basis states associated with the measuring device. The amplitude of the projection is rescaled so that the resulting state vector is of unit length. Note that the outcome of a measurement is probabilistic and changes the state of the system to the measured one.

To illustrate the idea, consider the case of an arbitrary state $|\mathbf{x}\rangle$ represented as a linear combination of a set of basis states $|0\rangle$ and $|1\rangle$: $|\mathbf{x}\rangle = \mathbf{c}_1|0\rangle + \mathbf{c}_2|1\rangle$ with $\mathbf{c}_1$ and $\mathbf{c}_2$ complex numbers, as depicted in Figure 10.5. The measurement of the state $|\mathbf{x}\rangle = \mathbf{c}_1|0\rangle + \mathbf{c}_2|1\rangle$ results in $|0\rangle$ with probability $|\mathbf{c}_1|^2$ and results in $|1\rangle$ with probability $|\mathbf{c}_2|^2$. In Figure 10.5 it can be observed that different bases associated with different measuring devices result in different outcomes. The key point is that measuring $|\mathbf{x}\rangle$ implies interacting with $|\mathbf{x}\rangle$, which has the effect of collapsing the state of the system onto one of the basis states. Therefore, by the Heisenberg uncertainty principle, measuring the state of a quantum system will 'disturb' its state.

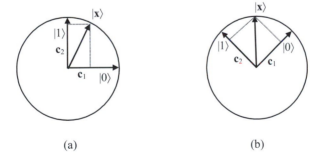

(a)   (b)

Figure 10.5: Pairs of basis associated with different measurement devices and the projection of $|\mathbf{x}\rangle$ onto the basis states. (a) Basis: $\{|0\rangle,|1\rangle\} = \{(1,0)^T,(0,1)\}$. (b) Basis: $\{|0\rangle,|1\rangle\} = \{(1/2,1/2)^T,(-1/2,1/2)\}$.

10.3.7. No-Cloning Theorem

The No-Cloning Theorem, introduced simultaneously but independently by Wootters and Zurek (1982) and Dieks (1982), states that it is not possible to create identical copies of an arbitrary unknown quantum state; that is, there is no quantum *copymachine*.

To prove the theorem, assume there is a quantum system A in a two-dimensional Hilbert space H^2 with basis states $|0\rangle_A$ and $|1\rangle_A$. An arbitrary state $|x\rangle_A$ of this quantum system can thus be written as a linear combination of the basis states: $|x\rangle_A = c_1|0\rangle_A + c_2|1\rangle_A$. Suppose we want to copy the state $|x\rangle_A$. The quantum copymachine will receive as input the state $|x\rangle_A$ and deliver as output $|x\rangle_A$ back again and one copy of it. Therefore, one system is inserted into the copymachine and two identical systems are retrieved. In order to make the copy, a system B with an identical Hilbert space and an arbitrary initial state $|s\rangle_B$ is taken. The initial state $|s\rangle_B$ of system B will be transformed into the state to be copied. The following initial state results in the copymachine:

$$|x\rangle_A|s\rangle_B \tag{10.8}$$

This composite system can be either observed, what would collapse the system into one of the basis states of the measuring device, or be subjected to an arbitrary unitary transformation $\mathbf{A}$ that would be responsible for copying the state $|x\rangle_A$:

$$\mathbf{A}(|x\rangle_A|s\rangle_B) = |x\rangle_A|x\rangle_B =$$
$$= (c_1|0\rangle_A + c_2|1\rangle_A)(c_1|0\rangle_B + c_2|1\rangle_B) = \tag{10.9}$$
$$= (c_1^2|0\rangle_A|0\rangle_B + c_1c_2|0\rangle_A|1\rangle_B + c_2c_1|1\rangle_A|0\rangle_B + c_2^2|1\rangle_A|1\rangle_B)$$

Let us take the particular case in which we want to copy the basis states of a two-dimensional quantum system:

$$\mathbf{A}(|0\rangle_A|s\rangle_B) = |0\rangle_A|0\rangle_B$$
$$\mathbf{A}(|1\rangle_A|s\rangle_B) = |1\rangle_A|1\rangle_B \tag{10.10}$$

Applying the copymachine to the generic state $|x\rangle$, assuming the linearity of transformation $\mathbf{A}$, and taking into account Equation (10.10), results in

$$\mathbf{A}(|x\rangle_A|s\rangle_B) = \mathbf{A}(c_1|0\rangle_A + c_2|1\rangle_A)|s\rangle_B =$$
$$= \mathbf{A}(c_1|0\rangle_A|s\rangle_B + c_2|1\rangle_A|s\rangle_B) = \tag{10.11}$$
$$= \mathbf{A}(c_1|0\rangle_A|s\rangle_B) + \mathbf{A}(c_2|1\rangle_A|s\rangle_B) =$$
$$= c_1|0\rangle_A|0\rangle_B + c_2|1\rangle_A|1\rangle_B$$

Note that the result presented in Equation (10.11) is different from the one obtained in Equation (10.9). Therefore, generally $\mathbf{A}(|x\rangle_A|s\rangle_B) \neq |x\rangle_A|x\rangle_B$. The no-cloning theorem implies that it is not possible to reliably clone an unknown state; it is possible to clone a known quantum state.

10.4 QUANTUM INFORMATION

To introduce the foundations of quantum information processing we will first define the basic information unit, describe means of processing these units, and then extend to the cases of multiple combined information units. The quantum analogues to the basic building blocks of standard computers will be presented: quantum bits, quantum gates, and quantum circuits.

10.4.1. Bits and Quantum Bits (Qubits)

In computing and information theory, the most basic information unit is the *binary digit* or *bit*, which is an abstract entity that can assume either the value '0' or '1'. Information is physically stored as bits in standard computers, and the devices realizing a bit can be a combination of transistors and other integrated circuit elements with a distribution of charge determining the state of the bit. Bits can be combined into *n*-dimensional bitstrings to represent more information, and the bitstrings can be manipulated to perform computation; for instance, implement algorithms and perform mathematical operations. It is possible to access a certain memory address of a standard computer and observe (read) it without affecting its contents.

When information is stored at atomic scales, quantum effects take place and the result is a completely different scenario. In this case, a *quantum bit*, or *qubit* for short, can assume both values '0' and '1' simultaneously, differently from the classical bits that are discrete and can only assume one of these values at any given time. Qubits are mathematical objects (abstract entities) with specific properties and correspond to the most basic units of information in quantum computing. They are represented by ideal two-state quantum systems, such as polarized photons, electrons, atoms, ions, or nuclear spins (Figure 10.6). Measurements and manipulations affect the contents of a qubit and can be modeled as matrix operations.

Quantum information is described by a state in a two-level quantum mechanical system with two basic states conventionally labeled $|0\rangle$ and $|1\rangle$. These particular states form an orthonormal basis for a two-dimensional complex vector space (a Hilbert space H^2) and, thus, are known as *computational basis states*. As such, it is possible to form linear combinations of the basis states, often called *superpositions*, which correspond to a *pure* qubit state:

Figure 10.6: Representation of an atom with three discrete energy levels according to quantum mechanics. An atom depicted with only the two inner circumferences (two energy levels) can represent a quantum bit.

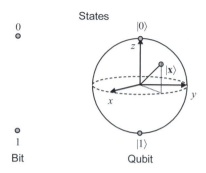

States

0
○

○
1

Bit

$|0\rangle$

$|\mathbf{x}\rangle$

$|1\rangle$

Qubit

Figure 10.7: Geometric interpretation of a bit and the Bloch sphere representation of a qubit.

$$|\mathbf{x}\rangle = \mathbf{c}_1|0\rangle + \mathbf{c}_2|1\rangle \qquad (10.12)$$

where $\mathbf{c}_1$ and $\mathbf{c}_2$ are complex numbers.

In practice, the quantum superposition of the basis states means that an infinite amount of information can potentially be encoded in a single qubit by appropriately defining the amplitudes $\mathbf{c}_1$ and $\mathbf{c}_2$. To visualize the difference between a bit and a qubit, one can think geometrically about their state spaces. The possible states of a bit correspond to two points ('0' or '1'), and the possible states (pure states) of a qubit correspond to the surface of a unit sphere with the logical states being the poles of the sphere and any other state being a point in the surface of the sphere (Figure 10.7). The sphere presented in Figure 10.7 is known as *Bloch sphere*. It corresponds to a geometrical representation of the pure state space of a bi-dimensional quantum mechanical system. Thus, a qubit can exist in a continuum of states between $|0\rangle$ and $|1\rangle$, until it is observed as will be discussed below.

As the basis states form a basis of the space, any other state $|\mathbf{x}\rangle$ can be written as a linear combination of the basis states. Despite the fact that a qubit can be found in infinite superposition states, a single classical bit can be extracted from a qubit. Thus, when a qubit is measured (observed), either the result '0' is obtained with probability $|\mathbf{c}_1|^2$ or the result '1' is obtained with probability $|\mathbf{c}_2|^2$. Like all probabilities, $|\mathbf{c}_1|^2 + |\mathbf{c}_2|^2 = 1$, what has the geometric interpretation that the qubit state is normalized to one and, thus, corresponds only to a direction in a Hilbert space. For example, the measurement of a qubit in the state

$$1/2 \, |0\rangle + \sqrt{3}/2 \, |1\rangle$$

has probability $(\tfrac{1}{2})^2$ of giving '0' as result, and probability $(\sqrt{3}/2)^2$ of giving '1' as result.

In the example representation of a qubit provided in Figure 10.6, when the atom changes energy level it either emits or absorbs light in quanta (packets) of energy, named photons. The movements of electrons are described by *wave-*

functions that provide statistical information about the possible outcomes of an observation, though the actual outcome is not known in advance. Due to the uncertainty principle, an observation (measurement) changes the wavefunction and, as a consequence, the probability distribution of future observations. The outcome is that every measurement results in only one of two states.

10.4.2. Multiple Bits and Qubits

In standard computing, n bits can be used to represent 2^n different states. For instance, a string of 2 bits allows the states 00, 01, 10, and 11. Correspondingly, an n-qubit system has n computational basis states denoted as

$$|y\rangle = \sum_{x=00\ldots0}^{11\ldots1} c_x \, |x\rangle \tag{10.13}$$

where c_x are complex numbers such that $\sum_x |c_x|^2 = 1$. Therefore, n qubits can represent any complex unit vector in a Hilbert space of dimension 2^n, having one dimension for each classical state. That is, any vector in this 2^n-dimensional space can be represented as a linear combination of the basis states. This is an exponential growth of the number of possible states that a quantum system may have in comparison with that of classical systems, what suggests a possible exponential speed-up of computation on quantum computers. An ordered system of n qubits is also known as a *quantum register*.

Quantum states are combined through tensor products, and this is what leads to the exponential growth of the number of possible states. To illustrate the idea, consider the state space for two qubits, each of which with the standard computational basis $\{|0\rangle,|1\rangle\}$. The basis of this state space is $\{|0\rangle\otimes|0\rangle, |0\rangle\otimes|1\rangle, |1\rangle\otimes|0\rangle, |1\rangle\otimes|1\rangle\}$, and can also be represented as $\{|00\rangle, |01\rangle, |10\rangle, |11\rangle\}$. For the standard linear algebra basis in H^2, $|0\rangle = (1,0)^T$ and $|1\rangle = (0,1)^T$, the basis in H^4 becomes:

$$|00\rangle = \begin{bmatrix} 1 \\ 0 \\ 0 \\ 0 \end{bmatrix}, \quad |01\rangle = \begin{bmatrix} 0 \\ 1 \\ 0 \\ 0 \end{bmatrix}, \quad |10\rangle = \begin{bmatrix} 0 \\ 0 \\ 1 \\ 0 \end{bmatrix}, \quad |11\rangle = \begin{bmatrix} 0 \\ 0 \\ 0 \\ 1 \end{bmatrix}$$

10.4.3. Gates and Quantum Gates

The elementary operations that can be used to manipulate bits in standard computers are called *logic functions* (*operations*). A *logic gate* is an electronic device used to perform a simple logic function and to construct larger logic systems. A logic gate may have one or more inputs and a single bit output. Common gates are the NOT, which operates on a single bit, and the AND and OR gates that operate on two or more inputs. Figure 10.8(a) illustrates some typical logic gates and Figure 10.8(b) presents their corresponding truth tables.

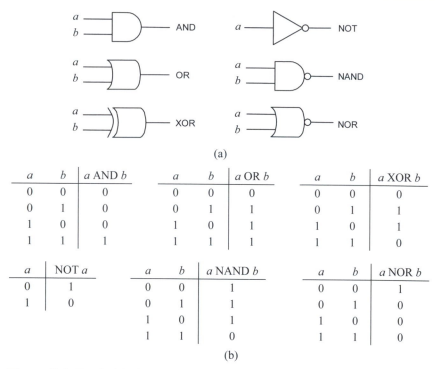

(a)

a	b	a AND b
0	0	0
0	1	0
1	0	0
1	1	1

a	b	a OR b
0	0	0
0	1	1
1	0	1
1	1	1

a	b	a XOR b
0	0	0
0	1	1
1	0	1
1	1	0

a	NOT a
0	1
1	0

a	b	a NAND b
0	0	1
0	1	1
1	0	1
1	1	0

a	b	a NOR b
0	0	1
0	1	0
1	0	0
1	1	0

(b)

Figure 10.8: Standard logic gates: AND, NOT, OR, NAND, XOR, NOR. (a) Pictoric representation. (b) Corresponding truth tables.

There are small, well-defined sets of standard logic gates considered *universal*, meaning that any function can be performed by combining these universal gates. Examples are the NAND gate, the {AND, OR, NOT} set of gates, the {AND, NOT} gates, and others. Most classical gates, such as the OR and XOR gates, have the property that the operations they perform on the inputs are not invertible. This is easy to see by observing that one cannot say for sure what were the input values to these gates by having only their outputs and knowing the transformation they perform on the inputs. This is not the case for *quantum gates*.

Quantum gates, the analogues of logic gates in standard computers, are the basic units of quantum algorithms and, thus, quantum computers. The simplest example of a quantum gate acting on a single bit is the quantum NOT gate. While the classical NOT gate maps $1 \rightarrow 0$ and $0 \rightarrow 1$, the quantum NOT gate maps the state $|0\rangle \rightarrow |1\rangle$ and $|1\rangle \rightarrow |0\rangle$. This operation is suitably represented by matrix $\mathbf{X}$ defined as follows:

$$\text{NOT} |0\rangle = \mathbf{X}|0\rangle = \begin{bmatrix} 0 & 1 \\ 1 & 0 \end{bmatrix}\begin{bmatrix} 1 \\ 0 \end{bmatrix} = \begin{bmatrix} 0 \\ 1 \end{bmatrix} = |1\rangle \qquad (10.14)$$

$$\text{NOT }|1\rangle = \mathbf{X}|1\rangle = \begin{bmatrix} 0 & 1 \\ 1 & 0 \end{bmatrix}\begin{bmatrix} 0 \\ 1 \end{bmatrix} = \begin{bmatrix} 1 \\ 0 \end{bmatrix} = |0\rangle$$

Any unitary matrix transformation may be used as a quantum gate, what implies that, in contrast to standard gates, quantum gates are *reversible* (*invertible*). The unitary constraint comes from the fact that an arbitrary quantum state $|\mathbf{y}'\rangle = \mathbf{c}_1'|0\rangle + \mathbf{c}_2'|1\rangle$, which is a result of the application of a quantum gate to the state $|\mathbf{y}\rangle = \mathbf{c}_1|0\rangle + \mathbf{c}_2|1\rangle$, has to satisfy the condition that $|\mathbf{c}_1'|^2 + |\mathbf{c}_2'|^2 = 1$. The reversible nature of quantum gates means that, for instance, if $|\mathbf{y}'\rangle$ is the result of $\mathbf{X}|\mathbf{y}\rangle$, then one can fully determine $|\mathbf{y}\rangle$ by applying $\mathbf{X}^\dagger$ (the adjoint of $\mathbf{X}$) to $|\mathbf{y}'\rangle$.

The Pauli matrices, presented in Equation (10.6), correspond to well known quantum gates. For instance, matrix $\mathbf{I}$ is the so-called *identity gate* and matrix $\mathbf{X}$ is the *NOT gate* presented above. The identity gate is so-called because it maintains the current state of the system, and the $\mathbf{Z}$ gate leaves $|0\rangle$ unchanged and maps $|1\rangle \rightarrow -|1\rangle$. When exponentiated, the Pauli matrices also give rise to three important classes of unitary transformations, known as the *rotation operators*, that rotate the state about the x, y and z axes:

$$\mathbf{R}_x(\theta) \equiv e^{-i\theta\mathbf{X}/2} = \begin{bmatrix} \cos\dfrac{\theta}{2} & -i\sin\dfrac{\theta}{2} \\ -i\sin\dfrac{\theta}{2} & \cos\dfrac{\theta}{2} \end{bmatrix}$$

$$\mathbf{R}_y(\theta) \equiv e^{-i\theta\mathbf{Y}/2} = \begin{bmatrix} \cos\dfrac{\theta}{2} & -\sin\dfrac{\theta}{2} \\ \sin\dfrac{\theta}{2} & \cos\dfrac{\theta}{2} \end{bmatrix} \tag{10.15}$$

$$\mathbf{R}_z(\theta) \equiv e^{-i\theta\mathbf{Z}/2} = \begin{bmatrix} e^{-i\theta/2} & 0 \\ 0 & e^{i\theta/2} \end{bmatrix}$$

There is another common gate known as the *phase* or *shift* gate represented by matrix, $\mathbf{S}$, which also leaves $|0\rangle$ unchanged, but maps $|1\rangle \rightarrow i|1\rangle$:

$$\mathbf{S} = \begin{bmatrix} 1 & 0 \\ 0 & i \end{bmatrix} \tag{10.16}$$

D. Deutsch (1992) introduced the square-root of the NOT gate as

$$\sqrt{\text{NOT}}\,|0\rangle = \tfrac{1}{2}(1+i)|0\rangle + \tfrac{1}{2}(1-i)|1\rangle$$

$$\sqrt{\text{NOT}}\,|1\rangle = \tfrac{1}{2}(1-i)|0\rangle + \tfrac{1}{2}(1+i)|1\rangle$$

$$\sqrt{\text{NOT}} = \frac{1}{2}\begin{bmatrix} 1+i & 1-i \\ 1-i & 1+i \end{bmatrix} \tag{10.17}$$

$$\sqrt{\text{NOT}} \cdot \sqrt{\text{NOT}} = \text{NOT}$$

Another transformation that is commonly used in quantum computing is the *Hadamard gate*

$$H = \frac{1}{\sqrt{2}} \begin{bmatrix} 1 & 1 \\ 1 & -1 \end{bmatrix} \tag{10.18}$$

When applied to n bits individually, the Hadamard gate H generates a superposition of all 2^n possible states. For example, when applied to the basis states, the Hadamard gate creates a superposition of states.

$$H|0\rangle = \frac{1}{\sqrt{2}} \begin{bmatrix} 1 & 1 \\ 1 & -1 \end{bmatrix} \begin{bmatrix} 1 \\ 0 \end{bmatrix} = \frac{1}{\sqrt{2}} \begin{bmatrix} 1 \\ 1 \end{bmatrix} = \frac{1}{\sqrt{2}} (|0\rangle + |1\rangle)$$

$$H|1\rangle = \frac{1}{\sqrt{2}} \begin{bmatrix} 1 & 1 \\ 1 & -1 \end{bmatrix} \begin{bmatrix} 0 \\ 1 \end{bmatrix} = \frac{1}{\sqrt{2}} \begin{bmatrix} 1 \\ -1 \end{bmatrix} = \frac{1}{\sqrt{2}} (|0\rangle - |1\rangle)$$

$$\tag{10.19}$$

Similarly to standard gates, quantum gates can also act upon more than one input. A transformation commonly applied on $H^2 \otimes H^2$ is the *controlled-NOT gate*, **CNOT**, defined as

$$\mathbf{CNOT} = \begin{bmatrix} 1 & 0 & 0 & 0 \\ 0 & 1 & 0 & 0 \\ 0 & 0 & 0 & 1 \\ 0 & 0 & 1 & 0 \end{bmatrix} \tag{10.20}$$

The **CNOT** gate works as follows. Given an input state $|x_1 x_2\rangle$, $x_1, x_2 \in \{0,1\}$, the output produced by the **CNOT** gate is $|x_1 x_3\rangle$, $x_3 = x_1 \oplus x_2$, where $\oplus$ is addition modulo two, which corresponds to the logical exclusive-OR operation.

The **CNOT** gate is so named because the first qubit is known as the *control qubit* and the second one is the *target qubit*. Whenever the control qubit is set to 0, the target qubit does not change. By contrast, when the control qubit is set to 1, the target qubit is flipped, as illustrated below:

$$\mathbf{CNOT} |00\rangle = |00\rangle$$
$$\mathbf{CNOT} |01\rangle = |01\rangle$$
$$\mathbf{CNOT} |10\rangle = |11\rangle$$
$$\mathbf{CNOT} |11\rangle = |10\rangle$$

$$\tag{10.21}$$

An important feature of the controlled-NOT gate is that it can be used to create entanglement. Take, for instance, the case illustrated below

$$\mathbf{CNOT} \frac{1}{\sqrt{2}} (|00\rangle + |10\rangle) = \frac{1}{\sqrt{2}} (|00\rangle + |11\rangle)$$

Transformation **CNOT** is clearly unitary, since $\mathbf{CNOT} = \mathbf{CNOT}^{\dagger}$.

Similarly to the **CNOT** transformation, it is possible to define the controlled-controlled-NOT (**CCNOT**) gate that operates on three qubits. **CCNOT** negates the rightmost bit if and only if the first two bits are equal to 1 simultaneously. This gate is called the *Toffoli gate* (Toffoli, 1980):

$$\mathbf{CCNOT}\,|000\rangle = |000\rangle$$
$$\mathbf{CCNOT}\,|001\rangle = |001\rangle$$
$$\mathbf{CCNOT}\,|010\rangle = |010\rangle$$
$$\mathbf{CCNOT}\,|011\rangle = |011\rangle$$
$$\mathbf{CCNOT}\,|100\rangle = |100\rangle$$
$$\mathbf{CCNOT}\,|101\rangle = |101\rangle$$
$$\mathbf{CCNOT}\,|110\rangle = |111\rangle$$
$$\mathbf{CCNOT}\,|111\rangle = |110\rangle$$

$$\mathbf{CCNOT} = \begin{bmatrix} 1 & 0 & 0 & 0 & 0 & 0 & 0 & 0 \\ 0 & 1 & 0 & 0 & 0 & 0 & 0 & 0 \\ 0 & 0 & 1 & 0 & 0 & 0 & 0 & 0 \\ 0 & 0 & 0 & 1 & 0 & 0 & 0 & 0 \\ 0 & 0 & 0 & 0 & 1 & 0 & 0 & 0 \\ 0 & 0 & 0 & 0 & 0 & 1 & 0 & 0 \\ 0 & 0 & 0 & 0 & 0 & 0 & 0 & 1 \\ 0 & 0 & 0 & 0 & 0 & 0 & 1 & 0 \end{bmatrix} \tag{10.22}$$

Another important transformation that operates on three bits (or qubits) is the *Fredkin gate*, $\mathbf{F_R}$, which works as a type of *controlled swap gate* (Fredkin and Toffoli, 1982). As with the **CNOT** gate, this gate has a control line (bit or qubit) whose value dictates the output of the gate. If the control line is set, then the other two values are swapped. Assuming the first line is the control line, the Fredkin gate works as follows:

$$\mathbf{F_R}\,|000\rangle = |000\rangle$$
$$\mathbf{F_R}\,|001\rangle = |001\rangle$$
$$\mathbf{F_R}\,|010\rangle = |010\rangle$$
$$\mathbf{F_R}\,|011\rangle = |011\rangle$$
$$\mathbf{F_R}\,|100\rangle = |100\rangle$$
$$\mathbf{F_R}\,|101\rangle = |110\rangle$$
$$\mathbf{F_R}\,|110\rangle = |101\rangle$$
$$\mathbf{F_R}\,|111\rangle = |111\rangle$$

$$\mathbf{F_R} = \begin{bmatrix} 1 & 0 & 0 & 0 & 0 & 0 & 0 & 0 \\ 0 & 1 & 0 & 0 & 0 & 0 & 0 & 0 \\ 0 & 0 & 1 & 0 & 0 & 0 & 0 & 0 \\ 0 & 0 & 0 & 1 & 0 & 0 & 0 & 0 \\ 0 & 0 & 0 & 0 & 1 & 0 & 0 & 0 \\ 0 & 0 & 0 & 0 & 0 & 0 & 1 & 0 \\ 0 & 0 & 0 & 0 & 0 & 1 & 0 & 0 \\ 0 & 0 & 0 & 0 & 0 & 0 & 0 & 1 \end{bmatrix} \tag{10.23}$$

Generalizations of the Hadamard Gate

The application of the Hadamard gate to a generic state and to n input states is very important for the development of several quantum algorithms. In order to find an expression for the application of the Hadamard gate to a generic state $|x\rangle$, rewrite the result of its application to $|0\rangle$ and $|1\rangle$ presented in Equation (10.19), as follows:

$$\mathbf{H}\,|0\rangle = \frac{1}{\sqrt{2}}(|0\rangle + |1\rangle) = \frac{1}{\sqrt{2}}\left((-1)^{0 \cdot 0}\,|0\rangle + (-1)^{0 \cdot 1}\,|1\rangle\right)$$
$$\mathbf{H}\,|1\rangle = \frac{1}{\sqrt{2}}(|0\rangle - |1\rangle) = \frac{1}{\sqrt{2}}\left((-1)^{1 \cdot 0}\,|0\rangle - (-1)^{1 \cdot 1}\,|1\rangle\right) \tag{10.24}$$

What results in

$$\mathbf{H}\,|\mathbf{x}\rangle = \frac{1}{\sqrt{2}}\sum_{y=0}^{1}(-1)^{\mathbf{x} \cdot \mathbf{y}}\,|\mathbf{y}\rangle \tag{10.25}$$

The parallel application of the Hadamard gate on a system of n qubits, represented by $\mathbf{H}^{\otimes n}\,|0\rangle^{\otimes n}$, prepared in the $|0\rangle$ state results in

$$\mathbf{H}^{\otimes n}\,|\,0\rangle^{\otimes n} = \frac{1}{2^{n/2}}\sum_{x=0}^{2^{n}-1}|\,\mathbf{x}\rangle \tag{10.26}$$

which is a superposition of all integers from 0 to $2^{n}-1$. Note that the Hadamard gate leads to a superposition of all computational basis states.

Finally, the application of Equation (10.25) to the tensor product of n arbitrary qubits results in Equation (10.27).

$$\mathbf{H}\,|\,\mathbf{x}_1\rangle \otimes \mathbf{H}\,|\,\mathbf{x}_2\rangle \otimes \dots \otimes \mathbf{H}\,|\,\mathbf{x}_n\rangle = \frac{1}{2^{n/2}}\sum_{y=0}^{2^{n}-1}(-1)^{\mathbf{x}\cdot\mathbf{y}}\,|\,\mathbf{y}\rangle \tag{10.27}$$

where $\mathbf{x}\cdot\mathbf{y}$ denotes the bitwise AND (or mod 2 scalar product): $\mathbf{x}\cdot\mathbf{y} = (x_1 \wedge y_1) \oplus (x_2 \wedge y_2) \oplus \dots \oplus (x_n \wedge y_n)$.

10.4.4.   Quantum Circuits

Graphical descriptions of a quantum transformation, called *quantum circuits*, are a useful means of representing and combining quantum state transformations (quantum gates). The notation is similar to those of many layouts of classical circuits: information flow is from left to right; horizontal lines represent wires; vertical lines connecting two or more qubits represent coupling terms; initial conditions are represented by quantum states written to the left of the circuit; and squared boxes together with special symbols placed on the horizontal lines correspond to gates.

To illustrate the main features of a quantum circuit, let us first consider the quantum circuit associated with the controlled-NOT gate (Figure 10.9). Each line in the circuit corresponds to a *wire*, and the direction for reading the circuit is from left to right. In the controlled-NOT gate circuit, the top wire is associated with the control qubit and the bottom wire is related to the target qubit.

Among the many possibilities of combining the controlled-NOT circuit, a useful circuit is the one that exchanges the states of the two input qubits, as illustrated in Figure 10.10.

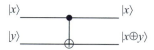

Figure 10.9: Quantum circuit for the controlled-NOT gate.

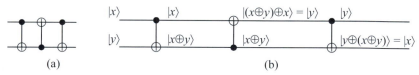

Figure 10.10: Quantum circuit that exchanges the states of the input qubits. (a) Circuit representation. (b) Sequence of steps on a computational basis $|x,y\rangle$.

Figure 10.11: Simplified swap circuit.

The representation of the bit swap circuit can be simplified, as illustrated in Figure 10.11. The controlled-controlled-NOT gate can also be represented by a quantum circuit, as depicted in Figure 10.12. Note that, in this case, the first two qubits control the output of the circuit.

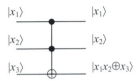

Figure 10.12: Quantum circuit for the Toffoli gate.

By combining multiple Toffoli gates the whole set of Boolean logic gates can be simulated. For instance, by setting the control bits $x_1 = x_2 = 1$, the Toffoli (**CCNOT**) gate simulates the NOT gate for input x_3: $\mathbf{CCNOT}|1,1,x_3\rangle = |1,1,\neg x_3\rangle$; and by setting $x_3 = 0$, **CCNOT** simulates the AND gate between x_1 and x_2: $\mathbf{CCNOT}|x_1,x_2,0\rangle = |x_1,x_2,x_1 \wedge x_2\rangle$.

The Fredkin gate can be represented by a quantum circuit as depicted in Figure 10.13. In this representation, the first qubit controls the output of the circuit: if $x_1 = 1$, then $x_2' = x_3$ and $x_3' = x_2$. The Fredkin gate is a reversible universal logic gate, and can be used to simulate the AND and NOT functions, among others.

Some particular gates can be represented by a simple circuit composed of input/output wires and an appropriately labeled box. For instance, the Hadamard gate can be represented by a box with an **H** inside, the phase shift gate can be represented by a box with an **S** inside, and the Toffoli gate can be represented by a box with a **T** inside, as illustrated in Figure 10.14.

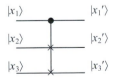

Figure 10.13: Quantum circuit for the Fredkin gate.

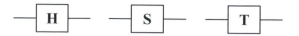

Figure 10.14: Quantum circuits for the Hadamard (**H**), shift (**S**), and Toffoli's (**T**) gate.

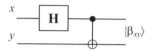

Figure 10.15: Quantum circuit to generate Bell states.

The Bell states, also called *EPR states* or *EPR pairs* were presented in Equation (10.4). They can be generated by a combination of a Hadamard gate to the first qubit followed by a controlled-NOT gate, as illustrated in Figure 10.15. The Hadamard gate generates a superposition of the top qubit, which is used as a control qubit to the **CNOT** gate, as summarized in Table 10.1.

Table 10.1: Application of the Hadamard gate to the first qubit followed by the application of the **CNOT** gate using the output of the Hadamard gate as the control qubit.

Initial state	H(first qubit, x)	CNOT(first qubit, x)
$|00\rangle$	$1/\sqrt{2}\,(|0\rangle+|1\rangle)|0\rangle$	$1/\sqrt{2}\,(|00\rangle+|11\rangle)$
$|01\rangle$	$1/\sqrt{2}\,(|1\rangle+|0\rangle)|1\rangle$	$1/\sqrt{2}\,(|10\rangle+|01\rangle)$
$|10\rangle$	$1/\sqrt{2}\,(|0\rangle-|1\rangle)|0\rangle$	$1/\sqrt{2}\,(|00\rangle-|11\rangle)$
$|11\rangle$	$1/\sqrt{2}\,(|0\rangle-|1\rangle)|1\rangle$	$1/\sqrt{2}\,(|01\rangle-|10\rangle)$

10.4.5.   Quantum Parallelism

As discussed previously, the Toffoli gate is a universal gate and can, thus, be used to implement any arbitrary classical function f with m input bits and k output bits, $f(\mathbf{x}) : \{0,1\}^{m} \rightarrow \{0,1\}^{k}$. Assume the existence of an $m+k$-bit *quantum gate array* $\mathbf{U}_f$ (Figure 10.16) that implements f as follows $\mathbf{U}_f :$ $|\mathbf{x}_1 \otimes \mathbf{x}_2\rangle \rightarrow |\mathbf{x}_1 \otimes (\mathbf{x}_2 \oplus f(\mathbf{x}_1))\rangle$. To compute $f(\mathbf{x}_1)$ it is necessary to apply $\mathbf{U}_f$ to $|\mathbf{x}_1\rangle$ tensored with k zeros $|\mathbf{x}_1 \otimes \mathbf{0}\rangle$. This transformation flips the second qubit if $f(\cdot)$ acting on the first qubit is 1, and does nothing if $f(\cdot)$ acting on the first qubit is 0.

When the transformation $\mathbf{U}_f$ is applied to a superposed state, it is applied to all basis vectors of the superposition simultaneously and will generate a superposition of the result. Therefore, it is possible to compute $f(\mathbf{x}_i)$, $i = 1, \ldots, n$, simultaneously with a single application of $\mathbf{U}_f$; a phenomenon called *quantum parallelism*.

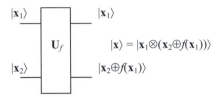

Figure 10.16: Graphical representation of the quantum gate array $\mathbf{U}_f$.

To illustrate quantum parallelism, assume $|\mathbf{x}_1\rangle = 1/\sqrt{2}(|0\rangle - |1\rangle)$, which can be obtained by applying a Hadamard gate on $|1\rangle$. For $|\mathbf{x}_2\rangle = 0$, the application of $\mathbf{U}_f$ on $|\mathbf{x}_1\rangle = 1/\sqrt{2}(|0\rangle - |1\rangle)$ results in

$$1/\sqrt{2}\,(|0\otimes f(0)\rangle - |1\otimes f(1)\rangle) \tag{10.28}$$

What is interesting about the state presented in Equation (10.23) is the fact that its first term contains information about $f(0)$ and its second term contains information about $f(1)$. Thus, a single application of $\mathbf{U}_f$ allows us to evaluate $f(\mathbf{x})$ for two values of $\mathbf{x}$ simultaneously.

10.4.6. Examples of Applications

Dense coding and *teleportation* serve to illustrate the use of simple quantum gates. Both processes have the same initial scenario: Alice and Bob have never communicated and now wish to communicate. To do so, they use an apparatus that generates entangled pairs of qubits (Bell states), such as:

$$\beta_{00} = |\,\mathbf{x}_0\rangle = \frac{1}{\sqrt{2}}(|\,00\rangle + |\,11\rangle) \tag{10.29}$$

One qubit is sent to Alice and another to Bob. Alice can only perform transformations on her qubit and Bob can only perform transformations on his until a particle is transmitted.

Dense Coding

Dense coding corresponds to the method by which Alice can communicate two classical bits by sending a single qubit to Bob (Bennett and Wiesner, 1992). Two quantum bits are involved, but Alice only sees one of them. She receives two classical bits and encodes the numbers 0, 1, 2, and 3. These numbers will correspond to one of the unitary transformations $\{\mathbf{I}, \mathbf{X}, \mathbf{Y}, \mathbf{Z}\}$, presented in Equation (10.30), that will act on Alice's first bit of the entangled pair $\mathbf{x}_0$.

$$\mathbf{I} = \begin{bmatrix} 1 & 0 \\ 0 & 1 \end{bmatrix} \qquad \mathbf{X} = \begin{bmatrix} 0 & 1 \\ 1 & 0 \end{bmatrix} \tag{10.30}$$

$$Y = \begin{bmatrix} 0 & 1 \\ -1 & 0 \end{bmatrix} \qquad Z = \begin{bmatrix} 1 & 0 \\ 0 & -1 \end{bmatrix}$$

The result of applying these transformations to the entangled pair is presented in Table 10.2.

Table 10.2: Application of transformations $\{I,X,Y,Z\}$ to the initial entangled state $|x_0\rangle$.

Encoded value	Initial state	Transformation	New state
0		$(I \otimes I)\,|x_0\rangle$	$1/\sqrt{2}\,(|00\rangle + |11\rangle)$
1	$|x_0\rangle = \dfrac{1}{\sqrt{2}}(|00\rangle + |11\rangle)$	$(X \otimes I)\,|x_0\rangle$	$1/\sqrt{2}\,(|10\rangle + |01\rangle)$
2		$(Y \otimes I)\,|x_0\rangle$	$1/\sqrt{2}\,(-|10\rangle + |01\rangle)$
3		$(Z \otimes I)\,|x_0\rangle$	$1/\sqrt{2}\,(|00\rangle - |11\rangle)$

Bob then applies a **CNOT** gate to the resulting entangled pairs obtaining a set of separable (decomposable) pairs. Bob can now measure the second qubit without disturbing the quantum state. If the measurement results $|0\rangle$, then the encoded value is either 0 or 3, otherwise it is either 1 or 2. To finally identify a unique encoded value, Bob applies a Hadamard gate (Equation (10.18)) to the first qubit resulting from the application of the **CNOT** gate. If the results of the first and second measurements are $|0\rangle$ and $|0\rangle$, respectively, then the encoded value is 0; if the results are $|0\rangle$ and $|1\rangle$, respectively, then the encoded value is 3; and so on. This part of the process is summarized in Table 10.3.

Dense coding can be performed by the circuit presented in Figure 10.17. On the left hand side of the picture the circuit to generate EPR pairs (Figure 10.15) is presented, followed by the transformation to be applied to the classical bits, **CNOT** gate, Hadamard gate and two measurement operators.

Table 10.3: Application of the **CNOT** gate to the input entangled state $|x_i\rangle$ and then application of the Hadamard gate to the first qubit of the resulting state.

Initial state	CNOT $|x_i\rangle$	H(first qubit)
$|x_0\rangle = 1/\sqrt{2}\,(|00\rangle + |11\rangle)$	$1/\sqrt{2}\,(|00\rangle + |10\rangle) = 1/\sqrt{2}\,(|0\rangle + |1\rangle)|0\rangle$	$|0\rangle$
$|x_1\rangle = 1/\sqrt{2}\,(|10\rangle + |01\rangle)$	$1/\sqrt{2}\,(|11\rangle + |01\rangle) = 1/\sqrt{2}\,(|1\rangle + |0\rangle)|1\rangle$	$|0\rangle$
$|x_2\rangle = 1/\sqrt{2}\,(-|10\rangle + |01\rangle)$	$1/\sqrt{2}\,(-|11\rangle + |01\rangle) = 1/\sqrt{2}\,(-|1\rangle + |0\rangle)|1\rangle$	$|1\rangle$
$|x_3\rangle = 1/\sqrt{2}\,(|00\rangle - |11\rangle)$	$1/\sqrt{2}\,(|00\rangle - |10\rangle) = 1/\sqrt{2}\,(|0\rangle - |1\rangle)|0\rangle$	$|1\rangle$

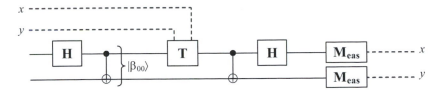

Figure 10.17: Quantum circuit to perform dense coding. $T \in \{I,X,Y,Z\}$.

Quantum Teleportation

In our Alice-Bob scenario, assume Alice wants to communicate an unknown qubit $|x\rangle$ to Bob. Due to the no-cloning theorem, quantum states cannot be copied and transmitted without being destroyed. Furthermore, if $|x\rangle$ is unknown any measurement she performs on it may change its state. Thus, the only way to transmit $|x\rangle$ to Bob seems to be by sending him a physical qubit. *Quantum teleportation* is a way around this apparent limitation (Bennett et al., 1993). It corresponds to the process by which Alice wants to communicate a single, unknown, qubit $|x\rangle$ to Bob, by sending him only classical information. Thus, the goal is to transmit the quantum state of a particle using classical bits (a classical channel) and reconstruct the state at the receiver.

The whole process can be summarized as follows. Alice and Bob are both in possession of one qubit of an EPR pair. Assume Alice controls the first half of the EPR pair and Bob the second half. Alice interacts the qubit to be transmitted $|x\rangle$ with her half of the EPR pair, and then measures the two qubits she holds, obtaining one out of the four possible classical results $\{00,01,10,11\}$. This information is then sent through a classical channel to Bob, who performs one out of four operations $\{I,X,Y,Z\}$ on his half of the EPR pair, what allows him to recover the transmitted information $|x\rangle$.

Let $|x\rangle = c_1|0\rangle + c_2|1\rangle$, $|c_1|^2 + |c_2|^2 = 1$, be the unknown quantum state Alice wants to communicate with Bob, and $1/\sqrt{2}(|00\rangle + |11\rangle)$ the EPR pair they possess. The input state to the system is the tensor product between $|x\rangle$ and $|x_0\rangle = 1/\sqrt{2}(|00\rangle + |11\rangle)$:

$$|x_0\rangle|x\rangle = (c_1|0\rangle + c_2|1\rangle)1/\sqrt{2}(|00\rangle + |11\rangle) =$$

$$1/\sqrt{2}(c_1|0\rangle(|00\rangle + |11\rangle) + c_2|1\rangle(|00\rangle + |11\rangle)) = \qquad (10.31)$$

$$1/\sqrt{2}(c_1|000\rangle + c_1|011\rangle + c_2|100\rangle + c_2|111\rangle).$$

Alice now applies the decoding step of dense coding - a controlled-NOT gate followed by a Hadamard gate - to the unknown qubit (first qubit) and to her member of the entangled pair; what corresponds to the transformation $(H \otimes I \otimes I) \cdot (CNOT \otimes I)$:

Table 10.4: Bits received, their corresponding quantum state, and the transformation Bob has to apply to the received qubit in order to place it in the transmitted state $|\mathbf{x}\rangle = \mathbf{c}_1|0\rangle + \mathbf{c}_2|1\rangle$.

Bits received	State	Transformation
00	$\mathbf{c}_1|0\rangle + \mathbf{c}_2|1\rangle$	**I**
01	$\mathbf{c}_1|1\rangle + \mathbf{c}_2|0\rangle$	**X**
10	$\mathbf{c}_1|0\rangle - \mathbf{c}_2|1\rangle$	**Z**
11	$\mathbf{c}_1|1\rangle - \mathbf{c}_2|0\rangle$	**Y**

$$(\mathbf{H}\otimes\mathbf{I}\otimes\mathbf{I})(\mathbf{CNOT}\otimes\mathbf{I})(|\mathbf{x}_0\rangle|\mathbf{x}\rangle) =$$

$$(\mathbf{H}\otimes\mathbf{I}\otimes\mathbf{I})\Big[1/\sqrt{2}\,\big(\mathbf{c}_1\,|000\rangle+\mathbf{c}_1\,|011\rangle+\mathbf{c}_2\,|110\rangle+\mathbf{c}_2\,|101\rangle\big)\Big] \;=$$

$$(\mathbf{H}\otimes\mathbf{I}\otimes\mathbf{I})\Big[1/\sqrt{2}\,\big(\mathbf{c}_1\,|0\rangle(|00\rangle+|11\rangle)+\mathbf{c}_2\,|1\rangle(|10\rangle+|01\rangle)\big)\Big] \;= \qquad (10.32)$$

$$1/2\big[\mathbf{c}_1(|0\rangle+|1\rangle)(|00\rangle+|11\rangle)+\mathbf{c}_2(|0\rangle-|1\rangle)(|10\rangle+|01\rangle)\big] \;=$$

$$1/2\big[|00\rangle(\mathbf{c}_1\,|0\rangle+\mathbf{c}_2\,|1\rangle)+|01\rangle(\mathbf{c}_1\,|1\rangle+\mathbf{c}_2\,|0\rangle)+$$
$$+|10\rangle(\mathbf{c}_1\,|0\rangle-\mathbf{c}_2\,|1\rangle)+|11\rangle(\mathbf{c}_1\,|1\rangle-\mathbf{c}_2\,|0\rangle)\big]$$

Alice then measures her two qubits, collapsing the state onto one of the four different possibilities {00,01,10,11}. This result is sent to Bob, who uses them to find out which of the transformations {**I**,**X**,**Y**,**Z**} he must apply to his qubit in order to place it in the superposed state $|\mathbf{x}\rangle = \mathbf{c}_1|0\rangle + \mathbf{c}_2|1\rangle$, as summarized in Table 10.4.

The whole process can be performed by the quantum teleportation circuit presented in Figure 10.18, where gate **T** corresponds to one of the transformations {**I**,**X**,**Y**,**Z**}.

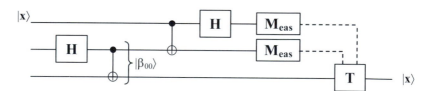

Figure 10.18: Quantum circuit to perform teleportation.

A Note on Teleportation

Teleportation, or *teletransportation*, is a term from science fiction that corresponds to the process of moving objects from one place to another by encoding information about the objects, transmitting this information to a different place, and then creating a copy of the original objects in the new location. Quantum teleportation is so named because it brings some features in common with the science fiction notion of teleportation: 1) it 'encodes' quantum bits as classical bits; 2) the information to be transmitted is destroyed in the process (since measurement places the transmitted qubit in one of the basis states associated with the measurement device); 3) and then the transmitted qubit is reconstructed from the classical bits and the EPR pair. The quantum information received is complete, for it is the complete description of the transmitted qubit.

10.5 UNIVERSAL QUANTUM COMPUTERS

The development of quantum computing theory has been marked by several attempts to formally describe a Universal Quantum Computer Q. One of the first attempts was made by P. Benioff (1980) with his proposal of building a quantum mechanical model of a Turing machine. He showed that a reversible unitary transformation was enough to realize a Turing machine. Later on, R. Feynman (1982, 1986) introduced a universal quantum simulator based on quantum analogues of logic circuits (i.e., quantum circuits). He was the first to suggest that quantum mechanics may be more computationally powerful than Turing machines. It is attributed to D. Deutsch (1985), however, the proposal of the first true model of a quantum computer; a quantum Turing machine.

10.5.1. Benioff's Computer

P. Benioff (1980, 1982) introduced a microscopic quantum mechanical model of Turing machines. To design this model, he assembled together specific representations for the main components of a Turing machine (TM). In the description of a TM presented in Appendix B.3.2, its main components are a *tape*, a *head*, a *state register*, and an *instruction table* (*program*). Benioff described a Turing machine as the union of three main components: a *tape T*, a head *H*, and a machine *M*; TM = $T \cup H \cup M$. Thus, this representation is related to the one presented in Appendix B.3.2 by assuming that *M* includes the state register and the program.

He represented the tape as a lattice of quantum spin systems, qubits, of finite length. The tape alphabet was assumed to be in a one to one correspondence with the set of possible spin projections that each of the spin systems is capable of assuming. An expression of length *n* (e.g., a string of *n* symbols) thus corresponded to a lattice system of length *n*, which was used to represent all tape expressions of length *n* or less.

The machine head was represented by a single spinless system that moved along the tape lattice, and the machine corresponded to a single system of spin fixed at a given lattice position. The spin system should be large enough so as to allow a full description of the configuration of a Turing machine. The quantum states of the system were assumed to lie in a Hilbert space of dimension 2, H^2, and were in a one to one correspondence with the spin projections of the system.

The state change of the system was performed by unitary operators applied to the spin projections.

One step of computation was performed by measuring an observable and then, depending on the value found, carrying out a given operation described by a unitary transformation. The measurement at the end of a step promoted a collapse of the quantum state onto one of the basis states. The computation was thus performed in steps of fixed duration, such that at the end of each step the tape was back in one of the basis states, but during a computational step the machine could be in a superposition of spin states. As a result, quantum superposition (interference) was only partially used and a classical Turing machine could be employed to simulate such computer.

10.5.2.   Feynman's Computer

Feynman (1982) starts his work discussing the main feature of universal computers: it does not matter how it is manufactured, anything computable should be computed by a universal computer. The question he posed was "can physics be simulated by a universal computer?". Knowing that the physical world is quantum mechanical, this problem becomes that of simulating quantum physics. He suggested that the best form of tackling this problem is by designing a quantum mechanical computer obeying quantum mechanical laws.

Feynman (1986) introduced his universal quantum mechanical computer by proposing some quantum analogues of classical logic gates. He considered the two logical primitives FAN OUT and EXCHANGE in addition to the AND and NOT gates, as illustrated in Figure 10.19. The FAN OUT creates a copy of the input and the EXCHANGE crosses the inputs.

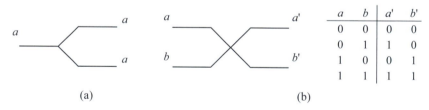

a	b	a'	b'
0	0	0	0
0	1	1	0
1	0	0	1
1	1	1	1

(a)   (b)

Figure 10.19: Additional primitive elements: (a) FAN OUT; and (b) EXCHANGE.

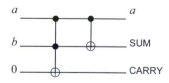

Figure 10.20: Circuit for an Adder.

Feynman (1986) then described three reversible primitives that could be used to make a universal machine, the first one being the NOT gate. The other primitive he introduced was the CONTROLLED NOT gate, which can be used to realize FAN OUT by setting its second wire to '0'. The EXCHANGE operation can be performed by the circuit presented in Figure 10.10. The CONTROLLED CONTROLLED NOT gate was introduced so as to make the set of gates complete; that is, so that any arbitrary logical function could be simulated by combining these circuits. To illustrate, he showed how to use these circuits to make an Adder, as depicted in Figure 10.20. Note that, in all these cases, only classical inputs are being discussed here, meaning that these gates can be used to compute with quantum information and also with classical bits.

To design the universal quantum computer, Feynman assumed a two-state system, called an *atom*, and an *n*-bit number, called a *register*, represented by a state of a set of *n* two-state systems. The state of an atom was represented using the bracket notation, $|0\rangle$ and $|1\rangle$, and the change of state in a system was performed by applying reversible operations on it, following Equation (10.5). He proposed four basic operators: 1) an operator *a* that *annihilates* if the point is occupied; 2) an operator *a** (the conjugate of *a*) that does the opposite, i.e., *creates* if the point is unoccupied; 3) an operator *n*, *number*, that asks if there is something there; 4) and an operator **I** corresponding to the identity operator to complete the set:

$$a = \begin{bmatrix} 0 & 0 \\ 1 & 0 \end{bmatrix}, \quad a^* = \begin{bmatrix} 0 & 1 \\ 0 & 0 \end{bmatrix}, \quad n = \begin{bmatrix} 1 & 0 \\ 0 & 0 \end{bmatrix}, \quad \mathbf{I} = \begin{bmatrix} 1 & 0 \\ 0 & 1 \end{bmatrix} \quad (10.33)$$

10.5.3. Deutsch's Computer

Deutsch (1985) proposed a reinterpretation of the Church-Turing thesis in which the computable functions are those that, in principle, can be computed by a natural system, in particular, a real physical system. To do so, he defined a computing machine $\aleph$ capable of perfectly simulating a physical system $\wp$, and stated a physical version of the Church-Turing thesis:

> "Every finitely realizable physical system can be perfectly simulated by a universal model computing machine operating by finite means." (Deutsch, 1985; p. 99)

According to Deutsch (1985), this formulation is better defined and more physical than Turing's proposal, because it refers exclusively to objective concepts such as measurement, preparation and physical system. In this proposal, a machine that operates by finite means is the one that proceeds in a sequence of steps. The reinterpretation of the classical Church-Turing thesis is stronger than the original one in the sense that it is not satisfied by a Turing machine τ in classical physics. This is because of the continuity of the classical dynamics, which is in contrast to the discreteness of τ. Furthermore, there is no reason why physical laws should respect the limitations of the functions (algorithms) that can be computed by a Turing machine.

Deutsch (1985) then introduced a universal quantum computer Q capable of simulating every finite, realizable physical system, including all other instances of quantum computers and simulators. His model quantum computer is composed of a finite *processor* and an infinite *memory* (*tape*), of which only a portion is used. The processor consists of M 2-state observables, $\{\check{n}_i\}$, $i \in Z^M$ (Z^M is the set of integers from 0 to $M - 1$), and a tape of infinite length of 2-state observables $\{\check{s}_i\}$, $i \in Z$. The head position of a Turing machine is associated with an observable $\check{o} \in Z$. The state of a universal quantum computer Q is defined as a unit vector in the Hilbert space H spanned by the eigenvectors of $\check{o}$, $\check{n}$, and $\check{s}$ labeled by the corresponding eigenvalues o, $\mathbf{n}$, and $\mathbf{s}$, called the *computational basis states* of Q:

$$|o;\mathbf{n};\mathbf{s}\rangle = |\, o;\, n_0,\, n_1,\, \ldots,\, n_{M-1};\, \ldots,\, s_{-1},\, s_0,\, s_1,\ldots\rangle \qquad \textbf{(10.34)}$$

The dynamics of Q is controlled by a constant unitary operator $\mathbf{A} \in H$, which specifies the evolution of any state $|\mathbf{x}\rangle \in H$, following Equation (10.5).

The universal quantum computer Q can simulate not only any Turing machine, but also any other quantum computer with arbitrary precision. Furthermore, Q can simulate several physical systems, real and theoretical, which are beyond the scope of the universal Turing machine.

Some important properties of the quantum computer Q is that two consecutive states of Q can never be identical after a non-trivial computation, Q must not be observed before the computation ends (for this would alter its relative state), and the dynamics of Q is necessarily reversible (due to the unitary condition of the transformation $\mathbf{A}$).

10.6 QUANTUM ALGORITHMS

The design of novel computing paradigms always raises the question as to how efficient they are. To demonstrate the efficiency of quantum computers, several authors proposed quantum algorithms capable of solving test and real world problems more efficiently than any classical computer. Quantum algorithms can be constructed using quantum gates. In the literature, the word *oracle* has been used to refer to a circuit for determining the property of a function.

One of the oldest and simplest quantum algorithms that demonstrated a superior power of quantum computers when compared to standard computers was proposed by Deutsch (1985) and later generalized by Deutsch and JoSza (1992). Although the Deutsch and Jozsa algorithm was capable of solving a problem more efficiently than a classical computer, it is of no practical relevance. Their proposal, however, served as a motivation for the development of other quantum algorithms, such as Simon's and Shor's work on the factoring and discrete logarithm problems, respectively.

10.6.1. Deutsch-Jozsa Algorithm

Deutsch (1985) and Deutsch and Jozsa (1992) explored quantum parallelism to demonstrate that a quantum computer can be much more efficient than a classical computer. They hypothesized the problem of distinguishing two different classes of Boolean functions that map $\{0,1\}^n$ into $\{0,1\}$, $f(\mathbf{x})$: $\{0,1\}^n \rightarrow \{0,1\}$. The functions belonging to one of the classes are *constant*, while the functions of the other class are *balanced*; that is, the goal is to verify if $f(\mathbf{x})$ is constant for all values of $\mathbf{x}$, or equal to one for exactly half of all possible $\mathbf{x}$.

In a classical computer, given an arbitrary function, $f(\mathbf{x})$, at best two function evaluations and at worst $2^{n-1} + 1$ evaluations would have to be performed so as to determine with certainty the class to which function $f(\mathbf{x})$ belongs. For large n, it may not be possible to find a result with a classical computer, while on a quantum computer a single function call is necessary.

Before going into the general case, let us assume first the simplest case in which $f(\mathbf{x})$: $\{0,1\} \rightarrow \{0,1\}$. The quantum circuit presented in Figure 10.21 can be used to implement the Deutsch (1985) algorithm, which is a particular case of the Deutsch-Jozsa (1992) algorithm and serves to classify function $f(\mathbf{x})$ above as constant or balanced.

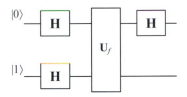

Figure 10.21: Quantum circuit to implement the Deutsch algorithm.

Initially, one Hadamard gate is applied to each of the input states $|0\rangle$ and $|1\rangle$, generating the following inputs to the U_f gate

$$\mathbf{H}|0\rangle = 1/\sqrt{2}\,(|0\rangle + |1\rangle)$$
$$\mathbf{H}|1\rangle = 1/\sqrt{2}\,(|0\rangle - |1\rangle) \tag{10.35}$$

We know from Section 10.4.5 that

$$\mathbf{U}_f(|\mathbf{x}\rangle \otimes (|0\rangle - |1\rangle)) = |\mathbf{x}\rangle \otimes ((|0\rangle - |1\rangle) \oplus f(\mathbf{x})) \tag{10.36}$$

Thus, if $f(\mathbf{x}) = 0$, then

$$(|0\rangle - |1\rangle) \oplus f(\mathbf{x}) = |0\rangle - |1\rangle = (-1)^0\,(|0\rangle - |1\rangle) = (-1)^{f(\mathbf{x})}\,(|0\rangle - |1\rangle) \tag{10.37}$$

And if $f(\mathbf{x}) = 1$, then

$$(|0\rangle - |1\rangle) \oplus f(\mathbf{x}) = |1\rangle - |0\rangle = (-1)^1\,(|0\rangle - |1\rangle) = (-1)^{f(\mathbf{x})}\,(|0\rangle - |1\rangle) \tag{10.38}$$

```
procedure [z] = Deutsch-Jozsa(U_f, |0⟩^⊗n, |1⟩)
    initialize |0⟩^⊗n|1⟩
    superpose states: H^⊗n|0⟩^⊗n,  H|1⟩
    eval(f(x),U_f)
    apply Hadamard on the top n qubits
    measure to obtain output z
end procedure
```

Algorithm 10.1: Main steps of the Deutsch-Jozsa algorithm.

Therefore,

$$\mathbf{U}_f(|\mathbf{x}\rangle \otimes (|0\rangle - |1\rangle)) = (-1)^{f(\mathbf{x})}|\mathbf{x}\rangle \otimes (|0\rangle - |1\rangle) \qquad (10.39)$$

(Note that the constant term $1/\sqrt{2}$ was suppressed.)

Applying $\mathbf{U}_f$ to the input state and using the result of Equation (10.39)

$$\mathbf{U}_f(\tfrac{1}{2}(|0\rangle + |1\rangle) \otimes (|0\rangle - |1\rangle)) =$$
$$= \tfrac{1}{2}((-1)^{f(0)}|0\rangle + (-1)^{f(1)}|1\rangle) \otimes (|0\rangle - |1\rangle) \qquad (10.40)$$

Finally, the Hadamard gate is applied to the first qubit

$$\tfrac{1}{2}((-1)^{f(0)}\mathbf{H}|0\rangle + (-1)^{f(1)}\mathbf{H}|1\rangle) \otimes (|0\rangle - |1\rangle) =$$
$$= \tfrac{1}{2}((-1)^{f(0)}1/\sqrt{2}(|0\rangle + |1\rangle) + (-1)^{f(1)}1/\sqrt{2}(|0\rangle - |1\rangle)) \otimes (|0\rangle - |1\rangle) =$$
$$= \frac{1}{2\sqrt{2}}(|0\rangle((-1)^{f(0)} + (-1)^{f(1)}) + |1\rangle((-1)^{f(0)} - (-1)^{f(1)})) \otimes (|0\rangle - |1\rangle) \qquad (10.41)$$

An analysis of Equation (10.41) allows us to conclude that if $f(\mathbf{x})$ is constant, then the top qubit evaluates to $\pm|0\rangle$. Otherwise, if $f(\mathbf{x})$ is balanced, then the top qubit evaluates to $\pm|1\rangle$. Thus, a single evaluation of $f(\mathbf{x})$ allows us to classify the function into balanced or constant.

The Deutsch-Jozsa algorithm is a generalization of Deutsch's algorithm for $f(\mathbf{x})$: $\{0,1\}^n \rightarrow \{0,1\}$ and will be left as an exercise to the reader. Algorithm 10.1 summarizes the main steps of the generic Deutsch-Jozsa algorithm, where $|0\rangle^{\otimes n}$ is an n-bit length string. Figure 10.22 presents a circuit to implement the Deutsch-Jozsa algorithm.

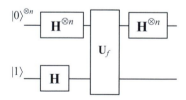

Figure 10.22: Quantum circuit to implement the Deutsch-Jozsa algorithm.

10.6.2. Simon's Algorithm

Assume a function $f(\mathbf{x})$: $\{0,1\}^n \rightarrow \{0,1\}^n$, and the conjecture that f is either 1-to-1 or there exists a non-trivial $\mathbf{a}$ such that $\forall \mathbf{x} \neq \mathbf{y}$, $f(\mathbf{x}) = f(\mathbf{y}) \Leftrightarrow \mathbf{y} = \mathbf{x} \oplus \mathbf{a}$, where the period $\mathbf{a}$ is an n-bit string. Simon's problem (Simon, 1997) involves two goals: i) determining which of the two conditions hold; and ii) finding $\mathbf{a}$.

Figure 10.23 illustrates the circuit used to implement Simon's algorithm. The analysis of the algorithm can be summarized as presented in the following sequence of steps:

1. Apply the Hadamard gate to the top n lines, $|0\rangle^{\otimes n}$, resulting in Equation (10.26). The state of the whole circuit right before $\mathbf{U}_f$ is applied is the superposition

$$\mathbf{H}^{\otimes n} |0\rangle^{\otimes n} |0\rangle^{\otimes n} = \frac{1}{2^{n/2}} \sum_{x=0}^{2^n-1} |\mathbf{x}\rangle |0\rangle^{\otimes n} \tag{10.42}$$

2. The application of the gate $\mathbf{U}_f$, $\mathbf{U}_f|\mathbf{x}\rangle|\mathbf{y}\rangle = |\mathbf{x}\rangle|\mathbf{y} \oplus f(\mathbf{x})\rangle$, to the result above leads to

$$\frac{1}{2^{n/2}} \sum_{x=0}^{2^n-1} |\mathbf{x}\rangle | f(\mathbf{x})\rangle \tag{10.43}$$

3. Measuring the bottom n lines results in either $f(\mathbf{x}_0)$ or $f(\mathbf{x}_0 \oplus \mathbf{a})$

$$\frac{1}{2^{1/2}} (|\mathbf{x}_0\rangle + |\mathbf{x}_0 \oplus \mathbf{a}\rangle) | f(\mathbf{x}_0)\rangle \tag{10.44}$$

4. Applying the Hadamard gate to the top n lines results in

$$\frac{1}{2^{(n+1)/2}} \sum_{y=0}^{2^n-1} \left[(-1)^{\mathbf{x}_0 \cdot \mathbf{y}} + (-1)^{(\mathbf{x}_0 \oplus \mathbf{a}) \cdot \mathbf{y}} \right] |\mathbf{y}\rangle \tag{10.45}$$

5. Measuring the first n lines results in a $\mathbf{y}_1$ such that $\mathbf{a} \cdot \mathbf{y}_1 = 0$ with probability $1/2^{n-1}$.

6. By repeating the procedure above until n linearly independent $\mathbf{y}_1, \dots, \mathbf{y}_n$ are obtained, and solving all the equations $\mathbf{y}_i \cdot \mathbf{a} = 0$, $i = 0, \dots, n$, $\mathbf{a}$ can be obtained.

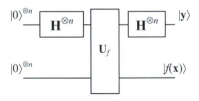

Figure 10.23: Quantum circuit to implement Simon's algorithm.

In summary, there are three key aspects of Simon's algorithm that make it a powerful approach (Meglicki, 2002). First, the application of the Hadamard gate to the top n lines generates a superposition of all possible numbers from 0 to 2^{n-1}, which is then fed to the second register that evaluates function $f(x)$ for all its arguments simultaneously (Equation (10.42) and Equation (10.43)). In standard computers, 2^{n-1} separate steps would be necessary to replace this single quantum step. Second, the use of the $\mathbf{U}_f$ gate promotes the entanglement of the whole computer. When the bottom n are subjected to measurement, the top n lines are forced into a superposition of some $|x_0\rangle$ and $|x_0 \oplus a\rangle$. Thus, **a** appears mixed with x_0 in the top n lines.

10.6.3. Shor's Algorithm

Due to their difficulty, finding discrete logarithms and factoring integers are two problems extensively studied in number theory. The hardness of these problems is the core of several important techniques for securing information, known as *cryptographic systems*, such as the RSA public key cryptosystem (Rivest et al., 1978). RSA uses a public key N that is the product of two large prime numbers p and q, $N = pq$. A form of breaking the RSA encryption is by factoring N; that is, finding p and q. However, factoring is a problem that becomes increasingly more time consuming when N grows.

Inspired by the work of Simon, P. W. Shor (1994, 1997) proposed algorithms for factoring integers and finding discrete logarithms in polynomial time using a hypothetical quantum computer. His proposal is composed of two parts: i) a reduction of the factoring problem to that of finding the period of a function; and ii) the development of a quantum algorithm to solve the problem of order-finding. The latter is based on the calculation of a *quantum Fourier transform* (QFT) of the function.

Quantum Fourier Transform

The quantum Fourier transform, which is the quantum analogue of the discrete Fourier transform (Appendix B.4.7), is defined as a linear operator that works as follows. Given an orthonormal basis $|0\rangle, \ldots, |2^n - 1\rangle$, where n is the number of qubits the QFT performs the following action on the basis states:

$$|x'\rangle = \text{QFT} \, |x\rangle = \frac{1}{2^{n/2}} \sum_{y=0}^{2^n-1} e^{2\pi i xy/2^n} \, |y\rangle \qquad (10.46)$$

Equivalently, the QFT takes the state vector $|y\rangle$ of Equation (10.47) (cf. Equation (10.13)) and transforms it into $|y'\rangle$

$$|y\rangle = \sum_{x=0}^{2^n-1} c_x \, |x\rangle \qquad (10.47)$$

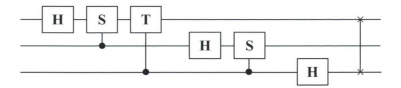

Figure 10.24: Quantum circuit to perform a three-qubit QFT.

$$|\mathbf{y'}\rangle = \text{QFT} \, |\mathbf{y}\rangle = \frac{1}{2^{n/2}} \sum_{x=0}^{2^n-1} \sum_{y=0}^{2^n-1} c_x e^{2\pi i x y / 2^n} \, |\mathbf{y}\rangle \qquad (10.48)$$

where n is the number of qubits.

The QFT can also be represented by a square matrix of dimension $n \times n$ and whose entries are $2^{-n/2}\omega$, where $\omega = e^{2\pi i / 2^n}$:

$$\frac{1}{2^{n/2}} \begin{bmatrix} 1 & 1 & 1 & \cdots & 1 \\ 1 & \omega & \omega^2 & \cdots & \omega^{2^n-1} \\ 1 & \omega^2 & \omega^4 & \cdots & \omega^{2(2^n-1)} \\ \vdots & \vdots & \vdots & \ddots & \vdots \\ 1 & \omega^{2^n-1} & \omega^{2(2^n-1)} & \cdots & \omega^{(2^n-1)(2^n-1)} \end{bmatrix} \qquad (10.49)$$

The discrete Fourier transform can be implemented as a quantum circuit consisting of Hadamard gates and controlled phase shift gates. Figure 10.24 depicts the quantum circuit to implement the QFT for a system with three qubits.

Factorization

The first part of Shor's algorithm for factoring is the reduction of the factoring problem to the order-finding problem that can be performed using the following sequence of steps on a classical computer:

1. Take a pseudo-random number $x < N$.

2. Calculate the greatest common divisor of integers x and N, $gcd(x,N)$, using, for instance, the Euclidean algorithm.

3. Test if $gcd(x,N) \neq 1$

   3.1. Then, Stop. There is a nontrivial factor for N.

   3.2. Else, use the period finding sub-routine to find r, the period of the following function: $f = x^a \bmod N$, such that $f(x + r) = f(x)$.

4. If r is odd, then return to Step 1.

5. If $x^{r/2} = -1 \bmod N$, then return to Step 1.

6. The factors of N are $gcd(x^{r/2} \pm 1, N)$.

The period-finding sub-routine can be implemented using a circuit that is very similar to that of Simon (Figure 10.23), but replacing the Hadamard gates in the top right side by a quantum Fourier transform (QFT) gate (Figure 10.25).

The analysis of the circuit presented in Figure 10.25 leads to the following sequence of steps (Wikipedia.org; Meglicki, 2002).

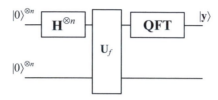

Figure 10.25: Quantum circuit to implement Shor's factoring algorithm.

1. Apply the Hadamard gate to the top n lines, $|0\rangle^{\otimes n}$. The state of the whole circuit right before U_f is applied is the superposition:

$$\mathbf{H}^{\otimes n}\,|0\rangle^{\otimes n}\,|0\rangle^{\otimes n} = \frac{1}{2^{n/2}}\sum_{x=0}^{2^n-1}|\,\mathbf{x}\rangle\,|0\rangle^{\otimes n} \tag{10.50}$$

2. The application of the gate U_f, $U_f|\mathbf{x}\rangle|\mathbf{y}\rangle = |\mathbf{x}\rangle|\mathbf{y}\oplus f(\mathbf{x})\rangle$, to the result above leads to

$$\frac{1}{2^{n/2}}\sum_{x=0}^{2^n-1}|\,\mathbf{x}\rangle\,|\,f(\mathbf{x})\rangle \tag{10.51}$$

3. Apply the quantum Fourier transform on the input register:

$$\frac{1}{2^n}\sum_{x=0}^{2^n-1}\sum_{y=0}^{2^n-1}e^{2\pi i xy/2^n}\,|\,\mathbf{y}\rangle\,|\,f(\mathbf{x})\rangle \tag{10.52}$$

4. As f is periodic, a measurement in the input register will result in $|\mathbf{y}\rangle$ with probability:

$$\frac{1}{2^n}\left|\sum_{x:f(x)=f(x_0)}e^{2\pi i xy/2^n}\right|^2 = \frac{1}{2^n}\left|\sum_b e^{2\pi i(x_0+rb)y/2^n}\right| \tag{10.53}$$

The closer $\mathbf{y}r/2^n$ is to an integer, the higher the probability.

5. Turn $\mathbf{y}/2^n$ into an irreducible fraction and extract the denominator r' as a candidate for r.

6. If $f(\mathbf{x}+r')=f(\mathbf{x})$:

    6.1.   Then Stop. The period has been determined and corresponds to r'.

6.2.   Else, find more candidates to r by using values close to $\mathbf{y}$ or multiples of r'. If any candidate satisfies Step 6, go to Step 6.1. Else, go to Step 1.

10.6.4.   Grover's Algorithm

L. K. Grover (1997a,b) presented a quantum mechanical algorithm to perform search faster than any classical algorithm. As a benchmark he used the problem of retrieving an item satisfying a given criterion from an unsorted database containing N items. In order to solve this problem classically; that is, using a standard algorithm, each datum in the database has to be examined. If an item satisfies the desired condition, then stop; otherwise, remove the tested datum from the database and continue the search. On average, this algorithm examines half the database: $O(N/2)$.

Grover (1997a,b) demonstrated that with a quantum computer it is possible to have the input and output in superpositions of states and thus find an object in a time that is proportional to the square root of the number of items: $O(N^{1/2})$. The quantum mechanical algorithm proposed is a sequence of three unitary transformations on a pure state followed by measurement. The unitary transformations used were: i) the creation of a superposition with equal amplitude for all the N possible basic states of the system; ii) the Hadamard transformation; and iii) the selective rotation of the phases of states.

The description to be presented here follows the one used in (Grover, 1997a). Other, more general, descriptions can be found in many texts from the literature, such as (Nielsen and Chuang, 2000; Rieffel and Polak, 2000; Pittenger, 2001; Meglicki, 2002; Perry, 2004; among others).

Assume a system with $N = 2^n$ states labelled $S_1, S_2, \ldots, S_N$ and represented as bitstrings of length n. The goal is to identify a single state S_v such that $C(S_v) = 1$; for all other states $C(S_i) = 0$, $\forall i \neq v$. The algorithm works as follows.

1.   Using the Hadamard transform initialize the system in the equal superposition state:

$$\frac{1}{2^{n/2}} \sum_{x=0}^{2^n-1} | \mathbf{x} \rangle \tag{10.54}$$

2.   Repeat the following unitary transformations $O(N^{1/2})$ times:

   a.   Let the system be in any state S_i, $i = 1, \ldots, N$. If $C(S) = 1$ rotate the phase by π radians; else, leave the system unaltered.

   b.   Apply a diffusion transformation $\mathbf{D}$ defined as: $d_{ij} = 2/N$ if $i \neq j$, and $d_{ii} = -1 + 2/N$.

3.   Measure the resulting state, which will be state S_v, such that $C(S_v) = 1$, with probability at least $\frac{1}{2}$.

10.7 PHYSICAL REALIZATIONS OF QUANTUM COMPUTERS: A BRIEF DESCRIPTION

The theoretical developments in DNA and quantum computing are almost always performed with a view of a biological or physical implementation of the respective natural computers. In the particular case of quantum computing, several physical devices have been proposed to perform quantum computation. However, the problem of building a quantum computer is a very difficult one. The most difficult obstacles appear to be the measurement of quantum states and the implementation of quantum state transformations with sufficient precision so as to provide accurate results after a large number of computational steps. These difficulties are aggravatted when the size of the computer grows.

Nielsen and Chuang (2000) and DiVincenzo (2000) discuss a set of basic criteria for the realization of quantum computing:

- *A scalable physical system to implement qubits*: The first requirement of a quantum computer is a physical system with a set of qubits. A generic qubit state can be written as a linear combination of a set of basis states (Equation (10.13)). A qubit can be physically realized using spin ½ particles, atoms, polarized photons, etc. Some of these proposals will be briefly reviewed below.

- *Qubit initialization*: Before starting any computation, it is necessary to initialize the quantum registers. The two main approaches to set qubits to desired states are by physically cooling them into their ground state or by measuring them, what will result in a projection of the system into the desired state or a state that can be rotated into it.

- *Definition of a universal set of quantum gates*: A quantum algorithm is usually specified as a sequence of applications of unitary transformations, each one acting on a small number of qubits. The most direct transcription of this into a physical implementation is to identify Hamiltonians capable of generating these unitary transformations and then designing a physical system that switches on and off each of the Hamiltonians at specified times. A problem that arises with the physical implementation of quantum gates is that they cannot be implemented perfectly. Thus, error correction techniques have to be devised in order for us to have a reliable computation.

- *Decoherence times longer than the gate operation time*: Decoherence corresponds to the appearance of quantum correlations of a system with its environment. These correlations result in entirely new properties and behavior compared to that shown by isolated objects (Joos, 1999). It corresponds to the main mechanism responsible for the appearance of classical behavior in a quantum system. Thus, if decoherence acts for very long it may lead to a quantum computer behaving like a classical one.

- *Measurement of the output*: The output of a good quantum algorithm is a state that provides a useful answer when a measurement is performed. Reading the final result of a quantum computation requires the ability to measure specific qubits. An important feature of measurement is the wavefunction collapse that describes what happens when a projective measurement is performed.

The following sections present a brief description of some important physical implementations of quantum computers. Further details can be found in the cited literature and in the book by Nielsen and Chuang (2000).

10.7.1. Ion Traps

Cirac and Zoller (1995) showed that a quantum computer could be implemented using a set of N cold ions moving in a linear trap and interacting with laser beams (Figure 10.26). According to the authors, the distinctive features of their proposal were the possibility of implementing n-bit quantum gates, the minimization of decoherence effects, and the efficiency of measurement. The qubits are the ions themselves, whose states can be identified with two of the internal states of the ion. The independent manipulation of each individual qubit is performed by directing distinct laser beams to each ion. A CNOT gate can be implemented by exciting the collective quantized motion of the ions with lasers. In an ion trap decoherence is an outcome of the spontaneous decay of the internal atomic states and damping of the motion of ions.

Figure 10.26: Ions in a linear trap interacting with different laser beams.

The decoherence time in the ion trap can be much longer than the time spent by many quantum gates to act. Finally, quantum measurement can be performed efficiently using a technique called *quantum jumps*. Using the ideas proposed by Cirac and Zoller (1995), Monroe et al. (1995) demonstrated the first fundamental logic gates that operate on prepared quantum states.

10.7.2. Cavity Quantum Electrodynamics (CQED)

In the proposal presented by Turchette et al. (1995), the authors focused on the implementation of quantum computers using large optical nonlinearities realizable in *cavity quantum electrodynamics* (CQED). In CQED systems, the mutual coupling of individual photons to a single intracavity atom allows them to have strong interactions. Depending on some parameters it is possible to strongly couple a single atom to a cavity mode so as to create a one-dimensional atom. Thus, the atom-cavity system may be viewed as a quantum-optical device. The dynamics in the proposed system is a result of the nonlinear optical response of a Cesium atom coupled to the cavity field. The interactions between photons in their atom-cavity device are characterized by the transmission of coherent state pump and probe beams. To summarize, the proposed quantum computer is composed of a CQED filled with Cesium atoms and an arrangement of lasers, polarisers, mirrors, and phase shift detectors.

10.7.3. Nuclear Magnetic Resonance (NMR)

When the nuclei of some atoms are immersed in a static magnetic field and subjected to another oscillating magnetic field, some nuclei experience a phenomenon called *nuclear magnetic resonance* (NMR). NMR spectroscopy has been widely employed to obtain chemical, physical, strucutural and electronic information about a molecule. Although NMR spectroscopy has been known for some time, only in 1997 N. A. Gershenfeld and I. L. Chuang (1997), and Cory et al. (1997) independently noticed that it could be used to realize a quantum computer. In their proposal each molecule can be viewed as a single computer whose state is determined by the orientations of its spins, and quantum logic gates correspond to sequences of pulses that manipulate the orientations of the spins, thus performing unitary transformations on the state. The system is started with molecules in known configurations and spin couplings, and carefully chosen NMR signals are applied in order to produce an output that depends solely on the quantum computation performed.

10.7.4. Quantum Dots

Another approach that has been investigated for the realization of quantum computers is the use of *quantum dots* (Loss and DiVincenzo, 1998; Sherwin et al., 1999). A quantum dot is a semiconductor crystal that has a diameter of a few nanometers and is, thus, also called a nanocrystal. It works like a potential well, confining electrons to regions of the order of the electron's wavelength, what results in discrete quantized energy levels. Loss and DiVincenzo (1998) proposed the implementation of a universal set of quantum gates based on the spin states of coupled single-electron quantum dots. The operations are performed by the gating of the tunneling barrier between neighboring dots. A set of initial states could be prepared using a uniform magnetic field to cool down the system. The authors discussed two possibilities for qubit measurement: the use of a switchable tunneling into a supercooled paramagnetic dot, and the use of a spin-dependent, switchable 'spin valve' tunnel barrier.

10.8 SCOPE OF QUANTUM COMPUTING

Quantum mechanics has been shown capable of speeding up several algorithms. For instance, Shor (1997) proposed a quantum search algorithm for factoring that is exponentially faster than any classical algorithm known to date, and Grover (1997) developed a quantum algorithm for search that is polinomially faster than any classical algorithm. However, the importance and usefulness of quantum systems go beyond the speeding up of some information processing tasks. In quantum communication, for instance, qualitative and quantitative improvements can be gained, such as the reduction of the amount of the communicated data required. Furthermore, there are some tasks that can be performed in quantum systems, but that cannot be accomplished using classical physical systems. An example of such a task is quantum cryptography, which provides an absolute secrecy of communication.

The physical implementation of quantum computers can bring benefits to many sciences, including biotechnology and physics, and to human life as well. Quantum computers could be used to perform real experiments in physics in place of the many *gedanken experiments* that have been proposed to date; it could be used in applied physics, such as generating high precision spectroscopy and atomic clocks; it could be used to simulate quantum systems, to perform super-fast code breaking, to detect eavesdroppers, to perform teleportation, and to improve control and communications. Quantum computing can also be applied to game theory, and a quantum game may provide nondeterministic behaviors, as required in some applications.

There are already commercial products based on quantum systems. A company called MagiQ Technologies sells systems based on quantum cryptography and communications. Quantum cryptographic key is already for sale, using the photon-by-photon encryption. According to MagiQ officials, the primary markets for these quantum encryption keys are financial institutions and government agencies (Pappalardo, 2005).

10.9 FROM CLASSICAL TO QUANTUM COMPUTING

This chapter introduced several important concepts from quantum computing and constrasted it with classical computing. Bits were compared with qubits, gates with quantum gates and circuits with quantum circuits. Table 10.5 provides a brief summary of the main similarities and differences between classical and quantum computing.

10.10 SUMMARY AND DISCUSSION

The power of quantum computers to perform calculations across a multitude of parallel universes gives it the ability to quickly perform tasks that classical computers will never be able to achieve. This power, however, can only be harnessed if a correct type of algorithm is designed, but this is usually a very difficult task. Examples of these algorithms are already available, such as Shor's and Grover's algorithms discussed here. These have proved major implications on some well-known fields of computing, like search and cryptography, because they enable the most commonly used cryptographic techniques to be easily broken.

Table 10.5: Comparison between classical and quantum computing.

Classical Computing	Quantum Computing
Information is represented as a bitstring, i.e., string of either '0' or '1'. Also denoted by $\lvert x \rangle^n$, $0 \leq x < 2^n$	Information is represented as a qubit, i.e., a superposition of '0' and '1'. Denoted by $\sum_x c_x \lvert x \rangle^n$, $\sum_x \lvert c_x \rvert^2 = 1$
Sequential computation	Parallel computation
Computation performed by moving bits through logic gates	Computation performed by altering the state of atoms
Information can be copied and read without being disturbed	Information cannot be copied or read without being disturbed
Information can be transmitted forward in time	Information can be transmitted in any direction in time
Information can be measured (read) without being destroyed	Information cannot be measured without being destroyed
It is always possible to predict with certainty the state of the system after measurement	It is generally only possible to predict probabilistically the state of the system after measurement

As is the case with DNA computing, the most important finding of quantum computing is that quantum computers can perform some tasks much more efficiently than classical computers. For instance, it was discussed that the prime factoring of an *n*-bit integer requires exponential time in classical computers, whilst quantum computers can solve the same problem in a quadratic time complexity class. Another outcome of quantum computing, which is quantum communication, allows information to be sent without eavesdroppers listening undetected.

This chapter started by reviewing some physical phenomena to illustrate why and how quantum and classical phenomena are different. It then followed with a brief introduction to quantum information and a description of quantum computation making a trade-off with classical computing. A major difficulty in the construction of a quantum computer is to interface it with the user. The system has to be initialized and run allowing the reading and writing of information, requiring the interaction with a user. Some of the main proposals of physically implementing a quantum computer together with a discussion of the scope of quantum computing concluded the chapter.

10.11 EXERCISES

10.11.1. Questions

1. Planck's proposal of the quantum theory was based on a phenomenon called *ultraviolet catastrophe*. Explain what it means and Planck's explanation to the catastrophe.

2. Explain the interference pattern for waves using two sinusoids with differences in their phases. What happens if the waves are in phase? What if they are out of phase?

3. Is it possible to have a single-input/single-output standard binary logic gate that satisfies $\sqrt{NOT} \cdot \sqrt{NOT} = NOT$? Justify your answer.

10.11.2. Exercises

1. The momentum p of a particle is given by its mass m times its velocity v: $p = mv$. Calculate the following momentums and compare:
   o A photon at light speed
   o A walking ant
   o A supersonic airplane
   o A walking person

2. Let a pair of basis vectors $\{|0\rangle, |1\rangle\} = \{(1,0)^T, (0,1)^T\}$. Determine, in terms of the basis vectors, the transformation $\mathbf{A}$ that exchanges $|0\rangle$ into $|1\rangle$.

3. In a two-qubit system of the form $|y\rangle = c_1|00\rangle + c_2|01\rangle + c_3|10\rangle + c_4|11\rangle$, what is the probability that measuring the first qubit alone results in '1'? What is the postmeasurement state?

4. Show that the two-qubit states below are not entangled.

   $1/\sqrt{2}(|10\rangle + |11\rangle)$.

   $\frac{1}{2}(|00\rangle_{AB} + i|01\rangle_{AB} - |10\rangle_{AB} - |11\rangle_{AB})$

5. Show that the Bell states $|x_1\rangle$ and $|x_2\rangle$, presented in Equation (10.4), are entangled; that is, $|x_1\rangle \neq |x\rangle|y\rangle$ and $|x_2\rangle \neq |x\rangle|y\rangle$ for all single qubit states $|x\rangle$ and $|y\rangle$.

6. Verify that the application of two or more unitary transformations is not commutative.

7. Show that the controlled-NOT gate cannot be written as a tensor product of two operators.

8. Show that the Hadamard gate performs a rotation of the Bloch sphere $90°$ about the y-axis followed by a rotation of $180°$ about the x-axis.

9. Show that the application of Equation (10.25) to the tensor product of n qubits results in Equation (10.27).

10. Show that a Toffoli gate can be used to simulate a NAND gate and to do FANOUT, thus proving that a quantum computer can be used to perform classical, deterministic, computing.

11. Show that a Fredkin gate can be used to simulate the following gates: AND, NOT, CROSSOVER, and FANOUT.

12. Demonstrate that the transformation $\mathbf{U}_f$ is unitary. (Show that $\mathbf{U}_f \mathbf{U}_f = \mathbf{I}$ and $\mathbf{U}_f^\dagger = \mathbf{U}_f$.)

13. Based on the set of gates introduced by Feynman to make a universal classical computer (Section 10.5.2), design a Full Adder circuit; that is, an Adder that takes a carry element c from some previous addition and adds it to the two elements a and b, and has an additional line d with a 0 input that will serve to provide as output the CARRY of the Full Adder.

14. Detail the steps of the Deutsch-Jozsa algorithm, which is a generalization of Deutsch's algorithm for $f(\mathbf{x})$: $\{0,1\}^n \rightarrow \{0,1\}^n$.

15. Show that the function $f(00) = 0$, $f(01) = 0$, $f(10) = 1$, and $f(11) = 1$, is balanced. Tip: apply the Deutsch-Josza algorithm.

16. Calculate the quantum Fourier transform of the following states:

a) $|\mathbf{x}\rangle = \dfrac{2^{1/2}}{3^{1/2}} |0\rangle + \dfrac{1}{3^{1/2}} |1\rangle$

b) $|\mathbf{x}\rangle = \dfrac{1}{2^{1/2}} |10\rangle + \dfrac{1}{2^{1/2}} |01\rangle$

17. Show that the diffusion transformation $\mathbf{D}$ defined as: $d_{ij} = 2/N$ if $i \neq j$, and $d_{ii} = -1 + 2/N$ can be implemented as a product of three elementary matrices (unitary transformations).

10.11.3. Thought Exercises

1. Assume a classical computer that performs 10^9 instructions per second (1GHz). Knowing that the light speed is approximately 3×10^8 m/s, what is the distance traversed by the light during the time spent by the execution of one instruction on such computer? Based on this result, explain the need for miniaturizing computer technology and the potential of optical computers (i.e., computers that perform computation with light, e.g., laser).

2. Show that any unitary operator can be realized in the form $\mathbf{U} = \exp(i\mathbf{K})$ for a Hermitian operator $\mathbf{K}$.

3. Based on the no-cloning theorem (Section 10.3), show that a copymachine can only clone states that are orthogonal to one another.

4. Show that any N-qubit gate can be performed as a sequence of arbitrary 1-qubit gates and CNOT gates.

5. Describe how the Toffoli gate can be used to simulate a probabilistic classical computer.

10.11.4. Projects and Challenges

1. Show that any multiple qubit gate can be decomposed from the CNOT gate and single qubit gates.

2. Show that the quantum Fourier transform (QFT) can be performed using only single-bit gates and measurements of single bits.

10.12 REFERENCES

[1]   Benioff, P. (1980), "The Computer as a Physical System: A Microscopic Quantum Mechanical Hamiltonian Model of Computers as Represented by Turing Machines", *Journal of Statistical Physics*, **22**(5), pp. 563–591.

[2]   Benioff, P. (1982), "Quantum Mechanical Hamiltonian Models of Turing Machines", *Journal of Statistical Physics*, **29**(3), pp. 515–546.

[3]   Bennett, C. H. and Wiesner, S. J. (1992), "Communication via one- and two-particle Operations on Eisen-Podolsky-Rosen States", *Phys. Rev. Lett.*, **69**, pp. 2881–2884.

[4]   Bennett, C. H., Brassard, G., Crépeau, C., Jozsa, R., Peres, A. and Wooters, W. K. (1993), "Teleporting an Unknown Quantum State via Dual Classical and Einstein-Podolsky-Rosen Channels", *Phys. Rev. Lett.*, **70**, pp. 1895–1898.

[5]   Cirac, J. I. and Zoller, P. (1995), "Quantum Computations with Cold Trapped Ions", *Physical Review Letters*, **74**(20), pp. 4091–4094.

[6]   Cory, D., Fahmy, A. and Havel, T. (1997), "Ensemble Quantum Computing by NMR Spectroscopy", *Proc. of the Nat. Ac. of Sci. U.S.A.*, **94**(5), pp. 1634–1639.

[7]   Deutsch, D. (1985), "Quantum Theory, the Church-Turing Principle, and the Universal Quantum Computer", *Proc. of the Royal Soc. of London A*, **400**, pp. 97–119.

[8]   Deutsch, D. (1992), "Quantum Computation", *Physics World*, **5**, pp. 57–61.

[9]   Deutsch, D. and Jozsa, R. (1992), "Rapid Solutions of Problems by Quantum Computation", *Proc. of the Royal Soc. of London A*, **439**, pp. 553–558.

[10]   Dieks, D. (1982), "Communication by EPR Devices", *Physics Letters A*, **92**(6), pp. 271–272.

[11]   Dirac, P. (1982), *The Principles of Quantum Mechanics*, 4th Ed., Oxford University Press.

[12]   DiVincenzo, D. P. (2000), "The Physical Implementation of Quantum Computation", *Fortschr. Phys.*, **48**, pp. 771–783.

[13]   Feynman, R. P. (1982), "Simulating Physics with Computers", *Int. Journal of Theor. Physics*, **21**(6/7), pp. 467–488.

[14]   Feynman, R. P. (1985), "Simulating Physics with Computers", *Int. Journal of Theor. Physics*, **11**, pp. 11–20.

[15]   Feynman, R. P. (1986), "Quantum Mechanical Computers", *Foundations of Physics*, **16**(6), pp. 507–531.

[16]   Fredkin, E. and Toffoli, T. (1982), "Conservative Logic", *Int. Journal of Theor. Physics*, **21**, pp. 219–253.

[17]   Gershenfeld, N. and Chuang, I. (1997), "Bulk Spin-Resonance Quantum Computation", *Science*, **275**, pp. 350–356.

[18]   Grover, L.K. (1997a), "Quantum Mechanics Helps in Searching for a Needle in a Haystack", *Physical Review Letters*, **79**(2), 325–328.

[19]   Grover, L. K. (1997b), "Quantum Computers Can Search Arbitrarily Large Databases by a Single Query", *Physical Review Letters*, **79**(23), pp. 4709–4712.

[20]   Hey, T. and Walters, P. (1987), *The Quantum Universe*, Cambridge University Press.

[21]   Hirvensalo, M. (2001), *Quantum Computing*, Natural Computing Series, Springer.

[22]   Joos, E. (1999), "Elements of Environmental Decoherence", In P. Blanchard, D. Giulini, E. Joos, C. Kiefer and I.-O. Stamatescu (Eds.), *Decoherence: Theoretical, Experimental, and Conceptual Problems*, Springer, pp. 1–17.

[23]   Loss, D. and DiVincenzo, D. P. (1998), "Quantum Computation with Quantum Dots", *Physical Review A*, **57**(1), pp. 120–126.

[24]   Meglicki, Z. (2002), "Introduction to Quantum Computing (M743)", *Lecture Notes from the Topological Quantum Computing Home Page*, [On Line] http://www.tqc.iu.edu/M743/Default.htm .

[25]   Monroe, C., Meekhof, D. M., King, B. E., Itano, W. M. and Wineland, D. J. (1995), "Demonstration of a Fundamental Quantum Logic Gate", *Physical Review Letters*, **75**(25), pp. 4714–4717.

[26]   Nielsen, M. A. and Chuang, I. L. (2000), *Quantum Computation and Quantum Information*, Cambridge University Press.

[27]   Pappalardo, J. (2005), "Researchers Cast Wary Eye on Atomic-Level Computing", *NDIA's Business and Technology Magazine*, March.

[28]   Pellizzari, T., Gardiner, S. A., Cirac, J. I. and Zoller P. (1995), "Decoherence, Continuous Observation, and Quantum Computing: A Cavity QED Model", *Physical Review Letters*, **75**(21), pp. 3788–3791.

[29]   Perry, R. T. (2004), *The Temple of Quantum Computing*, On Line Book, [On Line] http://www.toqc.com/.

[30]   Pittenger, A. O. (2001), *An Introduction to Quantum Computing Algorithms*, Birkhäuser.

[31]   Polkinghorne, J. (2002), *Quantum Theory: A Very Short Introduction*, Oxford University Press.

[32]   Rieffel, E. and Polak, W. (2000), "An Introduction to Quantum Computing for Non-Physicists", *ACM Computing Surveys*, **32**(3), pp. 300–335.

[33]   Rivest, R. L., Shamir, A. and Adleman, L. M. (1978), "A Method for Obtaining Digital Signatures and Public-Key Cryptosystems", *Communications of the ACM*, **21**(2), pp. 120–126.

[34]   Sherwin, M. S., Imamoglu, A. and Montroy, T. (1999), "Quantum Computation with Quantum Dots and Terahertz Cavity Quantum Electrodynamics", *Physical Review A*, **60**(5), pp. 3508–3514.

[35]   Shor, P. W. (1994), "Algorithms for Quantum Computation: Discrete Logarithms and Factoring", *Proc. of the 35th IEEE Annual Symposium on Foundations of Comp. Science*, pp. 124–134.

[36]   Shor, P. W. (1997), "Polynomial-Time Algorithms for Prime Factorization and Discrete Logarithms on a Quantum Computer", *SIAM Journal on Computing*, **26**(5), pp. 1484–1509.

[37]   Simon, D. (1997), "On the Power of Quantum Computation", *SIAM Journal on Computing*, **26**(5), pp. 1474–1483.

[38]   Toffoli, T. (1980), "Reversible Computing", In J. W. de Bakker and J. van Leeuwen (Eds.), *Proc. of the 7th Colloquium on Automata, Languages and Programming, Lecture Notes in Computer Science*, **84**, pp. 632–644.

[39]   Turchette, Q. A., Hood, C. J., Lange, W., Mabuchi, H. and Kimble, H. J. (1995), "Measurement of Conditional Phase Shifts for Quantum Logic", *Physical Review Letters*, **75**(25), pp. 4710–4713.

[40]   Wootters, W. K. and Zurek, W. H.   (1982), "A Single Quantum Cannot be Cloned", *Nature*, 299, pp. 802–803.

CHAPTER 11

AFTERWORDS

"Clearly, we are on the threshold of yet another revolution. Human knowledge is doubling every ten years. In the past decade, more scientific knowledge has been created than in all of human history. Computer power is doubling every eighteen months. The Internet is doubling every year. The number of DNA sequences we can analyze is doubling every two years. Almost daily, the headlines herald new advances in computers, telecommunications, biotechnology, and space exploration. In the wake of this technological upheaval, entire industries and lifestyles are being overturned, only to give rise to entirely new ones. But these rapid, bewildering changes are not just quantitative. They mark the birth pangs of a new era."

(M. Kaku, Visions: How Science will Revolutionize the 21st Century and Beyond, Oxford University Press, 1998, p. 4)

11.1 NEW PROSPECTS

Natural computing has taught us to *think 'naturally' about computation* and also to think *computationally about nature*. You can check it by yourself just by going back to Chapter 1 and comparing the answers you gave to the questions posed there with those provided throughout the text. It is very much likely that your answers were substantially different from those presented in this book, but this does not mean you did not get it right. It just indicates how diverse we are and how creative we can be. Your proposals may be novel and useful, it is now time to implement them and investigate the outcomes. But before you start this journey, let us just say a few more words about all that has been presented in this book.

We have discovered that all natural computing approaches yield novel and exciting capabilities for science as a whole, mainly computer science, engineering, philosophy, and the biosciences. Scientists and engineers have been gifted with a very rich paradigm for exploration; the combination of nature and computing. In a broad sense, natural computing has taught us that any (model of a) natural phenomenon may be used as a basis for the development of novel algorithmic tools for problem solving; a vast array of natural phenomena can be simulated/emulated in computers; and natural materials can be used for computation. The fruits of these explorations are continuously becoming new technological solutions and explanations to old and recent problems, and the full potential is far from being reached.

The interdisciplinary and multidisciplinary spirit fostered as a result of doing natural computing research is one of the most pleasant and remarkable features of this field. The excitement and freshness of the field build up the prospects for creativity, invention, innovation, and discovery in this endeavor. Most researchers of natural computing agree that the approaches described in this text represent not only the most accessible and exploitable openings to a large set of computational systems - the opening wedge into a new field that is likely to provide a satisfying synthesis of the worlds of Nature and Computing.

The fact that natural computing techniques can solve some problems that are intractable for classical computers raises some important questions: 1) What other problems can be more efficiently solved by these methods that are not tractable by classical computers? 2) What other patterns and behaviors can be synthesized using natural computing? 3) When will we effectively have some of these new technologies (e.g., DNA computers) actually implemented (in wetware or hardware)? Bear in mind that many natural computing techniques are already implemented; e.g., in washing machines, air-conditioners, car brakes, and so on.

This brief concluding chapter starts by summarizing why natural computing has developed so quickly over the past years, follows with a discussion of some lessons from natural computing, and then makes a distinction between different terminologies used to refer to, in many cases, the same techniques.

11.2 THE GROWTH OF NATURAL COMPUTING

Over the last two decades, natural computing has experienced an astonishing growth. Its frontiers have been expanding in many directions simultaneously, from nature-inspired computing to the 'infant' computing with natural materials. This rapid growth is due to a number of factors:

- Computing power and memory storage capacity have increased dramatically in these last decades. Fast processors combined with large memories have made it possible to implement complex adaptive systems of various sorts, to design efficient and near real-time search and optimization algorithms, to simulate and emulate life in computers, to simulate bio-computers *in silico*, and to tackle large-scale real world problems that may involve millions of samples in high-dimensional spaces; feats that were not feasibly performed a few decades ago.

- The vast number of problems with features such as nonlinearity, high-dimensionality, difficulty in modeling and finding derivatives, etc., make it unfeasible the use of traditional techniques, such as linear, nonlinear, dynamic, integer programming, and others. Examples of these problems are pervasive in areas such as robotics, data mining (including web mining and bioinformatics), pattern recognition, function approximation, search, and optimization. These problems constitute a great challenge to computing and engineering and have fostered a major interest by natural computing approaches.

- Increasing interaction and collaboration among different disciplines and fields of investigation, like computer science, engineering, biology, medicine, physics, ecology, neuroscience, mathematics, and statistics. These multidisciplinary and interdisciplinary efforts have resulted in new ideas, techniques, and paradigms, which embody the vast field of natural computing.

11.3 SOME LESSONS FROM NATURAL COMPUTING

Histories of intellectual and technological development come in different ways and at different times. One may be the result of an eureka moment that causes a major change in the world. Another way is a paradigm shift, where researchers and even ordinary people find that completely new types of science are available. Many of these developments, such as Newton's mechanics, are somehow related to nature. Natural computing is another science that is emerging guided by the same principles: observe, measure, model, qualify, quantify, implement, test, and evaluate. The objectives may range from problem solving to modeling to computing, but nature and computation come together at some point to result in natural computing.

All the natural systems studied in this text have many things to teach us. And some of these are of striking importance for computing. First, is that, at least by analogy, they serve the purpose of illustrating how nature works. Any system of units that lack identity or agency, whose behavior arises from the interactions of these components with themselves and the environment, has something in common with many of the systems reviewed. It was observed that completely distinct systems might have similar patterns of behavior.

Another lesson is that nature abounds with problem solving and computing techniques. Evolution can be used to design any type of structure; ants solve routing, clustering, navigation, and other problems; immune systems identify the self, eliminate foreigners and malfunctioning self, and integrate with other bodily systems; the nervous system reasons, take decisions, and coordinates actions; life-like behaviors and patterns allow the simulation, mimicking, and understanding of life; and molecules and atoms lead to novel types of computing machines.

In most of these systems, processes, and methods, however, the overall behavior is not encapsulated in any of its units individually. It is intriguing to realize that the individual entities are usually very simple and with very limited capabilities. The resultant complex behaviors are thus an emergent property of a number of simple units interacting with themselves and the environment. To see how the components produce the response of the whole system, these interactions must be tracked dynamically.

By looking the other way around, natural computing allows us to conclude that many phenomena in nature can be detached from the matter. For instance, the coordinated behavior of birds, fishes, and other animals can be mimmicked by simulated agents moving according to a finite and highly simplified set of rules; the development and flowering of a plant can be studied using a simple system based on production rules; and a few molecules can be used for computation.

11.4 ARTIFICIAL INTELLIGENCE AND NATURAL COMPUTING

The many readers who have ever heard of or are familiar with *artificial intelligence* (AI), or *computational intelligence* (CI), might be wondering what differs natural computing from AI and CI, and also what differs CI from AI. Let us thus take some time to discuss these fields of research, before concluding this book.

11.4.1.   The Birth of Artificial Intelligence

On the campus of the Dartmouth College in Hanover, New Hampshire, during the summer of 1956, several researchers and students joined to discuss the possibilities of creating 'intelligent' computational tools. Among the participants, some of them came to play leading roles in the development of the field of *artificial intelligence* (AI), such as John McCarthy, Marvin Minsky, Herbert Simon, and Allen Newell.

The development of AI happened only a few weeks before the symposium named 'Cerebral Mechanisms in Behavior', which is said to have given birth to the cognitive sciences. It is interesting to note that these symposia had some participants and ideas in common. While the cognitive sciences were mostly concerned with the analysis of human knowledge and intelligence, the field of AI was interested in their synthesis.

After a number of early attempts to develop systems capable of solving a wide range of problems, scientists realized some of the limitations of their approaches and research became more focused on goal-directed problem solving, instead of the development of general problem solvers or complete models of nervous cognition. The most traditional AI techniques were symbolic in nature, which proposed that an algorithmic manipulation of symbolic structures is necessary and sufficient for general intelligence. This symbolic tradition also encompasses approaches based on logic, in which the symbols are used to represent objects and relations among objects, and symbol structures are used to represent known facts. One aspect that marked the traditional AI concerned the way used to build AI systems. The procedural view held that systems could be designed by encoding expertise in domain-specific algorithms, generically known as knowledge-based systems. A classical example is a knowledge-based system for medical diagnosis: the symptoms typically associated with various illnesses are provided by an expert and, combined with a reasoning mechanism, the knowledge-based system gives the diagnosis for new patients.

11.4.2.   The Divorce Between AI and CI

In the mid sixties, new systems started being developed by looking at some forms of 'intelligent behavior' other than human brains, and also by trying to model the uncertainties of the natural language. As results, artificial neural networks (Chapter 4), evolutionary algorithms (Chapter 3), and fuzzy systems were developed. Even more recently, in the mid nineties, algorithms inspired by the social behavior of insects and the simulation of social processes have been developed; these are termed swarm intelligence (Chapter 5). Last, but not least, artificial immune systems (Chapter 6) were developed inspired by theoretical and empirical immunology.

Some of the disagreements between the traditional symbolic and knowledge-based artificial intelligence approaches and mainly artificial neural networks were a result of disputes for funding, and individual lack of success in providing their promised end results. For instance, leading figures in traditional AI, such as Marvin Minsky and Seymour Papert published a book (Minsky and Papert, 1968) demonstrating the limitations of one of the first neural network models, known as *perceptron*, introduced by Frank Rosenblatt. This resulted in an almost complete cessation of funding for research on artificial neural networks for around twenty years. This continued until Rumelhart and collaborators released their book on parallel-distributed processors (Rumelhart et al., 1986) presenting ways of overcoming the limitations of the perceptrons.

After this and other disputes among fields of research, there was a desire to dissociate the research on neural networks from that on knowledge-based systems. The older paradigms based mainly on symbolic, top-down, and expert processing remained as part of artificial intelligence, while the newer paradigms of neural networks, evolutionary computing, and fuzzy systems composed the younger field of computational intelligence (Marks II and Zurada, 1994).

11.4.3. Natural Computing and the Other Nomenclatures

It is common to suggest names for new fields of research. In addition to the two well-known names of artificial intelligence and computational intelligence discussed above, there are also new terminologies, such as *soft computing* (SC), *complex systems, biologically inspired computing* (BIC), and other names used to describe and refer to more or less the same or related topics. However, as we have just seen for AI and CI, there are not only historical reasons for the use of different names, but there are also differences, even if slight, in approaches in these fields. Instead of trying to find crisp boundaries among the nomenclatures, this section briefly describes the meaning (focus) of each terminology.

Soft computing is usually viewed as a fusion of the computational intelligence approaches, which in turn provides foundations for the conception, design, and development of, in most cases, hybrid systems (Bonissone, 1997; Novák, 1998). Thus, the difference between soft computing and computational intelligence is that the former is mainly concerned with hybrid systems. But SC is not just a mixture of approaches; rather it is a partnership, in which each strategy contributes with a distinct feature for addressing problems in its domain. It is also important to remind that soft computing was a term originally coined by L. Zadeh to refer mainly to *fuzzy systems*.

Complex systems, by contrast, is related to the science, mathematics and engineering of systems with simple components that exhibit complex behaviors (www.complex-systems.com). It is a new field of research that investigates how the parts of a system interact with one another and the environment, giving rise to emergent collective behaviors. The focus is on questions about the components, wholes, and their relationships. The natural systems that inspired the development of most (if not all) natural computing approaches are complex systems. The resultant natural computing techniques are also in most cases complex systems by themselves.

Biologically inspired computing is the field of research that takes inspiration from bio-systems or their theoretical models to design algorithms or systems for solving problems (de Castro and Von Zuben, 2004). The field is also termed *computing with biological metaphors*, *bio-inspired computing*, or *biologically motivated computing*. In most cases, biology has a broad meaning, incorporating not only biological, but also chemical and physical systems. Therefore, bio-inspired computing boils down to nature inspired computing; the first branch of natural computing.

To summarize, natural computing is the term used here to encompass all computational intelligence approaches based on some inspiration from nature, soft computing, man-made complex systems, biologically inspired computing, and novel computing paradigms rooted in nature in a higher or lesser degree.

11.5 VISIONS

With classical computers quickly approaching their limit, natural computing promises to deliver a novel computing paradigm. With it comes a brand new theory of computation that incorporates molecules or quantum devices as information storage and processing means. Quantum computers, for instance, have the theoretical capability of simulating any finite physical system and may even hold the enigma to creating 'strong' artificial intelligence and 'strong' artificial life. The influence of natural computing in our lives, and in particular in computing, may be so striking, that even some nomenclature we created to describe computationally hard problems may disappear. Once a molecular or a quantum computer is built, the problems we currently know as hard may not be hard any more.

Nature really has many things to teach us, including a revisitation of the current computing approaches, theory, and technology. Nature computes in a different way than doing arithmetic operations. The scientific community is witnessing the appearance and establishment of alternative ways of computing: *natural computing*. But natural computing also poses several challenges to researchers, such as how to implement it efficiently, how to correct errors, and how do these systems scale. It is often subtle what it may result in the near and far future.

...

Once upon a time, all machines were built, propelled, or controlled by mechanisms designed by living organisms, such as human beings. Wagons were pulled by horses, which, unlike modern vehicles, could steer themselves, refuel themselves, and even reproduce themselves. Automation has replaced most of these subtle creatures with strong but stupid, inflexible slaves. It has come the time for a revolution though. Natural computing is providing new forms of studying, synthesizing, looking at, using, and understanding the natural world so as to produce increasingly more powerful problem-solving techniques, life-like artificial beings and patterns, and computing paradigms.

My taste in science is even more indulgent, however, as perhaps you would expect from an engineer, but I do have my reasons. I personally look forward to the day when we can put life into technology - not quasi-intelligent, monolithic artificial intelligence, but fully thinking, caring, behaving, and even reproducing organisms that are more than the sum of their parts: artificially synthesized complex adaptive systems. Much of the techniques and examples described here may pave the ground towards these more ambitious goals.

11.6 REFERENCES

[1]   Bonissone, P. P. (1997), "Soft Computing: The Convergence of Emerging Reasoning Technologies", *Soft Computing*, **1**(1), pp. 6–18.

[2]   Novák, V. (1998), "Towards Formal Theory of Soft Computing", *Soft Computing*, **2**(1), pp. 4–6.

[3]   de Castro, L. N. and Von Zuben, F. J. (2004), *Recent Developments in Biologically Inspired Computing*, Idea Group Publishing.

[4]   Minsky, M. L. and Papert, S. A. (1969), *Perceptrons*, MIT Press.

[5]   Rumelhart, D. E., McClelland, J. L. and the PDP Research Group (1986), *Parallel Distributed Processing: Explorations in the Microstructure of Cognition, Volume 1: Foundations*, The MIT Press.

[6]   Marks II, R. J. and Zurada, J. M. (1994), *Computational Intelligence: Imitating Life*, IEEE Press.

APPENDIX A

GLOSSARY OF TERMS

The purpose of this glossary is to provide a quick reference for the biological, physical, and chemical terminology used within this book, plus a few other key concepts. The bibliography used to gather this list is available at the end of the appendix. Some terms defined here may have various meanings under different contexts, but the description presented is biased by the perspective on natural computing introduced in this text.

A

A: adenine.

Accessory cell: cell required to initiate an immune response. A term often used to describe antigen-presenting cells (APC).

Acquired immunity: see *adaptive immune response*.

Action potential: electric signal propagated over long distances by excitable cells, e.g., nerve and muscle; it is characterized by an all-or-none reversal of the membrane potential in which the inside of the cell temporarily becomes positive relative to the outside; has a threshold and is conducted without decrement. Also known as *nerve impulse*.

Adaptive immune response: immune response highly specific to antigens, including those associated with microbes. It has an associated lag time and is capable of developing a memory, i.e., to remember previously encountered antigens.

Adenine: purine base, $C_5H_5N_5$, that is the constituent involved in base pairing with thymine in DNA and with uracil in RNA.

Adrenal gland: either of two small, dissimilarly shaped endocrine glands, one located above each kidney, consisting of the cortex, which secretes several steroid hormones, and the medulla, which secretes epinephrine. It is also called suprarenal gland.

Affine: of or relating to a transformation of coordinates that is equivalent to a linear transformation followed by a translation.

Affinity: measure or tightness of the binding between an antigen combining site and an antigenic determinant; the stronger the binding, the higher the affinity. Antibodies produced in a secondary (memory) immune response usually present higher affinities than those of the primary response.

Affinity landscape: representation of the space of all possible antigen binding sites (antibodies or TCRs) along with their affinities.

Affinity maturation: the increase in antibody affinity frequently seen during a secondary immune response.

Agarose: polysaccharide polymer material; linear plymer of carbohydrates, produced by seaweed, that form a gel.

Agent: autonomous entity that senses and acts upon the environment.

Alfa helix: helix configuration assumed by some polypeptide chains; one of the most common types of secondary structures in protein.

Allele: one member of a pair or series of genes that occupy a specific position (or locus) on a specific chromosome.

Allergen: a substance (antigen) that causes an allergy by inducing IgE synthesis.

Allergy: an inappropriate and harmful response of the immune system to normally harmless (nonpathogenic) substances. Used as a synonym to *hypersensitivity*.

Amine: one of a class of strongly basic substances derived from ammonia by replacement of one or more hydrogen atoms by a basic atom or radical.

Amino acids: organic compounds containing an amino group (NH_2), a carboxylic acid group (COOH), and any of various side groups that link together by peptide bonds to form *proteins* or that function as chemical messengers and as intermediaries in metabolism. They form short polymer chains called *peptides* or *polypeptides*, which, in turn, form proteins.

Amino group: the chemical group NH_2, consisting of a nitrogen atom attached by single bonds to hydrogen atoms, alkyl groups, aryl groups, or a combination of them; also called *amino radical*.

AMP: adenosine monophosphate.

Amplification, gene: production of multiple copies of a DNA segment so as to increase the expression level of a gene taken by the segment. See *Polymerase chain reaction (PCR)*.

Anergy: lack of normal immune function, either generalized or antigen specific.

Anneal: to pair DNA or RNA by hydrogen bonds to a complementary sequence, forming a double-stranded polynucleotide. Often used to refer to the binding of a DNA probe, or the binding of a primer to a DNA strand during a polymerase chain reaction.

Annealing: subject a material, such as glass and metal, to a process of heating and slow cooling in order to toughen and reduce brittleness.

Antibody (**Ab**): a soluble (serum) protein molecule produced and secreted by B-cells in response to an antigen. Antibodies are usually defined in terms of their specific binding to an antigen.

Antigen (**Ag**): any substance that when introduced into the body, is capable of inducing an immune response.

Antigen binding site: the part of an antibody molecule that binds specifically to an antigen.

Antigen presentation: process by which certain cells in the body, named antigen presenting cells - APCs, express antigen on their surfaces in a form recognizable by lymphocytes.

Antigen-presenting cells (**APC**): B-cells, cells of the monocyte lineage (including macrophages as well as dendritic cells), and various other body cells that "present" antigen in a form that B- and T-cells can recognize.

Antigen processing: the conversion of an antigen into a form in which it can be recognized by lymphocytes.

Antigen receptor: specific antigen-binding site on B- and T-cells.

Antigenic determinant: chemical structure (small compared to the macromolecule) recognized by the V-region of an antibody. It determines the specificity of antibody-antigen interaction.

Apoptosis: programmed cell death in which a cell brings about its own death and lysis, signaled from outside or programmed in its genes, by systematically degrading its own macromolecules.

Artificial selection: selection by humans of a consciously chosen trait or combination of traits in a(usually captive) population; differing from natural selection in that the criterion for survival and reproduction is the trait chosen, rather than fitness as determined by the entire genotype.

Asexual reproduction: simplest form of reproduction and does not involve meiosis, gamete formation, or fertilization; any reproduction process that does not involve the fusion of cells.

Associative memory: also named content addressable memory (CAM) and is a kind of storage device which includes comparison with each element of storage. A data value is broadcast to all elements of storage and compared with the values there. Those that match are flagged in some way. Subsequent operations can then work on flagged elements, e.g., read them out one at a time or write to certain positions in all of them. A CAM can thus operate as a data parallel processor.

Astrocyte (astroglia): a star-shaped cell, especially a neuroglial cell of nervous tissue that supports the neurons.

Atom: smallest particle of matter with the properties of an element; composed of protons, neutrons, and electrons.

Attenuate: to weaken the effects of a pathogenic microorganism while retaining its viability.

Auto immunity: the production of antibody against your own tissues.

Autoantibody: an antibody that reacts against a person's own tissue or components.

Autoantigen: a molecule that behaves as a self-antigen.

Autocatalytic reaction: reaction in which a product molecule acts as a catalyst for the same reaction; the reagent molecule, thus, catalyses its own reaction.

Autoimmune disease: a disease that arises when the immune system mistakenly attacks the body's own tissues or components. It is the result of a breakdown in self-tolerance. Factors predisposing or contributing to the development of autoimmune diseases include age, genetics, gender, infections, and the nature of the *autoantigen*. Reumathoid arthritis and systemic lupus erythematosus are examples of autoimmune diseases.

Autonomic nervous system: part of the vertebrate nervous system that regulates involuntary action, such as the intestines, heart, and glands. It is divided into the sympathetic nervous system and the parasympathetic nervous system.

Avidity: the summation of multiple affinities.

Axon: the usually long process of a nerve fiber that generally conducts impulses away from the body of the nerve cell.

Axon hillock: anatomical part of a neuron that connects the cell body to the axon.

B

Bacterium: unicellular prokaryotic microorganism of the class schizomycetes which vary in terms of morphology, oxygen and nutritional requirements, and motility. It may be free-living, saprophytic, or pathogenic in plants or animals.

Base pair: structure with hydrogen bonds formed between two complementary nucleotides; the standard base pairs are A-T and C-G.

Base pairing rules: the requirement that adenine must always form a base pair with thymine (or uracil) and guanine with cytosine, in a nucleic acid double helix.

B-cell: small white blood cells expressing immunoglobulin molecules on its surface. Also known as B-lymphocytes, they are derived from the bone marrow and develop into plasma cells that are the main antibody secretors.

B-cell receptor (**BCR**): immunoglobulin molecule on the surface of B-cells. It is composed of four polypeptide chains: two identical heavy (H) and two identical light (L) chains.

Bit: in computing and information theory it corresponds to the fundamental unit of information. It can only assume one of two values: 0 or 1.

Bitstring: ordered sequence of bits.

Blood-brain barrier (**BBB**): a physiological mechanism that alters the permeability of brain capillaries, so that some substances, such as certain drugs, are prevented from entering brain tissue, while other substances are allowed to enter freely.

Binding site: the crevice or pocket on a protein in which a ligand binds.

Biomolecule: organic compound normally present as an essential component of living organisms.

Bond: a uniting force, tie, link, ligation; to join securely.

Bone marrow: soft tissue located in the cavity of the bones. It is the source of all blood cells.

Bony spine: the spinal column of a vertebrate.

Bottom-up: an approach to a problem or study that begins with details and works up to the highest conceptual level; of or relating to a structure or process that progresses from detailed small subunits to a large, basic unit.

Brain: portion of the vertebrate central nervous system that is enclosed within the cranium, continuous with the spinal cord, and composed of gray matter and white matter. It is the primary center for the regulation and control of bodily activities, receiving and interpreting sensory impulses, and transmitting information to the muscles and body organs. It is also the seat of consciousness, thought, memory, and emotion.

Brainstem: portion of the brain, consisting of the medulla oblongata, pons Varolii, and midbrain, that connects the spinal cord to the forebrain and cerebrum.

Bra-ket notation: standard notation for describing quantum states in the theory of quantum mechanics.

Brownian motion: physical phenomenon that minute particles immersed in a fluid move about randomly; mathematical models used to describe those random movements.

C

C: cytosine.

cAMP: see *cyclic AMP*.

Cancer: the name given to a group of diseases characterized by uncontrolled cell growth.

Carbon: chemical element that has the symbol **C** and atomic number 6. It is a naturally abundant nonmetallic element, which occurs in all organic and in many inorganic compounds, capable of chemical self-bonding to form an enormous number of biologically, chemically, and commercially important molecules

Carboxyl group: the univalent radical -COOH present in and characteristic of organic acids.

Cell: the fundamental unit of life; the smallest body, surrounded by a membrane, capable of independent reproduction.

Cell body: the portion of a nerve cell that contains the nucleus but does not incorporate the dendrites or axon. Also called *soma*.

Cell culture: culture of cells *in vitro*.

Cell differentiation: process by which the initially identical cells present during the earliest stages of development not only undergo anatomical alteration but also acquire specialized functional properties, e.g., the antibody production and secretion in high volumes by the plasma cells.

Cellular immunity: immune protection provided by the direct action of immune cells.

Central lymphoid organs: lymphoid organs primarily involved in the production and maturation of immune cells. They include the bone marrow and thymus.

Central nervous system: the portion of the vertebrate nervous system consisting of the brain and spinal cord.

Central tolerance: process whereby immature T- and B-cells acquire tolerance to self-antigens during maturation within the primary lymphoid organs (thymus and bone marrow, respectively). It involves the elimination of cells with receptors for self-antigens.

Centromere: specialized site within a chromosome that serves as the attachment point for the mitotic or meiotic spindle.

Cerebellum: trilobed structure of the brain lying posterior to the pons and medulla oblongata and inferior to the occipital lobes of the cerebral hemispheres. It is responsible for the regulation and coordination of complex voluntary muscular movement as well as the maintenance of posture and balance.

Cerebrum: large rounded structure of the brain occupying most of the cranial cavity, divided into two cerebral hemispheres that are joined at the bottom by the corpus callosum. It controls and integrates motor, sensory, and higher mental functions, such as thought, reason, emotion, and memory.

Chaos: mathematically chaos means an aperiodic deterministic behavior that is very sensitive to its initial conditions.

Chemokines: cytokines that direct cell migration or activate cells.

Chemotaxis: attraction of leukocytes or other cells by chemicals.

Chickenpox: acute contagious disease, primarily of children. It is caused by the varicella-zoster virus and characterized by skin eruptions, slight fever, and malaise. Also called *varicella*.

Chromatin: complex of DNA, RNA and protein that composes the eukaryotic chromosomes.

Chromatides: the two daughter strands of a duplicated chromosome that are still joined by a single centromere.

Chromatography: process in which complex mixtures of molecules are separated by many repeated partitionings between a flowing phase and a stationary phase.

Chromosome: a threadlike linear strand of DNA and associated proteins in the nucleus of eukaryotic cells that carries the genes and functions in the transmission of hereditary information. Each human cell has 23 pairs of chromosomes.

Clonal deletion: the loss of lymphocytes of a particular specificity due mainly to contact with self-antigens.

Clonal selection principle: the prevalent theory stating that the specificity and diversity of an immune response are the result of selection by antigen of specifically reactive clones from a large repertoire of preformed lymphocytes, each with individual specificities.

Clone: (n.) a group of genetically identical cells or organisms descended from a single common ancestor; (v.) to reproduce multiple identical copies.

Codon: sequence of three adjacent nucleotides in a nucleic acid that codes for a specific amino acid.

Coevolution: strictly, the joint evolution of two or more ecologically interacting species, each of which evolves in response to selection imposed by the other. Sometimes used loosely to refer to evolution of one species caused by its interaction with another.

Combinatorial joining: the joining of DNA segments to generate essentially new genetic information, as occurs with BCR and TCR genes during the development of B- and T-cells. Combinatorial joining, or assembly, allows multiple opportunities for a few sets of genes to combine in several different ways.

Complement: a complex series of blood serum proteins, which on sequential activation may mediate protection against microbial infection and contribute to the inflammatory response. They are synthesized by hapatocytes and monocytes and help (complement) antibody responses through a wide spectrum of activities, including a pivotal role in innate defense mechanisms. It might be activated by either the classical (through antibody) or alternative pathway (innate).

Complement cascade: sequence of events usually triggered by an antigen-antibody complex, in which each component of the complement system is activated in turn.

Complementary base sequences: polynucleotide sequences that are related by the base-pairings rules.

Complementary structures: two structures, each of which defines the other; for instance, the two strands of a DNA helix.

Connected: joined or fastened together by some type of connection, which may be physical, chemical or by means of another type of affinity.

Cortex: outer layer of gray matter that covers the surface of the cerebral hemisphere.

Co-stimulation: the delivery of a second signal from an APC to a T-cell. The second signal rescues the activated T-cell from anergy, allowing it to produce the lymphokines necessary for the growth of additional T-cells.

Covalent bond: chemical bonds that involve sharing of electron pairs. It produces a mutual attraction that holds the resultant molecule together.

Cranium: the part of the skull enclosing the brain; the braincase.

Crossover: exchange of genetic material in sexual reproduction. In each one of the parents, genes are exchanged between pairs of homologous chromosomes to form a single one.

Cross-reactivity: the ability of an antibody, specific for one antigen, to react with a slightly different antigen; a measure of relatedness between two different antigenic substances.

Cyclic AMP: molecule important in many biological processes; second messenger within cells whose formation by adenylyl cyclase is stimulated by certain hormones or other molecular signals, used for intracellular signal transduction.

Cytokine: small molecule that signal between cells, inducing growth, differentiation, chemotaxis, activation, enhanced cytotoxicity or regulation of immunity. They are referred to as lymphokines if produced by lymphocytes, interleukins if produced by leukocytes, and monokines if produced by monocytes and macrophages.

Cytoplasm: the total content of a cell excluding the nucleus.

Cytosine: a pyramidine base that occurs in DNA and RNA that binds only with guanine in DNA; one of the five main nucleobases found in nucleic acids.

D

Denaturation (melting): loss of the native configuration of a macromolecule resulting, for instance, from heat treatment, extreme PH changes, chemical treatment, or other denaturing agents. It is usually accompanied by loss of biological activity; double helix separation into single strands, possibly, but not exclusively, by heating.

Dendrite: short fiber that conducts information toward the cell body of the neuron.

Dendritic cells: set of antigen-presenting cells (APCs) present in lymph nodes, spleen and at low levels in blood, which are particularly active in stimulating T-cells.

Deoxyribonuclease: enzyme that breaks a DNA polynucleotide cleaving the phosphodiester bonds.

Deoxyribonucleotide: compound that consists of a purine or pyrimidine base bonded to a sugar, which in turn is bound to a phosphate group; molecule composed of a nitrogenous base attached to the five-carbon sugar deoxyribose, which also has a phosphate group attached to it.

Determinant: part of the antigen molecule that binds to an antibody-combining site or to a receptor on a T-cell. It is also termed *epitope*.

Determinism: proposition that every event and action is causally determined by an unbroken chain of antecedents. No mysterious thing or random events occur.

Development: process by which a single cell becomes a differentiated organism.

Diploid: cell with a full set of genetic material, consisting of chromosomes in homologous pairs and thus having two copies of each autosomal genetic locus. A diploid cell has one chromosome from each parental set.

Disease: alteration in the state of the body or of some of its organs, interrupting or disturbing the performance of the vital functions, and causing or threatening pain and weakness. Also termed *illness* or *sickness*.

Distributed: spread, scattered or diffused over the space or time.

DNA (deoxyribonucleic acid): nucleic acid that carries the genetic information in the cell and is capable of self-replication and synthesis of RNA. DNA consists of two long chains of nucleotides twisted into a double helix and joined by hydrogen bonds between the complementary bases *adenine* and *thymine* or *cytosine* and *guanine*. The sequence of nucleotides determines individual hereditary characteristics.

DNA hybridization: formation of a double helix of DNA or RNA from complementary single strands.

DNA ligase: enzyme that covalently links DNA backbone chains.

DNA microarray: collection of DNA sequences immobilized on a solid surface, with individual sequences laid out in patterned arrays that can be produced by hybridization.

DNA polymerase: an enzyme that catalyzes template-dependent synthesis of DNA from its deoxyribonuclotide 5′-triphosphate precursors.

DNA replication: synthesis of new DNA from an existing DNA.

Domain: a compact segment of an immunoglobulin or TCR chain, made up of amino acids.

Double helix: structure with base pairs containing two polynucleotides; the natural shape of DNA in the cells; coiled conformation of two complementary, antiparallel DNA chains.

Duality: in physics duality corresponds to the property of matter and electromagnetic radiation characterized by the fact that some properties can be explained best by wave theory and others by particle theory.

E

Ecological niche: range of combination of all relevant environmental variables under which a species or population can persist; often more loosely used to describe the 'role' of a species, or the resources it utilizes.

Ecology: science that studies the relationship of organisms to each other and to their environment.

Ecosystem: a system formed by the interaction of a community of organisms with their physical environment.

Electromagnetic: pertaining to or exhibiting magnetism produced by electric charge in motion.

Electron: elementary subatomic particle consisting of a charge of negative electricity.

Electrophoresis: movement of charged solutes in response to an electrical field; often used to separate mixtures of ions, proteins, or nucleic acids.

Electroencephalogram (**EEG**): record of electrical activity of the brain obtained from scalp electrodes. Also called *encephalogram*.

Emergence: complex pattern formation from simpler rules. For a phenomenon to be termed emergent it should generally be unpredictable from a lower level description.

Emulation: endeavor to equal or to excel another in qualities or actions. A software emulator allows computer programs to run on a platform (e.g., operating system or computer architecture) different from the one for which they were originally written.

Endocrine system: the system of glands that produce endocrine secretions (hormones), which help to control bodily metabolic activity.

Endonuclease: an enzyme that makes internal cuts in DNA backbone chains; an enzyme that breaks the phosphodiester bonds of a DNA molecule.

Entanglement: phenomenon in quantum mechanics by which the quantum states of two or more objects are described in relation to each other, even though the individual objects may be spatially separated.

Entity: something that has a distinct, separate existence, though it needs not be a material existence.

Entropy: measure of the degree of disorder of a system.

Environment: totality of the factors that influence the activities, achievements, and ultimate fate of an organism (e.g., animal, plant, or an artificial agent).

Enzyme: any of numerous proteins or conjugated proteins produced by living organisms and functioning as biochemical catalysts.

Epigenesis: theory that development is a process of gradual increase in complexity as opposed to the preformationist view which supposed that a mere increase in size is sufficient to produce adult from embryo.

Epistemology: science (branch of philosophy) that studies the nature of knowledge, its presuppositions and foundations, and its extent and validity.

Epitope: a unique shape, or marker, carried on an antigen's surface, which triggers a corresponding antibody response.

EPR paradox: thought experiment that demonstrates that the result of a measurement performed on one part of a quantum system can have an instantaneous effect on the result of a measurement performed on another part, independently of the distance separating the two parts.

Eukaryote: organism with cells that have nuclear membranes.

Evolution: gradual change in an organism or one of its parts as a result of reproduction, followed by genetic variation and natural selection.

Exon: sequence of DNA that codes information for protein synthesis that is transcribed to messenger RNA.

Exonuclease: enzyme that digests DNA from the ends of the strands; it removes sequentially the nucleotides from the ends of a nucleic acid molecule.

Experiment: procedure performed in a controlled environment for the purpose of gathering observations, data, or facts, demonstrating known facts or theories, or testing hypothesis or theories.

F

False negative: incorrect negative test result that occurs when the attribute for which the subject is being tested actually exists in that subject; test result that is read as negative but actually is positive. Also called a *miss*.

False positive: incorrect positive test result that occurs when a subject does not have the attributes for which the test is being conducted; test result that is read as positive but actually is negative. Also called *false alarm*.

Feedback: process whereby some proportion or in general, function, of the output signal of a system is passed (fed back) to the input. It is usually performed intentionally to control the dynamics of the system.

Fitness: extent to which an organism is adapted to, suitable, or able to produce offspring in a particular environment.

5'-end: end of a nucleic acid that lacks a nucleotide bound at the 5' position of the terminal residue.

Forebrain: anterior of the three principal divisions of the brain.

Fractal: a geometric pattern that is repeated at ever smaller scales to produce irregular shapes and surfaces that cannot be represented by classical geometry. Fractals are used especially in computer modeling of irregular patterns and structures in nature. In many cases a fractal can be generated by a repeating pattern, in a typically recursive or iterative process.

Fungus: general term used to denote a group of eukaryotic protists, including mushrooms, yeasts, rusts, moulds, smuts, etc., which are characterized by the absence of chlorophyll and by the presence of a rigid cell wall composed of chitin, mannans, and sometimes cellulose. They are usually of simple morphological form or show some reversible cellular specialization, such as the formation of pseudoparenchymatous tissue in the fruiting body of a mushroom. The dimorphic fungi grow, according to environmental conditions, as molds or yeasts.

G

G: guanine.

Gamete: cell involved in sexual reproduction, which has half the genetic makeup of the parent cell.

Ganglia: general term for a group of nerve cell bodies located outside the central nervous system, occasionally applied to certain nuclear groups within the brain or spinal cord, e.g., basal ganglia.

Gate: in computing a gate is a low-level digital logic component that performs Boolean functions, store bits of data, and connect and disconnect various parts of the overall circuit to control the flow of data.

Gel electrophoresis: electrophoresis performed in a gel. Electrophoresis is a method of separating substances, especially proteins, and analyzing molecular structure based on the rate of movement of each component in a colloidal suspension while under the influence of an electric field.

Gene: a hereditary unit consisting of a sequence of DNA that occupies a specific locus on a chromosome and determines a particular characteristic in an organism. Genes undergo mutation when their DNA sequence changes.

Gene expression: process by which the biological information carried by a gene is released and made available to the cell, by transcription followed by translation.

Genetic code: set of triplet code words in DNA (or mRNA) coding for the amino acids of proteins.

Genetic engineering: any alteration of genetic material, as in agriculture, to make them capable of producing new substances or performing new functions; process of manipulating genes in an organism, usually outside the organism's normal reproductive process. It often involves the isolation, manipulation, and reintroduction of DNA into model organisms. The aim is to introduce new characteristics to an organism in order to increase its usefulness, such as increasing the yield of a crop species, introducing a novel characteristic, or producing a new protein or enzyme.

Genetic information: information contained in a sequence of nucleotide bases in a DNA (or RNA) molecule.

Genetic recombination: formation of new combination of alleles in offspring as a result of exchange of DNA sequences between molecules. It occurs naturally in the crossing over between homologous chromosomes in meiosis.

Genetic variation: phenotypic variation resulting from the presence of different genotypes in the population.

Genetics: branch of biology dedicated to the study of the genes.

Generalization: formulation of general concepts/knowledge by abstracting/extracting common properties of known instances of a similar situation.

Genome: the complete collection of genetic material carried by an organism or cell.

Genotype: genetic constitution of an organism or cell, which is distinct from its expressed features or phenotype. In practice, it usually refers to the particular alleles present at the loci in question.

Germ line: refers to the genes in germ cells as opposed to somatic cells. In immunology it refers to the genes in their unrearranged state rather than those rearranged for production of immunoglobulin or TCR molecules.

Germ line cells: type of animal cell formed early in embryogenesis and may multiply by mitosis or may produce, by meiosis, cells that develop into gametes.

Gland: organ that produces a secretion for use elsewhere in the body or in a body cavity or for elimination from the body.

Glia: sustentacular tissue that surrounds and supports neurons in the central nervous system; glial and neural cells together compose the tissue of the central nervous system. Also named *neuroglia*, glial cells do not conduct electrical impulses, unlike neurons.

Guanine: purine base that occurs in DNA and RNA that binds only with cytosine in DNA; one of the five main nucleobases found in nucleic acids.

H

Haploid: nucleus, cell or organism possessing a single set of unpaired chromosomes.

Hapten: small molecule that reacts with a specific antibody but cannot induce the formation of antibodies unless bound to a carrier protein or other large antigenic molecule.

Helix: spiral structure with a repeating pattern described by two simultaneous operations - rotation and translation. It is the natural conformation of many regular biological polymers.

Heterozygote: individual organism that possesses different alleles at a locus.

Histamine: physiologically active amine, $C_5H_9N_3$, found in plant and animal tissue and released from mast cells as part of an allergic reaction in humans. It stimulates gastric secretion and causes dilation of capillaries, constriction of bronchial smooth muscle, and decreased blood pressure.

Histocompatibility: if tissues of two organisms are histocompatible, then grafts between the organisms will not be rejected. If however, major histocompatibility antigens are different, then an immune response will be mounted against the foreign tissue.

Homeostasis: tendency to stability in the normal body states (internal environment) of the organism. It is basically achieved by a system of feedback regulatory (control) mechanisms.

Homologous chromosome: chromosomes that pair during meiosis, have the same morphology, and contain genes governing the same characteristics.

Homozygote: cell or diploid organism containing two identical alleles for a certain gene.

Hormone: naturally occurring substance secreted by specialized cells which affect the metabolism or behavior of other cells possessing functional receptors for the hormone. Hormones may be hydrophilic like insulin, in which case the receptors are on the cell surface or lipophilic, like the steroids, where the receptor can be intracellular.

Humoral: pertaining to the extracellular fluids, including the serum and lymph.

Humoral immunity: immune reaction provided with immune fluids, i.e., soluble factors such as antibodies, which circulate in the body's fluids or "humors", primarily serum and lymph.

Hybrid: individual formed by mating between unlike forms, usually genetically differentiated populations or species.

Hydrogen bond: weak attractive force between one electronegative atom and a hydrogen atom that is covalently linked to a second electronegative atom.

Hydrolysis: cleavage of a covalent bond followed by the addition of water elements.

Hydroxyl: univalent radical OH that is a characteristic component of certain acids, bases, alcohols, phenols, carboxylic and sulfonic acids, and amphoteric compounds.

Hypothalamus: part of the brain that lies below the thalamus and functioning to regulate bodily temperature, certain metabolic processes, and other autonomic activities.

I

Idiotope: antigenic determinant (*epitope*) unique to a single clone of cells and located in the variable region of the immunoglobulin product of that clone. The idiotope forms part of the antigen-binding site. Any single immunoglobulin may have more than one idiotope.

Idiotype: the antigenic specificities defined by the unique sequences (idiotopes) of the antigen-combining site. Thus, anti-idiotype antibodies combined with those specific sequences may block immunological reactions and may resemble the epitope to which the first antibody reacts.

Immune: free from the possibility of acquiring a certain infectious disease.

Immune response: alteration in the reactivity of an organisms' immune system in response to an antigen. In vertebrates, this may involve antibody production, induction of cell-mediated immunity, complement activation, or development of immunological tolerance.

Immune system: integrated body system of organs, tissues, cells, and cell products such as antibodies that differentiates self from nonself and neutralizes potentially pathogenic microorganisms or substances.

Immunity: condition of being resistant to an infection, or *immune*.

Immunization: process that increases the reaction of the organisms to a given antigen and therefore improves its ability to resist or overcome infection. The term immunization also describes a technique used to induce immune resistance to a specific disease by exposing the individual to a weakened or died antigen in order to raise antibodies to that antigen.

Immunocompetent: capable of recognizing and acting against an antigen.

Immunoglobulin: general term for all antibody molecules. See *antibody*.

Immunology: science concerned mainly with the study of the structure and function of the immune system, innate and acquired immunity, the bodily distinction of self from nonself, and laboratory techniques involving the interaction of antigens with specific antibodies.

Individual: any specific object within a group, for instance, a person; something that exists as a distinct entity.

Infection: invasion and multiplication of microorganisms in body tissues, which may be not clinically apparent or may result in local cellular injury due to competitive metabolism, toxins, intracellular replication, or antigen-antibody response. The infection may remain localized, subclinical, and temporary if the body's defensive mechanisms are effective. A local infection may persist and spread by extension to become an acute, subacute or chronic clinical infection or disease state. A local infection may also become systemic when the microorganisms gain access to the lymphatic or vascular system.

Inflammatory response: a localized protective response elicited by injury or destruction of tissues, which serves to destroy, dilute or wall off (sequester) both the injurious agent and the injured tissue. It is characterized in the acute form by the classical signs of pain, heat, redness, swelling, and loss of function. Histologically, it involves a complex series of events, including dilatation of arterioles, capillaries and venules, with increased permeability and blood flow, exudation of fluids, including plasma proteins and leucocytic migration into the inflammatory focus.

Innate immune response: first immune response against infections. It works rapidly, gives rise to the acute inflammatory response, and has some specificity for microbes.

Inoculation: act or instance of inoculating, especially the introduction of an antigenic substance or vaccine into the body in order to produce immunity to a specific disease.

Interaction: strictly, the dependence of an outcome on a combination of causal factors, such that the outcome is not predictable from the average effects of the factors taken separately.

Interference: influence among waves when they come into contact; superposition of waves that results in a new wave pattern.

Interleukine (**IL**): glycoproteins secreted by a variety of leukocytes that have effects on other leukocytes.

Internal image: spatial configuration of the combining site of an anti-idiotype antibody that resembles the epitope to which the idiotype is directed.

In vitro: from the Latin *in glass*. Experiments done in a cell or organism free system.

In vivo: from the Latin *in life*. Experiments done in a system such that the organism remains intact, either at the level of the cell or at the level of the whole organism.

Intron: segment of DNA that does not code for protein. The intervening sequence of nucleotides between coding sequences or *exons*.

Ion: atom or group of atoms with a net electric charge acquired by gaining or losing one or more electrons.

K

Karyokinesis: see *mitosis*.

Killer T-cell: T-cell subset that can directly kill body cells infected by viruses or transformed by cancer. Also called *cytotoxic T-cell*.

L

Lag phase: period of time between the introduction of a microorganism (antigen) into the organism or a culture medium and the time B-cells start producing antibodies exponentially.

Lamarckism: theory that evolution is caused by inheritance of character changes acquired during the life of an individual, due to its behavior or to environmental influences.

Learning: act, process, or experience of gaining knowledge or skill; behavioral modification especially through experience or conditioning.

Lesion: any pathological or traumatic discontinuity of tissue or loss of function of a part.

Leukocyte: any of various blood cells that have a nucleus and cytoplasm, separate into a thin white layer when whole blood is centrifuged, and help protect the body from infection and disease. Include *neutrophils*, *eosinophils*, *basophils*, *lymphocytes*, and *monocytes*. Also called *white blood cells*.

Ligand: linking (or binding) molecule.

Ligase: see *DNA ligase*.

Ligation: binding.

Locus: position on a chromosome at which a particular gene is found.

Lymph: clear, watery, sometimes faintly yellowish fluid derived from body tissues that contains white blood cells and circulates throughout the lymphatic system, returning to the venous bloodstream through the thoracic duct. Lymph acts to remove bacteria and certain proteins from the tissues, transport fat from the small intestine, and supply mature lymphocytes to the blood.

Lymph nodes: small bean-shaped organs of the immune system, widely distributed throughout the body and linked by lymphatic vessels. The lymph nodes store special cells that can trap cancer cells or bacteria that are traveling through the body in lymph. Also called *lymph glands*.

Lymphatic vessels: a body wide network of channels, similar to the blood vessels, which remove cellular waste from the body by filtering through lymph nodes and eventually emptying into the blood stream. They carry the *lymph* and are also named *lymphatics*.

Lymphocyte: small white blood cell with virtually no cytoplasm, found in blood, tissue, and in lymphoid organs such as lymph nodes, spleen and Peyer's patches; bears antigen-specific receptors.

Lymphoid organs: the organs of the immune system where lymphocytes develop and congregate. They include the bone marrow, thymus, lymph nodes, spleen and various other clusters of lymphoid tissue. The blood vessels and lymphatic vessels can also be considered lymphoid organs.

Lymphokines: generic term for molecules other than antibodies that are involved in signaling between cells of the immune system and are produced by lymphocytes. These soluble molecules help and regulate the immune responses.

Lysis: dissolution or destruction of cells, such as blood cells or bacteria, as by the action of a specific lysine that disrupts the cell membrane and causes the loss of cytoplasm.

M

Macromolecule: molecule with molecular weight ranging from a few thousands to hundreds of millions.

Macrophage: a large and versatile immune cell derived from monocytes, which acts as a microbe-devouring phagocyte. It is also an antigen-presenting cell and an important source of immune secretions.

Major histocompatibility complex (MHC): a group of genes encoding polymorphic cell-surface molecules (MHC class I and II) that are involved in controlling several aspects of the immune response. MHC genes code for self-markers on all body cells and play a major role in transplantation rejection. Several other proteins are encoded in this region.

Maturation of the immune response: process by which the B-cell receptors (antibodies) increase their affinity in relation to the selective antigenic stimulus. It occurs through a combination of mutational and selective events.

Measurement: quantity, dimension, or capacity determined by measuring; process of estimating or determining the ratio of a magnitude of a quantitative property or relation to a unit of the same type of quantitative property or relation.

Medulla: the inner portion of an organ.

Meiosis: process of cell division in sexually reproducing organisms that reduces the number of chromosomes in reproductive cells from diploid to haploid, leading to the production of gametes in animals and spores in plants.

Membrane potential: electric potential difference across the membrane.

Memory: faculty of retaining and recalling past experience.

Memory cell: cell that presents an active state of immunity to a specific antigen, such that a second encounter with that antigen leads to an enhanced and faster response.

Messenger RNA (mRNA): RNA that serves as a template for protein synthesis.

Metabolism: act or process, by which living tissues or cells take up and convert into their own proper substance the nutritive material brought to them by the blood, or by which they transform their cell protoplasm into simpler substances, which are fitted either for excretion or for some special purpose, as in the manufacture of the digestive ferments. Hence, metabolism may be either constructive (*anabolism*), or destructive (*catabolism*).

Metaphor: one thing conceived as representing another; application of a concept to which it is not literally applicable but which suggests a resemblance and invites comparison.

Microbes: microscopic living organisms. The term is particularly applied to pathogenic organisms, such as bacteria, viruses, fungi, and protozoa. Also named *microorganism*.

Microbiology: branch of science that studies microorganisms.

Microorganism: microscopic organism. It usually refers to disease causing organisms, such as viruses, bacteria, and fungi.

Midbrain: portion of the vertebrate brain that develops from the middle section of the embryonic brain. It is usually involved in unconscious body function.

Mitogen: substance that stimulates the mitosis of certain cells.

Mitosis: process in cell division by which the nucleus divides, typically consisting of four stages, prophase, metaphase, anaphase, and telophase, and normally resulting in two new nuclei, each of which contains a complete copy of the parental chromosomes. Also called *karyokinesis*.

Model: theoretical construct that represents physical, biological, chemical, or social processes, with a set of variables and a set of logical and quantitative relationships between them.

Molecule: smallest particle of a substance that still maintains its chemical and physical properties and is composed of two or more atoms.

Molecular weight: the sum of the atomic weights of the constituent atoms in a molecule.

Molecule: smallest particle of a substance that retains its chemical and physical properties.

Monoclonal: literally, coming from a single clone. In immunology, monoclonal usually describes a preparation of antibody or T-cells that are homogenous; derived from a clone of cells with the same specificity toward an epitope.

Monocyte: large, circulating, phagocytic white blood cell, having a single well-defined nucleus and very fine granulation in the cytoplasm. They emigrate into tissue and differentiate into a *macrophage*.

Monokine: soluble factor secreted by monocytes and macrophages that act on other cells and help to direct and regulate the immune response.

Monomer: small molecule that may become chemically bonded to other monomers to form a *polymer*.

Monospecificity: of single specificity. Refers to cells whose receptors present the capability of recognizing a single antigenic pattern.

Multispecificity: of multiple specificity. Refers to cells whose receptors present the capability of recognizing different antigenic patterns, as far as a minimal amount of interactions occur.

Mutagen: any agent that provokes mutation.

Mutation: change of the DNA sequence within a gene or chromosome of an organism resulting in the creation of a new character or trait not found in the parental type. Also an individual exhibiting such a change.

Myelin sheath: insulating envelope of myelin that surrounds the core of a nerve fiber or axon and facilitates the transmission of nerve impulses. In the peripheral nervous system, the sheath is formed from the cell membrane of the Schwann cell and in the central nervous system, from oligodendrocytes. Also called *medullary sheath*.

N

Naïve lymphocyte: lymphocyte that has not been involved in an immune response.

Natural selection: process in nature by which, according to Darwin's theory of evolution, only the organisms most adapted to their environment tend to survive and transmit their genetic characteristics (material) in increasing numbers to succeeding generations while those less adapted tend to be eliminated.

Negative feedback: type of feedback in which a system responds so as to reverse the direction of change; usually enhances stability.

Negative selection: process that prevents self-specific lymphocytes from becoming auto-aggressive.

Nerve impulse: see *action potential*.

Nervous system: the specialized coordinating system of cells, tissues, and organs that endows animals with sensation and volition. In vertebrates, it is often divided into two systems: the central (brain and spinal cord), and the peripheral (somatic and autonomic nervous system).

Nest: container or shelter made by an animal or insect to hold or grow eggs or offspring. Usually a whole in the ground, tree, rock or building or made of organic material, such as mud, twig, leaves, grass.

Neuroglia: network of branched cells and fibers that supports the tissue of the central nervous system. It is also called *glia*.

Neuron: impulse-conducting cells that constitute the brain, spinal column, and nerves, consisting of a nucleated cell body with one or more dendrites and a single axon. Also known as *nerve cell*.

Neurotransmitters: any of a group of substances that is released on excitation from the axon terminal of a presynaptic neuron of the central or peripheral nervous system and travel across the synaptic cleft to either excite or inhibit the target cell.

Nitrogen bases: type of molecule that forms an important part of nucleic acids, composed of ring structures containing nitrogen. Hydrogen bonds between them bind two single DNA strands into a double helix.

Nuclease: enzyme that cleaves the phosphodiester bonds between the nucleotide subunits of nucleic acids; enzyme that degrades a nucleic acid molecule.

Nucleic acid: see *polynucleotide*.

Nucleotide: the basic structural unit of nucleic acids (DNA or RNA); molecule composed of a nitrogen base, a sugar, and a phosphate.

Nucleus: large, membrane-bound, usually spherical protoplasmic structure within a living cell, containing the hereditary material of the cell and controlling its metabolism, growth, and reproduction.

O

Offspring: progeny or descendants.

Oligonucleotide: short polymer of nucleotides (usually less than 50).

Ontogeny: origin and development of an individual organism from embryo to adult; history of the individual development of an organism, or of the evolution of the germ. Also called *ontogenesis*.

Organ: combination of different types of tissues.

Organelle: differentiated structure within an eukaryote cell, such as mitochondria, ribosome or lisosome, that performs a specific function.

Organism: individual form of life, such as a plant, animal, bacterium, protist, or fungus; a body made up of organs, organelles, or other parts that work together to carry on the various processes of life.

Ovaries: the usually paired female or hermaphroditic reproductive organ that produces ova and, in vertebrates, secrete female hormones estrogen, which develops and maintains female characteristics, and progesterone, which prepares the uterus for pregnancy.

P

Pairing: the sideways attachment of two homologous chromosomes prior to crossover.

Parasite: organism that grows, feeds, and is sheltered on or in a different organism while contributing nothing to the survival of its host.

Paratope: an antibody-combining site that is complementary to an epitope.

Parsimony: economy in the use of means to an end.

Particle: basic unit of matter or energy; small piece or part; a tiny portion or speck.

Pathogen: a microorganism that causes disease.

Pathogenic: disease-causing.

Pathology: (study of) the nature of disease and its causes, processes, development, and consequences; anatomic or functional manifestation of a disease.

Peptide: any of various natural or synthetic compounds containing two or more amino acids linked by the carboxyl group of one amino acid to the amino group of another.

Peptide bond: a covalent bond between two amino acids in which the amino group of one amino acid is bonded to the carboxyl group of the other with the elimination of H_2O.

Peyer's patches: lymphoid organs located in the sub mucosal tissue of the mammalian gut containing very high proportions of cells capable of secreting a specific type of antibody. The patches have B- and T-dependent regions and germinal centers. A specialized epithelium lies between the patch and the intestine. It is involved in gut associated immunity.

pH: negative logarithm of the hydrogen ion concentration of an aqueous solution.

Phagocyte: large white blood cells that contribute to the immune defenses by ingesting and digesting waste material, microbes or other cells and foreign particles.

Phagocytosis: process by which cells engulf material and enclose it within a vacuole (phagosome) in the cytoplasm.

Phenotype: the physical or biochemical expression of an individuals' genotype; observable expressed traits, such as eye and skin color.

Pheromone: chemical produced and secreted by a living organism, especially insects, to transmit a message to other members of the same species. There are different types of pheromone, such as sex, food trail, alarm, and others.

Phosphate: radical consisting of one phosphorus atom and four oxygen, PO_4; salt or an ester of phosphoric acid.

Phosphodiester bond: chemical ligation that unites adjacent nucleotides in a polynucleotide.

Photoelectric effect: emission of electrons from matter by incidence of electromagnetic radiation, such as light or ultraviolet radiation.

Photon: quantum of the electromagnetic field, for instance light, originally called energy quantum; quantum of electromagnetic energy, regarded as a discrete particle of no electric charge, zero mass, and an indefinitely long lifetime.

Physiology: (study of) the functions of living organisms and their parts

Plasma: fluid part of the circulating blood.

Plasma cells: terminally differentiated antibody-producing cells that develop from B-cells.

Polymer: regular, covalently bonded arrangement of basic subunits (monomers) that is produced by repetitive application of one or a few chemical reactions; generic term used to describe a substantially long molecule.

Polymerase: see *DNA polymerase*.

Polymerase chain reaction (PCR): technique that allows generating multiple copies of a DNA molecule generated by amplifying a target DNA.

Polynucleotide: nucleotide polymer; linear sequence of nucleotides in which the 3′ position of the sugar of one nucleotide is linked through a phosphate group to the 5′ position on the sugar of the adjacent nucleotide. Also called *nucleic acid*.

Polypeptide: a peptide, such as a small protein, containing many molecules of amino acids.

Pons: band of nerve fibers on the ventral surface of the brain stem that links the medulla oblongata and the cerebellum with upper portions of the brain. It is also called *pons Varolii*.

Positive feedback: in this type of feedback the response of the system is to change a changing variable even more in the same direction. If uncontrolled it may promote a snow ball effect.

Positive selection: serves the purpose of avoiding the accumulation of useless lymphocytes with either no receptor at all or with receptors that are unproductive for the organism.

Primary lymphoid organs: organs mainly responsible for the production and maturation of lymphocytes.

Primary response: immune response as a consequence of the first encounter with a given antigen. It is generally weak, has a long lag phase, and generates immune memory.

Primer: short olygonucleotide that binds to a single stranded DNA to provide a site for DNA replication; short DNA or RNA that can act as the start point for the growth of a chain in 3′ when bound to a template.

Prokaryote: organisms whose cells are characterized by the absence of a delimited nucleus or by the general absence of a membrane.

Protein: organic compound made up of amino acids, which is one of the major constituents of plants and animal cells.

Purine: one of the types of nitrogen bases that compose the nucleotides. The DNA purine bases are adenine and guanine.

Pyramidine: of the types of nitrogen bases that compose the nucleotides. The DNA pyramidine bases are thymine and cytosine, and in RNA is uracil.

Q

Quantization: state of being constrained or act of constraining to a set of discrete values, rather than varying continuously.

Quantum: smallest amount of a physical quantity that can exist independently, especially a discrete quantity of electromagnetic radiation; refers to an indivisible and perhaps elementary entity. See *photon*.

Quantum bit: basic unit of information in quantum computing and information. Differently from the bits in classical systems, a quantum bit may have more than two possible states: 0, 1, and a combination of the two states obeying the superposition principle.

Quantum computer: type of computer (device) that uses the principles (e.g., entanglement and superposition) of quantum systems, like a collection of atoms, to compute.

Quantum gate: simple quantum circuit that operates on a small number of quantum bits.

Quantum information: information held in the 'state' of a quantum system.

Quantum mechanics: mathematical framework that describes quantum physics.

Quantum physics: theory from physics that explains the behavior of objects with atomic scales.

Quantum register: the analogue in quantum computing of a classical processor register.

Quantum states: possible states that can be assumed by a quantum mechanical system.

Qubit: see *quantum bit*.

R

Realization: making real or giving the appearance of reality; literal, material model that implements certain functions of the original.

Receptor: cell surface molecule that binds specifically to particular proteins or peptides.

Recombination: appearance in the offspring of traits that were not found together in either of the parents; genetic event that occurs during the formation of sperm and egg. See *cross-over*.

Recognition: process by which an immune cell or molecule specifically identifies and matches (bind) with a given antigen.

Recruitment: behavioral mechanism that enables an ant colony to assemble rapidly a large number of foragers at a desirable food source and to perform efficient decision making, such as the selection of the most profitable food source or the choice of the shortest path between the nest and the food source.

Red blood cell: cell in the blood of vertebrates that transports oxygen and carbon dioxide to and from the tissues. In mammals, the red blood cell is disk-shaped and biconcave, contains hemoglobin, and lacks a nucleus. Also called *erythrocyte, red cell, red corpuscle*.

Reductionism: attempt or tendency to explain a complex set of facts, entities, phenomena, or structures by another, simpler set; theory that suggests that the nature of complex things can always be reduced to (explained by) simpler or more fundamental things.

Reflex: involuntary action or response, such as a sneeze, blink, or hiccup. It is also used to describe muscle responses.

Register: in terms of computer processor it corresponds to the part used as a storage location, often aimed at speeding up the execution of programs by providing fast access to frequently used values.

Regulatory genes: genes whose primary function is to control the rate of synthesis of the products of other genes.

Reinforcement: strengthening of a response; increase in likelihood of something.

Renaturation: return of a protein or nucleic acid from a denatured state to its native configuration; refolding of an unfolded (denatured) protein so as to restore the native structure and protein function.

Repertoire: set of cells or molecules in the immune system. Used as a synonym to *population*.

Resilience: for what disturbance the system might flip out of it stability domain; the better the resilience the larger the disturbance required to get the system out of its stability domain.

Restriction enzyme: enzyme that cuts double-stranded DNA at specific short nucleotide sequences. Variations in this sequence within a population results in variation in DNA sequence lengths after treatment with a restriction enzyme.

Ribonucleotide: nucleotide in which a purine or pyrimidine base is linked to a ribose molecule. The base may be adenine, guanine, cytosine, or uracil.

Ribonuclease: enzyme that degrades RNA.

Ribosome: organelles or sites in which proteins are synthesized.

RNA (ribonucleic acid): polymer of ribonucleotides.

RNA polymerase: enzyme that catalyzes the formation of RNA from ribonucleoside triphosphates using DNA as a template.

S

Scavenger cells: any of a diverse group of cells that have the capacity to engulf and destroy foreign material, dead tissues, or other cells.

Seaweed: any of a large number of marine benthic algae that are multicellular, large-bodied, and thus differentiated from most algae that tend toward microscopic size.

Second signal: the delivery of a co-stimulatory signal from an APC to a T-cell. The co-stimulatory signal rescues the activated T-cell from anergy, allowing it to produce the lymphokines necessary for the growth of additional T-cells.

Secondary lymphoid organs: organs where the immune cells interact with the antigenic stimuli, thus initiating adaptive immune responses.

Secondary response: immune response that follows a second or subsequent encounter with a particular antigen.

Selection: nonrandom differential survival (choice) or reproduction of an entity.

Self-organization: organize yourself/itself; process in which the organization of a system increases automatically without being guided or managed by an outside source. Self-organizing systems usually present emergent properties.

Self-similarity: object that is exactly or approximately similar to part(s) of itself.

Serum: the clear fluid that is obtained upon separating whole blood into its solid and liquid components after it has been allowed to clot. Blood serum from the tissues of immunized animals contains antibodies that are used to transfer immunity to another individual.

Sexual reproduction: production of offspring whose genetic constitution is a mixture of that of two potentially genetically different gametes.

Simulation: imitation of something that attempts to represent certain features of the behavior of the system to be simulated by the behavior of another system; representation of the operation or features of one process or system through the use of another. It can be used in various contexts, including the modeling of nature.

Skull: bony or cartilaginous framework of the head of vertebrates made up of the bones of the braincase and face. It is also named *cranium*.

Sociobiology: scientific discipline that investigates the biological determinants and (evolutionary) advantages of social behavior.

Soma: the neuron cell body that contains the nucleus.

Somatic cell: cell not involved in reproduction; any cell other than a germ cell.

Somatic mutation: process occurring during B-cell clonal expansion and affecting the antibody gene region, which, together with selection, permits refinement of the antibody specificity with relation to the selective antigen.

Speciation: evolution of reproductive isolation within an ancestral species, resulting in two or more descendant species.

Species: in the sense of biological species, the members of a group of populations that interbreed or potentially interbreed with each other under natural conditions; a fundamental taxonomic category to which individual specimens are assigned, which often but not always corresponds to the biological species.

Spinal cord: thick, whitish cord of nerve tissue that extends from the medulla oblongata down through the spinal column and from which the spinal nerves branch off to various parts of the body.

Spleen: large, highly vascular lymphoid organ, lying in the human body to the left of the stomach below the diaphragm, serving to store blood, disintegrate old blood cells and filter foreign substances from the blood.

Splice: to join together or insert (segments of DNA or RNA) so as to form new genetic combinations or alter a genetic structure.

Stem cell: cell that gives rise to a lineage of cells. Particularly used to describe the most primitive cells in the bone marrow from which all the various types of blood cells are derived.

Stigmergy: method of communication in which the individuals of the system communicate with each other by modifying their local environment.

Stimulus: a factor that can be detected by a receptor, which in turn produces a response.

Strand: single filament that can be twisted together to form a double-strand.

Sugar: carbohydrate which contains the functional group $(CH_2O)_n$.

Superposition: state of being superposed or act of superposing; property of certain physical quantities in which the net result caused by two or more phenomena is the (weighted) sum of the results obtained by each phenomenon independently.

Suppression: mechanism for producing a specific state of immunologic unresponsiveness by which one cell or its products act on another cell.

Swarm: aggregate of organisms (e.g., animals and insects), especially when in motion.

Synapse: junction across which a nerve impulse passes from an axon terminal to a neuron, muscle cell, or gland cell.

Synaptic cleft: narrow space between the presynaptic cell and the postsynaptic cell in a chemical synapse, across which the neurotransmitter diffuses.

T

T: thymine.

T-cell: small white blood cell that orchestrate and/or directly participate in the immune defenses. Also known as T lymphocyte, it maturates in the thymus and secretes lymphokines.

Teleportation: process of moving matter or information by dematerializing, usually instantaneously, at one point and recreating it at another. Also called *teletransportation*.

Teletransportation: See *teleportation*.

Template: structure in some direct physical process that can cause the patterning of a second structure, usually complementary to it in some sense.

Template strand: strand of nucleic acid used by a polymerase as a template to synthesize a complementary strand.

Thalamus: large ovoid mass of grey matter situated in the posterior part of the forebrain that relays sensory and motor impulses to the cerebral cortex.

3' end: end of a nucleic acid that lacks a nucleotide bond at the 3' position of the terminal residue.

Thymine: pyramidine basis that occurs in DNA and that binds with adenine; one of the five main nucleobases found in nucleic acids.

Thymus: small glandular primary lymphoid organ situated behind the top of the breastbone; site where T-cells proliferate and maturate. It is also considered to be an endocrine organ.

Tissue: organization of differentiated cells of of a similar type. There are four basic types of tissue: muscle, nerve, epidermal, and connective

Tolerance: a state of nonresponsiveness to a particular antigen or group of antigens.

Tonsils and adenoids: prominent oval masses of lymphoid tissues on either sides of the throat. They are primarily associated with the protection of the respiratory system.

Top-down: an approach to a problem that begins at the highest conceptual level and works down to the details; of or relating to a structure or process that progresses from a large, basic unit to smaller, detailed subunits

Toxins: poisonous substance, especially a protein, produced by living cells or organisms, normally very damaging to mammalian cells, that can be delivered directly to target cells by linking them to antibodies or lymphokines. They are different from the simple chemical poisons by their high molecular weight and antigenicity.

Trail: path, mark, scent, course, or trace left by a moving organism.

Trait: genetically determined characteristic or condition.

Transcription: process by which messenger RNA is synthesized from a DNA template resulting in the transfer of genetic information from the DNA molecule to the messenger RNA.

Transduction: transfer of genetic material (and its expressed phenotype) from a bacterium cell to another through bacteriophages.

Transformation: in mathematics it corresponds to invertible functions; mapping of one space onto itself or onto another.

Translation: process by which messenger RNA directs the amino acid sequence of a growing polypeptide during protein synthesis.

True negative: correct negative test result that occurs when the attribute for which the subject is being tested actually exists in that subject; test result that is read as negative when it is really negative.

True positive: correct positive test result that occurs when a subject does have the attributes for which the test is being conducted; test result that is read as positive when it really is positive. Also called a *hit*.

Tumor: abnormal growth of tissue resulting from uncontrolled, progressive multiplication of cells and serving no physiological function; may be benign (not cancerous) or malignant. It is also called a *neoplasm*.

U

U: uracil.

Unresponsiveness: inability to respond to an antigenic stimulus. It may be specific for a particular antigen (see *tolerance*), or broadly nonspecific as a result of damage to the entire immune system, e.g., after whole-body irradiation.

Uracil: pyramidine basis that occurs in RNA replacing thymine as found in DNA, and that binds with adenine; one of the five main nucleobases found in nucleic acids.

V

Vaccine: preparation of a weakened or killed pathogen, such as a bacterium or virus, or of a portion of the pathogen's structure that upon administration stimulates antibody production or cellular immunity against the pathogen but is incapable of causing severe infection. By stimulating an immune response (but not disease), it protects against subsequent infection by that organism.

Vaccination: process of inoculating an individual with a vaccine in order to protect against a particular disease.

Vertebrates: animals having a vertebral column, members of the phylum chordata, subphylum vertebrata comprising mammals, birds, reptiles, amphibians, and fishes.

Virus: obligate intracellular parasite that often causes disease and that consist essentially of a core of RNA or DNA surrounded by a protein coat.

W

Watson-Crick complementarity: process in DNA bonding where the nucleotide A bonds with T and G bonds with C.

Wavefunction: in quantum mechanics a wave function is a mathematical function used to describe the propagation of the wave associated with any particle or group of particles.

White blood cell: any of various blood cells that have a nucleus and cytoplasm, separate into a thin white layer when whole blood is centrifuged, and help protect the body from infection and disease. White blood cells include *neutrophils, eosinophils, basophils, lymphocytes*, and *monocytes*. It is also called *leukocytes, white cells* and *white corpuscles*.

Z

Zygote: the result of the union of the male and female sex cells. The zygote therefore has d diploid number of chromosomes.

A.1 BIBLIOGRAPHY

[1] Abbas, A. K., Lichtman, A. H. and Pober, J. S. (1998), *Cellular and Molecular Immunology*, W. B. Saunders Company.

[2] Barret, J. T. (1983), *Textbook of Immunology An Introduction to Immunochemistry and Immunobiology*, 4th Ed., The C. V. Mosby Company.

[3] Benjamini, E., Sunshine, G. and Leskowitz, S. (1996), *Immunology A Short Course*, 3rd Ed., Wiley-Liss.

[4] Biology Online: *Information in the Biological Sciences*, [On Line] http://www. biology-online.org/default.htm

[5] Brown, T. A. (1998), *Genetics: A Molecular Approach*, 3rd Ed., Stanley Thornes.

[6] Bullock, J., Boyle, J. and Wang, M. B. (1984), *Physiology*, New York: J. Wiley, Pennsylvania: Harwal Pub.

[7] Burns, G. W. and Bottino, P. J. (1989), *The Science of Genetics*, 6th Ed., Macmillan Publishing Company.

[8] Burton, G. R. W. and Engelkirk, P. G. (1996), *Microbiology for the Health Sciences*, Lippincott-Raven Publishers.

[9] Cancer WEB Online Dictionary: *On-line Medical Dictionary, © 1997-98 Academic Medical Publishing and CancerWEB* [On Line] http://cancerweb.ncl.ac.uk/.

[10] Coleman, R. M., Lombard, M. F. and Sicard, R. E. (1992), *Fundamental Immunology*, 2nd Ed., Wm. C. Brown Publishers.

[11] de Castro, L. N. and Timmis, J. (2002), *Artificial Immune Systems: A New Computational Intelligence Approach*, Springer-Verlag.

[12] Futuyma, D. J., (1998), *Evolutionary Biology*, 3rd Ed., Sinauer Associates, Inc.

[13] Griffiths, A. J. F., Miller, J. H., Suzuki, D. T., Lewontin, R. C. and Gelbart, W. M. (2000), *Introduction to Genetics*, 7th Ed., W. H. Freeman and Company.

[14] Guyton, A. C. (1991), *Textbook of Medical Physiology*, 8th Ed., W. B. Saunders Company.

[15] Hood, L. E., Weissman, I. L., Wood, W. B. and Wilson, J. H. (1984), *Immunology*, 2nd Ed., The Benjamin/Cummings Publishing Company, Inc.

[16] Janeway, C. A., Travers, P., Walport, M. and Capra, J. D. (1999), *Immunobiology: The Immune System in Health and Disease*, 4th Ed., Garland Publishing.

[17] Klein, J. (1990), *Immunology*, Blackwell Scientific Publications.

[18] Krebs, C. J. (1994), *Ecology The Experimental Analysis of Distribution and Abundance*, 4th Ed., Harper Collins College Publishers.

[19] Lehninger, A. L. (2000), *Lehninger Principles of Biochemistry*, 3rd Ed., Worth Publishers.

[20] Lydyard, P. M., Whelan, A. and Fanger, M. W. (2000), *Instant Notes in Immunology*, BIOS Scientific Publishers Limited.

[21] Mackenna, B. R. and Callander, R. (1998), *Illustrated Physiology*, Churchill Livingstone.

[22] Voet, D., Voet, J. G. and Pratt, C. W. (1999), *Fundamentals of Biochemistry*, John Wiley & Sons.

[23] Watson, J. D. (2003), *Molecular Biology of the Gene*, 5th Ed., Benjamin Cummings.

[24] WEB Online Dictionary,*Dictionary.com Copyright © 2005, Lexico Publishing Group, LLC*, [On line] http://dictionary.com.

[25] WEB Wikipedia, The Free Encyclopedia. *Wikipedia Foundation*, [On Line] http://www.wikipedia.org.

APPENDIX B

THEORETICAL BACKGROUND

The collection of concepts, theorems, and axioms to be presented in the following are basic for a thorough understanding of natural computing. Specific chapters will require more of some concepts than others, and this is emphasized in each chapter/section as appropriate and also in the Preface. The inclusion of these topics as an Appendix to the book aims at making it a self-contained volume on the subjects studied. Even those readers well versed in mathematics are invited to go through this appendix or parts of it when directed. Only fundamental concepts are presented, and a list of relevant textbooks in which to search for a more complete coverage of each topic is provided as a bibliography at the end of the appendix. Three main topics are reviewed here: linear algebra, statistics, and theory of computation. Other concepts, such as optimization, logic of propositions, theory of nonlinear dynamical systems, graph theory, data clustering, affine transformations, and Fourier transforms are briefly reviewed as well.

B.1 LINEAR ALGEBRA

Linear algebra is the branch of mathematics concerned with the study of vectors, matrices, vector spaces, linear transformations, and systems of linear equations. All these concepts are central for a good understanding of natural computing.

B.1.1. Sets and Set Operations

Sets

A *set* may be defined as an aggregation or collection of objects, called the elements or members of the set. Sets are denoted here by italic capital letters of the Roman alphabet. Some special sets are described by specific symbols:

- $\aleph$: set of natural numbers
- $\Re$: set of real numbers
- c: set of complex numbers

The logic state or association of an element $\mathbf{x}$ to a set X is represented by

$$\mathbf{x} \in X: \mathbf{x} \text{ belongs to } X$$

$$\mathbf{x} \notin X: \mathbf{x} \text{ does not belong to } X$$

A set may be specified by listing its elements

$$X = \{\mathbf{x}_1, \mathbf{x}_2, ..., \mathbf{x}_n\}$$

or by stating the properties that characterize its elements:

$$X_2 = \{\mathbf{x} \in X_1 \text{ such that } P(\mathbf{x}) \text{ is true}\} \text{ or } X_2 = \{\mathbf{x} \in X_1\colon P(\mathbf{x})\}$$

Set Operations

The main set operations are:

- *Union*: $X_1 \cup X_2 = \{\mathbf{x}\colon \mathbf{x} \in X_1 \text{ or } \mathbf{x} \in X_2\}$

- *Intersection*: $X_1 \cap X_2 = \{\mathbf{x}\colon \mathbf{x} \in X_1 \ e \ \mathbf{x} \in X_2\}$

  $X_1 \cap X_2 = \varnothing$ (empty set) if X_1 and X_2 are *disjoint sets*.

The *complement* of a set X is represented by $\overline{X}$ and is defined as

$$\overline{X} = \{\mathbf{x}\colon \mathbf{x} \notin X\}$$

S is a *subset* of X if every element in S is also an element of X. In this case, we say that S is *contained* in X $(S \subset X)$ or that X *contains* S $(X \supset S)$.

B.1.2. Vectors and Vector Spaces

Scalar

A *scalar* is a variable that assumes values on the axis of real numbers. They are described here by lowercase italic letters of the Roman alphabet. The set of all scalars is represented by $\Re$.

The magnitude of a real scalar x is given by $|x| = \begin{cases} x & \text{if } x \geq 0 \\ -x & \text{if } x < 0 \end{cases}$.

Vector

An ordered array of n scalars $x_i \in \Re$ $(i = 1, 2, \dots, n)$ is termed a *vector* of dimension n. In other words, an n-tuple of real numbers is a vector (of length n). Vectors are described here by lowercase boldface letters of the Roman alphabet, and are placed columnwise:

$$\mathbf{x} = \begin{bmatrix} x_1 \\ x_2 \\ \vdots \\ x_n \end{bmatrix}$$

The numbers $x_1, \dots, x_n$ are called the *coordinates, components, entries* or *elements* of $\mathbf{x}$. The set of all real-valued vectors of length n is represented by $\Re^n$.

- A scalar is a vector of length $n = 1$.
- Vector $\mathbf{0}_n$ is the zero vector of length n, with all elements equal to 0.
- Vector $\mathbf{1}_n$ is the vector of length n, with all elements equal to 1.

Vector $\mathbf{e}_i$ is a unit norm vector of length n (indicated by the context) with all elements equal to 0, with the exception of the i-th element, which is equal to 1 $(1 \leq i \leq n)$.

Linear Vector Space

A *linear vector space* $(X,\Im)$ consists of a set of vectors X and a field $\Im$ on which two operations are defined:

Addition (+): a mapping $X \times X \to X$ such that $\forall$ **x**, **y** $\in X$, (**x**,**y**) $\to$ (**x** + **y**) results in **x** + **y** $\in X$.

Scalar multiplication ($\cdot$): a mapping $\Re \times X \to X$ such that $\forall$ **x** $\in X$ and $\forall$ $a \in \Im$, (a,**x**) $\to$ ($a\cdot$**x**) results in $a\cdot$**x** $\in X$.

A number of axioms may still be deduced to the space $(X,\Im)$:

(1)	**x**+**y** = **y**+**x**	(commutativity)
(2)	(**x**+**y**)+**z** = **x**+(**y**+**z**)	(associativity)
(3)	$a\cdot$(**x**+**y**) = $a\cdot$**x** + $a\cdot$**y**	(distributivity)
(4)	**x**+**0** = **x**	(zero vector)
(5)	(ab)$\cdot$**x** = $a\cdot$($b\cdot$**x**)	(associativity)
(6)	(a+b)$\cdot$**x** = $a\cdot$**x**+ $b\cdot$**x**	(distributivity)
(7)	$1\cdot$**x** = **x**	(zero vector)

Examples: $(\Re,\Re)$, $(\Re^n,\Re)$, $(\Re^{m\times n},\Re)$, where n and m are positive integers.

Linear Vector Subspace

Let $(X,\Im)$ be a linear vector space and S a subset of X. We say that $(S,\Im)$ is a linear vector subspace of $(X,\Im)$ if S forms a linear space over $\Im$ with respect to the same operations defined on $(X,\Im)$.

Linear Variety

The *translation* of a subspace is called *linear variety*. A linear variety may be described as:

$$V = \mathbf{x} + S$$

where $(S,\Im)$ is a linear subspace of $(X,\Im)$ and **x** $\in X$.

Convex Set

A set X in a linear vector space is a *convex set* if $\forall$ $\mathbf{x_1},\mathbf{x_2} \in X$, all elements such that $a\mathbf{x_1} + (1 - a)\mathbf{x_2}$, $0 \leq a \leq 1$, also belong to X (Figure B.1). The following are convex sets:

- Empty set.
- Linear varieties.
- Subspaces.
- Intersections of convex sets.

(a)   (b)

Figure B.1: Convexity. (a) Convex. (b) Non convex.

Linear Combinations, Spanning Sets and Convex Combinations

Let $S = \{\mathbf{x}_1, \mathbf{x}_2, \dots, \mathbf{x}_n\}$ be a subset of a linear vector space $(X,\mathfrak{I})$. A *linear combination* $\mathbf{v}$ of the elements of S is formed by

$$\mathbf{v} = a_1\mathbf{x}_1 + a_2\mathbf{x}_2 + \dots + a_n\mathbf{x}_n$$

where $a_1, a_2, \dots, a_n \in \mathfrak{I}$. Alternatively, vector $\mathbf{v}$ is a linear combination of vectors $\mathbf{x}_1, \mathbf{x}_2, \dots, \mathbf{x}_n$ if there is a solution to the equation above, where $a_1, a_2, \dots, a_n$ are unknown constants.

Vectors $\mathbf{x}_1, \mathbf{x}_2, \dots, \mathbf{x}_n$ in $(X,\mathfrak{I})$ are said to *span* $(X,\mathfrak{I})$ or to form a *spanning set* of $(X,\mathfrak{I})$ if every $\mathbf{v} \in V$ is a linear combination of the vectors $\mathbf{x}_1, \mathbf{x}_2, \dots, \mathbf{x}_n$; that is, if there exist scalars $a_1, a_2, \dots, a_n \in \mathfrak{I}$ such that

$$\mathbf{v} = a_1\mathbf{x}_1 + a_2\mathbf{x}_2 + \dots + a_n\mathbf{x}_n$$

If the scalars $a_1, a_2, \dots, a_n \in \mathfrak{I}$ are such that $a_i \geq 0$ ($i=1, 2, \dots, n$) and $\sum_{i=1}^{n} a_i = 1$, then the linear combination is called *convex combination* of the elements $\mathbf{x}_1, \mathbf{x}_2, \dots, \mathbf{x}_n \in X$.

Linear Dependence and Independence

A vector $\mathbf{x}$ is said to be *linearly dependent* in relation to a set of vectors S if $\mathbf{x}$ can be written as a linear combination of the elements of S; that is, if $\mathbf{x} \in [S]$. Otherwise, $\mathbf{x}$ is said to be *linearly independent* of the set S. A set of vectors is linearly independent if and only if $a_i = 0$, $i = 1, 2, \dots, n$ is the single solution to equation $\sum_{i=1}^{n} a_i\mathbf{x}_i = \mathbf{0}$.

Basis and Dimension of a Linear Vector Space

A set $S = \{\mathbf{x}_1, \mathbf{x}_2, \dots, \mathbf{x}_n\}$ of vectors is a *basis* of $(X,\mathfrak{I})$ if: 1) S is linearly independent and 2) S spans $(X,\mathfrak{I})$.

If $S = \{\mathbf{x}_1, \mathbf{x}_2, \dots, \mathbf{x}_n\}$ is a basis to $(X,\mathfrak{I})$, then the vector space $(X,\mathfrak{I})$ has dimension n. Any other basis to the same space $(X,\mathfrak{I})$ has the same number of elements.

Any vector $\mathbf{y} \in X$ may be expressed in the form:

$$y = b_1 \mathbf{x}_1 + b_2 \mathbf{x}_2 + \cdots + b_n \mathbf{x}_n = \begin{bmatrix} \mathbf{x}_1 & \mathbf{x}_2 & \cdots & \mathbf{x}_n \end{bmatrix} \cdot \begin{bmatrix} b_1 \\ b_2 \\ \vdots \\ b_n \end{bmatrix}$$

where $\mathbf{b} = [b_1 \; b_2 \; \cdots \; b_n]^T$ is the representation of $\mathbf{y}$ in the basis S (the superscript T represents transposition). In other words, any vector $\mathbf{y} \in X$ can be written as a linear combination of the basis vectors. The representation of a vector in a given basis is unique.

If n is finite, the vector space $(X, \Im)$ has finite dimension. For an infinite n, $(X, \Im)$ is said to have an infinite dimension.

Dot (Inner) Product

The *dot or inner product* between two vectors is a function $f_{dot}: \Re^n \times \Re^n \to \Re$ that associates to each pair of vectors a scalar $\mathbf{x}, \mathbf{y} \in \Re^n$ given by

$$\langle \mathbf{x}, \mathbf{y} \rangle = \sum_{i=1}^{n} x_i y_i = \mathbf{x}^T \mathbf{y} = x_1 y_1 + x_2 y_2 + \cdots + x_n y_n$$

where $\mathbf{x} = [x_1 \; x_2 \; \cdots \; x_n]^T$ and $\mathbf{y} = [y_1 \; y_2 \; \cdots \; y_n]^T$.

The dot product is thus a real value that satisfies the following criteria:

- $\langle \mathbf{x}, \mathbf{y} \rangle = \langle \mathbf{y}, \mathbf{x} \rangle$
- $\langle \mathbf{x} + \mathbf{y}, \mathbf{z} \rangle = \langle \mathbf{x}, \mathbf{z} \rangle + \langle \mathbf{y}, \mathbf{z} \rangle$
- $\langle \alpha \cdot \mathbf{x}, \mathbf{y} \rangle = \alpha \cdot \langle \mathbf{x}, \mathbf{y} \rangle$
- $\langle \mathbf{x}, \mathbf{x} \rangle \geq 0 \; \forall \; \mathbf{x} \in X$, e $\langle \mathbf{x}, \mathbf{x} \rangle = 0 \Leftrightarrow \mathbf{x} = \mathbf{0}$

Outer Product

The *outer product* between two vectors is a function $f_{outer}: \Re^m \times \Re^n \to \Re^{m \times n}$ that associates to each pair of vectors $\mathbf{x} \in \Re^m$, $\mathbf{y} \in \Re^n$ a matrix of dimension $m \times n$ defined as:

$$\rangle \mathbf{x}, \mathbf{y} \langle = \mathbf{x} \mathbf{y}^T = \begin{bmatrix} x_1 y_1 & x_1 y_2 & \cdots & x_1 y_n \\ x_2 y_1 & \ddots & & \vdots \\ \vdots & & & \\ x_m y_1 & \cdots & & x_m y_n \end{bmatrix}$$

In contrast to inner products, the outer product can involve vectors of different dimensions.

B.1.3. Norms, Projections, and Orthogonality

Norms, Semi-Norms and Quasi-Norms

By introducing the concept of *norm* (*distance measure*), it is possible to study topological properties, such as continuity and convergence. The *norm* is a function f_N: $\mathfrak{R}^n \to \mathfrak{R}$ that associates to each vector $\mathbf{x} \in \mathfrak{R}^n$ a scalar represented by $\|\mathbf{x}\|$. The norm satisfies the following axioms:

- $\|\mathbf{x}\| \geq 0, \ \forall \mathbf{x} \in X; \ \|\mathbf{x}\| = 0 \Leftrightarrow \mathbf{x} = 0$

- $\|\mathbf{x} + \mathbf{y}\| \leq \|\mathbf{x}\| + \|\mathbf{y}\|, \ \forall \mathbf{x}, \mathbf{y} \in X$ (triangular inequality)

- $\|a\mathbf{x}\| = |a|\|\mathbf{x}\|, \ \forall \mathbf{x} \in X, \ \forall a \in \mathfrak{I}$

When a norm is associated with a vector space $(X, \mathfrak{I})$, there is a *normed vector space*.

Examples of functions f_N that satisfy the axioms for a norm are:

- $\|\mathbf{x}\|_p = \left(\displaystyle\sum_{i=1}^{n} |x_i|^p \right)^{1/p}$ (*p* norm)

- For $p = 1$: $\|\mathbf{x}\|_1 = \displaystyle\sum_{i=1}^{n} |x_i|$

- For $p = 2$: $\|\mathbf{x}\|_2 = \langle \mathbf{x}, \mathbf{x} \rangle^{1/2} = \left(\mathbf{x}^T \mathbf{x} \right)^{1/2} = \left(\displaystyle\sum_{i=1}^{n} |x_i|^2 \right)^{1/2}$ (Euclidean norm)

- For $p = \infty$: $\|\mathbf{x}\|_\infty = \max_i |x_i|$

A *semi-norm* is a function f_{SN}: $\mathfrak{R}^n \to \mathfrak{R}$ that satisfies all the properties of a norm, with the exception of the first axiom. The linear subspace $X_0 \subset \mathfrak{R}^n$ where $\|\mathbf{x}\| = 0$ is termed the null space of the semi-norm.

A *quasi-norm* is a function f_{QN}: $\mathfrak{R}^n \to \mathfrak{R}$ that satisfies all properties of a norm, with the exception of the second axiom (triangular inequality), which assumes the form

- $\|\mathbf{x} + \mathbf{y}\| \leq b(\|\mathbf{x}\| + \|\mathbf{y}\|), \ \forall \mathbf{x}, \mathbf{y} \in X, \ b \in \mathfrak{R}$

These concepts are very important for the definition of topological properties (Figure B.2).

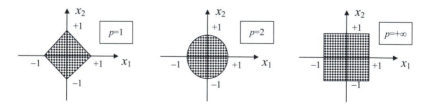

Figure B.2: Topological view of norms in two-dimensions ($n = 2$): $p = 1$, $p = 2$ and $p = +\infty$. (Courtesy of © Fernando J. Von Zuben.)

There is a relationship between the Euclidean norm and the inner product between two vectors as follows (Cauchy-Schwartz-Buniakowsky inequality):

$$\left|\langle \mathbf{x}, \mathbf{y} \rangle\right|^2 \leq \langle \mathbf{x}, \mathbf{x} \rangle \langle \mathbf{y}, \mathbf{y} \rangle \Rightarrow \left|\langle \mathbf{x}, \mathbf{y} \rangle\right| \leq \|\mathbf{x}\|_2 \cdot \|\mathbf{y}\|_2$$

Orthogonal and Orthonormal Vectors

Two vectors $\mathbf{x}, \mathbf{y} \in \mathfrak{R}^n$ are orthogonal, $\mathbf{x} \perp \mathbf{y}$, if their inner product is zero:

$$\mathbf{x} \perp \mathbf{y} \Rightarrow \langle \mathbf{x}, \mathbf{y} \rangle = 0$$

Furthermore, if $\langle \mathbf{x}, \mathbf{x} \rangle = \langle \mathbf{y}, \mathbf{y} \rangle = 1$, then the vectors $\mathbf{x}, \mathbf{y} \in \mathfrak{R}^n$ are *orthonormal*.

A vector $\mathbf{x}$ is orthogonal to a set S ($\mathbf{x} \perp S$) if $\langle \mathbf{x}, \mathbf{s} \rangle = 0$, $\forall \, \mathbf{s} \in S$.

Projecting a Vector along a Given Direction

Given a linear vector space X, let $\mathbf{y} \in X$ be a vector that provides a certain direction. The projection of any vector $\mathbf{x} \in X$ in a direction $\mathbf{y}$ (projection of $\mathbf{x}$ along $\mathbf{y}$ - Figure B.3) is given by

$$\mathrm{proj}_{\mathbf{y}}(\mathbf{x}) = \frac{\langle \mathbf{x}, \mathbf{y} \rangle}{\langle \mathbf{y}, \mathbf{y} \rangle^{1/2}} \cdot \frac{\mathbf{y}}{\langle \mathbf{y}, \mathbf{y} \rangle^{1/2}} = \frac{\mathbf{x}^T \mathbf{y}}{\mathbf{y}^T \mathbf{y}} \cdot \mathbf{y}$$

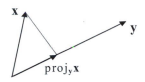

Figure B.3: Projection of $\mathbf{x}$ along $\mathbf{y}$.

Orthonormal Vectors Generated from Linearly Independent Vectors

Although all sets of non-zero orthonormal vectors are linearly independent, not all sets of linearly independent vectors is orthonormal, but these can orthonormalized. Given a set of m linearly independent n-dimensional vectors ($m \leq n$), $\{\mathbf{x}_1, \mathbf{x}_2, \dots, \mathbf{x}_m\}$, it is always possible to make a linear combination of these vectors that results in m orthogonal n-dimensional vectors $\{\mathbf{u}_1, \mathbf{u}_2, \dots, \mathbf{u}_m\}$ that span the same space. Besides, if the vectors $\mathbf{u}_i$ ($i=1, \dots, m$) have unit norm, then they are orthonormal.

A set of orthonormal vectors $\{\mathbf{u}_1, \mathbf{u}_2, \dots, \mathbf{u}_m\}$ may be obtained from a set of linearly independent vectors $\{\mathbf{x}_1, \mathbf{x}_2, \dots, \mathbf{x}_m\}$ using the *Gram-Schmidt orthogonalization process*, which works as follows:

Step (1) $\mathbf{y}_1 = \mathbf{x}_1$

$$\mathbf{y}_i = \mathbf{x}_i - \sum_{j=1}^{i-1} \frac{\langle \mathbf{x}_i, \mathbf{y}_j \rangle}{\langle \mathbf{y}_j, \mathbf{y}_j \rangle} \mathbf{y}_j, \quad i = 2, \dots, m$$

Step (2) $$\mathbf{u}_i = \frac{\mathbf{y}_i}{\langle \mathbf{y}_i, \mathbf{y}_i \rangle^{1/2}}, \quad i = 1, \dots, m$$

Then,

$$\mathbf{u}_1 = a_{11}\mathbf{x}_1$$
$$\mathbf{u}_2 = a_{21}\mathbf{x}_1 + a_{22}\mathbf{x}_2$$
$$\mathbf{u}_3 = a_{31}\mathbf{x}_1 + a_{32}\mathbf{x}_2 + a_{33}\mathbf{x}_3$$
$$\vdots \qquad \vdots \qquad \vdots \qquad \ddots$$

where $a_{ii} > 0$ ($i = 1, \dots, m$). It is important to remark that there are other orthonormalization processes that are more general than this one and do not impose any constraint on the coefficients a_{ij} ($i \geq j$; $i,j = 1, \dots, m$).

B.1.4. Matrices and Their Properties

Matrix

An ordered array of $m.n$ scalars x_{ij} ($i=1, 2, \dots, m$; $j = 1, 2, \dots, n$) is called a *matrix* of dimension $m \times n$. Matrices are described here by boldface capital letters of the Roman alphabet and are commonly presented as a rectangular array of the form

$$\mathbf{X} = \begin{bmatrix} x_{11} & x_{12} & \cdots & x_{1n} \\ x_{21} & x_{22} & \cdots & x_{2n} \\ \vdots & \vdots & \ddots & \vdots \\ x_{m1} & x_{m2} & \cdots & x_{mn} \end{bmatrix}.$$

The set of all matrices $m \times n$ with real-valued elements is represented by $\Re^{m \times n}$.

- The columns of matrix $\mathbf{X}$ are column vectors described by

$$\mathbf{x}_i = \begin{bmatrix} x_{1i} \\ x_{2i} \\ \vdots \\ x_{mi} \end{bmatrix}, \quad i=1,\dots,n$$

- The rows of matrix $\mathbf{X}$ are row vectors described by $\mathbf{x}_{(j)} = [x_{j1}\ x_{j2} \cdots x_{jn}]$, $j=1, \dots, m$.

- A vector is a matrix with a unitary number of rows or columns.

Basic Operations Involving Vectors and Matrices

Given two vectors $\mathbf{x}, \mathbf{y} \in \Re^m$, $\mathbf{x} = [x_1, x_2, \dots, x_m]^T$ and $\mathbf{y} = [y_1, y_2, \dots, y_m]^T$, their addition and subtraction (Figure B.4) are performed as follows:

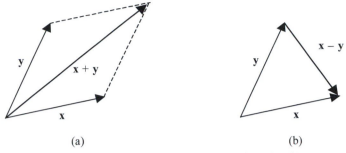

(a)   (b)

Figure B.4: Vector addition (a) and subtraction (b).

Vector addition: $\mathbf{x} + \mathbf{y} = [x_1 + y_1, x_2 + y_2, \ldots, x_m + y_m]^T$

Vector subtraction: $\mathbf{x} - \mathbf{y} = [x_1 - y_1, x_2 - y_2, \ldots, x_m - y_m]^T$

The multiplication of a vector $\mathbf{x} \in \mathfrak{R}^m$ by a scalar $a \in \mathfrak{R}$ is performed as follows:

Scalar multiplication: $a.\mathbf{x} = [a.x_1, a.x_2, \ldots, a.x_m]^T$

Matrix multiplication: a matrix $\mathbf{A} \in \mathfrak{R}^{m \times n}$ can be multiplied by another matrix $\mathbf{B} \in \mathfrak{R}^{r \times l}$ if and only if $n = r$, and the result is a matrix $\mathbf{C} \in \mathfrak{R}^{m \times l}$. Given $\mathbf{A}$ and $\mathbf{B}$ below:

$$\mathbf{A} = \begin{bmatrix} a_{11} & a_{12} & \cdots & a_{1n} \\ a_{21} & a_{22} & \cdots & a_{2n} \\ \vdots & \vdots & \ddots & \vdots \\ a_{m1} & a_{m2} & \cdots & a_{mn} \end{bmatrix}, \; \mathbf{B} = \begin{bmatrix} b_{11} & b_{12} & \cdots & b_{1l} \\ b_{21} & b_{22} & \cdots & b_{2l} \\ \vdots & \vdots & \ddots & \vdots \\ b_{r1} & b_{r2} & \cdots & b_{rl} \end{bmatrix}$$

each element c_{ij} of matrix $\mathbf{C} = \mathbf{A}.\mathbf{B}$, $\mathbf{C} \in \mathfrak{R}^{m \times l}$, is obtained as follows:

$$c_{ij} = \sum_{k=1}^{n} a_{ik} b_{kj}, i = 1, \ldots, m; j = 1, \ldots, l$$

The multiplication of a matrix $\mathbf{A} \in \mathfrak{R}^{m \times n}$ by a scalar $b \in \mathfrak{R}$ is performed by simply multiplying each element of matrix $\mathbf{A}$ by the scalar, similarly to the vector multiplication by a scalar.

Transpose and Square Matrices

The transpose of a matrix $\mathbf{A}$, written $\mathbf{A}^T$, is the matrix obtained by writing the columns of $\mathbf{A}$, in order, as rows. In other words, if $\mathbf{A} = [a_{ij}]$ is an $m \times n$ matrix, then $\mathbf{A}^T = [b_{ij}]$ is the $n \times m$ matrix where $b_{ij} = a_{ji}$. As discussed previously, the transpose of a row vector is a column vector.

If the matrix has the same number of columns and rows ($m = n$), then the matrix is *square*.

Trace

Given a matrix $\mathbf{A}$ of dimension $n \times n$, the trace of $\mathbf{A}$, represented by $\mathrm{tr}(\mathbf{A})$, is the sum of the elements of the main diagonal of $\mathbf{A}$ ($a_{ii} \; \forall i$):

$$\mathrm{tr}(\mathbf{A}) = \sum_{i=1}^{n} a_{ii}$$

Range and Rank

The range, $R(\mathbf{A})$, of a matrix $\mathbf{A}$ of dimension $m \times n$ is a subspace of $\Re^m$ defined as

$$R(\mathbf{A}) = \left\{ \mathbf{y} \in \Re^m : \mathbf{y} = \mathbf{Ax}, \forall \, \mathbf{x} \in \Re^n \right\}$$

Therefore, $R(\mathbf{A})$ is given by the set of all possible linear combinations of the columns of $\mathbf{A}$. The dimension of $R(\mathbf{A})$ is termed the rank of matrix $\mathbf{A}$ and is represented by $\rho(\mathbf{A})$.

The rank of a matrix is equivalent to the number of columns linearly independent of $\mathbf{A}$. Since $\rho(\mathbf{A}) = \rho(\mathbf{A}^T)$, the rank may also be equivalent to the number of rows linearly independent of $\mathbf{A}$. If $n \geq m$ and $\rho(\mathbf{A}) = m$, then $R(\mathbf{A}) \equiv \Re^m$, and it is possible to conclude that:

$$\dim\{R(\mathbf{A})\} = \rho(\mathbf{A}) = \rho(\mathbf{A}^T) \leq \min(m,n)$$

Symmetry

A matrix $\mathbf{A} = \{a_{ij}\}$ of dimension $n \times n$ is symmetric if $a_{ij} = a_{ji}$, $i,j = 1, 2, \ldots, n$. In other words, a symmetric matrix is a square matrix $\mathbf{A}$ that satisfies the following condition: $\mathbf{A}^T = \mathbf{A}$, where ^{T} is the transposing operator.

Inversion

Given a matrix $\mathbf{A}$ of dimension $n \times n$, if there exists a matrix $\mathbf{B}$ of same dimension such that $\mathbf{AB} = \mathbf{BA} = \mathbf{I}$, then $\mathbf{B}$ is unique and is termed the inverse of $\mathbf{A}$; that is, $\mathbf{B} = \mathbf{A}^{-1}$.

Pseudo-inversion

The pseudo-inverse of a given matrix $\mathbf{A} \in \Re^{m \times n}$ is a matrix $\mathbf{B} \in \Re^{n \times m}$ that satisfies the following criteria:

- $\mathbf{ABA} = \mathbf{A}$;   - $\mathbf{BAB} = \mathbf{B}$;   - $(\mathbf{AB})^T = \mathbf{AB}$;   - $(\mathbf{BA})^T = \mathbf{BA}$

Matrix $\mathbf{B}$ is usually denoted by $\mathbf{A}^+$, and can be demonstrated to be unique. Matrix $\mathbf{A}^+$ can be directly obtained by single value decomposition.

If matrix $\mathbf{A}$ has full rank; that is, if $\rho(\mathbf{A}) = \min(m,n)$, its pseudo-inverse is given by

- If $m < n$, $\rho(\mathbf{A}) = m$ and $\mathbf{A}^+ = \mathbf{A}^T(\mathbf{AA}^T)^{-1}$
- If $m > n$, $\rho(\mathbf{A}) = n$ and $\mathbf{A}^+ = (\mathbf{A}^T\mathbf{A})^{-1}\mathbf{A}^T$
- If $m = n$, $\rho(\mathbf{A}) = m = n$ and $\mathbf{A}^+ = \mathbf{A}^{-1}$

Therefore, if **A** is invertible (square and with full rank), its pseudo-inverse is equal to its inverse.

Cofactor

Given a matrix **A** of dimension $n \times n$, the *cofactor* of element a_{ij} ($i,j = 1, 2, \ldots, n$) is given by

$$c_{ij} = (-1)^{i+j} m_{ij}$$

where m_{ij} is the determinant of matrix **A** formed by eliminating its i-th row and j-th column.

Determinant

Given a matrix **A** of dimension $n \times n$, the determinant of **A**, $det(\mathbf{A})$ or $|\mathbf{A}|$, is given by

$$|\mathbf{A}| = \det(\mathbf{A}) = \begin{cases} \sum_{i=1}^{n} a_{ij} c_{ij}, \forall i \\ \text{or} \\ \sum_{j=1}^{n} a_{ij} c_{ij}, \forall j \end{cases}$$

where c_{ij} is the cofactor of a_{ij}.

Let $\mathbf{A} = [\mathbf{a}_1 \ \mathbf{a}_2 \ \cdots \ \mathbf{a}_n]^T$ ($1 \leq j \leq n$). The determinant of **A** has the following properties:

- Invariance: $\det([\mathbf{a}_1 \ \mathbf{a}_2 \ \cdots \ \mathbf{a}_n]) = \det([\mathbf{a}_1 \ \cdots \ \mathbf{a}_j + \mathbf{a}_k \ \cdots \ \mathbf{a}_n])$, $j \neq k$, $1 \leq j,k \leq n$
- Homogeneity: $\det([\mathbf{a}_1 \ \cdots \ b\mathbf{a}_j \ \cdots \ \mathbf{a}_n]) = b.\det([\mathbf{a}_1 \ \cdots \ \mathbf{a}_j \ \cdots \ \mathbf{a}_n])$

Thus, if **B** is a matrix obtained from **A** by exchanging two of its columns, then $\det(\mathbf{B}) = -\det(\mathbf{A})$.

Adjoint

Given a matrix **A** of dimension $n \times n$, the adjoint of **A**, represented by adj(**A**), is given by

$$\text{adj}(\mathbf{A}) = \{ a'_{ij} \}$$

where $a'_{ij} = c_{ji}$ is the cofactor of a_{ji}.

The following equalities hold: $\begin{cases} \mathbf{A}.\text{adj}(\mathbf{A}) = \det(\mathbf{A}).\mathbf{I} \\ \text{adj}(\mathbf{A}).\mathbf{A} = \det(\mathbf{A}).\mathbf{I} \end{cases} \Rightarrow \mathbf{A}^{-1} = \dfrac{\text{adj}(\mathbf{A})}{\det(\mathbf{A})}$

Singularity

A matrix **A** of dimension $n \times n$ is singular if $\det(\mathbf{A}) = 0$. Otherwise, **A** is said to be nonsingular. Since $\mathbf{A}^{-1} = \text{adj}(\mathbf{A})/\det(\mathbf{A})$, **A** admits inverse if and only if $\det(\mathbf{A}) \neq 0$; that is, if **A** is nonsingular.

Nullity

The *null space* of a matrix $\mathbf{A}$ of dimension $m \times n$ is a subspace of $\mathfrak{R}^n$ defined as

$$N(\mathbf{A}) \triangleq \left\{ \mathbf{x} \in \mathfrak{R}^n : \mathbf{A}\mathbf{x} = \mathbf{0} \right\}$$

The dimension of $N(\mathbf{A})$ is termed the *nullity* of matrix $\mathbf{A}$ and is represented by $\nu(\mathbf{A})$. The nullity and rank are related by

$$\nu(\mathbf{A}) + \rho(\mathbf{A}) = n$$

Important properties: $N(\mathbf{A}) \perp R(\mathbf{A}^T)$ in $\mathfrak{R}^n$ $N(\mathbf{A}^T) \perp R(\mathbf{A})$ in $\mathfrak{R}^m$

Eigenvalues and Eigenvectors

Let $\mathbf{A}$ be a matrix of dimension $n \times n$. A scalar $\lambda \in \mathbb{C}$ (set of complex numbers) is an *eigenvalue* of $\mathbf{A}$ if there exists a nonzero vector $\mathbf{x} \in \mathbb{C}^n$, called the associated *eigenvector*, such that

$$\mathbf{A}\mathbf{x} = \lambda \mathbf{x}$$

- $\mathbf{A}\mathbf{x} = \lambda\mathbf{x}$ may be rewritten as $(\lambda\mathbf{I} - \mathbf{A})\mathbf{x} = \mathbf{0}$

- $\exists \, \mathbf{x} \in \mathbb{C}^n, \mathbf{x} \neq \mathbf{0}$ such that $(\lambda\mathbf{I} - \mathbf{A})\mathbf{x} = \mathbf{0}$ if and only if $\det(\lambda\mathbf{I} - \mathbf{A}) = 0$

- $\Delta(\lambda) \triangleq \det(\lambda\mathbf{I} - \mathbf{A})$ is the *characteristic polynomial* of $\mathbf{A}$

- Since the degree of $\Delta(\lambda)$ is n, matrix $\mathbf{A}$ has n eigenvalues

Positivity

A matrix $\mathbf{A}$ of dimension $n \times n$ is said to be *positive definite* if and only if

$$\langle \mathbf{x},\mathbf{A}\mathbf{x} \rangle = \mathbf{x}^T\mathbf{A}\mathbf{x} > 0, \ \forall \ \mathbf{x} \in \mathfrak{R}^n, \mathbf{x} \neq \mathbf{0},$$

or equivalently, if all its eigenvalues are positive.

A matrix $\mathbf{A}$ of dimension $n \times n$ is *positive semi-definite* if and only if

$$\langle \mathbf{x},\mathbf{A}\mathbf{x} \rangle = \mathbf{x}^T\mathbf{A}\mathbf{x} \geq 0, \ \forall \ \mathbf{x} \in \mathfrak{R}^n$$

or equivalently, if all its eigenvalues are non-negative.

A matrix $\mathbf{A}$ of dimension $n \times n$ is said to be *negative definite* (*negative semi-definite*) if $-\mathbf{A}$ is *positive definite* (*positive semi-definite*).

B.1.5. Complex Numbers and Spaces

Complex Numbers

The set of complex numbers is denoted by $\mathbb{C}$. Formally, a complex number is an ordered pair (a,b) of real numbers where equality, addition and multiplication are defined as follows:

- $(a,b) = (c,d)$ if and only if $a = c$ and $b = d$

- $(a,b) + (c,d) = (a+c, b+d)$

- $(a,b) \cdot (c,d) = (ac-bd, \, ad+bc)$

The real number a can be identified with the complex number $(a,0)$, whilst the complex number $(0,1)$ is denoted by i. The number i has the important property that

$$i^2 = ii = (0,1).(0,1) = (-1,0) = -1 \quad \text{or} \quad i = \sqrt{-1}$$

Accordingly, any complex number $\mathbf{z} = (a,b)$ can be written as

$$\mathbf{z} = (a,b) = (a,0) + (0,b) = (a,0) + (b,0).(0,1) = a + bi.$$

The above notation $\mathbf{z} = a + bi$, where $a \equiv \mathrm{Re}(\mathbf{z})$ and $b \equiv \mathrm{Im}(\mathbf{z})$ are called the *real* and the *imaginary* part of $\mathbf{z}$, respectively, is more convenient than $\mathbf{z} = (a,b)$.

Complex Conjugate and Absolute Value

Let $\mathbf{z} = a + bi$ be a complex number. The *conjugate* of $\mathbf{z}$, $\mathbf{z}^*$ or $\overline{\mathbf{z}}$, is denoted and defined by

$$\mathbf{z}^* = \overline{\mathbf{z}} = \overline{a + bi} = a - bi$$

The *absolute value* of $\mathbf{z}$, $|\mathbf{z}|$, is defined as the nonnegative square root of $\mathbf{z}\mathbf{z}^*$, and is equal to the norm of the vector (a,b) in $\mathfrak{R}^2$:

$$|\mathbf{z}| = \sqrt{\mathbf{z}\overline{\mathbf{z}}} = \sqrt{a^2 + b^2}$$

Complex Plane

Analogously to the fact that real numbers can be represented by points on a line, complex numbers c can be represented by points in a plane (Figure B.5).

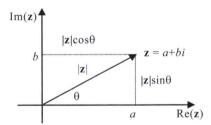

Figure B.5: Complex plane.

Polar Coordinates

It is also possible to express complex numbers in terms of *polar coordinates* as an ordered pair:

$$(|\mathbf{z}|,\theta) = |\mathbf{z}|(\cos\theta + i\sin\theta)$$

where $|\mathbf{z}|$ is the absolute value of $\mathbf{z}$ and θ is the angle relative to the positive real coordinate axis.

Conversion from rectangular to polar coordinates is as follows:

$$|\mathbf{z}| = \sqrt{a^2 + b^2}, \quad \theta = \tan^{-1}(b/a)$$

Conversion from polar to rectangular coordinates is as follows:

$$a = |\mathbf{z}|\cos\theta \qquad b = |\mathbf{z}|\sin\theta$$

Exponential Form

Knowing that $e^{i\theta} = \cos\theta + i\sin\theta$, which is known as the Euler Formula, and $e^{-i\theta} = \cos\theta - i\sin\theta$, complex numbers can also be represented in exponential form, as follows:

$$\mathbf{z} = |\mathbf{z}|e^{i\theta}$$

Vectors in c^n

The *complex n-space*, c^n, is the set of all *n*-tuples of complex numbers. The elements of c^n are called *points* or *vectors*.

Dot (Inner) Product in c^n

Let $\mathbf{u} = [u_1, \ldots, u_n]$ and $\mathbf{v} = [v_1, \ldots, v_n]$ be two vectors in c^n. The *dot* or *inner* product between $\mathbf{u}$ and $\mathbf{v}$ is defined by:

$$\mathbf{u}.\mathbf{v} = \langle \mathbf{u}, \mathbf{v} \rangle \equiv \sum_{i=1}^{n} u(i)v(i)*$$

The function $c^n \times c^n \to c$ is an inner product if it satisfies:

- $\langle \cdot, \cdot \rangle$ is linear in the second argument
- $\langle \mathbf{u}, \mathbf{v} \rangle = \langle \mathbf{v}, \mathbf{u} \rangle *$
- $\langle \mathbf{u}, \mathbf{u} \rangle \geq 0$, with equality only if $\mathbf{u} = 0$.

Complex Matrices

A *complex matrix* is a matrix with complex entries. The *conjugate* of a complex matrix, $\mathbf{A}*$, is obtained from $\mathbf{A}$ by taking the conjugate of each entry in $\mathbf{A}$:

If $\mathbf{A} = [a_{ij}]$, then $\mathbf{A}* = [b_{ij}]$, where $b_{ij} = a_{ij}*$.

Special Complex Matrices: Self-Adjoint (Hermitian), Unitary

A matrix $\mathbf{A}$ for which

$$\mathbf{A}^\dagger = (\mathbf{A}*)^T = \mathbf{A}$$

where the conjugate transpose is denoted $\mathbf{A}^\dagger$, $\mathbf{A}^T$ is the transpose, and $\mathbf{A}*$ is the complex conjugate, is said to be *self-adjoint*. If a matrix is self-adjoint, it is said to be *Hermitian*. This is equivalent to saying that $a_{ij} = a_{ji}*$.

A complex matrix $\mathbf{A}$ is *unitary* if $\mathbf{A}^\dagger \mathbf{A}^{-1} = \mathbf{A}^{-1}\mathbf{A}^\dagger = \mathbf{I}$; that is, $\mathbf{A}^\dagger = \mathbf{A}^{-1}$. In other words, a *unitary matrix* satisfies the condition $\mathbf{A}^\dagger \mathbf{A} = \mathbf{A}\mathbf{A}^\dagger = \mathbf{I}$. A complex matrix is unitary if and only if its rows (columns) form an orthonormal set relative to the dot product of complex vectors.

There are several examples of Hermitian matrices that are relevant for the study of Natural Computing, in particular the *Pauli Matrices* are often seen in quantum computing (Chapter 10):

$$\sigma_0 = \mathbf{I} = \begin{bmatrix} 1 & 0 \\ 0 & 1 \end{bmatrix} \qquad \sigma_1 = \mathbf{X} = \begin{bmatrix} 0 & 1 \\ 1 & 0 \end{bmatrix}$$

$$\sigma_2 = \mathbf{Y} = \begin{bmatrix} 0 & i \\ -i & 0 \end{bmatrix} \qquad \sigma_3 = \mathbf{Z} = \begin{bmatrix} 1 & 0 \\ 0 & -1 \end{bmatrix}$$

Hilbert Spaces

A finite-dimensional *Hilbert space H* is a complete vector space over the complex numbers that is equipped with an inner product $H \times H \to c$, $(\mathbf{x},\mathbf{y}) \to \langle \mathbf{x},\mathbf{y} \rangle$.

A vector space H is complete if for each vector sequence $\mathbf{x}_i$ such that

$$\lim_{m,n \to \infty} \|\mathbf{x}_m - \mathbf{x}\| = 0$$

there exists a vector $\mathbf{x} \in H$ such that

$$\lim_{n \to \infty} \|\mathbf{x}_n - \mathbf{x}\| = 0$$

Tensor Products

The *tensor product*, also called *Kronecker product*, may be applied in different contexts to matrices, vectors, vector spaces, etc. In the particular case to be studied in this volume, it is a way of putting vector spaces together to form larger vector spaces in quantum systems. The tensor product between two matrices $\mathbf{A}$ of dimension $r \times s$ and $\mathbf{B}$ of dimension $t \times u$ is denoted by $\mathbf{A} \otimes \mathbf{B}$ and is given by

$$\mathbf{A} \otimes \mathbf{B} = \begin{bmatrix} a_{11}\mathbf{B} & a_{12}\mathbf{B} & \cdots & a_{1s}\mathbf{B} \\ a_{21}\mathbf{B} & a_{22}\mathbf{B} & \cdots & a_{21}\mathbf{B} \\ \vdots & \vdots & \ddots & \vdots \\ a_{r1}\mathbf{B} & a_{r2}\mathbf{B} & \cdots & a_{rs}\mathbf{B} \end{bmatrix}$$

and has dimension $rt \times su$.

B.2 STATISTICS

Statistics can be understood as a collection of methods used for planning *experiments* (i.e., any process which leads to a well-defined outcome), and collecting, organizing, summarizing, analyzing, and interpreting data so as to extract some conclusions from them.

B.2.1. Elementary Concepts

Population, Sample, Variables

A *population* is the complete collection of elements (e.g., values, measures, cities, etc.) to be studied. A *sample* is a subset of elements extracted from the population. An *attribute* or *parameter* is a measure that describes a feature of a population. Each of these features is called a *variable*, and can be classified into *quantitative* (numeric) or *qualitative* (not numeric) variable (Figure B.6).

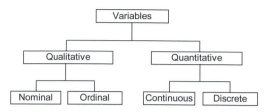

Figure B.6: Types of variables.

Quantitative variables correspond to numbers that represent countings or measurements, and can be divided into *discrete variables* (with a finite or countable number of possible values) or *continuous variables* (with an infinite number of possible values that can be associated with a continuous scale). Qualitative variables can be divided into *nominal* (consisting only of names, labels, or categories) and *ordinal* (involving variables that can be ordered).

Branches of Statistics

It is possible to divide statistics into three main branches: *descriptive statistics*, *probability*, and *inferential statistics*. Descriptive statistics aims at describing and summarizing the data obtained from a sample or experiment, such as determining a measure of certainty for the mean or other descriptive values obtained, i.e., confidence limits. Probability is the mathematical theory used to study *uncertainty* from random events. Statistical inference is the branch of statistics that concern the generalization, from samples to populations, of the information and conclusions obtained.

B.2.2. Probability

Event and Sample Space

Let an *event* be a collection of results from an experiment. A *simple event* is the one that does not allow any further decomposition. The *sample space* (Ω) of an experiment consists of all possible simple events; that is, the sample space consists of all results that do not allow any further decomposition.

Probability

Probability is a function $P(\cdot)$ that attributes numeric values to the events on a sample space, according to the following definition - A function $P(\cdot)$ is a probability if it satisfies the following conditions:

$0 \leq P(A) \leq 1, \forall A \subset \Omega.$

$P(\Omega) = 1.$

$P\left(\bigcup_{j=1}^{n} A_j\right) = \sum_{j=1}^{n} P(A_j)$, where A_j's are disjoint events.

The probability of occurrence of a certain event can be approximated, or estimated, by its relative frequency; that is, realize an experiment a large number of times and count how many times the event effectively occurred:

$$P(A) = \frac{\text{number of occurrences of event } A}{\text{number of repetitions of the experiment}}$$

Assume that an experiment has n simple different events, each of which with the same chance of occurrence. If event A can occur in s among the n forms, then

$$P(A) = \frac{\text{number of ways } A \text{ can occur}}{\text{number of different simple events}} = \frac{s}{n}$$

The complement $\bar{A}$ of an event A consists of all results in which the event A does *not* occur.

The probability of the union of events A and B, $P(A \cup B)$; that is, the probability of either A or B or both events occurring in a given experiment, can be determined by the *law of addition*:

$$P(A \cup B) = P(A) + P(B) - P(A \cap B)$$

where $P(A \cap B)$ corresponds to the probability of both events occurring at the same time in a given experiment.

Conditional Probability

In many situations, the random process we are dealing with can be divided into stages, and the information concerning what happened in a certain stage may influence the probabilities of occurrence of the other stages. The *conditional probability*, written $P(A|B)$ and read "the probability of A given B", is the probability of some event A, assuming event B has already occurred:

$$P(A \mid B) = \frac{P(A \cap B)}{P(B)}, P(B) > 0$$

From the definition of conditional probability we can deduce the *multiplication law of probabilities*:

$$P(A \cap B) = P(A \mid B) P(B), \text{ with } P(B) > 0$$

Two events A and B are said to be *independent* if the information about the occurrence or not of B does not alter the probability of occurrence of A and vice-versa; that is,

$$P(A \mid B) = P(A), P(B) > 0$$

Or, equivalently,

$$P(A \cap B) = P(A)P(B)$$

Bayes Theorem

One of the most important relations involving conditional probabilites is given by the *Bayes Theorem*, that expresses a conditional probability in terms of other conditional and marginal probabilities (i.e., the probability of one event ignoring the others). Assume that the events $C_1, C_2, \ldots, C_k$, form a partition of Ω and that their probabilities are known. Assume, also, that for an event A the probabilities $P(A \mid C_i)$ are known for all $i = 1, 2, \ldots, k$. Then, for any j,

$$P(C_j \mid A) = \frac{P(A \mid C_j)P(C_j)}{\sum_{i=1}^{k} P(A \mid C_i)P(C_i)}, j = 1, 2, ..., k$$

Counting

Given two events, the first one happening in m distinct forms and the second one happening in n different forms. Both events together may happen in $m.n$ distinct forms.

A *permutation* is an *ordered list without repetitions*; that is, an ordered sequence with no two elements the same, drawn from a fixed set of symbols, and of maximum length. Thus, the essential difference between a *permutation* and a *set* is that the elements of a permutation are arranged in a specified order. The number of possible permutations, also called *arrangements*, of k elements chosen among n elements, $P(n,k)$, without repetition is:

$$P(n,k) = \frac{n!}{(n-k)!}$$

where the symbol "!" corresponds to the factorial function: $n! = n \times (n-1) \times (n-2) \times ... \times 1$, and, by definition, $0! = 1$.

If there are repetitions; that is, there are n elements with n_1 the same, n_2 the same, ..., n_k the same, then the number of permutations is:

$$P = \frac{n!}{n_1! n_2! ... n_k!}$$

The number of combinations of k elements extracted from a set of n distinct elements is the binomial coefficient "n choose k",

$$C(n,k) = \binom{n}{k} = \frac{n!}{(n-k)!k!}$$

B.2.3. Discrete Random Variables

Random Variable

A *random variable* is a variable, usually represented by x, which has a single numeric value (randomly determined) for each result of an experiment. This terminology is used to describe the value that corresponds to the result of a certain experiment. The word *random* indicates that, usually, we can only know its value after the experiment has been performed; thus, to each possible value of the variable there is an associated probability of occurrence.

Discrete Random Variable

A *discrete random variable* either admits a finite number of values or has an enumerable number of values. By contrast, a *continuous random variable* can assume an infinite number of values, and these can be associated with a continuous (without gaps or interruptions) scale.

Probability Distributions

A *discrete probability distribution*, also called *discrete probability function*, gives the probability of occurrence of each value of a discrete random variable. Any probability distribution P must satisfy the following conditions:

$$\sum P(x) = 1, \forall x; \text{ and } 0 \leq P(x) \leq 1, \forall x$$

Some random variables appear quite often in practical situations, motivating a deeper study. In these cases, the probability distribution may be written in a compact form; that is, there is a rule to attribute probabilities. Some examples are the *Uniform distribution*, the *Binomial distribution*, and the *Poisson distribution*. A random variable x_i, $i = 1, \ldots, k$, follows a Uniform distribution if the same probability, $1/k$, is attributed to each of the k values of variable x_i. A random variable has a Binomial distribution if it obeys the following distribution: $C(n,k).p^k.(1-p)^{n-k}$, where $C(n,k)$ is the Binomial coefficient, n is the number of (Bernoulli) trials and p is the probability of success. Finally, a random variable x has a Poisson distribution with parameter $\lambda > 0$ if its probability function is given by $(e^{-\lambda}\lambda^k)/k!$, $k = 1, 2, \ldots$.

B.2.4. Summary and Association Measures

Central Tendency and Dispersion Measures

Some measures can be used to summarize the information available about the probability distribution of a random variable or contained in a data set. Two types of measures are of major importance: *central tendency measures* and *dispersion measures*. The most common measures of central tendency are the *mean*, *median* and *mode*, summarized in Table B.1.

While the central tendency measures provide us with information about the behavior of data within a data set (variable values), they do not carry information about the variability of the data (variable values); this is provided by the dispersion measures, summarized in Table B.2. For instance, the variance of a given data set corresponds to how much the data vary in relation to the mean value of the data set.

Table B.1: Measures of central tendency for data sets and discrete random variables: mean, median, and mode. *Md* is the mode.

	Data set	**Random Variable**
Mean	$\bar{x} = \dfrac{1}{N}\sum\limits_{i=1}^{N} x_i$	$\mu = \dfrac{1}{N}\sum\limits_{i=1}^{N} p_i x_i$
Median	Central value	$P(x \geq Md) \geq 0.5$ and $P(x \leq Md) \geq 0.5$
Mode	Most frequent value	Most probable value

Table B.2: Dispersion measures for data sets and discrete random variables: variance and standard deviation.

	Data set	Random Variable
Variance	$\mathrm{var}(x) = \dfrac{1}{N}\displaystyle\sum_{i=1}^{N}(x_i - \bar{x})^2$	$\mathrm{var}(x) = \displaystyle\sum_{i=1}^{N} p_i(x_i - \mu)^2$
Standard deviation	$\sigma(x) = \sqrt{\mathrm{var}(x)}$	$\sigma(x) = \sqrt{\mathrm{var}(x)}$

Association Measures

For two different (random) variables it is possible to determine if there is any dependence between them by using some *association measures*. For instance, if two variables deviate from their respective means in a similar way, one can say that they are *covariant*, i.e., there is a statistical correlation of their fluctuations.

Assume a raw data set with N pairs of values (x_i, y_i), $i = 1, 2, \dots, N$. The *covariance*, $\mathrm{cov}(x,y)$, between the two random variables x and y is given by:

$$\mathrm{cov}(x, y) = \frac{1}{N}\sum_{i=1}^{N}(x_i - \bar{x})(y_i - \bar{y})$$

As the covariance value depends on the scale used to measure x and y, it is hard to use it as a standard to compare the statistical association degree of different pairs of variables. To solve this problem, a scale factor can be applied dividing each individual deviation by the standard deviation of the respective variable, resulting in the *correlation coefficient*:

$$\rho(x, y) = \frac{\mathrm{cov}(x, y)}{\sigma(x).\sigma(y)}$$

The *correlation coefficient*, $\rho(x,y)$, measures the linear dependence between the variables. In other words, it determines if there is a (linear) relationship between the variables. It can be written in terms of the covariance given above or explicitly as below:

$$\rho(x, y) = \frac{\displaystyle\sum_{i=1}^{N}(x_i - \bar{x})(y_i - \bar{y})}{\sqrt{\displaystyle\sum_{j=1}^{N}(x_j - \bar{x})^2} \cdot \sqrt{\displaystyle\sum_{j=1}^{N}(y_j - \bar{y})^2}}$$

Note that the correlation coefficient between variables x and y can assume values in the range $[-1, +1]$.

B.2.5. Estimation and Sample Sizes

Point and Interval Estimators

An *estimator* is a sample statistics used to obtain an approximation of an unknown parameter. An *estimate* is a specific value or an interval of values corresponding to the result of the application of an estimator to a particular set of data. A *point estimator* is a single value used to approximate a population parameter, while an *interval estimator* is an interval of values that have a probability of containing the true population value.

Confidence Interval

A *confidence interval* is associated with a confidence degree corresponding to a measure of our certainty that the interval contains a population parameter. The *confidence degree* is the probability $(1 - \alpha)$, usually expressed as a percentile value, that the confidence interval contains the true value of the population parameter. Common choices for the confidence degree are 90%, 95%, and 99%, because these provide a good equilibrium between precision and confidence.

For sample sizes greater than 30 (big sample) and assuming a normal distribution, the confidence interval for the population mean μ is given by:

$$\bar{x} - z\frac{\sigma}{\sqrt{N}} < \mu < \bar{x} + z\frac{\sigma}{\sqrt{N}}$$

where $\bar{x}$ is the sample average, and $E = z\dfrac{\sigma}{\sqrt{N}}$ is the error margin (the maximum difference probable between the sample average and the population average). Parameter z is called the *critical value* and assumes some standard values based on the desired confidence degree: 90% $\rightarrow z = 1.645$; 95% $\rightarrow z = 1.96$; and 99% $\rightarrow z = 2.575$. Parameter σ is the population standard deviation, and can be replaced by the sample standard deviation when the population value is not known.

It is possible to determine the minimal sample size to estimate a given parameter, such as the population mean, from the error margin equation:

$$N = \left[\frac{z\sigma}{E}\right]^2$$

B.3 THEORY OF COMPUTATION AND COMPLEXITY

Intuitively, a *computer* is a physical system whose dynamics takes it from one set of input states to one of a set of output states. The *theory of computation* began early before computers were designed. It constitutes a branch of computer science and mathematics that deals formally (mathematically) with the problem of computing. It concerns the study of formal models of computation and also the limits of computing. This section reviews three of the most important concepts from the theory of computation for the understanding of natural computing.

B.3.1. Production Systems and Grammars

Lindenmayer was concerned with finding developmental instructions with which known kinds of organisms could be generated. Inspired by the way theoretical linguists were concerned with production or transformation rules by which certain types of words or sentences can be generated, his formalism was based on the generation of constructs using a specific *grammar*. The formalism proposed thus contained elements of *production systems* and grammars.

A production system, also known as *rule-based system*, uses *implications* as its primary representation. Typically, it consists of a set of *production rules*, a *working memory* and a *recognize-act phase*. In the context of L-systems, we will be restricted to the *production rules*, simply called *productions*. A production is a condition-action pair of inference rules. The condition part of the rule is a pattern that determines when that rule may be applied, and the action part defines the associated action to be taken. The example given in Chapter 5, "**If** *it starts to rain* **then** *open your umbrella*" can be expressed as $a \rightarrow b$, meaning that "**If** *a* **then** *b*", where *a* corresponds to "*it starts to rain*" and *b* corresponds to "*open your umbrella*".

An *alphabet V* is a finite set of symbols. A *word* or *string w* on the set *V* is a finite sequence of its elements. The set of all finite strings formed from members, called *letters*, of the alphabet *V* is defined as V^*. The *length* of a word *w*, denoted by $|w|$ or $l(w)$, is the number of elements in its sequence of letters. The empty string *e*, the string of zero length, is also a member of V^*. The set of strings over *V* without the empty string is denoted by V^+, thus $V^+ = V^* - \{e\}$. If an alphabet consists of symbols 0 and 1, $G = \{0,1\}$, then G^* is the set of all finitely long sequences of 0's and 1's plus the empty string, thus $G^* = \{e,0,1,00,01,10,11,000,001,...\}$. A *language L* over an alphabet *V* is any subset (collection of words) of V^*, or, *L* is a language over *V* if and only if $L \subseteq V^*$.

A *grammar G* is a construct consisting of four parts, $G = \langle V,T,P,S \rangle$, where *V* is a finite set (alphabet, also called *vocabulary*); *T* is a set whose elements are called terminals $(T \subset S)$, the elements of $N = S \setminus T$ are called nonterminals $(N \subset S)$; *P* is the set of productions, and *S* is a nonterminal symbol designated as the starting symbol. Thus, a *grammar* is a set of production rules *P* over a set of nonterminal symbols *N* and a set of terminal symbols *T* of an alphabet, with a certain nonterminal symbol *S*. Each production in *P* must contain at least one nonterminal symbol on its left side.

B.3.2. Universal Turing Machines

A *Turing Machine* (TM) is an abstract computing machine with the power of both real computers and of other mathematical definitions of what can be computable. It is an idealized computing device introduced in 1936 by A. Turing (1936) to give a mathematical definition of an *algorithm* or "mechanical procedure". Turing believed that if one could write down a set of rules describing a computation, then this idealized machine could carry it out. Thus, a Turing machine can carry out all kinds of computations on numbers and symbols. If something (e.g., a problem) is computable, then it can be solved by a Turing machine.

The *Church-Turing thesis* states, in its most common form, that every effective computation or algorithm can be carried out by a Turing machine. A TM capable of simulating any other Turing machine is called a *Universal Turing machine*. A Turing machine is composed of: 1) a *tape* for storing information; 2) a *head* for reading and writing symbols on the tape and moving left and right; 3) a *state register* for storing the state of the machine; and 4) an *instruction table* (or *program*) for controlling the machine; that is, indicating what symbol to write, how to move the head, and what is the next state.

The tape is divided into squares (or cells), each square containing a symbol from some finite alphabet (e.g., the binary digits '0' and '1'). This tape is the general-purpose storage medium of the machine, serving both as the vehicle for input and output, and as a working memory for storing the results of intermediate steps of the computation. It is assumed to be arbitrarily extendible to the left and to the right; thus, the Turing machine is always supplied with as much tape as it needs for its computation. The main reason for the unlimited length of the tape was because Turing wanted to show that there are problems that cannot be solved by Turing machines, even when unlimited working memory is supplied and unlimited time is allowed for computing. Note that every part of the machine is finite, but it is the potentially unlimited amount of tape that gives it an unbounded amount of storage space.

The read/write head is programmable. It is helpful to think of the operation of programming as consisting of altering the head's internal wiring by means of a plug board arrangement. To compute with the device, you program it, write the input on the tape using the alphabet chosen, place the head over the cell containing the leftmost input symbol, and set the machine in motion. Once the computation is completed, the machine will come to a *halt* state with the head positioned at any programmed cell or over the cell containing the leftmost symbol of the output.

Turing's machine gave two major contributions to the development of the digital computers known nowadays: 1) the idea of controlling the workings of a computing device by storing a program of encoded instructions in a memory, and 2) the proof that a single machine - a universal Turing machine - can carry out every type of computation carried out by any other Turing machine.

A *deterministic Turing machine* (DTM) is the one with a transition rule $\delta(q,x) = (q',x',d)$ that specifies for a given current state of the head and computer (q,x) a single instruction (q',x',d), where x' is the symbol to be written by the head, q' is the next state of the computer, and d is the direction (left or right - $\{L,R\}$) in which to step. A *non-deterministic Turing machine* (NDTM) differs in that, rather than a single instruction triplet, the transition rule may specify a number of alternative instructions. At each step of the computation, the computer can choose one among a number of possible instructions. Therefore, whereas a DTM has a single 'path' to follow, an NDTM has a number of possible 'paths' to follow.

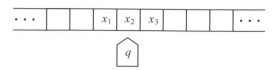

Figure B.7: Picture of a Turing machine M in state q reading the symbol x_2 and with $x_1qx_2x_3$ written on the tape.

More formally, a deterministic Turing machine can be defined as a quadruple $M = (Q, \Sigma, \delta, s_0)$, where Q is the set of states (including the *halting* state), Σ is a finite alphabet, $\delta : (Q \times \Sigma) \rightarrow (Q \times \Sigma \times \{L,R,N\})$ is the state transition function – $\{L,R,N\}$ indicates the movement of the head to the Left, Right or No movement, and s_0 is the start state. Without loss of generality, the Turing machine can be assumed to work on a single-sided tape, and the input $a_1, a_2, \ldots$, a_n, occupying the first n tape cells. Each Turing computation can be represented by a sequence of *configurations* of the machine, where each configuration $C = (x_1 \ldots x_{k-1} q x_k x_{k+1} \ldots x_m)$ encodes a single step in the computation. The characters $x_1 \ldots x_m$ represent the significant contents of the tape, q is the state of the machine, and k is the position of the head. The state q of the machine is written before the symbol x_k that the machine is reading. Figure B.7 illustrates a Turing machine in state q reading the symbol x_2 and with the word $x_1qx_2x_3$ written on the tape.

B.3.3. Complexity Theory

Computational complexity, also known as *complexity theory*, is a central field of computer science concerned with the study of the intrinsic complexity of computational problems. It focuses on the computational resources and considers the effect of limiting these resources on the class of problems that can be solved (see Figure B.8). Researchers in this area define computational models, such as Turing machines and Boolean circuits, and resource measures, such as *space* (how much memory does it take to solve a problem) and *time* (how many steps does it take to solve a problem). A complexity class is then the set of problems solvable by a particular model under particular resource constraints. This section provides an elementary introduction to complexity theory.

A *problem* is a set of finite-length questions, where each question is a finite-length *string* (sequence of entities - e.g., bits, DNA molecules, etc.) with associated finite-length answers. For example, we have studied, in several chapters of this book, the *travelling salesman problem* (TSP): given a set of cities to be visited, find the shortest path that passes through every city exactly once. A particular question is called an *instance*. In the TSP problem for example, an instance with 12 cities has already been solved using an evolutionary algorithm, an ant colony optimization algorithm, and an immune algorithm. Many other instances of the TSP problem are available in the literature and on-line (cf. the TSPBIB available at: http://www.densis.fee.unicamp.br/~moscato/TSPBIB_home.html).

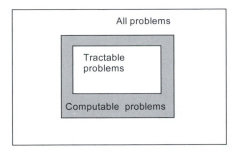

Figure B.8: Within the set of all problems there are the computable ones, and within this smaller set there are the tractable problems.

A *decision problem* is a problem where all the answers are YES or NO. Decision problems are often considered because an arbitrary problem can always be reduced to a decision problem. As an example of a decision problem, consider the *Hamiltonian path problem* stated as follows. Given a graph, is there a path that starts at a given node and ends at another given node passing through all nodes exactly once? This problem is a decision problem known as the Hamiltonian path problem and will be further studied using a DNA computing technique.

An *algorithm* can be defined as a list of instructions (unambiguous description of steps) to perform some task, and the task can be anything that has a recognizable end-point or result, such as solving a given (instance of a) problem. It is important to stress that an algorithm can be implemented using a standard computer (as we are accustomed to) or any other means, such as using DNA molecules and manipulation techniques or Quantum bits, as will be discussed later in this chapter and in the next chapter, respectively.

Some of the specific steps in an algorithm may be repeated until the task is accomplished. Normally, there are different algorithms for the same task, and some may perform better than others. In computer science, the quality of an algorithm is evaluated by investigating its cost; that is, by analyzing its complexity in terms of resources required to solve a problem. In most cases, the *time* and *space* complexity of an algorithm are of major interest. As an algorithm cannot have a quantity of data higher than the amount it has time to analyze, the *time complexity* of an algorithm poses an upper bound on its *space complexity*.

Time and space complexity are measured with respect to the size of the problem to be solved. For example, consider a problem P and an algorithm A that solves P. If n is the size of an instance of P, then the time complexity of A, T_A, is a function of n, $T_A(n)$. This time complexity indicates how the execution time of the algorithm grows with n. Consider now the following two cases for $T_A(n)$:

$$T_A(n) \propto k^n$$
$$T_A(n) \propto n^k$$

where k is a constant greater than 1 in both cases. In the first case, the problem has an *exponential (time) complexity*, and in the second case, the problem has a *polynomial (time) complexity*. Problems for which a polynomial time algorithm is known are said to be *tractable*, while problems for which only exponential time algorithms are known are *intractable*.

The class of problems known as P is the set of decision problems that can be solved by a deterministic algorithm (machine) in polynomial time. The class NP is the set of decision problems that can be solved by a nondeterministic algorithm (machine) in polynomial time. As deterministic algorithms may be seen as particular cases of nondeterministic algorithms, the class P is often assumed to be contained in NP. On the other side, NP problems cannot have polynomial time deterministic algorithms solving them, and they are thus intractable. The question of whether P is the same set as NP is thus a major question in theoretical computer science. There is a US$1,000,000 prize for the one who answers this question:

http://www.claymath.org/Millennium_Prize_Problems/ P_vs_NP/

This difficulty in distinguishing P from NP problems motivated the concepts of *hard* and *complete*. A set of problems X is hard for a set of problems Y if every problem in Y can be transformed "easily" into some problem in X with the same answer. The most important hard set is *NP-hard*. Informally, an NP-hard problem is a decision problem that is at least as hard as any problem in NP. In other words, a problem is NP-hard if solving it in polynomial time would make it possible to solve all problems in class NP in polynomial time. Set X is complete for Y if it is hard for Y and is also a subset of Y. The most important complete set is *NP-complete*. NP-complete problems are the hardest problems in NP in the sense that they are the ones most likely not to be in P. If one solves an NP-complete problem, then the algorithm used to solve this problem can be used to solve all NP problems. S. Cook (1971) defined the first NP-complete problem, namely, the problem of the satisfiability of a Boolean formula, also known as the SAT problem.

B.4 OTHER CONCEPTS

This last section reviews several other concepts and ideas important in natural computing. In particular, optimization, logic of propositions, theory of nonlinear dynamical systems, graph theory, data clustering, and affine transformations, are briefly reviewed here.

B.4.1. Optimization

Optimization refers to concepts, methods and applications related to the determination of the *best* (*optimal*) solutions to a given problem. It involves the study of optimality conditions, the development and analysis of algorithms, computer experimentation, and applications.

Before applying an *optimization algorithm* to solve a given problem, it is first necessary to develop a *mathematical formulation* of the problem, which describes all relevant aspects of the problem, including the objective(s) to be optimized and a set of constraints or restrictions that have to be satisfied. Therefore, the mathematical formulation of an optimization problem requires the definition of an *objective function* to be optimized, either *maximized* or *minimized*, and a set of solutions, called *feasible solutions*, that satisfy the *constraints* of the problem.

For instance, consider an optimization problem formulated as follows:

$$\text{Minimize} \qquad f(\mathbf{x})$$

$$\text{Subject to} \qquad g_i(\mathbf{x}) \leq 0, \, i = 1, \ldots, m \qquad (1)$$

$$h_j(\mathbf{x}) = 0, \, j = 1, \ldots, l \qquad (2)$$

$$\mathbf{x} \in X.$$

where functions $f, g_i, h_j \in \Re$ ($i = 1, \ldots, m; j = 1, \ldots, l$), $X \subseteq \Re$, and $\mathbf{x} \in \Re^n$.

This problem must be solved for the values of the variables $x_1, x_2, \ldots, x_n$, that satisfy constraints (1) and (2), while minimizing function $f(\mathbf{x})$.

Function f is usually called the *objective function*, the *cost function* or the *criterion function*. A vector $\mathbf{x} \in \Re$ satisfying all constraints, (1) and (2), is called a *feasible solution* to the problem. The set of all possible solutions to a problem is called *search space*.

Depending on the problem features and thus formulation, its solution can be obtained algebraically, in a closed form. However, most optimization problems have features (e.g., nonlinearities) that avoid the determination of analytic solutions, requiring the application of iterative search procedures from a given initial condition. The presence of constraints complicates even further the problem, dividing the space of possible solutions into *feasible solutions* and *unfeasible solutions*.

Another aspect that has to be accounted for is the *convergence* behavior of the iterative search process. Three conditions are desirable: *guarantee of convergence*, *speed of convergence*, and *convergence to a good quality solution*, that is, a satisfying solution.

Solving the minimization problem formulated above corresponds to finding a feasible solution $\mathbf{x}^*$ such that $f(\mathbf{x}^*) \leq f(\mathbf{x})$, $\forall \mathbf{x} \in \Re^n$. Vector $\mathbf{x}^*$ is called a *minimum* or *optimum* of the problem. The solution $\mathbf{x}$ that satisfies this condition is called a *global solution, global minimum* or *global optimum*. In contrast to the global optimum, a *local optimum* is a potential feasible solution $\mathbf{x}^*$ in respect to a neighborhood N of a point $\mathbf{y}$, if and only if $f(\mathbf{x}^*) \leq f(\mathbf{y})$, $\forall \mathbf{y} \in N(\mathbf{x}^*)$, where $N(x) = \{y \in F : dist(x,y) \leq \varepsilon\}$, *dist* is a function that determines the distance between x and y, and ε is a positive constant. Figure B.9 illustrates a function with several local optima (minima) solutions and a single global optimum (minimum).

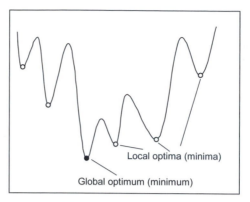

Figure B.9: Illustration of global and local minima of an arbitrary function. Dark circle: global minimum; White circles: local minima.

B.4.2. Logic of Propositions

Several proofs in mathematics and many algorithms in computer science use *logical expressions* such as:

"**If** *it starts to rain* **then** *open your umbrella*"

This expression can be generically written as:

"**If** *a* **then** *b*"

It is usually necessary to know the cases in which a given logical expression is TRUE (T) or FALSE (F), referred to as the *truth-value* of such expressions.

A *proposition* or *statement* is a declarative sentence that can be either true or false, but not both. For instance, consider the following sentences:

"Brazil is in South America." It is a true proposition.

"Mars is the Blue Planet." It is a false proposition.

"What are you doing tonight?" It is not a proposition.

Many propositions are *composite*; that is, composed of sub-propositions and various *connectives*, thus forming *compound propositions*. The fundamental property of a compound proposition is that its truth-value is completely determined by the truth-value of its sub-propositions together with the way they are connected to form the compound propositions.

There are three basic logical operations or *connectives*: conjunction ("AND"), disjunction ("OR"), and negation ("NOT"). These are also called Boolean function. Let *a* and *b* be two propositions, the definition for each of the three basic logical operations are:

- *a* AND *b*: If *a* and *b* are both true, then *a* AND *b* is true; otherwise *a* AND *b* is false.

- *a* OR *b*: If *a* and *b* are both false, then *a* OR *b* is false; otherwise *a* OR *b* is true.

- NOT *a*: If *a* is true, then NOT *a* is false; otherwise NOT *a* is true.

A	b	a AND b		a	b	a OR b		a	NOT a
0	0	0		0	0	0		0	1
0	1	0		0	1	1		1	0
1	0	0		1	0	1			
1	1	1		1	1	1			

Figure B.10: Truth tables for the connectives AND, OR, and NOT.

A simple and concise way to show the relationship between the truth-value of a compound proposition and the truth-value of its sub-propositions is through a *truth table*. Figure B.10 depicts the truth tables for the connectives AND, OR, and NOT; the value '0' corresponds to FALSE, and '1' corresponds to TRUE.

B.4.3. Theory of Nonlinear Dynamical Systems

Dynamical systems embody all systems whose behavior varies with time, more formally, whose *state* varies with time. A suitable model to study dynamical neural networks is the *state space model*, which takes into account state variables whose values, at any instant of time, are supposed to contain sufficient information to predict the future behavior of the system.

Dynamical systems are composed of two main parts: *state* and *dynamics*. The *state* describes the current condition of the system under the form of a vector parameterized in relation to time. The set of all possible states is termed the *state space* of the system. The *dynamics* describes how the state of the system varies with time, and the sequence of states of a dynamical system is termed its *trajectory* in the state space.

Assume a nonlinear dynamical system whose state variables $s_1(t)$, $s_2(t)$, ... , $s_m(t)$, where t is the continuous time (independent variable) and m is the order of the system, can be concatenated into a vector $\mathbf{s}(t) = [s_1(t), s_2(t), \ldots , s_m(t)]^T$ named the state vector of the system. Assume also nonlinear dynamical systems whose dynamics can be expressed as a first-order differential equation:

$$\frac{d}{dt}\mathbf{s}(t) = \mathbf{F}(\mathbf{s}(t))$$

Note that the function $\mathbf{F}(\cdot)$ in the equation above is diagonal; that means:

$$\mathbf{F}(\mathbf{s}) = \begin{bmatrix} f_1(s_1) & 0 & \cdots & 0 \\ 0 & f_2(s_2) & \cdots & 0 \\ \vdots & \vdots & \ddots & \vdots \\ 0 & 0 & \cdots & f_m(s_m) \end{bmatrix}$$

The system described in differential equation above is said to be *autonomous* because there is no explicit dependence of $\mathbf{F}(\cdot)$ with relation to time. The state space proposed in the equation above can be viewed as describing the motion of a point in an m-dimensional state space. This is important because it provides a visual and conceptual tool to analyze the dynamics of nonlinear systems. At a particular instant of time the state of the system is represented by a single point in the *m*-dimensional state space.

A nonlinear dynamical system can present a variety of behaviors. Consider an autonomous dynamical system described by the state space euqation above. A given constant vector **s*** is said to be an *equilibrium* or *stationary state* of the system if $F(s^*) = 0$, where **0** is the null vector. Therefore, **s*** is a solution to the differential equation presented.

An *attractor* is a region of the state space to where the trajectories converge given some initial condition. The region of the state space from which the system evolves toward the attractor is termed its *basin of attraction*. There are different types of attractors, for instance, *point attractors* and *limit cycles*. A point attractor is a single equilibrium point to which the trajectory of the system converges, and the limit cycle is a periodic orbit to which the trajectory converges. Therefore, a system initialized in the basin of attraction of a given attractor is going to converge to such attractor. The notions of attractors and basins of attraction are illustrated in Figure B.11.

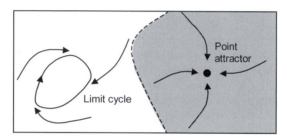

Figure B.11: Attractors and basins of attraction. There is an imaginary boundary (dashed line) that separates the basin of attraction of each attractor. The left-hand side attractor is a limit cycle, while the right-hand one is a point attractor.

B.4.4. Graph Theory

Graph theory is a rich source of decision problems with surprisingly many practical applications. In the context of natural computing, for instance, most ant colony optimization algorithms are used to solve combinatorial optimization problems represented as a graph problem. In addition, ant algorithms have also been applied to problems such as graph partitioning. Therefore, a brief review of some elementary *graph theory* is appropriate.

Consider the map illustrated in Figure B.12. It depicts a number of cities in Brazil and their possible flight connections. Each city on this picture corresponds to a *vertex*, a *point*, or a *node*, and each arc (link) connecting the nodes is termed an *edge*. This type of object is called a *graph*, and it denotes a set of nodes possibly connected by edges. A graph *G* can thus be defined by a set of vertices (nodes) *V* and a set of edges *E* - pairs of elements of *V*: $G = (V, E)$. Therefore, $V = \{v_0, v_1, \ldots, v_N\}$ is the set of nodes of *G*, and $E = \{(v_i, v_j) : i \neq j\}$ is the set of edges of *G*.

Figure B.12: Map of Brazil with some cities depicted as *nodes* on a graph and some *edges* (dashed lines) corresponding to the main flight connections among the cities.

A *path* in a graph G consists of an alternating sequence of vertices and edges. When there is no ambiguity, a path can be denoted by its sequence of vertices $(v_0, v_1, \ldots, v_N)$. The path is said to be *closed* if $v_0 = v_N$. A graph is said to be *connected* when there is at least one edge connecting each of its nodes, i.e., there is a path between any two of its vertices. It is said to be *directed*, when one can "walk through" the graph in a pre-specified direction, i.e., the edges are one-way. A graph G is said to be weighted if to each edge $e \in E$ is assigned a non-negative number $w(e) \geq 0$, called the *weight* or *length* of e. For an unweighted graph, the *length* of a path is given by the number of edges in the path. In the case of weighted graphs however, the length or weight of a path is determined by summing all the weights of the edges in the path. Two vertices v_i and v_j of a graph are said to be *adjacent* if there is an edge $e \in G$ connecting them, i.e., $e = \{v_i, v_j\}$. A *cycle* is a closed path of length 3 or more in which all vertices are distinct except v_0 and v_N.

Figure B.13 illustrates the concepts of graph theory discussed. For the graph G_1 the sets V_1 and E_1 are, respectively, $V_1 = \{A,B,C,D,E\}$ and $E_1 = \{(A,B),(A,C),(C,E),(E,D)\}$, and for the graph G_2 the sets V_2 and E_2 are, respectively, $V_2 = \{A,B,C\}$ and $E_2 = \{(A,B),(B,C)\}$. The graphs G_1 and G_2 are both not weighted and undirected, while graph G_3 is directed but not weighted, and G_4 is directed and weighted. Only graph G_3 is not connected. The weight (length) of the path (D,E,B,F,C) of G_4 is 10.

If a graph G is weighted, this can be represented as $G = (V,E,w)$, where $w(e)$ is the weight of edge e, also represented as $w(i,j)$: the weight of the edge (i,j) connecting nodes i and j. For the example of graph G_4 in Figure B.13, $w(D,E) = 5$, $w(E,B) = 2$, and so forth.

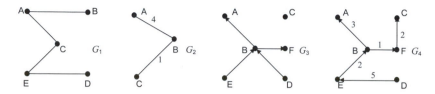

Figure B.13: Examples of graphs. G_1: connected, undirected, unweighted. G_2: connected, undirected, weighted. G_3: disconnected, directed, unweighted. G_4: connected, directed, and weighted.

The simple definition of graph presented above makes graph theory the appropriate language for discussing several topics of interest, including topological properties such as connectivity, paths, cycles, and distances in graphs. There are a significant number of graph-theoretic topics that are the object of complexity studies in computation (e.g., the traveling salesman problem, and sorting algorithms). The theory also extends to directed, labeled, or multiply connected graphs.

A graph T is called a *tree* if it is connected and has no cycles. A tree is a *spanning tree* of a graph if it is a sub-graph containing all the vertices of the graph. A *minimal spanning tree* of a graph is a spanning tree with minimum weight, where the weight of a tree is defined as the sum of the weights of its constituent edges, assuming the tree is weighted (Leclerc, 1995). Figure B.14 illustrates the concept of a tree, spanning trees, and a minimal spanning tree.

Several algorithms can be used to draw the MST of a graph, such as the Prim's algorithm (Prim, 1957). After drawing the minimal spanning tree of aiNet, it is still necessary to define how to prune connections from the MST so as to define the clusters in the network. Zahn (1971) introduced the concept of edge inconsistency as follows. *An MST edge (i,k) whose weight s_{ik} is significantly larger than the average of nearby edge weights on both sides of the edge (i,k) should be deleted. This edge is called inconsistent.* One natural way to measure the significance referred to is to calculate the *factor* or *ratio* (r) between s_{ik} and the respective averages.

The parameter r can be determined as follows. After building the MST for a given set of points, for each edge of the tree, its two end points are analyzed with a depth p.

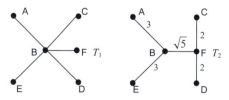

Figure B.14: Examples of trees. T_1 is a spanning tree and T_2 is a minimal spanning tree.

This means that the average (*avg*) and the standard deviation (σ) of the length of all edges which are within p steps of each end point are calculated. An edge is considered inconsistent if its length (l) is greater than the average plus r standard deviations. That is, if $l > (avg + r.\sigma)$, then the edge is considered inconsistent.

B.4.5. Data Clustering

The process of data analysis underlies many computing applications, either in a design phase or as part of their on-line operation. Data analysis procedures may be of one of two types: exploratory or confirmatory. Both are aimed at grouping or classifying data (measurements) based on either natural groupings (termed clusters) revealed by analysis (exploratory data analysis), or goodness-of-fit to a postulated model (confirmatory data analysis).

Cluster analysis or *data clustering* can thus be defined as the organization or separation of a collection of patterns - the data - into *clusters* or groups, based on *similarity* or *dissimilarity*. The data are usually represented as a vector of measurements, or a point in a multidimensional space. Intuitively, patterns within a cluster are more similar to each other than they are to patterns outside the cluster, i.e., to patterns belonging to other clusters. An example of clustering is shown in Figure B.15. The input patterns are shown in Figure B.15(a), and the desired clusters are shown in Figure B.15(b), where the patterns belonging to the same cluster are given the same shape.

The large number of techniques to represent data, to measure data proximity (similarity), and to group data, has produced a number of data clustering techniques. Assume the following notation. A *datum* **x** (also termed *pattern, item, feature vector,* or *observation*), is a single data item used by the clustering algorithm. It typically consists of a vector of L measurements: $\mathbf{x} = (x_1, \ldots, x_L)$, where each component x_i of **x** is called a *feature* or an *attribute* of **x**, and L is the *dimensionality* of the pattern space. A data-clustering problem can thus be formally defined as follows:

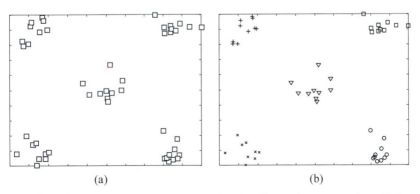

(a)   (b)

Figure B.15: Data clustering. (a) Input patterns. (b) Clustered patterns, where different clusters are represented by objects with different shapes.

*Given a set **X** of N data samples or items, **X** = {**x**₁,...,**x**ₙ}, each of which has a dimensionality L (each item is measured on L variables), devise a discrimination scheme for grouping (clustering) the objects into c classes (clusters). The number of classes and the characteristics of the classes are to be determined.*

To devise a discrimination scheme to cluster data, it is necessary to define a measure, usually a *distance measure*, that quantifies the degree of dissimilarity of two patterns drawn from the same feature space. As there are a variety of feature types and scales, the choice of a distance measure must be made carefully. There are a number of metrics to be used for determining the dissimilarity between a pair of items. For instance, the most popular metric for continuous spaces is the Euclidean distance:

$$D_2(\mathbf{x}_i, \mathbf{x}_j) = \left(\sum_k (x_{i,k} - x_{j,k})^2\right)^{1/2} = \|\mathbf{x}_i - \mathbf{x}_j\|_2$$

which is a special case of the Minkowski metrics

$$D_p(\mathbf{x}_i, \mathbf{x}_j) = \left(\sum_k (x_{i,k} - x_{j,k})^p\right)^{1/p} = \|\mathbf{x}_i - \mathbf{x}_j\|_p$$

It has been shown previously that the Hamming distance is an appropriate metric when the feature space is binary.

It is important to discuss, at this stage, the difference between *clustering* also termed *unsupervised classification*, and *discriminant analysis*, also named *supervised classification*. Clustering involves the grouping of unlabelled data into meaningful clusters that is, the raw data are presented to the clustering algorithm and it is expected that, given a similarity metric, these data be grouped into clusters. By contrast, in supervised classification a set of labeled (pre-classified) data is provided and the problem of classification involves the labeling of a newly, yet unlabelled, pattern. Typically, the labeled patterns are used to *learn* the descriptions of classes that, in turn, are used to label a new pattern. In the previous chapter we have seen that supervised learning approaches can be used to design classification and function approximation techniques (e.g., Hebb net, perceptron, Adaline, and MLP), while unsupervised learning techniques are used to devise clustering methods (e.g., self-organizing map).

Clustering problems arise in several contexts and have been addressed by researchers in various disciplines. The application of clustering techniques is of wide range, from image segmentation and object recognition to information retrieval. Therefore, there is a broad appeal and usefulness of clustering techniques.

B.4.6. Affine Transformations

A *space* **X** is a set, and the *points* of the space are the elements of the set. A metric space (**X**,*d*) is a space **X** together with a real-valued function $d : \mathbf{X} \times \mathbf{X} \to \Re$, which measures the distance between pairs of points $x, y \in \mathbf{X}$. The function *d*, called a *metric*, has to obey the following axioms:

$$d(x,y) = d(y,x) \ \forall \ x,y \in \mathbf{X}$$

$$0 < d(x,y) < \infty \ \forall \ x,y \in \mathbf{X}, x \neq y$$

$$d(x,x) = 0 \ \forall \ x \in \mathbf{X}$$

$$d(x,y) \leq d(x,z) + d(z,y) \ \forall \ x,y,z \in \mathbf{X}$$

Let $(\mathbf{X},d)$ be a metric space. A transformation on $\mathbf{X}$ is a function $f : \mathbf{X} \rightarrow \mathbf{X}$ that assigns exactly one point $f(x) \in \mathbf{X}$ to each point $x \in \mathbf{X}$ (see Figure B.16).

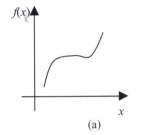
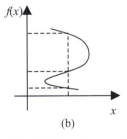

(a)   (b)

Figure B.16: Example of a transformation (a) and something that is not a transformation (b).

Let $f : \mathbf{X} \rightarrow \mathbf{X}$ be a transformation on a metric space. The forward iterates of f are transformations $f^{\circ n} : \mathbf{X} \rightarrow \mathbf{X}$ defined by $f^{\circ 0}(x) = x$, $f^{\circ 1}(x) = f(x)$, $f^{\circ(n+1)}(x) = f \circ f^{(n)}(x) = f(f^{(n)}(x))$ for $n = 0, 1, 2, \ldots$.

A transformation $w : \Re^2 \rightarrow \Re^2$ of the form

$$w(x_1,x_2) = (ax_1 + bx_2 + e, \ cx_1 + dx_2 + f)$$

where a, b, c, d, e, and f are real numbers, is called a (two-dimensional) *affine transformation*.

It is also possible to use the equivalent matrix notation for an affine transformation:

$$w(\mathbf{x}) = w \begin{bmatrix} x_1 \\ x_2 \end{bmatrix} = \begin{bmatrix} a & b \\ c & d \end{bmatrix} \begin{bmatrix} x_1 \\ x_2 \end{bmatrix} + \begin{bmatrix} e \\ f \end{bmatrix} = \mathbf{A}\mathbf{x} + \mathbf{t}$$

where $\mathbf{A} = \begin{bmatrix} a & b \\ c & d \end{bmatrix}$ and $\mathbf{t} = \begin{bmatrix} e \\ f \end{bmatrix}$.

These transformations have important geometric and algebraic properties. In order to work with fractal geometry, one needs to be familiar with the basic families of transformations and to know well the relationships between formulas for transformations and the geometric changes they cause in the metric spaces they act upon. It is more important to understand what the transformations do to sets than how they act on individual points. Thus, it is useful to know how an affine transformation in $\Re^2$ acts on a straight line, a rectangle, and a circle. The four main affine transformations are *translation, scaling, reflection,* and *rotation,* illustrated in Figure B.17.

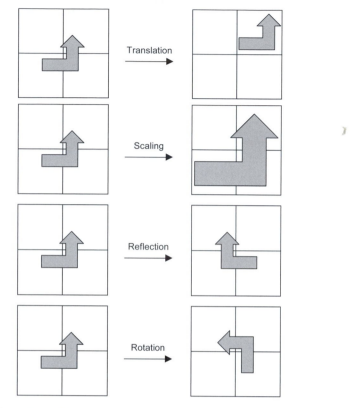

Figure B.17: Geometric representation of affine linear transformations.

Translation: this transformation moves a point on the plane along a straight line to some new location. It is just a simple vector addition operation, and can be described by

$$\begin{bmatrix} x_1 \\ x_2 \end{bmatrix} + \begin{bmatrix} e \\ f \end{bmatrix} = \begin{bmatrix} x_1 + e \\ x_2 + f \end{bmatrix}$$

Scaling: in the scaling of an image for example, the original image is changed so as to be stretched or squeezed into an area with a different width and height. In the most general case, a bi-dimensional figure can be scaled by multiplying each of its coordinates by a scalar factor. In matrix form this corresponds to:

$$\begin{bmatrix} s_1 & 0 \\ 0 & s_2 \end{bmatrix} \begin{bmatrix} x_1 \\ x_2 \end{bmatrix} = \begin{bmatrix} s_1 x_1 \\ s_2 x_2 \end{bmatrix}$$

Reflection: the application of reflection is similar to the application of scaling, but one or more of the values used in the diagonal matrix that multiply the coordinate vector has to be negative:

$$\begin{bmatrix} -1 & 0 \\ 0 & 1 \end{bmatrix}\begin{bmatrix} x_1 \\ x_2 \end{bmatrix} = \begin{bmatrix} -x_1 \\ x_2 \end{bmatrix}$$

Rotation: the image can also be rotated clockwise or counterclockwise by multiplying the vector **x** by a matrix:

$$\begin{bmatrix} 0 & 1 \\ -1 & 0 \end{bmatrix}\begin{bmatrix} x_1 \\ x_2 \end{bmatrix} = \begin{bmatrix} x_2 \\ -x_1 \end{bmatrix}$$

Note that the affine transformation $w(\mathbf{x}) = \mathbf{Ax} + \mathbf{t}$ can perform combinations of these operations. For example, an image can be rotated, scaled, translated, and reflected at the same time.

B.4.7. Fourier Transforms

Periodic phenomena are those that repeat themselves after some definite *period*. They occur quite frequently in nature and are also very useful in computing, engineering and many other sciences as well. Motors, rotating machines, hearts (under normal conditions), motion of the Earth, etc., are all examples of periodic phenomena.

A *periodic function* $f(x)$, $x \in \Re$, is the one for which there is a positive number p such that

$$f(x + p) = f(x)$$

where p is called the period of the function.

Fourier series correspond to a practical form of representing periodic functions $f(x)$ in terms of cosine and sine functions. The Fourier series has the form

$$f(x) = a_0 + \sum_{n=1}^{\infty}\left(a_n \cos nx + b_n \sin nx\right)$$

Thus, the problem of modeling a periodic function is transformed into the problem of finding the coeficients a_i, b_i, $i = 0,1, \ldots, n$ and the number of cosines and sines in the series.

The *Fourier Transform* provides a means to transform a signal defined in the time domain into one defined in the frequency domain. The *Discrete Fourier Transform* (DFT) is used when both, time and frequency, variables are discrete. It performs a mapping from a discrete, periodic sequence x_k to a set of coeficients representing the frequencies of the discrete sequence. It takes as input and outputs an array of complex numbers. The number of elements in the array depends on the sampling rate and the length of the waveform. Formally, n complex numbers $\mathbf{x}_0, \mathbf{x}_1, \ldots, \mathbf{x}_{n-1}$ are transformed into n complex numbers $\mathbf{y}_0, \mathbf{y}_1, \ldots, \mathbf{y}_{n-1}$ using the DFT:

$$\mathbf{y}_j = \frac{1}{n^{1/2}}\sum_{k=0}^{n-1}\mathbf{x}_k e^{-\frac{2\pi ijk}{n}}, \quad j = 0, \ldots, n-1$$

B.5 BIBLIOGRAPHY

[1]   Anton, H. A. (2000), *Elementary Linear Algebra*, Wiley.

[2]   Arabie, P., Hubert, L. J. and De Soete, G. (1996), *Clustering and Classification*, World Scientific Publishing Company.

[3]   Bazaraa, M. S., Sherali, H. D. and Shetty, C. M. (1993), *Nonlinear Programming: Theory and Algorithms*, John Wiley & Sons, Inc., New York, 2nd Ed.

[4]   Bluman, A. G. (2003), Elementary Statistics: A Step by Step Approach, McGraw-Hill.

[5]   Bondy, J. A. and Murty, U. S. R. (1976), *Graph Theory with Applications*, Elsevier North-Holland.

[6]   Chen C.-T. (1984), *Linear Systems: Theory and Design*, New York: Holt, Rinehart & Winston.

[7]   Cook, S. (1971), "The Complexity of Theorem-Proving Procedures", In *Conf. Record of the Third Annual ACM Symposium on Theory of Computing*, pp. 151–158.

[8]   Diestel, R. (2000), *Graph Theory*, 2nd Edition, Springer-Verlag.

[9]   Everitt, B. S., Landau, S. and Leese, M. (2001), *Cluster Analysis*, Arnold Publishers, 4th Ed.

[10] Garey, M. and Johnson, D. (1979), *Computers and Intractability*, Freeman.

[11] Han, J. and Kamber, M. (2000), *Data Mining: Concepts and Techniques*, Morgan Kaufmann.

[12] Hopcroft, J. E., Motwani, R. and Ullman, J. D. (2001), *Introduction to Automata Theory, Languages, and Computation*, Addison Wesley, 2nd Ed.

[13] Jain, A. K., Murty, M. N. and Flynn, P. J. (1999), "Data Clustering: A Review", *ACM Computing Surveys*, **31**(3), pp. 264–323.

[14] Jordan, D. W. and Smith, P. (1999), Nonlinear Ordinary Differential Equations: An Introduction to Dynamical Systems, Oxford University Press.

[15] Katok, A., Hasselblatt, B. and Rota, G.-C. (1996), *Introduction to the Modern Theory of Dynamical Systems*, Cambridge University Press.

[16] Kreyszig, E. (1999), *Advanced Engineering Mathematics*, 8th Ed., John Wiley & Sons.

[17] Lay, D. C. (2002), *Linear Algebra and Its Applications*, Addison Wesley, 3rd Ed.

[18] Lewis, H. R. and Papadimitriou, C. H. (1981), *Elements of the Theory of Computation*, Prentice Hall Inc.

[19] Lipschutz, S. and Lipson, M. (2000), *Shaum's Outline of Linear Algebra*, McGraw-Hill.

[20] Luenberger, D. G. (1989), *Linear and Nonlinear Programming*, 2nd Ed, Kluwer Academic Publishers.

[21] Magalhães, M. N., and de Lima, A. C. P. (2004), *Notions about Probability and Statistics* (in Portuguese), EDUSP.

[22] Michalewicz, Z. and Fogel, D. B. (2000), *How to Solve It: Modern Heuristics*, Springer-Verlag.

[23] Naadimuthu, G. (1997), *Schaum's Outline of Operations Research*, Schaum, 2nd Ed.

[24] Nocedal, J. and Wright, S. J. (1999), *Numerical Optimization*, Springer-Verlag.

[25] Papadimitriou, C. H. (1994), *Computational Complexity*, Addison-Wesley.

[26] Papadimitriou, C. H. and Steiglitz, K. (1998), *Combinatorial Optimization: Algorithms and Complexity*, Dover Publications.

[27] Pospeset, H., *Introduction to Logic: Propositional Logic*, Prentice Hall, 3rd. Ed.

[28] Rardin, R. L. (1998), *Optimization in Operations Research*, Prentice Hall.

[29] Sipser, M. (1997), *Introduction to the Theory of Computation*, PWS Publishing Company.

[30] Triola, M. F. (1998), *Elementary Statistics*, Addison-Wesley.

[31] Turing, A. (1936), "On Computable Numbers, with an Application to the Entscheidungsproblem", *Proc. of the London Mathematical Society*, **42**(2), pp. 230–265.

[32] Tutte, W. T. (1984), *Graph Theory*, Encyclopedia of Mathematics and Its Applications, **21**, Addison-Wesley.

[33] Weiss, N. A. (2004), *Introductory Statistics*, Addison-Wesley, 7th Ed.

APPENDIX C

A QUICK GUIDE TO THE LITERATURE

Natural computing is a very broad field of investigation. It covers ideas from biology, chemistry, and physics with a view to computing. Each chapter contains its own set of references that were cited throughout the text. This appendix, thus, is aimed at complementing the already presented lists of references and guiding the reader in relation to where to publish or find material about all the subjects discussed here. The approach taken is to briefly comment on a few selected works cited and present a list of generic and specific conferences and journals to each field. The journals' descriptions include their websites, publishers, and ISSN numbers for the printed versions, but conferences are only listed here, mainly because the links change from one year to the other. This guide is divided by chapter names as follows: Section C.1 refers to Chapter 1, Section C.2 refers to Chapter 2, and so on.

Important Author's Note: the comments provided about specific books and papers are based solely on my own reading. As a consequence, the comments express a single opinion (that may be different from others!) and the list is unavoidably incomplete. I do apologize in advance to all those authors and researchers whose relevant works I did not list here. Those readers interested in a more complete list of references should refer to the end of each chapter.

C.1 INTRODUCTION

C.1.1. Comments on Selected Bibliography

- Sipper, M. (2002), *Machine Nature: The Coming Age of Bio-Inspired Computing*, McGraw-Hill.

- Bentley, P. (2002), *Digital Biology*, Simon & Schuster.

- Forbes, N. (2004), *Imitation of Life: How Biology is Inspiring Computing*, MIT Press.

  Although there are many specific popular science books on the topics covered here, these are the three more recent ones that make an effort to provide the reader with a general overview of natural computing, but using a different nomenclature, such as machine nature and digital biology. They are motivating and primarily written to newcomers.

- Flake, G. W. (2001), *The Computational Beauty of Nature: Computer Explorations of Fractals, Chaos, Complex Systems, and Adaptation*, MIT Press.
  Flake's book is the only book I know that brings some similarity with 'Fundamentals of Natural Computing' in the sense that it was written primarily

as a textbook to teach newcomers about specific subjects on computing inspired by nature. With good web support and source codes available for users, the book is a very good reading.

C.1.2. Main (General) Journals

- Adaptive Behavior (ISAB), ISSN 1059-7123, http://www.isab.org/journal/

- AI Magazine (AAAI), ISSN 0738-4602, http://www.aaai.org/Magazine/magazine.html

- Applied Artificial Intelligence (Taylor & Francis), ISSN 0883-9514, http://www.tandf.co.uk/journals/titles/08839514.asp

- Applied Soft-Computing (Elsevier), ISSN 1568-4946, http://www.elsevier.com/locate/asoc/

- Artificial Intelligence Review (Springer), ISSN 0269-2821, http://www.springer.com/sgw/cda/frontpage/0,11855,4-0-70-35740170-0,00.html?referer=www.wkap.nl

- BioSystems (Elsevier), ISSN 0303-2647, http://www.sciencedirect.com/science/journal/03032647

- Complex Systems (Complex Systems Publications), ISSN 0891-2513, http://www.complex-systems.com/

- Computational Intelligence (Blackwell Publishing), ISSN 0824-7935, http://www.blackwellpublishing.com/journal.asp?ref=0824-7935&site=1

- Computers in Biology and Medicine (Elsevier), ISSN 001-4825, http://www.elsevier.com/wps/product/cws_home/351

- Cybernetics and Systems (Taylor & Francis), ISSN 0196-9722, http://www.tandf.co.uk/journals/titles/01969722.asp

- Future Generation Computer Systems (Elsevier), ISSN 0167-739X, www.elsevier.com/locate/future

- IEEE/ACM Transactions on Computational Biology and Bioinformatics, ISSN 1545-5963, http://www.computer.org/tcbb/

- IEEE Expert (IEEE), ISSN 0885-9000, http://www.computer.org/expert/

- IEEE Intelligent Systems (IEEE), ISSN 1541-1672, http://www.computer.org/portal/site/intelligent/

- IEEE Transactions on Pattern Analysis and Machine Intelligence (IEEE), ISSN 0018-9340, http://www.computer.org/tpami/

- IEEE Transactions on Systems, Man, and Cybernetics (IEEE), http://www.ieeesmc.org/

- Informatica (Slovene Society Informatika), ISSN 0350-5596, http://ai.ijs.si/informatica/

- Information Sciences (Elsevier), ISSN 0020-0255, http://www.sciencedirect.com/science/journal/00200255

- Information Systems (Elsevier), ISSN 0306-4379,
  http://www.sciencedirect.com/science/journal/03064379

- Information Systems Journal (Blackwell Publishing), ISSN 1350-1917,
  http://www.blackwellpublishing.com/journal.asp?ref=1350-1917

- International Journal of Computational Intelligence and Applications
  (World Scientific), ISSN 1469-0268,
  http://www.worldscinet.com/journals/ijcia/ijcia.shtml

- International Journal of Computational Intelligence Research (Research
  India Publications), ISSN 0973-1873,
  http://falklands.globat.com/~softcomputing.net/ijcir/index.html

- International Journal of Intelligent Systems (Wiley Interscience), ISSN
  0884-8173, http://www3.interscience.wiley.com/cgi-bin/jhome/36062

- International Journal of Systems Science (Taylor & Francis), ISSN 0020-
  7721, http://www.tandf.co.uk/journals/titles/00207721.asp

- International Journal of Unconventional Computing (Old City Publishing),
  ISSN 1548-7199, http://www.oldcitypublishing.com/IJUC/IJUC.html

- Journal of Artificial Intelligence Research (AAAI), ISSN 1076-9757,
  http://www.cs.washington.edu/research/jair/home.html,
  http://www.aaai.org/Press/Journals/JAIR/jair.html

- Journal of Bioinformatics and Computational Biology (World Scientific),
  ISSN 0219-7200, http://www.worldscinet.com/jbcb/jbcb.shtml

- Journal of Computational Biology (Mary Ann Liebert), ISSN 1066-5277,
  http://www.liebertonline.com/loi/cmb

- Journal of Experimental and Theoretical Artificial Intelligence (Taylor &
  Francis), ISSN 0952-813X,
  http://www.tandf.co.uk/journals/titles/0952813x.asp

- Journal of Heuristics (Kluwer Academic Publishers), ISSN 1381-1231,
  www.kluweronline.com/issn/1381-1231/contents

- Journal of Intelligent Information Systems (Kluwer Academic Publishers),
  ISSN 0925-9902, http://www.kluweronline.com/issn/0925-9902/contents

- Journal of Intelligent Systems (Freund & Pettman), ISSN 0334-1860,
  http://www.brunel.ac.uk/~hssrjis/

- Knowledge and Information Systems (Springer), ISSN 0219-1377,
  http://www.cs.uvm.edu/~kais/

- Knowledge-Based Systems (Elsevier), ISSN 0950-7051,
  http://www.sciencedirect.com/science/journal/09507051

- Machine Learning (Springer), ISSN 0885-6125,
  http://www.wkap.nl/journalhome.htm/0885-6125

- Mathware and Soft Computing (DCSAI), ISSN 1134-5632, http://docto-
  si.ugr.es/Mathware/ENG/mathware.html

- Natural Computing (Kluwer Academic Publishers), ISSN 1567-7818, http://www.kluweronline.com/issn/1567-7818

- New Generation Computing (Ohmsha and Springer-Verlag), ISSN 0288-3635, http://www.ohmsha.co.jp/ngc/

- New Mathematics and Natural Computation (World Scientific), ISSN 1793-0057, http://www.worldscinet.com/nmnc/nmnc.shtml

- Physics of Life Reviews (Elsevier), ISSN 1571-0645, http://www.sciencedirect.com/science/journal/15710645

- PLoS Computational Biology (Public Library of Science), ISSN 1553-734X, http://www.ploscompbiol.org/

- Soft Computing (Springer), ISSN 1432-7643, http://link.springer.de/link/service/journals/00500/index.htm

C.1.3. Main Conferences

- ACM International Conference on Computing Frontiers

- Artificial Neural Networks in Engineering (ANNIE)

- Bio-Computing and Emergent Computation (BCEC)

- Computational Intelligence Methods and Applications (CIMA)

- Hybrid Intelligent Systems (HIS)

- IEEE International Conference on Systems, Man, and Cybernetics (SMC)

- IEEE Symposium on Computational Intelligence in Bioinformatics and Computational Biology (CIBCB)

- IEEE World Congress on Computational Intelligence (WCCI)

- Innovative Applications of Artificial Intelligence (IAAI)

- Intelligent Systems Design and Applications (ISDA)

- International Conference on Artificial Intelligence and Applications (AIA)

- International Conference on Artificial Intelligence and Soft Computing (ICAISC)

- International Conference on Artificial Neural Networks (ICANN)

- International Conference on Computational Intelligence (ICCI)

- International Conference on Computational Intelligence and Natural Computation (CINC)

- International Conference on Machine Intelligence (ICMI)

- International Conference on Machine Learning (ICML)

- International Conference on Machine Learning and Applications (ICMLA)

- International Conference on Natural Computation (ICNC)

- International Conference on Natural Intelligence (ICNI)

- International Conference on Neural Information Processing (ICONIP)

- International Conference on the Simulation of Adaptive Behavior (SAB)
- International Joint Conference on Artificial Intelligence (IJCAI)
- International Symposium on Bio-Inspired Computing (BIC)
- International Symposium on Computational Life Science (CompLife)
- Joint Conference on Information Sciences (JCIS)
- Parallel Problem Solving from Nature (PPSN)
- Recent Advances in Soft Computing (RASC)
- Workshop on Bio-Inspired Computing (WBIC)
- World Conference on Soft Computing (WSC)

C.2 CONCEPTUALIZATION

C.2.1. Comments on Selected Bibliography

- Camazine, S., Deneubourg, J.-L., Franks, N. R., Sneyd, J., Theraulaz, G. and Bonabeau, E. (2001), *Self-Organization in Biological Systems*, Princeton University Press.

  All authors of this book are well known for their great contributions to theoretical and experimental biology, focusing the case of social insects. Self-organization in Biological Systems is one of the best accounts of self-organization in nature, and also brings results of theoretical studies and models.

- Holland, J. H. (1998), *Emergence: From Chaos to Order*, Oxford University Press.

- Johnson, S. (2002), *Emergence: The Connected Lives of Ants, Brains, Cities and Software*, Penguin Books.

  Two 'must read' popular science books about emergence, the concept that the whole is more than the sum of its parts. The authors show through a number of examples that complex adaptive systems, such as brains, ant colonies, cities, and even the Internet, are all emergent systems governed by small sets of local rules. J. Holland is a bit more technical focusing on the proposal of a general setting for emergent systems, while S. Johnson provides an account linking brains with insect societies, software, and cities.

- Holland, J. H. (1995), *Hidden Order: How Adaptation Builds Complexity*, Addison-Wesley.

  In *Hidden Order* J. Holland proposes a general framework to study complex adaptive systems. This framework is based on four properties and three mechanisms of all CAS systems. He also introduces the classifier systems as tools to model agents for CAS systems.

- Nicolis, G. and Prigogine, I. (1989), *Exploring Complexity: An Introduction*, W. H. Freeman & Company.

- Waldrop, M. M. (1992), *Complexity: The Emerging Science at the Edge of Order and Chaos*, Simon & Schuster.

- Lewin, R. (1999), *Complexity: Life at the Edge of Chaos*, 2nd. Ed., Phoenix.

  Three popular science books with gentle introductions to the science of complexity and its main players. All authors focus on an intuitive understanding of complexity.

- *Science*, Special Issue on "Complex Systems", **284**(5411), 2nd April 1999, pp. 1–212.

  A special issue of the renowned science journal fully dedicated to complex systems. It covers complexity in several different domains such as chemistry, physics, weather, economy, and biology, with particular emphasis in the nervous system and ecosystems.

- Stewart, I. (1995), *Nature's Numbers*, Phoenix.

  I. Stewart is a very talented writer and researcher. Nature's Numbers impresses by the brevity with which the author was able to show us that patterns abound nature, and how to look at them. Although he does not talk much about fractal geometry, other books (e.g., B. Mandelbrot (1982), *The Fractal Geometry of Nature*, W.H. Freeman & Company) serve this purpose.

- Russell, S. J. and Norvig, P. (1995), *Artificial Intelligence: A Modern Approach*, Prentice Hall.

  A remarkably successful artificial intelligence textbook used in more than 200 colleges and universities only in 1995. It is considered to be the intelligent agent book, where the authors were particularly interested in software agents embodying AI techniques.

- Gleick, J. (1997), *Chaos: Making a New Science*, Vintage.

- Stewart, I. (1997), *Does God Play Dice?: The New Mathematics of Chaos*, Penguin Books.

  The two best popular science books I know of concerning chaos. They introduce the field to the layman without the more elaborate mathematical terminology.

- Solé, R. and Goodwin, B. (2002), *Signs of Life: How Complexity Pervades Biology*, Basic Books.

  The book is rich in drawings, tables, and photographs. It will appeal to most readers with an interest in the larger themes of biology, such as evolution, development, and inheritance. The authors have carefully segregated the toughest math in sidebars, making the text more easy reading.

- The Santa Fe Institute Website: www.santafe.edu.

The Santa Fe Institute is one of the leading interdisciplinary centers for complex systems theory research. It holds a number of dedicated workshops and publishes several volumes in complex systems.

C.3 EVOLUTIONARY COMPUTING

C.3.1. Comments on Selected Bibliography

- Bäck, T., Fogel, D. B. and Michalewicz, Z. (eds.) (2000a), *Evolutionary Computation 1: Basic Algorithms and Operators*, Institute of Physics Publishing.
- Bäck, T., Fogel, D. B. and Michalewicz, Z. (eds.) (2000b), *Evolutionary Computation 2: Advanced Algorithms and Operators*, Institute of Physics Publishing.
  These two edited volumes constitute the handbook of evolutionary computation. They provide a comprehensive and understandable introduction to the main topics in the field. The reader can easily find information about specific aspects of evolutionary algorithms, such as selection, recombination and mutation operators.

- Darwin, C. R. (1859), *The Origin of Species*, Wordsworth Editions Limited (1998).
  Darwin's book on the origin of species is one suh book anyone should read. Despite its more than 150 years of age, it is still up to date. It contains literally hundreds of examples supporting the natural selection theory. One very interesting feature of Darwin was his capability of finding possible flaws in his theory and discussing them in this book.

- Dawkins, R. (1986), *The Blind Watchmaker*, Penguin Books.
  Dawkins is a great writer. This book introduces the *biomorphs*, creatures evolved through a simple evolutionary algorithm (cf. Chapter 7) and gives several examples of why and how evolution occurs.

- Dennett, D. C. (1995), *Darwin's Dangerous Idea Evolution and the Meanings of Life*, Penguin Books.
  The philosopher D. Dennett provides one of the best philosophical discussions of the theory of evolution. His book has an interesting appeal for engineers and computer scientists as well because it describes evolution as an algorithmic process.

- Fogel, D. B. (ed.) (1998), *Evolutionary Computation: The Fossil Record*, The IEEE Press.
  The Fossil Record discusses the many faces of evolutionary computation and how it fits with other computational intelligence algorithms. It brings chapters on artificial life, hybrids of EAs and artificial neural networks, and hybrids of EAs and fuzzy systems. Discussions about evolution as an optimization process are also provided.

- Fogel, D. B. (2000), *Evolutionary Computation: Toward a New Philosophy of Machine Intelligence*, 2nd Ed., The IEEE Press.
  This book also discusses the various types of evolutionary algorithms and suggest they form a new approach for machine (artificial) intelligence.

- Futuyma, D. J. (1998), *Evolutionary Biology*, 3rd Ed., Sinauer Associates, Inc.
  A reference book on the topic. Well-written and full of information, Futuyma's book presents an impartial perspective of evolutionary biology in a language that even a layman can understand.

- Goldberg, D. E. (1989), *Genetic Algorithms in Search, Optimization and Machine Learning*, Addison-Wesley.
  This book contains a complete listing of a simple genetic algorithm in Pascal, which C programmers can easily understand. The book covers the basic topics in the field, including crossover, mutation, classifier systems, and fitness scaling, giving a novice enough information to implement a genetic algorithm.

- Holland, J. J. (1975), *Adaptation in Natural and Artificial Systems*, MIT Press.
  Holland's book is a classic in the field of complex adaptive systems. Holland is known as the father of genetic algorithms and classifier systems, the main subjects of this book. Drawing on ideas from the fields of genetics and economics, he shows how computer programs can evolve. The schema theory used to explain why genetic algorithms work is also presented in this text.

- Kirkpatrick, S., Gerlatt, C. D. Jr. and Vecchi, M. P. (1983), "Optimization by Simulated Annealing", *Science*, **220**, 671–680.
- Aarts, E. and Korst, J. (1989), *Simulated Annealing and Boltzmann Machines: A Stochastic Approach to Combinatorial Optimization and Neural Computing*, John Wiley & Sons.
  This is the paper that introduced the simulated annealing algorithm and one good introductory text on the subject, respectively.

- Koza, J. R. (1992), *Genetic Programming: On the Programming of Computers by means of Natural Selection*, MIT Press.
- Koza, J. R. (1994), *Genetic Programming II: Automatic Discovery of Reusable Programs*, MIT Press.
- Kinnear Jr., K. E. (ed.) (1994), *Advances in Genetic Programming*, MIT Press.
- Banzhaf, W., Nordin, P., Keller, R. E. and Francone, F. D. (1998), *Genetic Programming - An Introduction: On the Automatic Evolution of Computer Programs and Its Applications*, Morgan Kaufmann Publishers.
- Langdon, W. B. and Poli, R. (2002), *Foundations of Genetic Programming*, Springer.

All the books above are good introductions to genetic programming. The first two in particular were organized by Koza and contain didactic descriptions of how to implement and what types of problems can be solved using GP. The others, more recent ones, describe the theoretical background and some new advances in the field.

- Man, K. F., Tang, K. S. and Kwong, S. (1999), *Genetic Algorithms: Concepts and Designs*, Springer-Verlag.
  In addition to the basic introduction to genetic algorithms, this book presents some principles of multiobjective optimization and computing parallelism, and illustrate them with real-world applications. Also described is a hierarchical GA designed to address the problems in determining system topology.

- Michalewicz, Z. and Fogel, D. B. (2000), *How To Solve It: Modern Heuristics*, Springer-Verlag.
  This book provides a comprehensive, updated, and detailed discussion about problem solving using heuristics. It covers classic methods of optimization, such as dynamic programming, the simplex method, and gradient techniques, and recent proposals such as simulated annealing, tabu search, and evolutionary algorithms. It is written in a nice style and accompanies a number of puzzles and challenges to the reader.

- Michalewicz, Z. (1996), *Genetic Algorithms + Data Structures = Evolution Programs*, Springer-Verlag.
  This book brings a good discussion of constraint handling techniques for evolutionary algorithms and the use of float-point representation as data structures for GAs.

- Mitchell, M. (1996), *An Introduction to Genetic Algorithms*, The MIT Press.
  This book is ideal for newcomers to the field of genetic algorithms. The book covers background, history, motivation, and provides informative examples of applications.

C.3.2. Specific Journals

- IEEE Transactions on Evolutionary Computation (IEEE), ISSN 1089-778X, http://ieee-cis.org/pubs/tec/
- Evolutionary Computation Journal (MIT Press), ISSN 1063-6560, http://mitpress.mit.edu/catalog/item/default.asp?ttype=4&tid=25
- Genetic Programming and Evolvable Machines (Kluwer Academic Publishers), ISSN 1389-2576, www.kluweronline.com/issn/1389-2576

C.3.3. Specific Conferences

- Biological Applications of Genetic and Evolutionary Computation (Bio-GEC)

- European Conference on Genetic Programming (EuroGP)
- Evolutionary Optimization (EvoOpt)
- Foundations of Genetic Algorithms (FOGA)
- Genetic and Evolutionary Computation Conference (GECCO)
- IEEE International Conference on Evolutionary Computation (CEC)
- International Conference on Artificial Evolution (AE)
- International Conference on Evolvable Systems (ICES)
- International Conference on Evolvable Systems: From Biology to Hardware
- International Conference on Genetic Algorithms (ICGA)
- NASA/DoD Conference on Evolvable Hardware (EH2005)

Most journals and conferences on artificial intelligence, soft-computing, computational intelligence, bio-inspired computing, complex systems, machine-learning and natural computing accept and publish works involving evolutionary algorithms (cf. Section C.1). Further information can be found at the EC FAQ web site http://www.faqs.org/faqs/ai-faq/genetic/.

C.4 NEUROCOMPUTING

C.4.1. Comments on Selected Bibliography

- Anderson, J. A. (1995), *An Introduction to Neural Networks*, The MIT Press.
  Anderson presents a very good attempt at linking artificial neural networks with neuroscience. His book is very readable and discusses the biological plausibility and motivation of several neural network types.

- Churchland, P. S. and Sejnowski, T. J. (1994), *The Computational Brain*, MIT Press.
- O'Reilly, R. C. and Munakata, Y. (2000), *Computational Explorations in Cognitive Neuroscience: Understanding the Mind by Simulating the Brain*, The MIT Press.
- Dayan, P. and Abbot, L. F. (2001), *Theoretical Neuroscience: Computational and Mathematical Modeling of Neural Systems*, The MIT Press.
- Trappenberg, T. (2002), *Fundamentals of Computational Neuroscience*, Oxford University Press.
  The area of computational neuroscience is an emerging field of research that focuses on the understanding of how the brain works by using biologically-based computational models made up of networks of neuron-like units. The four books above are some of the most recent works that describe computational models that are more biologically plausible than the artificial neural networks described in this book. The focus is on biological modeling instead of problem-solving. However, all of them bring (extensive) dis-

cussions of the plausibility of ANNs as biological models. The importance of ANNs for neuroscience becomes evident.

- Fausett, L. (1994), *Fundamentals of Neural Networks: Architectures, Algorithms, and Applications*, Prentice Hall.
  Fausett's book is a very comprehensible textbook introducing the most basic neural network algorithms.

- Hagan, M. T., Demuth, H. B. and Beale, M. H. (1996), *Neural Network Design*, PWS Publishing Company.
  This book is particularly interesting for readers familiar with linear algebra and programming languages like Matlab® and Mathematica®. In particular, it was written by some of the contributors of the Matlab® neural network toolbox and accompanies a disk with several demos of neural net learning.

- Haykin, S. (1999), *Neural Networks: A Comprehensive Foundation*, 2nd Ed., Prentice Hall.
  Haykin's book is a must-read textbook about neural networks. It is written primarily for graduate students and provides detailed mathematical description of the hows, whys, and whens the models described work.

- Hebb, D. O. (1949), *The Organization of Behavior: A Neuropsychological Theory*, New York: Wiley.
  Hebb's describes one of the first learning rules for neural networks. The basic premise of this work is that behavior can be explained by the action of neurons.

- Hopfield, J. J. (1982), "Neural Networks and Physical Systems with Emergent Collective Computational Capabilities", *Proc. of the Nat. Acad. of Sci. U.S.A.*, **79**, pp. 2554–2558.
- Hopfield, J. J. (1984), "Neurons with Graded Response Have Collective Computational Properties like those of Two-State Neurons", *Proc. of the Nat. Acad. of Sci. U.S.A.*, **81**, pp. 3088–3092.
  These two pioneer papers describe how a discrete-time and continuous-time network, respectively, behave as content-addressable memories. Hopfield demonstrates the similarity of these models with some physical systems and discusses their biological plausibility under the perspective of neuroscience.
- Kohonen, T (1990), "The Self-Organizing Map", *Proceedings of the IEEE*, **78**(9), pp. 1464–1480.
- Kohonen, T. (1982), "Self-Organized Formation of Topologically Correct Feature Maps", *Biological Cybernetics*, **43**, pp. 59–69.
- Kohonen, T. (2000), *Self-Organizing Maps*, 3rd Ed., Springer-Verlag.
  Kohonen's works introduce competitive networks with particular emphasis on the self-organizing maps, and provide extensive analyses of linear associative networks.

- McCulloch, W. and Pitts, W. H. (1943), "A Logical Calculus of the Ideas Immanent in Nervous Activity", *Bulletin of Mathematical Biophysics*, **5**, pp. 115–133.

This paper introduces the first mathematical model of a neuron in which a weighted sum of the input signals is compared to a threshold in order to determine the firing of the neuron. It is considered to be the pioneer work in the field of ANN, and is an interesting reading because it allows us to see how ideas have matured from that time to now.

- Minsky, M. L. and Papert, S. A. (1969), *Perceptrons*, MIT Press.
  A book that was responsible for a massive abandonment of research and funding in artificial neural networks for almost two decades. It rigorously demonstrated the limitations of single layer perceptrons and speculated few hope for the proposal of learning algorithms for multi-layer networks (this is where the authors made a mistake!).

- Rumelhart, D. E., McClelland, J. L. and the PDP Research Group (1986), *Parallel Distributed Processing: Explorations in the Microstructure of Cognition, Volume 1: Foundations*, The MIT Press.
- McClelland, J. L., Rumelhart, D. E. and the PDP Research Group (1986), *Parallel Distributed Processing: Explorations in the Microstructure of Cognition, Volume 2: Psychological and Biological Models*, The MIT Press.
  These two volumes have for a long time been the most cited works in the field of artificial neural networks. They played a major role in the resurgence of interest in the field of ANN. Among the many aspects discussed, the backpropagation learning algorithm is fully described (vol. 1) and the biological plausibility of neural networks (vol. 2) was also presented.

- Artificial Neural Networks FAQ (Frequently Asked Questions): On line website [November 27th 2002] ftp://ftp.sas.com/pub/neural/FAQ.html
  The FAQ website for neural networks is worth looking at, mainly the part describing books about artificial neural networks. A variety of texts is presented along with some comments about their level and contents. It presents links to free and commercial software, hardware, data sets; most things one may need to understand, play with, and apply ANNs.

C.4.2. Specific Journals

- IEEE Transactions on Neural Networks (IEEE), ISSN 1045-9227, http://ieee-cis.org/pubs/tnn/
- Neural Networks (Elsevier), ISSN 0893-6080, http://www.inns.org/nnjournal.asp
- Neurocomputing (Elsevier), ISSN 0925-2312, http://www.sciencedirect.com/science/journal/09252312
- International Journal of Neural Systems (World Scientific Publishing), ISSN 0129-0657, http://www.sciencedirect.com/science/journal/09252312
- Neural Computation (MIT Press), ISSN 0899-7667, http://mitpress.mit.edu/catalog/item/default.asp?ttype=4&tid=31
- Neural Computing Surveys (Lawrence Erlbaum Associates), ISSN 1093-7609, http://www.cse.ucsc.edu/NCS/

- Neural Processing Letters (Kluwer Academic Publishers), ISSN 1370-4621, www.kluweronline.com/issn/1370-4621/contents
- Neural Network News (AIWeek Inc.), http://www.neoxi.com/NNR/Neural_Network_News.html
- Network: Computation in Neural Systems (Taylor & Francis), ISSN 0954-898X, http://www.tandf.co.uk/journals/titles/0954898X.asp
- Connection Science: Journal of Neural Computing (Taylor & Francis), ISSN 0954-0091, http://www.tandf.co.uk/journals/titles/09540091.asp
- Neural Computing and Applications (Springer-Verlag), ISSN 0941-0643, http://www.springerlink.com/link.asp?id=102827

C.4.3. Specific Conferences

- Artificial Neural Networks and Expert Systems (ANNES)

- Artificial Neural Networks in Engineering (ANNIE)

- European Symposium on Artificial Neural Networks (ESANN)

- IEEE International Joint Conference on Neural Networks (IJCNN)

- International Conference on Artificial Neural Networks (ICANN)

- International Conference on Cognitive and Neural Systems (ICCNS)

- International Conference on Neural Information Processing (ICONIP)

- International Congress on Computational Intelligence (ICCI)

Most journals and conferences on artificial intelligence, soft-computing, computational intelligence, bio-inspired computing, complex systems, machine-learning, and natural computing accept and publish works involving neural networks (cf. Section C.1). Further information can be found at the ANN FAQ web site ftp://ftp.sas.com/pub/neural/FAQ.html.

C.5 SWARM INTELLIGENCE

C.5.1. Comments on Selected Bibliography

- Deneubourg, J. -L., Goss, S., Franks, N., Sendova-Franks, A., Detrain, C. and Chrétien, L.. (1991), "The Dynamics of Collective Sorting: Robot-Like Ant and Ant-Like Robot", In J. A. Meyer and S. W. Wilson (eds.) *Simulation of Adaptive Behavior: From Animals to Animats*, pp. 356–365, Cambridge, MA, MIT Press/Bradford Books.
- Beckers, R., Holland, O. E. and Deneubourg, J. L. (1994), "From Local Actions to Global Tasks: Stigmergy and Collective Robotics", In R. A. Brooks and P. Maes (eds.) *Proc. of the 4th Int. Workshop on the Synthesis and Simulation of Life, Artificial Life IV*, pp. 181–189.
  The two papers above are some of the pioneer and most well-known works describing the implementation of robotic systems inspired by social insects, in particular ant colonies. Both are focused on the problem of clustering objects inspired by the clustering of dead bodies in ants. They showed that a

simple mechanism involving the modulation of the probability of dropping an object in a certain position of the arena is a function of the local density of objects.

- Bonabeau, E., Dorigo, M. and Théraulaz, G. (1999), *Swarm Intelligence from Natural to Artificial Systems*, Oxford University Press.
  Bonabeau and his collaborators extended the concept of swarm intelligence from the context of cellular robotic systems to the more general view of any system or algorithm inspired by the collective behavior of social systems. This is a pioneer book that starts discussing experimental results on several biological studies, describes models for these investigations in natural systems, and finally explains how these models and investigations can be used for the development of swarm intelligence.

- Cao, Y. U., Fukunaga, A. S. and Kahng, A. B. (1997), "Cooperative Mobile Robotics: Antecedents and Directions", *Autonomous Robots*, **4**, pp. 1–23.
  A very broad survey of collective robotic systems focusing cooperative behaviors. The paper also discusses open problems in the field of cooperative robotics emphasizing theoretical issues.

- Dorigo, M., Maniezzo, V. and Colorni, A. (1996), "The Ant System: Optimization by a Colony of Cooperating Agents", *IEEE Trans. on Systems, Man, and Cybernetics - Part B*, **26**(1), pp. 29–41.
- Dorigo, M., and Di Caro, G. (1999), "The Ant Colony Optimization Meta-Heuristic", In D. Corne, M. Dorigo, and F. Glover (eds.), *New Ideas in Optimization*, McGraw-Hill, pp. 13–49.
  The first of the two papers above introduces the ant colony optimization approach as originally proposed. And the second one reviews the current trends in ACO algorithms, including a review of applications.

- Dorigo, M., Stützle, T. (2004), *Ant Colony Optimization*, MIT Press.
  Dorigo and Stützle provide a comprehensive and gentle introduction to ant colony optimization. It starts with the biological motivation to the algorithm and follows with a description of the algorithm, together with many of its variations. The book also contains a brief history of the development of the algorithm.

- Gordon, D. (1999), *Ants at Work: How an Insect Society is Organized*, W. W. Norton, New York.
  A very accessible popular science book introducing several of the main concepts and behaviors involved in the intricate lives of ants. A delightful reading.

- Hölldobler, B. and E. O. Wilson, (1990), *The Ants*, Belknap Press.
  A remarkably well-written and comprehensive introduction to ants. Full of detailed pictures and photographs, the authors delve into the world of ants, and leave several research questions unanswered for investigation.

- Krieger, M. J. B., Billeter, J.-B. and Keller, L. (2000), "Ant-Task Allocation and Recruitment in Cooperative Robots", *Nature*, **406**, 31 August, pp. 992–995.

- Krieger M. B. and Billeter J.-B. (2000), "The Call of Duty: Self-Organised Task Allocation in a Population of Up to Twelve Mobile Robots". *Robotics and Autonomous Systems*, **30**(1-2), pp. 65–84.

- Kube, C. R. and Bonabeau, E. (2000), "Cooperative Transport by Ants and Robots", *Robotics and Autonomous Systems*, **30**(1-2), pp. 85–101.
  The three papers above are key readings in collective robotics and provide links in which to find videos for collective robotics experiments.

- Lumer, E. D. and Faieta, B. (1994), "Diversity and Adaptation in Populations of Clustering Ants", In D. Cliff, P. Husbands, J. A. Meyer and S.W. Wilson (eds.), *Proc. of the 3rd Int. Conf. on the Simulation of Adaptive Behavior: From Animals to Animats*, **3**, MIT Press, pp. 499–508.
  An influential paper describing the standard ant clustering algorithm.

- Kennedy, J. and Eberhart, R. (1995), "Particle Swarm Optimization", *Proc. of the IEEE Int. Conf. on Neural Networks*, Perth, Australia, **4**, pp. 1942–1948.
  The founding paper of the field introduces the particle swarm optimization algorithm as a model for simulating social behavior.

- Kennedy, J., Eberhart, R. and Shi. Y. (2001), *Swarm Intelligence*, Morgan Kaufmann Publishers.
  This book introduces the particle swarm algorithm, its scope, some variations, and analyses of its behavior. It can also be seen as a discussion about the mind and intelligence, and how they can emerge from the dynamics of social interaction. A sociocognitive theoretical view was presented referring to the special qualities of thoughts about social objects, with the special characteristics they have relative to other objects. This perspective is supported by the simple PS algorithm introduced.

C.5.2. Specific Journals

There is no specific journal dedicated to swarm intelligence yet. However, most works in the field are published in the evolutionary computing journals. Other possibilities are journals on artificial and computational intelligence, soft-computing, complex systems, bio-inspired, biosystems and natural computing. The following is a list of special issues on swarm intelligence.

- Annals of Mathematics and Artificial Intelligence, *From Ants to A(ge)nts: A Special Issue on Ant-Robotics*, I. A. Wagner and A. M. Bruckstein, **31**(1–4), 2001.

- Autonomous Robots, *Special Issue on Swarm Robotics*, M. Dorigo and E. Şahin (Eds.), **17**(2–3), 2004.

- Future Generation Computer Systems, *Special Issue on Ant Algorithms*, M. Dorigo, T. Stützle and G. Di Caro (Eds.), **16**(8), 2000.

- IEEE Transactions on Evolutionary Computation, *Special Issue on Particle Swarm Optimization*, R. C. Eberhart and Y. Shi (Eds.), **8**(3), 2004.
- IEEE Transactions on Evolutionary Computation, *Special Issue on Ant Algorithms and Swarm Intelligence*, M. Dorigo, L. M. Gambardella, M. Middendorf and T. Stützle (Eds.), **6**(4), 2002.
- Informatica, *Special Issue on Ant Colony and Multi-Agent Systems*, N. Nedjah and L. de M. Mourelle (Eds.), **29**(2), 2005.
- Lecture Notes in Computer Science, *Ant Colony Optimization and Swarm Intelligence*, M. Dorigo, M. Birattari, C. Blum, L. M. Gambardella, F. Mondada and T. Stützle (Eds.), **3172**, 2004, Springer-Verlag.
- Lecture Notes in Computer Science, *Ant Algorithms*, M. Dorigo, G. Di Caro and M. Sampels (Eds.), **2463**, 2002, Springer.
- Mathware and Soft Computing, *Special Issue on Ant Colony Optimization: Models and Applications*, O. Cordón, F. Herrera and T. Stützle (Eds.), **9**(2-3), 2002.

C.5.3. Specific Conferences

- IEEE Swarm Intelligence Symposium
- International Workshop on Ant Algorithms (ANTS)
- Workshop on Swarm Intelligence and Patterns (SIP)
- Special sessions and tracks at GECCO and CEC are also frequent

  Most journals and conferences on artificial intelligence, soft-computing, computational intelligence, bio-inspired computing, complex systems, machine-learning and natural computing accept and publish works involving swarm systems (cf. Section C.1). Further information can be found at the ANN FAQ web site given above.

C.6 IMMUNOCOMPUTING

C.6.1. Comments on Selected Bibliography

- Burnet, F. M. (1959), *The Clonal Selection Theory of Acquired Immunity*, Cambridge University Press.

  Describes the original proposal of the clonal selection and expansion of B- and T-cells.

- Dasgupta, D. (ed.) (1999), *Artificial Immune Systems and Their Applications*, Springer-Verlag.
- Dasgupta, D. and Attoh-Okine, N., (1997), "Immunity-Based Systems: A Survey", *Proc. of the IEEE SMC*, **1**, pp. 369–374.
- Dasgupta, D., Majumdar, N. and Niño, F. (2001), "Artificial Immune Systems: A Bibliography", CS Technical Report, CS-01-002 [On line] http://www.cs.memphis.edu/~dasgupta/AIS/AIS-bib.pdf

- de Castro, L. N. and Von Zuben, F. J. (2000a), "Artificial Immune Systems: Part II – A Survey of Applications", *Technical Report - RT DCA 02/00*, p. 65. [On line] http://www.dca.fee.unicamp.br/~lnunes

  All references above either cite or present a number of works on artificial immune systems.

- de Castro, L. N. and Timmis, J. I. (2002), *Artificial Immune Systems: A New Computational Intelligence Approach*, Springer-Verlag: London.

  The first textbook in English of the field of artificial immune systems. Some theoretical foundations are given; basic immune algorithms and how to design them is introduced; works from the literature are surveyed; and a comparison with other approaches is also provided.

- Farmer, J. D., Packard, N. H. and Perelson, A. S. (1986), "The Immune System, Adaptation, and Machine Learning", *Physica 22D*, pp. 187–204.

  A pioneer work on AIS. The text is very accessible and provides an immune network model that has been broadly applied to solve problems in various domains, such as autonomous navigation and classification.

- Janeway, C. A., P. Travers, Walport, M. and Capra, J. D. (1999), *Immunobiology: The Immune System in Health and Disease*, 4th Ed., Garland Publishing.

- Tizard, I. R. (1995), *Immunology An Introduction*, 4th Ed, Saunders College Publishing.

  These textbooks are two important introductory level texts for those more interested in the biological basis of artificial immune systems.

- Jerne, N. K. (1974), "Towards a Network Theory of the Immune System", *Ann. Immunol.* (Inst. Pasteur) **125C**, pp. 373–389.

  The pioneer work by Jerne where he introduces the immune network theory. The text is easy reading and also provides an introduction to the history of immunology.

- *Scientific American*, **269**(3), 1993.

  This issue of the Scientific American journal is fully dedicated to immunology. There are papers describing clonal selection, self/nonself discrimination, and some basics on how the immune system works.

- Paton, R. (ed.) (1994), *Computing with Biological Metaphors*, Chapman & Hall.

- Corne, D., Dorigo, M. and Glover F. (eds.) (1999), *New Ideas in Optimization*, McGraw-Hill.

- Zadeh, L. and Kacprzyk, J. (2000), *Intelligent Information Systems, Series: Advances in Soft Computing*.

- Bentley, P. (2001), *Digital Biology*, Headline.

- Abbass, H. A., Sarker, R. A. and Newton, C. S. (eds.) (2001), *Data Mining: A Heuristic Approach,* Idea Group Publishing.

  All the books above present chapters with artificial immune systems and their applications.

C.6.2. Specific Journals

There is no specific journal dedicated to AIS yet. However, most works in the field are published in the evolutionary computing journals. Other possibilities are journals on artificial and computational intelligence, soft-computing, complex systems, bio-inspired and natural computing. The following is a list of special issues on AIS.

- IEEE Transactions on Evolutionary Computation, *Special Issue on Artificial Immune Systems*, D. Dasgupta (Ed.), **6**(3), 2002.
- Genetic Programming and Evolvable Machines, *Special Issue on Artificial Immune Systems*, P. J. Bentley and J. Timmis, **4**(4), 2003.
- International Journal of Unconventional Computing, *Special Issue on Artificial Immune Systems*, J. Timmis, S. Stepney, P. J. Benley, G. Nicosia, V. Cutello (Eds.), **1**(3), 2005.
- Lecture Notes in Computer Science, *Proceedings of the 2nd International Conference on Artificial Immune Systems*, J. Timmis, P. J. Bentley and E. Hart (Eds.), **2787**, Springer, 2003.
- Lecture Notes in Computer Science, *Proceedings of the 3rd International Conference on Artificial Immune Systems*, G. Nicosia, V. Cutello, P. J. Bentley and J. Timmis (Eds.), **3239**, Springer, 2004.
- Lecture Notes in Computer Science, *Proceedings of the 4th International Conference on Artificial Immune Systems*, C. Jacob, M. L. Pilat, J. Timmis and P. J. Bentley (Eds.), **3627**, 2005.

C.6.3. Specific Conferences

- International Conference on Artificial Immune Systems (ICARIS)
- Special sessions and tracks at GECCO and CEC are also frequent

  Most journals and conferences on artificial intelligence, soft-computing, computational intelligence, bio-inspired computing, complex systems, machine-learning and natural computing accept and publish works involving artificial immune systems (cf. Section C.1).

C.7 FRACTAL GEOMETRY OF NATURE

C.7.1. Comments on Selected Bibliography

- Falconer, K. (2003), *Fractal Geometry: Mathematical Foundations and Applications*, John Wiley & Sons.

Falconer's book provides a good and accessible introduction to fractals, including several forms of measuring fractal dimension. It presents examples and applications of iterated function systems and Brownian motion.

- Barnsley, M. F. (1993), *Fractals Everywhere*, 2nd ed., Academic Press.
  The second edition of Barnsley's book provides an introductory description about fractals and how fractal geometry can be used to model real objects in the physical world, focusing iterated function systems. It contains quite a lot of mathematical definitions, some of which may require a little mathematical background.

- Barnsley, M. F. and Hard, L. P. (1993), *Fractal Image Compression*, A. K. Peters, Wellesley, Mass.
- Fisher, Y. (1995), *Fractal Image Compression: Theory and Applications*, Springer-Verlag.
  Fractal image compression is one of the major digital image compression techniques. These books present the logic, technology, and various uses of fractals by analyzing a complete, usable fractal image representation system.

- Bentley, P. J. and Corne, D. W. (2002), *Creative Evolutionary Systems*, Morgan Kaufmann.
- Bentley, P. J. (1999), *Evolutionary Design by Computers*, Morgan Kaufmann.
  These volumes provide an interesting perspective on the application of evolutionary algorithms to purposes essentially distinct from optimization. Applications of simulated evolution to artificial life, arts, music, and many other domains are presented.

- Lesmoir-Gordon, N., Rood, W. and Edney, R. (2000), *Introducing Fractal Geometry*, ICON Books U.K.
  This small volume is a very basic introduction to fractal geometry, from its historical roots to some basic concepts. Very easy reading and highly illustrated.

- Mandelbrot, B. (1983), *The Fractal Geometry of Nature*, W. H. Freemand and Company.
  The seminal book by Mandelbrot introduces the concept of fractal and fractal dimension. It describes chaos and dynamical systems as applied in the real world, and how fractals appear in nature.

- Meinhardt, H. (1982), *Models of Biological Pattern Formation*, Academic Press, London.
- Murray, J. D. (1989), *Mathematical Biology*, Springer-Verlag, Berlin.
  These are just some of the books that bring models of patterns and phenomena in nature.

- Peitgen, H.-O, Jürgens, H. and Saupe, D. (1992), *Chaos and Fractals: New Frontiers of Science*, Springer-Verlag.

- Peitgen, H.-O and Saupe, D. (eds.) (1988), *The Science of Fractal Images*, Springer-Verlag.
  These two bestsellers by Peitgen, Saupe, and Jürgens contain a lot of information about fractals and chaos in a language that is accessible to most readers. Both provide hundreds of illustrative figures and a number of codes for implementing fractals. The edited volume emphasizes computer graphics.

- Prusinkiewicz, P. and Lindenmayer, A. (1990), *The Algorithmic Beauty of Plants*, Springer-Verlag.
  A highly illustrated book that goes straight to its point: L-systems. It presents the basic L-systems, and some of its more sophisticated versions, such as stochastic and parametric L-systems. It also brings a discussion about the fractal nature of plants.

- Ilachinski, A. (2001), *Cellular Automata: A Discrete Universe*, World Scientific.
- Wolfram, S. (1994), *Cellular Automata and Complexity*, Perseus Books.
  Two very good books on cellular automata.

C.7.2. Specific Journals

- Artificial Life (MIT Press), ISSN 1064-5462,
  http://mitpress.mit.edu/catalog/item/default.asp?ttype=4&tid=41
- Communications of the ACM (ACM), ISSN 0001-0782,
  http://www.acm.org/pubs/cacm/
- Computer Graphics (ACM SIGGRAPH Proceedings), ISSN (various),
  http://www.siggraph.org/
- FRACTALIA: Applied Mathematics and Recreational Computing Quarterly Magazine, http://pages.codec.ro/fractalia/
- Fractals (World Scientific), ISSN 0218-348X,
  http://ejournals.wspc.com.sg/fractals/fractals.shtml
- IEEE Computer Graphics and Applications (IEEE Press), ISSN 0272-1716,
  www.computer.org/cga/edcal.htm
- International Journal of Image and Graphics (World Scientific), ISSN 0219-4678, http://ejournals.wspc.com.sg/ijig/ijig.shtml
- Journal of Artificial Societies and Social Simulations (JASSS), ISSN 1460-7425, http://jasss.soc.surrey.ac.uk/JASSS.html
- Journal of Recreational Mathematics (Baywood Publishing Company), ISSN 0022-412X, http://www.ashbacher.com/jrecmath.stm
- Journal of Theoretical Biology (Elsevier), ISSN 0022-5193,
  http://www.sciencedirect.com/science/journal/00225193

- Kybernetes: The International Journal of Systems and Cybernetics (Emerald Group Publishing), ISSN 0368-492X, http://www.emeraldinsight.com/0368-492X.htm

- Modeling & Simulation (Society for Computer Simulation), ISSN 1537-792, http://www.modelingandsimulation.org/

- Pattern Recognition (Elsevier), ISSN 0031-3203, www.sciencedirect.com/science/journal/00313203

- Pattern Recognition Letters (Elsevier), ISSN 0167-8655, http://www.sciencedirect.com/science/journal/01678655/

- Simulation (Society for Computer Simulation), ISSN 0037-5497, http://www.scs.org/pubs/simulation/simulation.html

C.7.3. Specific Conferences

- ACM Symposium on Solid and Physical Modeling
- Computer Animation Conference
- Computer Graphics and Image Processing International Conference
- Computational Aesthetics in Graphics, Visualization and Imaging
- European Conference on Computer Graphics (ECCG)
- Computer Graphics and Applications
- Symposium on Geometry Processing
- International Conference on Cyberworlds
- International Conference on Computer Vision Theory and Applications
- International Conference on Computer Graphics, Theory and Applications
- European Conference on Machine Learning (ECML)
- Fractal: Complexity and Fractals in Nature
- Graphics Interface
- International Conference of the Society for Adaptive Behavior
- International Conference on Shape Modeling and Applications
- International Conference on Artificial Life
- International Conference on Virtual Worlds
- International Conference in Central Europe on Computer Graphics, Visualization and Computer Vision (WSCG)
- International Conference on Computer Graphics and Artificial Intelligence
- International Workshop on Graph Grammars and Their Applications to Computer Science
- SIGGRAPH

These are the most well-known international periodicals and conferences of the field. Search for conferences on 'computer graphics', 'virtual reality', 'computational geometry', 'computer animation', etc., and you may retrieve several other events.

C.8 ARTIFICIAL LIFE

C.8.1. Comments on Selected Bibliography

- Adamatzky, A. and Komosinski, M. (2005), *Artificial Life Models in Software*, Springer-Verlag.
  This organized volume brings a collection of works describing several artificial life software developments. Most of the softwares described are free and can be downloaded from the internet.

- Colella, V., Klopfer, E. and Resnick, M. (2001), *Adventures in Modelling: Exploring Complex Dynamic Systems with StarLogo*, Teachers College Press.
- Resnick, M. (1994), *Turtles, Termites, and Traffic Jams: Explorations in Massively Parallel Microworlds*, MIT Press.
  These two books are good introductions to the StarLogo programming language and self-organized processes. They are easy reading and were written primarily for secondary students.

- Levy, S. (1992), "Artificial Life", Vintage Books.
- Ward, M. (2000), "Virtual Organisms: The Startling World of Artificial Life", Thomas Dunne Books.
  Two easy to read popular science books about ALife. Highly recommended for those that want to have a feeling of what is artificial life.

- Gardner, M. (1970), "Mathematical Games: The Fantastic Combinations of John Conway's New Solitaire Game 'Life'", *Scientific American*, **233**, pp. 120–123.
  This classical paper by Gardner is available on line and popularized Conway's Game of Life.

- Grand, S. (2000), *Creatures: Life and How to Make It*, Harvard University Press.
  From the author of the Creatures virtual toy, this popular science book describes, from the perspective of its creator, what constitutes the essence of existence and what is intelligence.

- Langton, C. (1988), "Artificial Life", in C. Langton (ed.), *Artificial Life*, Addison-Wesley, pp. 1–47.
  The introductory paper to artificial life. Anyone interested in starting on the field should read it.

- Packard, N. H. and Bedau, M. A. (2000), "Artificial Life", *Encyclopedia of Cognitive Science*, Macmillan Reference Ltd.
  Another good introductory text to the field of artificial life.

C.8.2. Specific Journals

- Artificial Life (MIT Press), ISSN 1064-5462,
  http://mitpress.mit.edu/catalog/item/default.asp?ttype=4&tid=41
- Artificial Life and Robotics (Springer), ISSN 1433-5298,
  http://www.springer.com/sgw/cda/frontpage/0,11855,4-40109-70-1121900-0,00.html
- Communications of the ACM (ACM), ISSN 0001-0782,
  http://www.acm.org/pubs/cacm/
- Computer Graphics (ACM SIGGRAPH Proceedings), ISSN (various),
  http://www.siggraph.org/

C.8.3. Specific Conferences

- Artificial Life, Proceedings Volume in the Santa Fe Institute Studies in the Sciences of Complexity, Addison-Wesley
- Computer Animation
- Emergent Computation
- Engineering Societies in the Agents World (ESAW)
- European Conference on Artificial Life (ECAL)
- From Animals to Animats: International Conference on the Simulation of Adaptive Behavior (SAB)
- International Conference on Autonomous Agents (AA)
- International Conference on Intelligent Autonomous Systems (IAS)
- International Conference on the Simulation and Synthesis of Living Systems (ALife)
- International Conference on Virtual Worlds (Virtual Worlds)
- International Symposium on Artificial Life and Robotics (AROB)
- International Workshop on Regulated Agent-Based Social Systems: Theories and Applications (RASTA)

  The series of Artificial Life Conference Proceedings, some volumes published in the Lecture Notes on Artificial Intelligence (e.g., vols. 929, 1674 and 2159) and the journal by the MIT Press provide a great collection of works and the state of the art on artificial life.

C.9 DNA COMPUTING

C.9.1. Comments on Selected Bibliography

- Adleman, L. M. (1994), "Molecular Computation of Solutions to Combinatorial Problems", *Science*, **226**, November, pp. 1021–1024.

- Lipton, R. J. (1995), "Using DNA to Solve NP-Complete Problems", *Science*, **268**, pp. 542–545.
  Adleman's and Lipton's works on the use of DNA molecules to solve combinatorial optimization problems were pioneer in the field and are easily understandable.

- Amos, M. (1997), *DNA Computation*, Ph.D. Thesis, Department of Computer Science, The University of Warwick.
- Amos, M. (2003), *Theoretical and Experimental DNA Computation*, Springer-Verlag.
  Amos' Ph.D. thesis is a good introduction to newcomers. It starts with a review of molecular biology and the experiments of Adleman, Lipton and the ones performed by himself. His book starts in a similar form, but extends the contents to more recent outcomes.

- Gramß, T., Bornholdt, S., Groß, M., Mitchell, M. and Pellizzari, T. (2001), *Non-Standard Computation: Molecular Computation - Cellular Automata - Evolutionary Algorithms - Quantum Computers*, Wiley-VCH.
- Maley, C. C. (1998), "DNA Computation: Theory, Practice, and Prospects", *Evolutionary Computation*, **6**(3), pp. 201–229.
- Păun, G., Rozenberg, G. and Saloma, A. (1998), *DNA Computing: New Computing Paradigms*, Springer-Verlag.
- Pisanti, N. (2000), "DNA Computing: A New Computational Paradigm Using Molecules", In R. Lupacchini and G. Tamburrini (eds.), *Grounding Effective Processes in Empirical Laws: Reflections on the Notions of Algorithms*, pp. 101–116.
- Zingel, T. (2000), "Formal Models of DNA Computing: A Survey", *Proc. of the Est. Ac. of Sci. Phys. Math.*, **49**(2), pp. 90–99.
  Although slightly outdated, all these books and surveys are good introductions to the field.

C.9.2. Specific Journals

There is no specific journal dedicated to DNA Computing yet. Many works in the field are published in the Discrete Mathematics and Theoretical Computer Science or in the Lecture Notes in Computer Science series devoted to the proceedings of the Unconventional Models of Computation conferences and DNA Based Computers conferences. The following is a nonexhaustive list of these volumes and special issues on DNA Computing.

- Unconventional Models of Computation, *Proceedings of the 1ˢᵗ International Conference on Unconventional Models of Computation*, C. Calude, J. Casti and M. J. Dinneen (Eds.), Springer-Verlag, 1998.

- Unconventional Models of Computation, *Proceedings of the 2ⁿᵈ International Conference on Unconventional Models of Computation*, I. Antoniu, C. Calude, and M. J. Dinneen (Eds.), Springer-Verlag, 2000.

- Lecture Notes in Computer Science, *Proceedings of the 3rd International Conference on Unconventional Models of Computation*, C. Calude, M. J. Dinneen and F. Peper (Eds.), **2509**, Springer-Verlag, 2002.

- Lecture Notes in Computer Science, *Proceedings of the 4th International Conference on Unconventional Computation*, C. Calude, M. J. Dinneen, G. Paun, M. J. Pérez-Jiménes and G. Rozenberg (Eds.), **3699**, Springer-Verlag, 2005.

- DNA Based Computers, *Proceedings of the 1st International Meeting on DNA Based Computers*, R. J. Lipton and B. Baum (Eds.), American Mathematical Society, **27**, 1996.

- DNA Based Computers II, *Proceedings of the 2nd International Meeting on DNA Based Computers*, L. F. Landweber and B. Baum (Eds.), American Mathematical Society, **44**, 1999.

- DNA Based Computers III, *Proceedings of the 3rd International Meeting on DNA Based Computers*, H. Rubin and D. H. Wood (Eds.), American Mathematical Society, **48**, 1999.

- DNA Based Computers III, *Proceedings of the 3rd International Meeting on DNA Based Computers*, H. Rubin and D. H. Wood (Eds.), American Mathematical Society, **48**, 1999.

- BioSystems, *Proceedings of the 4th International Meeting on DNA Based Computers*, **52**(1-3), 1999.

- DNA Based Computers V, *Proceedings of the 5th International Meeting on DNA Based Computers*, E. Winfree and D. K. Gifford (Eds.), American Mathematical Society, **54**, 2000.

- Lecture Notes in Computer Science, *Proceedings of the 6th International Meeting on DNA Based Computers*, A. Condon and G. Rozenberg (Eds.), **2054**, Springer-Verlag, 2001.

- Lecture Notes in Computer Science, *Proceedings of the 7th International Meeting on DNA Based Computers*, N. Jonoska and N. C. Seeman (Eds.), **2340**, Springer-Verlag, 2001.

- Lecture Notes in Computer Science, *Proceedings of the 8th International Meeting on DNA Based Computers*, M. Hagiya and A. Ohuchi (Eds.), **2568**, Springer-Verlag, 2003.

- Genetic Programming and Evolvable Machines, *Special Issue on Biomolecular Machines and Artificial Evolution*, M. H. Garzon (Ed.), **4**(2), 2003.

- Lecture Notes in Computer Science, *Proceedings of the 9th International Meeting on DNA Based Computers*, J. Chen and J. Reif (Eds.), **2943**, Springer-Verlag, 2004.

- Lecture Notes in Computer Science, *Proceedings of the 10th Workshop on DNA Computing*, C. Ferretti, G. Mauri and C. Zandron (Eds.), **3384**, Springer-Verlag, 2005.

- Computing in Science and Engineering, *Special Issue on Biocomputation*, S. Kumar and S. Sastry, **4**(1), 2002.

- New Generation Computing, *Special issue on Biomolecular Computing*, M. Yamamura, T. Head and M. Hagiya (Eds.), **20**(3), (2002).

C.9.3. Specific Conferences

- Annual Pacific Symposium on Biocomputing

- Annual Symposium on the Foundations of Computer Science

- Bio-Computing and Emergent Computation (BCEC)

- Intelligent Systems for Molecular Biology (ISMB)

- International Conference on Biocomputers (Biocomputers)

- International Conference on Research in Computational Molecular Biology

- International Meeting on DNA Computing (formerly International Meeting on DNA Based Computers)

- Pacific Symposium on Biocomputing

- Recent Advances in Biomolecular Computing

- SECABC Fall Workshop on Biocomputing

- Unconventional Models of Computation (UMC)

C.10 QUANTUM COMPUTING

C.10.1. Comments on Selected Bibliography

- Benioff, P. (1980), "The Computer as a Physical System: A Microscopic Quantum Mechanical Hamiltonian Model of Computers as Represented by Turing Machines", *Journal of Statistical Physics*, **22**(5), pp. 563–591.

- Feynman, R. P. (1982), "Simulating Physics with Computers", *Int. Journal of Theoretical Physics*, **21**(6/7), pp. 467–488.

- Deutsch, D. (1985), "Quantum Theory, the Church-Turing Principle, and the Universal Quantum Computer", *Proc. of the Royal Soc. of London A*, **400**, pp. 97–119.

  Some of the pioneer attempts to formalize a universal quantum computer.

- Dirac, P. (1982), *The Principles of Quantum Mechanics*, 4th Ed., Oxford University Press.

  A classic book on quantum mechanics. The first chapter is a very good introduction to the field, and presents the bra-ket notation used throughout Chapter 10 of 'Fundamentals of Natural Computing'.

- Hirvensalo, M. (2001), *Quantum Computing*, Natural Computing Series, Springer-Verlag.

- Nielsen, M. A. and Chuang, I. L. (2000), *Quantum Computation and Quantum Information*, Cambridge University Press.
- Pittenger, A. O. (2001), *An Introduction to Quantum Computing Algorithms*, Birkhäuser.

  These are some of the pioneer books on quantum computing. The most complete text among these is the one by Nielsen and Chuang, which contains a broad description of the principles of quantum computation and information.

- Perry, R. T. (2004), *The Temple of Quantum Computing*, On Line Book, [On Line] http://www.toqc.com/.
- Preskill, J. (1998), *Quantum Information and Computation*, On Line Book, [On Line] http://www.theory.caltech.edu/people/preskill/ph229/#lecture

  These are two on line texts on quantum computing for beginners and for those with little background on mathematics and physics. They are highly recommended for newcomers from computer science and engineering.

C.10.2. Specific Journals

- International Journal of Nanoscience (World Scientific), ISSN 0219-581X, http://www.worldscinet.com/ijn/ijn.shtml
- International Journal of Quantum Information (World Scientific), ISSN 0219-7499, http://www.worldscinet.com/ijqi/ijqi.shtml
- Natural Computing (Kluwer Academic Publishers), ISSN 1567-7818, http://www.kluweronline.com/issn/1567-7818
- Quantum Information and Computation (Rinton Press), ISSN 1533-7146, http://www.rintonpress.com/journals/qic/
- Quantum Information Processing (Kluwer Academic Publishers), ISSN 1570-0755, www.kluweronline.com/issn/1570-0755
- Virtual Journal of Quantum Information (American Institute of Physics and the American Physical Society), ISSN 1553-961X, http://www.vjquantuminfo.org/quantuminfo/?jsessionid=1207961085586897503

C.10.3. Specific Conferences

- Coding Theory and Quantum Computing
- Conference on the Theoretical and Experimental Foundations of Recent Quantum Technology
- Gordon Research Conference on Quantum Information Science (QIS)
- International Conference on Quantum Computing and Many-Body Systems
- International Conference on Quantum Dots (QD)
- Quantum Information and Computation

- Quantum Information Sciences Meetings
- Quantum Information, Computation and Complexity
- School and Workshop on Theory and Technology in Quantum Information, Communication, Computation and Cryptography
- Simons Conference on Quantum and Reversible Computation
- Unconventional Models of Computation (UMC)
- Workshop on Classical and Quantum Information Security
- Workshop on Decoherence, Entanglement and Information in Complex Systems (DEICS III)
- Workshop on Quantum Information Processing

INDEX